HUMAN DEVELOPMENT 87/88

Editor

Hiram E. Fitzgerald

Michigan State University

Hiram E. Fitzgerald is a professor and associate chairperson in the Department of Psychology at Michigan State University. He received a B.A. in 1962 from Lebanon Valley College, and an M.A. in 1964 and a Ph.D. in 1967 from the University of Denver. In addition to holding memberships in a variety of scientific associations, he is the Executive Director of the International Association for Infant Mental Health, and editor of the *Infant Mental Health Journal.* He has authored and edited over one hundred and twenty publications.

Editor

Michael G. Walraven

Jackson Community College

Michael G. Walraven is Dean of Instruction and Professor of Psychology at Jackson Community College. He received a B.A. from the University of Maryland in 1966, an M.A. from Western Michigan University in 1968, and a Ph.D. from Michigan State University in 1974. He is affiliated with the American Psychological Association, the Association for Behavior Analysis, and the Biofeedback Society of Michigan.

Cover illustration by Mike Eagle

Annual Editions

A Library of Information from the Public Press

The Dushkin Publishing Group, Inc.

Sluice Dock, Guilford, Connecticut 06437

The Annual Editions Series

Annual Editions is a series of over forty volumes designed to provide the reader with convenient, low-cost access to a wide range of current, carefully selected articles from some of the most important magazines, newspapers, and journals published today. Annual Editions are updated on an annual basis through a continuous monitoring of over 200 periodical sources. All Annual Editions have a number of features designed to make them particularly useful, including topic guides, annotated tables of contents, unit overviews, and indexes. For the teacher using Annual Editions in the classroom, an Instructor's Resource Guide with test questions is available for each volume.

PUBLISHED

Africa
Aging
American Government
American History, Pre-Civil War
American History, Post-Civil War
Anthropology
Biology
Business/Management
China
Comparative Politics
Computers in Education
Computers in Business
Computers in Society
Criminal Justice
Drugs, Society and Behavior
Early Childhood Education
Economics
Educating Exceptional Children
Education
Educational Psychology
Environment
Geography

Global Issues
Health
Human Development
Human Sexuality
Latin America
Macroeconomics
Marketing
Marriage and Family
Middle East and the Islamic World
Nutrition
Personal Growth and Behavior
Psychology
Social Problems
Sociology
Soviet Union and Eastern Europe
State and Local Government
Urban Society
Western Civilization,
   Pre-Reformation
Western Civilization,
   Post-Reformation
World Politics

FUTURE VOLUMES

Abnormal Psychology
Death and Dying
Congress
Energy
Ethnic Studies
Foreign Policy
Judiciary
Law and Society
Parenting
Philosophy

Political Science
Presidency
Religion
South Asia
Third World
Twentieth-Century American
   History
Western Europe
Women's Studies
World History

Library of Congress Cataloging in Publication Data
Main entry under title: Annual Editions: Human development.
   1. Child study—Addresses, essays, lectures.   2. Socialization—Addresses, essays,
lectures.   3. Old age—Addresses, essays, lectures.   I. Title: Human development.
HQ768.A55   155'.05   72-91973
ISBN 0-87967-668-X

Fifteenth Edition

Manufactured by The Banta Company, Menasha, Wisconsin 54952

Editors/ Advisory Board

EDITORS

Hiram E. Fitzgerald
Michigan State University

Michael G. Walraven
Jackson Community College

ADVISORY BOARD

Members of the Advisory Board are instrumental in the final selection of articles for each edition of Annual Editions. Their review of articles for content, level, currency, and appropriateness provides critical direction to the editor and staff. We think you'll find their careful consideration well reflected in this volume.

STAFF

To The Reader

In publishing ANNUAL EDITIONS we recognize the enormous role played by the magazines, newspapers, and journals of the *public press* in providing current, first-rate educational information in a broad spectrum of interest areas. Within the articles, the best scientists, practitioners, researchers, and commentators draw issues into new perspective as accepted theories and viewpoints are called into account by new events, recent discoveries change old facts, and fresh debate breaks out over important controversies.

Many of the articles resulting from this enormous editorial effort are appropriate for students, researchers, and professionals seeking accurate, current material to help bridge the gap between principles and theories and the real world. These articles, however, become more useful for study when those of lasting value are carefully *collected, organized, indexed,* and *reproduced* in a *low-cost format,* which provides easy and permanent access when the material is needed. That is the role played by *Annual Editions.* Under the direction of each volume's *Editor,* who is an expert in the subject area, and with the guidance of an *Advisory Board,* we seek each year to provide in each *ANNUAL EDITION* a current, well-balanced, carefully selected collection of the best of the public press for your study and enjoyment. We think you'll find this volume useful, and we hope you'll take a moment to let us know what you think.

Any history of the field of human development will reflect the contributions of many individuals responsible for crafting the topical domain of the discipline. For example, Binet launched the intelligence test movement, Freud focused attention on personality development, and Watson and Thorndike paved the way for the emergence of social learning theory. However, the philosophical principles that give definition to the field of human development have their direct ancestral roots in the evolutionary biology of Darwin, Wallace, and Spencer, and in the embryology of Preyer. Each of the two most influential developmental psychologists of the early twentieth century, James Mark Baldwin and G. Stanley Hall, was markedly influenced by questions about phylogeny (species' adaptation) and ontogeny (individual adaptation or fittingness). Baldwin's persuasive arguments challenged the assertion that species changes precede individual organism changes. Instead, Baldwin argued, ontogeny not only precedes phylogeny but is the process that shapes phylogeny. Thus, as Robert Cairns points out, developmental psychology has always been concerned with the study of the forces that guide and direct development. Early theories stressed that development was the unfolding of already formed or predetermined characteristics. Most contemporary students of human development stress the epigenetic principle which asserts that development is an emergent process of active, dynamic, reciprocal, and systemic change. This systems perspective forces one to think about the historical, social, cultural, interpersonal, and intrapersonal forces that shape the developmental process.

Thus, the study of human development embraces all fields of inquiry comprising the social, natural, and life sciences and professions. The need for depth and breadth of knowledge creates a paradox: while students are being advised to acquire a broad-based education, each discipline is becoming highly specialized. One way to combat specialization is to integrate the theories and findings from a variety of disciplines with those of the parent discipline. This, in effect, is the approach of *Annual Editions: Human Development 87/88.* The anthology includes articles that discuss the problems, issues, theories, and research findings from many fields of study; the common element is that they all address issues relevant to the study of human development. In most instances, the articles were not prepared for technical professional journals but were written specifically to communicate information about recent scientific findings or controversial issues to the general public. As a result, the articles tend to blend the history of a topic with the latest available information. In many instances, they challenge the reader to consider the personal and public implications of the topic. Thus, an article addressing the effects of televised violence on children's aggression can provide an in-depth analysis from a variety of perspectives. Similarly, a series of articles on genetic engineering and human reproduction can expand the textbook discussion of these topics, and focus student attention on such issues as genetic counseling, surrogate parenting, prenatal screening, and abortion—issues that many students will have to confront personally during the remainder of the decade.

Our selection of articles in this anthology was guided by the valued advice and recommendations from an advisory board consisting of faculty from community colleges, small liberal arts colleges, and large universities. Evaluations obtained from students, instructors, and advisory board members influenced the decision to retain or replace specific articles. Throughout the year we screen over one hundred articles for accuracy of information, interest value, writing style, and recency of information.

Human Development 87/88 is organized into six major units. Unit 1 focuses on conception and prenatal development. Unit 2 focuses on infant development. Unit 3 is divided into subsections addressing social and emotional development and cognitive and language development during childhood. Unit 4 also is subdivided with sections on family structure and child development, and school, culture, and child development. Units 5 and 6 cover the period of human development from adolescence to old age. In our experience this organization provides great flexibility for use of the anthology with any standard textbook. The units can be assigned sequentially, or instructors can devise any number of arrangements of individual articles to fit their specific needs. Use of the Topic Guide facilitates individualization of reading assignments. In large lecture classes this annual edition seems to work best as assigned readings to supplement the basic text. In smaller sections we have used the readings to stimulate instructor-student discussions, relying more heavily on guided individual instruction outside the classroom to help students achieve mastery of course content. Regardless of the instructional style used, it is our hope that our excitement over the study and teaching of human development will come through to you as you read the articles in this fifteenth edition of *Human Development.*

Hiram E. Fitzgerald

Michael G. Walraven

Editors

Contents

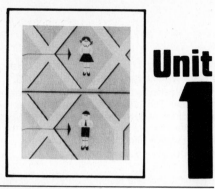

Unit 1

Development During the Prenatal Period

Five selections discuss development during the prenatal period with an emphasis on genetic influences, reproductive technology, prenatal development, and premature birth.

Unit 2

Development During Infancy

Six selections discuss development of the brain and development of communication, emotions, and cognitive systems during the first years of life.

The topics in italics are developed in the article. For further expansion please refer to the Topic Guide and the Index.

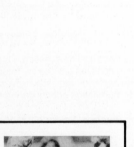

Unit 3

Development During Childhood

Thirteen selections examine human development
during childhood, paying specific attention to social
and emotional development and cognitive and
language development.

The topics in italics are developed in the article. For further expansion please refer to the Topic Guide and the Index.

The topics in italics are developed in the article. For further expansion please refer to the Topic Guide and the Index.

Unit
4

Family, School, and Cultural Influences on Child Development

Eight selections discuss the impact of home, school, and culture on child rearing and child development. The topics include discipline, parenting styles and family structure, as well as the role that our educational system plays in social and cognitive development of the child.

The topics in italics are developed in the article. For further expansion please refer to the Topic Guide and the Index.

Unit 5

Development During Adolescence and Early Adulthood

Seven selections examine some of the effects of social environment, sibling relationships, sex differences, twinship, suicide, and psychic surrender on human development during adolescence and early adulthood.

Unit 6

Development During Adulthood and Aging

Eight selections explore the relationship of family lifestyles, loneliness, and depression to development during adulthood as well as physical, cognitive, and social changes in the aged.

The topics in italics are developed in the article. For further expansion please refer to the Topic Guide and the Index.

The topics in italics are developed in the article. For further expansion please refer to the Topic Guide and the Index.

Topic Guide

This topic guide suggests how the selections in the book relate to topics of traditional concern to human development students and professionals. It is very useful in locating articles that relate to each other for reading and research. The guide is arranged alphabetically according to topic. Articles may, of course, treat topics that do not appear in the topic guide. In turn, entries in the topic guide do not necessarily constitute a comprehensive listing of all the contents of each selection.

TOPIC AREA	TREATED AS AN ISSUE IN:	TOPIC AREA	TREATED AS AN ISSUE IN:
Achievement	3. Are the Progeny Prodigies? 29. The Shame of American Education	Coping Skills	13. Stress and Coping in Children 17. Resilient Children
Adolescence/ Adolescent Development	32. Alienation and the Four Worlds of Childhood 33. Rites of Passage 38. Suicide	Creativity	3. Are the Progeny Prodigies?
		Crying	8. There's More to Crying Than Meets the Ear
Adulthood	27. Women at Odds 28. The Children of the '60s as Parents 35. Holocaust Twins 36. Men and Women 42. Toward an Understanding of Loneliness	Cultural Influences	13. Stress and Coping in Children 22. Intelligence
		Death	13. Stress and Coping in Children 23. Origins of Speech
Aggression/Violence	12. Children's Winning Ways 15. Aggression	Depression	16. Depression at an Early Age 38. Suicide 42. Toward an Understanding of Loneliness 43. The Good News About Depression
Aging	22. Intelligence 44. Aging 46. Never Too Late 47. The Reason of Age	Despair	16. Depression at an Early Age 38. Suicide 42. Toward an Understanding of Loneliness
Alienation	32. Alienation and the Four Worlds of Childhood 42. Toward an Understanding of Loneliness	Developmental Disabilities	1. Genes 3. Are the Progeny Prodigies? 8. There's More to Crying Than Meets the Ear 23. Origins of Speech
Artificial Insemination	2. The New Origins of Life 3. Are the Progeny Prodigies?		
Attachment	8. There's More to Crying Than Meets the Ear 26. The Importance of Fathering	Discipline	25. Punishment Versus Discipline 28. The Children of the '60s as Parents 30. Moral Education for Young Children
Behavior	8. There's More to Crying Than Meets the Ear 29. The Shame of American Education	Divorce	13. Stress and Coping in Children 26. The Importance of Fathering
Birth Defects	3. Are the Progeny Prodigies? 8. There's More to Crying Than Meets the Ear	Drugs	32. Alienation and the Four Worlds of Childhood 43. The Good News About Depression
Brain Organization/ Function/Chemistry	6. Making of a Mind 15. Aggression 21. Insights into Self-Deception 37. Male Brain, Female Brain 45. New Evidence Points to Growth of the Brain	Education/Educators	6. Making of a Mind 24. Why Children Talk to Themselves 26. The Importance of Fathering 29. The Shame of American Education 30. Moral Education for Young Children
		Emotional Development	13. Stress and Coping in Children 16. Depression at an Early Age 18. Face to Face 32. Alienation and the Four Worlds of Childhood 38. Suicide 43. The Good News About Depression
Bulimia	16. Depression at an Early Age	
Child Abuse	8. There's More to Crying Than Meets the Ear	
Childbirth	3. Are the Progeny Prodigies? 5. Before Their Time	Environmental Factors/ Stimulation	4. Life Before Birth 6. Making of a Mind 26. Alienation and the Four Worlds of Childhood 45. New Evidence Points to Growth of the Brain
Child Rearing/ Caregiving	10. Your Child's Self-Esteem 13. Stress and Coping in Children 25. Punishment Versus Discipline 26. The Importance of Fathering 28. The Children of the '60s as Parents		
		Evolutionary Theory	18. Face to Face 23. Origins of Speech
		Family Development	13. Stress and Coping in Children 28. The Children of the '60s as Parents 34. The Sibling Bond 46. Never Too Late
Chromosomes	1. Genes 3. Are the Progeny Prodigies?		
Cognitive Development	11. The Roots of Morality 14. Racists Are Made, Not Born 22. Intelligence 31. Rumors of Inferiority 37. Male Brain, Female Brain 47. The Reason of Age	Fathers	26. The Importance of Fathering
		Fertilization	2. The New Origins of Life 3. Are the Progeny Prodigies?
Competence	10. Your Child's Self-Esteem 29. The Shame of American Education		

Development During the Prenatal Period

In human reproduction, sexual intercourse brings together two living cells; a sperm, contributed by the male, and an ovum, contributed by the female. The unique union of these cells marks the beginning of development as well as the beginning of organism-environment transaction. From the moment of its conception, the newly formed organism exists in relationship to its environment. The quality of the prenatal environment and subsequent postnatal care received by the developing individual will lay a foundation for its development, but will not determine its developmental outcome. The articles selected for inclusion in this unit address issues related to genetic influences on development, prenatal development, premature birth, and various ethical issues related to advances in reproductive technology.

The first article, "Genes: Our Individual Programing System," explains that genes regulate hormones but are, in turn, influenced by many environmental events. A detailed presentation of genetic and hormonal influences on sex and sexual behavior is balanced by reference to the lack of any clear connection between hormonal levels and behavior.

Although couples may elect "artificial" methods for achieving fertilization, some of which are discussed in "The New Origins of Life," the sperm and ovum retain their dynamic characteristics; only the environment in which fertilization occurs or the techniques for inserting the sperm into the womb change. Nevertheless, reproductive technology has produced a variety of ethical, legal, and social issues that will not easily be resolved. One such issue involves selection of sperm based on known characteristics of the donor. Although sperm banks provide hope for many infertile couples and single women, selecting sperm based on the donor's athletic, intellectual, or artistic accomplishments may lead to unreasonable expectations for the child's subsequent achievements. "Are the Progeny Prodigies?" provides several anecdotal illustrations of infants conceived via sperm banks and addresses the issue of eugenics, the attempt to improve the hereditary characteristics of a species through selective breeding. During each of the periods of prenatal development, the fetus is vulnerable to a variety of biological and environmental stresses. Biological stresses include a variety of chromosomal anomalies and malformations. Environmental stresses include such factors as infectious disease, mal-nutrition, blood incompatibility, drugs, radiation, parental age, maternal emotional state, and chemical toxins. In addition to studies of the physical development of the fetus, new technologies are allowing investigators to study behavioral differentiation of the fetus as well. Much of this pioneering work is described in "Life Before Birth."

Recent technological advances have produced various techniques for assessing the developmental status of the fetus. Whereas each of these techniques contributes to a more complete evaluation of the structural and biological viability of the fetus, each technique also raises ethical and moral issues concerning decisions such as whether to retain or abort the fetus. Ethical issues, however, are not restricted to the question of abortion: they are also raised with respect to preterm birth. Thirty years ago, prematurity generally referred to infants born no more than about two or three months prior to the expected date. Today, infants born less than seven months gestational age are brought to term with the assistance of biomedical technology. Although survival rates of very low birth weight premature infants have increased, "Before Their Time" points out that such infants remain at risk for a variety of developmental problems, including those related to child rearing. Parent-infant interaction is difficult to establish when an infant is premature. Consequently, the quality of caregiving intervention received by prematurely born infants may not be equivalent to the quality of life-sustaining technology. It is clear, however, that without modern technology most premature infants weighing less than 1500 grams would die regardless of the quality of caregiving they receive.

The articles in this section discuss many of the discoveries that have been made about the early development of the fetus, and challenge the reader to consider the question, "When does life begin?" In addition, the articles review what is and what is not known about the consequences of preterm birth for the infant and for infant-caregiver interaction.

Looking Ahead: Challenge Questions

Does artificial insemination or in vitro fertilization alter the dynamic relationship between the ovum and sperm during fertilization? At what point in development would you feel secure in concluding that such alternatives to normal sexual intercourse have no long-term consequences?

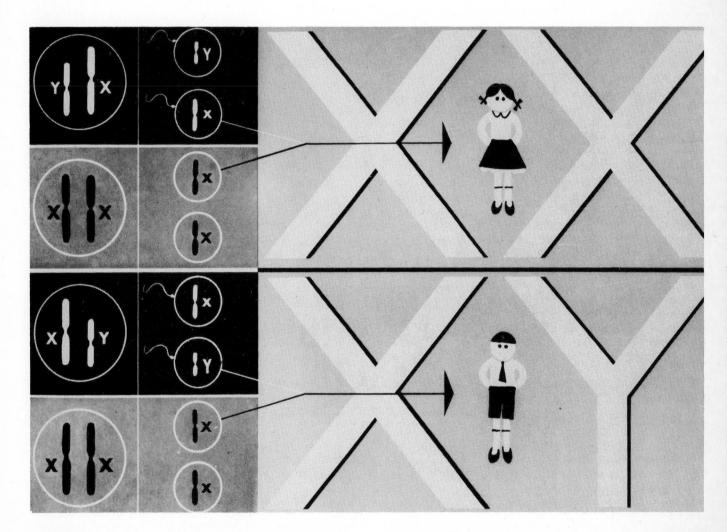

How is knowledge of the fetus' responsiveness to stimulation translated into prenatal care practices for the mother and fetus?

Consider your current beliefs about abortion, genetic engineering, and socialized medicine. How do you think these views would be challenged if you learned that your baby-to-be was expected to be profoundly retarded?

Would you demand efforts to save your five hundred gram premature infant if it had a seventy-five percent chance of dying and an equally high chance of having some form of severe developmental disability? Who should bear the financial burden associated with maintenance of neonatal intensive care, subsequent rehabilitation, and education?

If manipulation of genetic material can prevent the appearance of physical dysfunctions, might not similiar manipulations be used to engineer intellectual abilities, personality traits, or socially desirable behaviors? What factors would greatly constrain any society that attempted to actively and explicitly practice eugenics?

The hormonal constitution is as hereditary as eye color, body structure or any other physical trait carried by

Genes:

Our individual programing system

Nature occasionally produces tragic experiments in which disease or heredity disturbs the body's hormone balance. Unfortunate as these occurrences are, they nonetheless teach scientists a great deal about how the endocrine system works in health and disease.

New knowledge about the endocrine system comes slowly and often by chance; for it is difficult and often unethical to perform genetic research on human beings. People cannot be bred like laboratory animals. Affected individuals and their families are scattered around the world. The genetic conditions are rare. And few physicians are interested in research so that many instances simply go unrecorded.

Animal research on hormones is no substitute for human research. Animals' hormones, especially animal sex hormones and fertility cycles, are somewhat unlike those of human beings. Monkeys, it is true, most closely resemble human beings; but research with monkeys is expensive and difficult. And monkeys are still not the same as people.

Medical investigators therefore, must simply wait for those unlikely experiments of nature and be ready to learn from them whenever and wherever they occur. Despite the many problems, this research is beginning to show the connections between genes and hormones, sex and fertility.

From *Bostonia*, September/October 1984, pp. 26-33. Reprinted with the permission of Bostonia Magazine 1984.

Many factors influence the endocrine system

ONE FUNCTION OF GENES IS TO shuffle the biological deck repeatedly so that individuals of a species differ from each other. The genes responsible for controlling hormones are no exception to this rule. People differ as to how much of a particular hormone they produce, in the sensitivity and number of their receptors for a hormone on target tissues, and in the ability of target tissues to respond to their hormones' chemical signals.

Many genetic factors are involved in a programing system with so many variables. But the complexity of the endocrine system notwithstanding, each element must be the expression of programs written in the genes.

The genes are often overruled by forces that are not genetic. Genes simply give a program for how the body will function. Many cultural, social, environmental, nutritional and many other factors have cooperated to change the basic pattern. That women menstruate and go through menopause, for example, is determined by the genes that make a woman. Both menarche and menopause are the expression of hormones, in turn controlled by processes set up by the genes. Today 95 percent of young women in Europe and the United States have their first menstrual period (menarche) between the ages of 10 and 16.

Medical investigators must simply wait for unlikely experiments of nature in order to study genetic mistakes.

Although the genes have remained the same for the past several centuries, the average age of menarche has been coming down about three years a century for young women in industrialized countries.

Consequently, women are maturing faster than they did in the past. The average age of menarche is now stable in industrialized countries but is still coming down in the less developed countries, showing again the importance of social, environmental and nutritional factors on maturation. Menopause, in contrast, has been delayed by about three years a century and it is safe to assume that this, too, is because of environmental and nutritional factors.

Nonetheless, despite wide variations in human hormones, it is clear that genes must be at least partly responsible for the patterns of hormone actions, the types of hormones produced and the body's sensitivity to them. This is certain because:

• Certain hormonal problems run in families;

• Animals of different species have similar but not identical hormones;

• And, perhaps the most telling piece of evidence, males differ from females. The difference between males and females is genetically determined. Somehow in the genes are the instructions that make males and females produce different hormones or the same hormones in different amounts.

Genes determine hormones and hormones determine sex

GENES INHERITED FROM THE mother and the father establish a person's sex. By the time a human fertilized egg has divided three times, it is either male or female. Its sex is genetically determined at the moment of fertilization and the genetic pattern already makes a difference when the embryo is only eight cells large. There are no outward signs of sex yet, of course. These will not appear until the fetus has grown for another two months. But particularly from the moment of conception, the chemical potential is programed into the embryo's cells for the person to be of one sex or the other.

Before the eight-cell stage of the embryo, the X chromosome runs the show. In the earliest cell divisions, cells of the growing embryo seem to pay no attention to whether the other chromosome is an X (that is, the embryo is female) or a Y (that is, the embryo is

male).

If the embryo is genetically male, the Y chromosome begins to dominate the scene after the eight-cell stage and eventually directs the chemical processes that stimulate the growth of testes and male germ cells. The baby is a boy.

If the embryo does not have a Y chromosome, testes do not develop, and the fetus grows ovaries. The baby is a girl.

The genetics of sex

Every cell of a human being (except for the sperm and egg) has 23 pairs of chromosomes. Twenty-two pairs determine traits of the body. One pair determines the person's sex. Sex chromosomes, like all the other chromosomes a person has, come from the parents: one of a pair from the mother and one of a pair from the father. Sex chromosomes are either X or Y, so called because of their shape when examined under the microscope.

The first step that brings together the combination of genes for male or female determines most of the steps that follow. In the normal course of events, a female has two X chromosomes (XX) and a male has one X and one Y chromosome (XY). These combinations are dealt in the moment that a sperm (carrying either a single X or Y chromosome) fertilizes an egg (carrying one of the mother's X chromosomes). The genes carried on these chromosomes carry the family history. In the moment of fertilization, they project that history one more generation into the future.

Sex chromosomes

		Mother's Sex Chromosomes	
		X	X
Father's Sex Chromosomes	X	XX	XX = Female
	Y	XY	XY = Male

Male Sex Hormones

THE EMBRYO'S TESTES PRODUCE male hormones, notably testosterone, and a variety of other powerful substances that may be said loosely to "defeminize" and "masculinize" the growing male fetus. This is a gross oversimplification. But animal research and some work with human beings does seem to show that if nothing changes the original pattern, you get a girl. To get a boy, the fetus needs to change some of

Animal research and some work with human beings does seem to show that if nothing is done to change the original genetic pattern, you get a girl. To get a boy, the fetus needs to change.

the starting materials. And the agents of change are hormones.

French biologist Alfred Jost studied this process, mostly in rabbits, more than 30 years ago and found that the process of sexual differentiation is sequential, ordered and simple.

When Jost removed the testes from a genetically male (XY) rabbit fetus, he produced a female. She was sterile, of course, because she lacked ovaries; but she was in most other ways a female. When Jost implanted a testis in the neck of a genetically female (XX) rabbit fetus, he produced a male. He was sterile, too, because he lacked the male germ cells to produce sperm. But otherwise he was a male. Jost concluded that the difference between males and females is in the action of sex hormones in the fetus—hormones for which the genes carry the programs.

Once the genes have given their chemical instructions for the fetus to develop along the male model, the fetus takes over, helping to determine how it will look by producing more sex hormones.

In the early embryo, the developing testis secretes testosterone and a protein substance called Mullerian regression factor (MRF). MRF is responsible, in a sense, for "defeminizing" the male embryo. In the female embryo, a set of tubes called Mullerian ducts normally develops into the structures of the female reproductive system—Fallopian tubes, uterus and upper part of the vagina. In the male, almost nothing remains of the primitive Mullerian ducts because the protein messenger MRF instructs the embryo to absorb them.

A second hormone, human chorionic gonadotropin (hCG), starts the processes by which male sexual structures

develop. Produced by the placenta, hCG stimulates cells of the testis (Leydig cells) to synthesize testosterone, which gives the signal for another set of tubes called Wolffian ducts to develop in the male. Wolffian ducts eventually become seminiferous tubules and the vas deferens. Testosterone secreted by the Leydig cells also diffuses into areas of the embryo that will become the male external genitalia—penis, scrotum, prostate and urethra.

Once testosterone is produced, male development proceeds quickly because the little bit of testosterone produced stimulates the production of more. Blood vessels invade the tissues that will become the testes, the Mullerian ducts that have been characteristically female structures degenerate and male Wolffian ducts develop. All this happens within the first trimester. In contrast, ovaries do not develop until the second trimester.

This situation is not so simple, however. True, in the normal course of events, the presence of a Y chromosome in mammals is necessary to force development toward maleness. But it takes more than the presence of the Y chromosome and the enzymes it codes to make a male. There must also be a sufficient number of receptors on the appropriate tissues to receive the chemical signal from the Y chromosome. The tissue receptors must be sensitive enough to respond to the signal, and the small response of recognition on the part of the receptors must be capable of triggering a greater response by the cell.

There are cases of people with the normal XY genetic endowment who nevertheless appear female. Their Y chromosome produces the proper

Research has now found that one of the X chromosomes is actually inactivated in the female. Only one X chromosome is responsible for directing the course of female fetal development.

enzymes but their bodies lack the receptors to stimulate the growth and development of male tissues.

This is not the end of the story of X and Y. The female has two X chromosomes. For many years, research scientists wondered if both X chromosomes contributed to the development of the ovary and female characteristics. Research has now come up with the surprising finding that one of the X chromosomes is actually inactivated in the female. Only one X chromosome is responsible for directing the course of female fetal development. Even more surprising, the X chromosome is inactivated in males, too. In fact, having an activated X chromosome in males interferes with the normal processes by which germ cells divide and mature to be sperm cells. Men with the genetic anomaly, XXY, develop testes but because they do not inactivate their X chromosome, they do not usually produce sperm. Sometime later in development, a female takes advantage of having both X chromosomes in producing eggs.

The significance of these complex patterns of turning on and off genes and their enzyme products and the secrets of the silent X chromosome remain mysteries of evolutionary history.

The genetics of sterility

Hormones are not the entire story. In one kind of male infertility, testes are normal but they produce no sperm. The explanation is to be found in the chromosomes. Part of the Y chromosome near the center codes for enzymes that produce structures of the testis; another part of the Y chromosome, further out on one of the arms of the Y-shaped structure, codes for the messengers that produce the sperm. It is possible, therefore, to have a normal testis, producing all the right hormones, without having any sperm at all.

Gonads and the primitive cells that eventually mature to germ cells arise from two different sources in the embryo. The germ cells that will become sperm cells or eggs actually arise in the yolk sac, not strictly speaking part of the embryo itself. They then migrate to a part of the developing embryo known as the genital ridges. These migratory cells differentiate and mature to become eggs or sperms depending on which hormones act on them. The tissue of the ovaries or testes, however, comes from a different part of the growing embryo. The result is that it is possible for gonads to develop normally in response to the hormones triggered by the presence of a Y chromosome and still have no sperm or eggs.

Sometimes the message gets garbled

FOUR STEPS ARE REQUIRED FOR normal development of a male:
- Leydig cells must produce normal concentrations of testosterone;
- Synthesis of Mullerian regression factor (MRF) must be normal;
- Testosterone must be converted to the more biologically potent androgen, dihydrotestosterone; and
- Sex hormones must actually induce their effects within cells.

Sometimes, however, the message gets garbled. The appropriate androgens or Mullerian regression factor cannot exert their effects within the cells. If androgens are not effective, either because they are not being synthesized in high enough concentration or because something is wrong with the androgen receptors, an individual who is a genetic male —that is, whose genotype is XY— matures like a woman at puberty. The condition is known as testicular feminization.

If Mullerian regression factor (MRF) does not instruct the embryo to absorb the Mullerian ducts, genetic males may grow to have some of the sexual structures of females. Men may look normal but may actually have incompletely developed Fallopian tubes and uterus.

Do hormones shape behavior?

THE BODY PARTS THAT DISTINGUISH males from females are clearly shaped by genes and hormones. How much of human behavior—and especially behaviors directly and indirectly related to mating—is initiated and shaped by hormones? The answer is that no one knows if there is a connection between hormone levels and human behavior.

Males have more circulating testosterone than females do. Males may differ from females in activity level, preference for physical contact in sports and demonstrations of anger, assertiveness and aggressiveness. There are many problems with this research. Separating the effects of genetics from the effects of culture is almost impossible. Even if the differences in male and female behavior are real, whether the differences in behavior are related to differences in hormone

How much of human behavior—and especially behavior directly and indirectly related to mating—is initiated and shaped by hormones? The answer is that no one knows if there is a connection between hormone levels and human behavior.

levels simply cannot be determined.

For a long time it was an attractive hypothesis that men with the genetic endowment of XYY, that is, an extra Y chromosome capable of triggering the production of abnormally high levels of testosterone—might account for the abnormally high levels of violent and assaultive behavior in some men. XYY men may be impulsive, but endocrin-

---

Hormones and Homosexuality: Is There a Connection?

Homosexuality has intrigued researchers for decades. The notion that homosexuality could be biologically determined is not new, for example. As early as 1920, Freud suggested that hormones could ultimately prove to be the key to the mystery. And recent technical advancements in measuring blood hormone levels have focused new attention on whether varying levels of hormones in males or females, even before birth, can lead to adult homosexuality.

Investigators remain divided on the issue. In one early study launched by a California endocrinologist, ketosteroids in the urine of active male homosexuals and heterosexuals were analyzed and revealed differences in the endocrines that matched the participants' sexual preferences 90 percent of the time. In 1971, the Masters and Johnson Sex Research Institute reported that young homosexual males studied had lower levels of the male hormone testosterone than participating heterosexual men. These findings have remained unsubstantiated, however. Two later studies conducted by other investigators revealed *higher* testosterone levels among homosex-

uals than those found in heterosexual participants, and eight other studies reported no difference in testosterone level among the male participants. The findings of studies measuring gonadotropins and other sex hormones in the blood of male subjects also varied widely.

Even less research has been devoted to the relationship between hormones and female homosexuality. In the only reported study conducted with female participants, researchers found a higher level of plasma testosterone in the homosexual women than in heterosexual females, but noted that hormone measurements for all of the participants fell within healthy ranges.

Other questions persist, say scientists. The findings of plasma tests are also weakened by the results of early attempts to "treat" homosexuals—a reflection on the long period when homosexuality was widely viewed as an "illness." Administering testosterone to male homosexuals in these treatments increased their sex drive but did not change sexual orientation.

Other scientists have speculated that pre-

natal deficiencies or excess levels of sex hormones may lead to homosexual behavior later in life. Dr. Günter Dörner, director of the Institute for Experimental Endocrinology in Germany, has launched several studies to investigate the effect of estrogen injections on the pituitary production of LH hormones in adult homo- and heterosexual men. Dörner reported that homosexual males responded to the injections with an LH increase—a reaction more typical of females than heterosexual men. From his findings, he theorized that homosexual men may have a testosterone deficiency before birth, which enables the brain to develop in a primarily female manner. All sexual characteristics, according to Dörner —including identity and orientation—may be shaped by sex hormones, while in the womb.

Scientists and psychiatrists do agree that all hormonal theories about sexual orientation must be subject to more research before they can provide true insight into homo- or even heterosexuality.

CHERYL COLLINS

ologic studies have not shown a close relationship between the XYY genetic endowment and higher levels of testosterone in all XYY men.

Some medical and correctional authorities, for example, would like to show that the assaultive behaviors of violent criminals are due, in part, to their higher levels of testosterone or the abnormal sensitivity of tissues in their brain to circulating testosterone. If this were the case, dealing with the complex social, economic, psychological and other factors involved in criminal behavior would be reduced to the simple expedient of treating the criminal's hormone levels. But at this point, no one knows if hormones and behavior are so related.

Drugs have been used to lower the testosterone levels of violent sexual offenders in hopes of reducing their sexual urges and impulsive acts. It has turned out, however, that physicians must prescribe so much of the drug that the person's testosterone levels are reduced 50 to 70 percent below normal, not simply to normal levels. Drugs do not seem to reduce non-sexual violent behaviors; and when they do, it is more because they lower excitability in general and not because they have a specific effect.

There is no clear connection between levels of testosterone in normal men and their levels of resentment, hostility, assaultive behavior or irritability. On the other hand, there is some evidence to show that violent and assaultive prisoners have higher than normal levels of circulating testosterone, especially prisoners with long histories of violent behaviors. Separating cause from effect

The significance of these complex patterns of turning on and off genes and their enzyme products and the secrets of the silent X chromosome remain a mystery of evolutionary history.

is a problem; for it is not clear whether the environmental and social conditions provoking the violence raises the testosterone levels or whether the already raised testosterone levels predisposes the person to act violently.

Hormones and behavior

TESTOSTERONE HAS MANY EFFECTS on tissues of the body other than developing sexual tissues, including the brain. It would be surprising if hormones affected the brain and not behavior. The central nervous system plays such an important role in controlling mating behaviors, the selection of mates, the timing of mating, hormonal changes in pregnancy, rearing the young and so forth, it would be surprising indeed if the brain and hormones were not intimately related.

Anecdotal evidence and crude experiments have shown that this is the case. Recently, careful experiments to show precisely what parts of the brain have receptors for sex hormones have proved the case. Many male behaviors are controlled by centers in the brain prompted by androgens circulating in the blood.

As with so many of these experiments, the work has been done with animals, not people, and so it is difficult to generally apply the results to human beings. That genes affect hormones, hormones affect the brain, the brain affects behavior is a reasonable hypothesis, however. But human beings, perhaps unlike the rest of the animal kingdom, have powerful social and psychological determinants of behavior, as well as the chemical determinants. Hormones make a difference but they cannot account for the entire difference between men and women or between violent and less violent individuals.

Sex hormones do account for some of the sexual differences in nonsexual tissues. In some mammals (mice, for instance), the male kidney is larger than the female kidney because of the effects of testosterone. In many mammals, including human beings, there are sex-related differences in how the liver metabolizes steroids and drugs and in the proteins the liver secretes. Exposure to testosterone is responsible for the differences. Likewise, the red blood cell

produces hemoglobin in response to the protein erythropoietin but also in response to testosterone. But the most dramatic effect of testosterone on non-sexual tissue of the body is in the growth of muscle. Men as a group are larger and more muscular than women because of the greater concentrations of testosterone circulating in their blood streams.

Hormones make a difference but they cannot account for the entire difference between men and women or between violent and less violent individuals.

Testosterone also affects the brain. In animals, such "male" behaviors as courtship, aggressiveness, defense of territory and so on, are directly related to testosterone levels in the blood. In the brain, receptors recognize the presence of testosterone and trigger the appropriate reproductive and nonreproductive behaviors. In lower animals, testosterone is related to:

- Regulation of gonadotropin and prolactin;
- Courtship and other reproductive behaviors;
- Activity levels;
- Aggressiveness;
- Play;
- Taste preferences;
- Scent marking;
- Feeding and body weight;
- Learning;
- Circadian rhythms;
- Brain function.

In human beings, the role of hormones in determining behavior is not quite so clear. There is considerable evidence, however, that hormones do play a major role in shaping behavior.

There have been studies of the relationship between hormones and rearing in the development of sexual definition and gender identity. Most information has come from studies of children born without clear biological sexual definition.

Unclear sexual definition

Testicular feminization and congenital adrenal hyperplasia

For a variety of reasons, the genetic definition of sexual identity may not agree with the outward signs of sexual definition. When this happens, parents raise their children as the opposite of the child's biologic definition. Biologic boys are raised as girls in a condition known as testicular feminization, for example; and biologic girls are raised as boys in the condition known as congenital adrenal hyperplasia. In the first, males are born with normal testes and produce normal testosterone but the tissues of the rest of the body lack receptors (or somehow fail to get the message) for testosterone and fail to develop in the normal male pattern. The boy takes on feminine characteristics. In the second, a biologic female has an abnormal adrenal gland that produces testosterone rather than cortisone. As a result, the girl's tissues throughout the body respond to testosterone and she develops male characteristics.

Sexual definition and gender orientation

IF LEFT UNTREATED, TESTICULAR feminization and congenital adrenal hyperplasia create serious problems when biologic and gender identities conflict. It is an odd fact of human psychology, however, that gender identity is stronger than biologic sexual identity. Even when secondary sexual characteristics of the opposite sex begin at puberty, the child raised as a girl continues to think of herself as a girl; the child raised as a boy continues to think of himself as a boy.

When these conditions are treated early, gender identity and sexual identity are in harmony. (Hormones and sexual identity were in conflict only during the time of gestation.) The unfortunate condition nonetheless gives researchers the opportunity to study the relative contributions of hormones and rearing to sexual identity. Much research has now been devoted to the question of whether abnormally high or low androgens, progestogens or estrogens in fetal development have a continuing effect on the person's behavior. The hypothesis is that, even when the hormonal abnormalities are corrected after birth, prenatal exposure to high levels of androgens will produce "male" behavior and high levels of estrogens or low levels of androgens will

produce "female behaviors."

Researchers have studied several behaviors:

- Activity levels, outdoor play and athletic skills;
- Physical and verbal fighting;
- Play as rehearsal of the parent role;
- Preference in clothes, grooming, jewelry

The research shows that prenatal hormone levels do not affect gender identity. Even in children with abnormalities leading to opposing biologic and gender sex identities, the gender identity given by the child's rearing dominates. However, the research also shows that behaviors are determined by prenatal androgen levels.

Biologic females exposed to high levels of androgen before birth have been shown to prefer intense outdoor play, associate with boys for the most part, think of themselves and be labeled by others as tomboys. They do not play with dolls or rehearse parental roles.

> One function of genes is to shuffle the biological deck repeatedly so that individuals of a species differ from each other.

They are more likely to engage in body contact sports.

Boys can also have abnormalities that expose them to excessive levels of androgens before birth. Naturally these

---

The Nature vs. Nurture Debate

It is well-known that hormones play an essential role in the development of physical differences in males and females. But to what extent an individual's sexual identity is influenced by hormones versus his or her social environment has long been debated by scientists.

As in other areas of hormone research, studies are complicated by the difficulty of investigating the subtleties of physiology without harming human subjects. To understand healthy functioning, said Dr. Judith Vaitukaitis, professor of Medicine and Physiology at Boston University Medical Center, researchers often must work backward to find out what went wrong, and then infer what is "normal." In fact, it was in this way—the retrospective study of abnormalities of nature—that medical researchers recently provided insight into the "nature versus nurture" issue of gender identity.

In 1979, researchers from the Cornell University Medical College and the National University Pedro Henriquez Urena reported in the *New England Journal of Medicine* a study of a group of boys from the Dominican Republic who were born with "female-appearing" external genitalia and who were subsequently raised as girls. The condition was caused by deficiency of an

enzyme that in normal boys activates testosterone before birth to stimulate growth of male genitals.

The Dominican Republic boys, lacking male genitals, were raised as girls during childhood. The onset of puberty, however, resulted in normal increases in testosterone, which, in turn, stimulated the long-delayed development of a penis and scrotum. The boys consequently underwent an identity transition over the course of several years, passing through stages of "no longer feeling like girls, to feeling like men, to the conscious awareness that they were indeed men."

The researchers wrote that of 18 subjects "unambiguously" raised as girls, 16 changed to a "male-gender role," despite social pressure that included parental amazement and confusion. Because the boys made the transition with relatively little difficulty—in spite of a cultural environment that emphasizes a "definite socialization of children according to sex"— the researchers concluded that "environmental or sociocultural factors are not solely responsible for the formation of a male-gender identity. Androgens make a strong and definite contribution."

JON QUEIJO

Biologic females exposed to high levels of androgen before birth have been shown to prefer intense outdoor play, associate with boys for the most part, think of themselves and be labeled by others as tomboys.

═══════════

boys are physically and genetically male; there is no ambiguity as in the case of females exposed to androgens. But these boys differ from normally developing boys in being more active, more interested in physical contact sports and perhaps have higher levels of aggressiveness.

In contrast, boys with testicular feminization, whose androgen levels during prenatal development are normal but incapable of stimulating male development in tissues of the body, are stereotypically feminine. The problem with such research, however, is that these

biologic males are reared as females because they lack characteristic male appearance. There is no way to separate the effects of rearing from those of hormones.

Most researchers are quick to point out that sexual orientation is the result of complex and little understood psychological, social, environmental factors, as well as genetic and hormonal ones. Few people would argue that hormones alone strictly define sexual orientation. That they play a role, however, is clear.

Can you recognize the problem?

───────────

A woman and her husband have been trying to conceive a child for almost two years. She has had her period regularly every 28 to 32 days and is apparently in good health. Her gynecologist tested her to be certain that she had no infection and that her tubes were not blocked. Her husband was also examined and the tests showed that his sperm was entirely normal. At this point, she is told by her

physician that there is nothing wrong with her, she should just relax, and nature will take its course. Is the physician right?

A. Yes
B. No
C. Can't be certain

Answer: The answer is C, the physician cannot be certain, and the wife should be tested further if she and her husband want to have children. Approximately 15 to 20 percent of couples trying to conceive cannot although there is no immediately obvious reason for the difficulty. Nevertheless, in such cases a woman may be abnormal endocrinologically. She may have regular menstrual bleeding and appear normal but still have anovulatory cycles in which no mature egg is produced. More sophisticated testing is required before both husband and wife can be given a clean bill of health. Ironically, it is just these apparently normal cases that are the most difficult to diagnose and treat. The more marked the menstrual disturbance, the easier it is to correct, provided there are potentially fertilizable eggs in the follicles within her ovary.

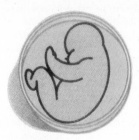

The age of the test-tube baby is fast developing. Already science has produced an array of artificial methods for creating life, offering solutions to the growing problem of infertility. In these stories, TIME explores the startling techniques, from laboratory conception to surrogate mothers, and examines the complex legal and ethical issues they raise.

The New Origins of Life

How the science of conception brings hope to childless couples

A group of women sit quietly chatting, their heads bowed over needlepoint and knitting, in the gracious parlor at Bourn Hall. The mansion's carved stone mantelpieces, rich wood paneling and crystal chandeliers give it an air of grandeur, a reflection of the days when it was the seat of the Earl De La Warr. In the well-kept gardens behind the house, Indian women in brilliant saris float on the arms of their husbands. The verdant meadows of Cambridgeshire lie serenely in the distance. To the casual observer, this stately home could be an elegant British country hotel. For the women and their husbands, however, it is a last resort.

Each has come to the Bourn Hall clinic to make a final stand against a cruel and unyielding enemy: infertility. They have come from around the globe to be treated by the world-renowned team of Obstetrician Patrick Steptoe and Reproductive Physiologist Robert Edwards, the men responsible for the birth of the world's first test-tube baby, Louise Brown, in 1978. Many of the patients have spent more than a decade trying to conceive a child, undergoing tests and surgery and taking fertility drugs. Most have waited more than a year just to be admitted to the clinic. Some have mortgaged their homes, sold their cars or borrowed from relatives to scrape together the $3,510 fee for foreign visitors to be treated at Bourn Hall (British citizens pay $2,340). All are brimming over with hope that their prayers will be answered by in-vitro fertilization (IVF), the mating of egg and sperm in a laboratory dish. "They depend on Mr. Steptoe utterly," observes the husband of one patient. "Knowing him is like dying and being a friend of St. Peter's."

In the six years that have passed since the birth of Louise Brown, some 700 test-tube babies have been born as a result of the work done at Bourn Hall and the approximately 200 other IVF clinics that have sprung up around the world. By year's end there will be about 1,000 such

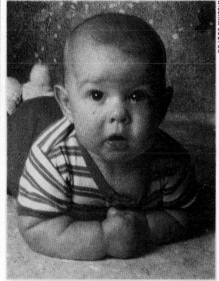

In-vitro Infant Daniel Brooks at four months

infants. Among their number are 56 pairs of test-tube twins, eight sets of triplets and two sets of quads.

New variations on the original technique are multiplying almost as fast as the test-tube population. Already it is possible for Reproductive Endocrinologist Martin Quigley of the Cleveland Clinic to speak of "old-fashioned IVF" (in which a woman's eggs are removed, fertilized with her husband's sperm and then placed in her uterus). "The modern way," he notes, "mixes and matches donors and recipients" *(see chart page 15)*. Thus a woman's egg may be fertilized with a donor's sperm, or a donor's egg may be fertilized with the husband's sperm, or, in yet another scenario, the husband and wife contribute their sperm and egg, but the resulting embryo is carried by a third party who is, in a sense, donating the use of her womb. "The possibilities are limited only by your imagination," observes Clifford Grobstein, professor of biological science and public policy at the University of California, San Diego. Says John Noonan,

professor of law at the University of California, Berkeley: "We really are plunging into the Brave New World."

Though the new technologies have raised all sorts of politically explosive ethical questions, the demand for them is rapidly growing. Reason: infertility, which now affects one in six American couples, is on the rise *(see box page 16)*. According to a study by the National Center for Health Statistics, the incidence of infertility among married women aged 20 to 24, normally the most fertile age group, jumped 177% between 1965 and 1982. At the same time, the increasing use of abortion to end unwanted pregnancies and the growing social acceptance of single motherhood have drastically reduced the availability of children for adoption. At Catholic Charities, for instance, couples must now wait seven years for a child. As a result, more and more couples are turning to IVF. Predicts Clifford Stratton, director of an in-vitro lab in Reno: "In five years, there will be a successful IVF clinic in every U.S. city."

It is a long, hard road that leads a couple to the in-vitro fertilization clinic, and the journey has been known to rock the soundest marriages. "If you want to illustrate your story on infertility, take a picture of a couple and tear it in half," says Cleveland Businessman James Popela, 36, speaking from bitter experience. "It is not just the pain and indignity of the medical tests and treatment," observes Betty Orlandino, who counsels infertile couples in Oak Park, Ill. "Infertility rips at the core of the couple's relationship; it affects sexuality, self-image and self-esteem. It stalls careers, devastates savings and damages associations with friends and family."

For women, the most common reason for infertility is a blockage or abnormality of the fallopian tubes. These thin, flexible structures, which convey the egg from the

THE BIG MOMENT: droplets containing sperm are added to waiting egg immersed in solution

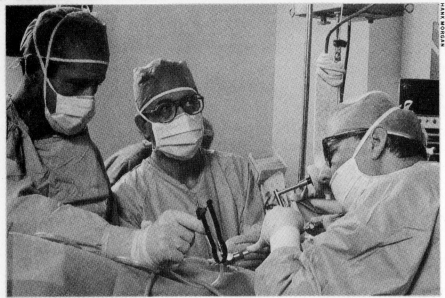

Doctors at the Jones Institute remove eggs from the ovary via a small incision

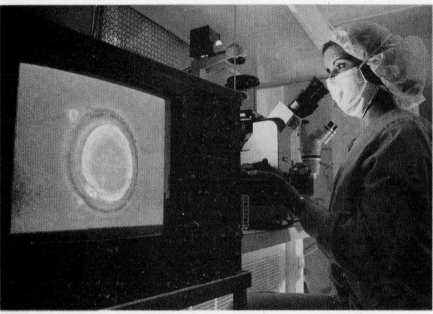

The extracted fluid containing the tiny ova is then examined under a microscope

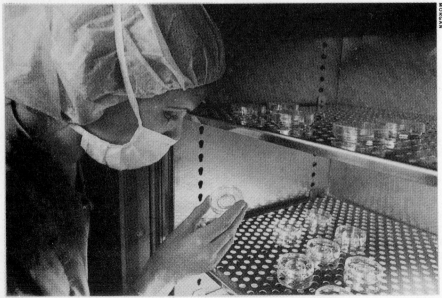

After the egg is fertilized, it is placed in an incubator for two to three days

ovaries to the uterus, are where fertilization normally occurs. If they are blocked or damaged or frozen in place by scar tissue, the egg will be unable to complete its journey. To examine the tubes, a doctor uses X rays or a telescope-like instrument called a laparoscope, which is inserted directly into the pelvic area through a small, abdominal incision. Delicate microsurgery, and, more recently, laser surgery, sometimes can repair the damage successfully. According to Beverly Freeman, executive director of Resolve, a national infertility-counseling organization, microsurgery can restore fertility in 70% of women with minor scarring around their tubes. But for those whose tubes are completely blocked, the chance of success ranges from 20% to zero. These women are the usual candidates for in-vitro fertilization.

Much has been learned about the technique since the pioneering days of Steptoe and Edwards. When the two Englishmen first started out, they assumed that the entire process must be carried out at breakneck speed: harvesting the egg the minute it is ripe and immediately adding the sperm. This was quite a challenge, given that the collaborators spent most of their time 155 miles apart, with Edwards teaching physiology at Cambridge and Steptoe practicing obstetrics in the northwestern mill town of Oldham. Sometimes, when one of Steptoe's patients was about to ovulate, the doctor would have to summon his partner by phone. Edwards would then jump into his car and charge down the old country roads to Oldham. Once there, the two would remove the egg and mate it with sperm without wasting a moment; by the time Lesley Brown became their patient, they could perform the procedure in two minutes flat. They believed that speed was the important factor in the conception of Louise Brown.

As it happens, they were wrong. Says Gynecologist Howard Jones, who, together with his wife, Endocrinologist Georgeanna Seegar Jones, founded the first American in-vitro program at Norfolk in 1978: "It turns out that if you get the sperm to the egg quickly, most often you inhibit the process." According to Jones, the pioneers of IVF made so many wrong assumptions that "the birth of Louise Brown now seems like a fortunate coincidence."

Essential to in-vitro fertilization, of course, is retrieval of the one egg normally produced in the ovaries each month. Today in-vitro clinics help nature along by administering such drugs as Clomid and Pergonal, which can result in the development of more than one egg at a time. By using hormonal stimulants, Howard Jones "harvests" an average of 5.8 eggs per patient; it is possible to obtain as many as 17. "I felt like a pumpkin ready to burst," recalls Loretto Leyland, 33, of Melbourne, who produced eleven eggs at an Australian clinic, one of which became her daughter Zoe.

According to Quigley, the chances for pregnancy are best when the eggs are retrieved during the three- to four-hour period when they are fully mature. At Bourn Hall, women remain on the premises, waiting for that moment to occur. Each morning, Steptoe, now 71 and walking with a cane, arrives on the ward to check their charts. The husband of one patient describes the scene: "Looking at a woman like an astonished owl, he'll say, 'Your estrogen is rising nicely.' The diffidence is his means of defense against desperate women. They think he can get them pregnant just by looking at them."

When blood tests and ultrasound monitoring indicate that the ova are ripe, the eggs are extracted in a delicate operation performed under general anesthesia. The surgeons first insert a laparoscope, which is about ⅓ in. in diameter, so that they can see the target: the small, bluish pocket, or follicle, inside the ovary, where each egg is produced. Then, a long, hollow needle is inserted through a second incision, and the eggs and the surrounding fluid are gently suctioned up. Some clinics are beginning to use ultrasound imaging instead of a laparoscope to guide the needle into the follicles. This procedure can be done in a doctor's office under local anesthesia; it is less expensive than laparoscopy but may be less reliable.

Once extracted, the follicular fluid is rushed to an adjoining laboratory and examined under a microscope to confirm that it contains an egg (the ovum measures only four-thousandths of an inch across). The ova are carefully washed, placed in petri dishes containing a solution of nutrients and then deposited in an incubator for four to eight hours. The husband, meanwhile, has produced a sperm sample. It is hardly a romantic moment, recalls Cleveland Businessman Popela, who made four trips to Cambridgeshire with his wife, each time without success. "You have to take the jar and walk past a group of people as you go into the designated room, where there's an old brass bed and a couple of *Playboy* magazines. They all know what you're doing and they're watching the clock, because there are several people behind you waiting their turn."

The sperm is prepared in a solution and then added to the dishes where the eggs are waiting. The transcendent moment of union, when a new life begins, occurs some time during the next 24 hours, in the twilight of an incubator set at body heat. If all goes well, several of the eggs will be fertilized and start to divide. When the embryo is at least two to eight cells in size, it is placed in the woman's uterus. During this procedure, which requires no anesthetic, Steptoe likes to have the husband present talking to his wife. "The skill of the person doing the replacement is very important," he says. "The womb doesn't like things being put into it. It contracts and tries to push things out. We try to do it with as little disturbance as possible."

The tension of the next two weeks, as

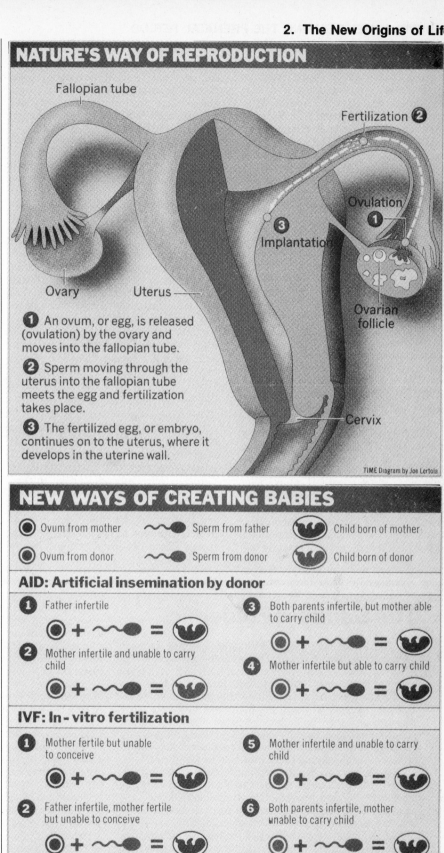

NATURE'S WAY OF REPRODUCTION

Fallopian tube
Fertilization ②
Ovulation ①
Implantation ③
Ovary
Uterus
Ovarian follicle
Cervix

① An ovum, or egg, is released (ovulation) by the ovary and moves into the fallopian tube.

② Sperm moving through the uterus into the fallopian tube meets the egg and fertilization takes place.

③ The fertilized egg, or embryo, continues on to the uterus, where it develops in the uterine wall.

TiME Diagram by Joe Lertola

NEW WAYS OF CREATING BABIES

⦿ Ovum from mother ~ Sperm from father ⚭ Child born of mother
⦿ Ovum from donor ~ Sperm from donor ⚭ Child born of donor

AID: Artificial insemination by donor

① Father infertile
⦿ + ~ = ⚭

② Mother infertile and unable to carry child
⦿ + ~ = ⚭

③ Both parents infertile, but mother able to carry child
⦿ + ~ = ⚭

④ Mother infertile but able to carry child
⦿ + ~ = ⚭

IVF: In-vitro fertilization

① Mother fertile but unable to conceive
⦿ + ~ = ⚭

② Father infertile, mother fertile but unable to conceive
⦿ + ~ = ⚭

③ Mother infertile but able to carry child
⦿ + ~ = ⚭

④ Both parents infertile, but mother able to carry child
⦿ + ~ = ⚭

⑤ Mother infertile and unable to carry child
⦿ + ~ = ⚭

⑥ Both parents infertile, mother unable to carry child
⦿ + ~ = ⚭

⑦ Mother unable to carry child, but both parents fertile
⦿ + ~ = ⚭

⑧ Mother fertile but unable to carry child, father infertile
⦿ + ~ = ⚭

the couple awaits the results of pregnancy tests, is agonizing. "Women have been known to break out in hives," reports Linda Bailey, nurse-coordinator at the IVF program at North Carolina Memorial Hospital in Chapel Hill. Success rates vary from clinic to clinic; some centers open and close without a single success. But even the best clinics offer little more than a 20% chance of pregnancy. Since tiny factors like water quality seem to affect results, both physicians and patients tend to become almost superstitious about what else might sway the odds. Said one doctor: "If someone told us that painting the ceiling pink would make a difference, we would do it."

In recent years, IVF practitioners have discovered a more reliable way of improving results: transferring more than one embryo at a time. At the Jones' clinic, which has one of the world's highest success rates, there is a 20% chance of pregnancy if one embryo is inserted, a 28% chance if two are used and a 38% chance with three. However, transferring more than one embryo also increases the likelihood of multiple births.

For couples who have struggled for years to have a child, the phrase "you are pregnant" is magical. "We thought we would never hear those words," sighs Risa Green, 35, of Framingham, Mass., now the mother of a month-old boy. But even if the news is good, the tension continues. One-third of IVF pregnancies spontaneously miscarry in the first three months, a perplexing problem that is currently un-der investigation. Says one veteran of Steptoe's program: "Every week you call for test results to see if the embryo is still there. Then you wait to see if your period comes." The return of menstruation is like a death in the family; often it is mourned by the entire clinic.

Many couples have a strong compulsion to try again immediately after in vitro fails. Popela of Cleveland compares it to a gambling addiction: "Each time you get more desperate, each time you say, 'Just one more time.' " In fact, the odds do improve with each successive try, as doctors learn more about the individual patient. But the stakes are high: in the U.S., each attempt costs between $3,000 and $5,000, not including travel costs and time away from work. Lynn Kellert, 31, and her husband Mitchell, 34, of New York City, who tried seven times at Norfolk before finally achieving pregnancy, figure the total cost was $80,000. Thus far, few insurance companies have been willing to foot the bill, arguing that IVF is still experimental. But, observes Grobstein of UCSD, "it's going to be increasingly difficult for them to maintain that position."

Second and third attempts will become easier and less costly with the wider use of cryopreservation, a process in which unused embryos are frozen in liquid nitrogen. The embryos can be thawed and then transferred to the woman's uterus, eliminating the need to repeat egg retrieval and fertilization. Some 30% to 50% of embryos do not survive the deep freeze. Those that do may actually have a better chance of successful implantation than do newly fertilized embryos. This is because the recipient has not been given hormones to stimulate ovulation, a treatment that may actually interfere with implantation.

Opinion is sharply divided as to how age affects the results of IVF. Although most clinics once rejected women over age 35, many now accept them. While one faction maintains that older women have a greater tendency to miscarry, Quigley, for one, insists that "age should not affect the success rate." Curiously, the Joneses in Norfolk have achieved their best results with women age 35 to 40. This year one of their patients, Barbara Brooks of Springfield, Va., had a test-tube son at age 41; she can hardly wait to try again.

Doctors are also beginning to use IVF as a solution to male infertility. Ordinarily, about 30 million sperm must be produced to give one a chance of penetrating and fertilizing the egg. In the laboratory, the chances for fertilization are good with only 50,000 sperm. "In vitro may be one of the most effective ways of treating men with a low sperm count or low sperm motility, problems that affect as many as 10 million American men," says Andrologist Wylie Hembree of Columbia-Presbyterian Medical Center in New York City.

While most clinics originally restricted IVF to couples who produced normal sperm and eggs, this too is changing. Today, when the husband cannot supply adequate sperm, most clinics are willing to use sperm from a donor, usually obtained from one of the nation's more than 20 sperm

The Saddest Epidemic

Richard and Diana Barger of Virginia could be a textbook case of an infertile couple. Diana's fallopian tubes and left ovary are blocked with scar tissue, ironically the result of an intrauterine device (I.U.D.) she used for three years. Even if an egg did manage to become fertilized, the embryo might be rejected by her uterus, which has been deformed since birth. Richard has his own difficulties: his sperm count is 6.7 million per milliliter, considerably below the number ordinarily required for fertilization under normal conditions. Says Diana: "I never thought getting pregnant would be so difficult."

The Bargers are victims of what Reproductive Endocrinologist Martin Quigley of the Cleveland Clinic calls "an epidemic" of infertility in the U.S. In the past 20 years, the incidence of barrenness has nearly tripled, so that today one in six American couples is designated as infertile, the scientific term for those who have tried to conceive for a year or more without success. More than a million of these desperate couples seek the help of doctors and clinics every year. Women no longer carry the sole blame for childless marriages. Research has found that male deficiencies are the cause 40% of the time, and problems with both members of the marriage account for 20% of reported cases of infertility.

Doctors place much of the blame for the epidemic on liberalized sexual attitudes, which in women have led to an increasing occurrence of genital infections known collectively as pelvic inflammatory disease. Such infections scar the delicate tissue of the fallopian tubes, ovaries and uterus. Half of these cases result from chlamydia, a common venereal disease, and 25% stem from gonorrhea.

Other attitudes are also at fault: by postponing childbirth until their mid- or even late 30s, women risk a barren future. A Yale University study of 40 childless women found that after 35 years of age, the time it takes to conceive lengthens from an average of six months to more than two years.

Other surveys have found that such athletic women as distance runners, dancers and joggers can suffer temporary infertility. The reason is that their body fat sometimes becomes too low for the production of the critical hormone estrogen. Stress can also suppress ovulation; women executives often miss two or three consecutive menstrual periods.

Infertility is easier to trace in men, but often much harder to treat. The commonest problems are low sperm counts and blocked sperm ducts. Among all men, 15% have varicose veins on the left testicle, which can reduce sperm production. Certain drugs and chemicals such as insecticides can also lower sperm counts. A man's fecundity also decreases with age, although not with the dramatic finality of female menopause. Happily, the source of infertility in couples can be diagnosed 95% of the time, and half of all these cases can be treated.

banks. An even more radical departure is the use of donor eggs, pioneered two years ago by Dr. Alan Trounson and Dr. Carl Wood of Melbourne's Monash University. The method can be used to bring about pregnancy in women who lack functioning ovaries. It is also being sought by women who are known carriers of genetic diseases. The donated eggs may come from a woman in the Monash IVF program who has produced more ova than she can use. Alternately, they could come from a relative or acquaintance of the recipient, providing that she is willing to go through the elaborate egg-retrieval process.

At Harbor Hospital in Torrance, Calif., which is affiliated with the UCLA School of Medicine, a team headed by Obstetrician John Buster has devised a variant method of egg donation. Instead of fertilizing the ova in a dish, doctors simply inseminate the donor with the husband's sperm. About five days later, the fertilized egg is washed out of the donor's uterus in a painless procedure called lavage. It is then placed in the recipient's womb. The process, which has to date produced two children, "has an advantage over IVF," says Buster, "because it is nonsurgical and can be easily repeated

until it works." But the technique also has its perils. If lavage fails to flush out the embryo, the donor faces an unwanted pregnancy.

The most controversial of the new methods of reproduction does not depend on advanced fertilization techniques. A growing number of couples are hiring surrogate mothers *(see box)* to bear their children. Surrogates are being used in cases where the husband is fertile, but his wife is unable to sustain pregnancy, perhaps because of illness or because she has had a hysterectomy. Usually, the hired woman is simply artificially inseminated with the

A Surrogate's Story

Valerie is a New Jersey mother of two boys, age two and three, whom she describes as "little monsters full of mischief." Her husband works as a truck driver, and money is tight. The family of four is living with her mother while they save for an apartment of their own. One day last March, Valerie, 23, who prefers to remain anonymous, saw the following advertisement in a local New Jersey paper: *Surrogate mother wanted. Couple unable to have child willing to pay $10,000 fee and expenses to woman to carry husband's child. Conception by artificial insemination. All replies strictly confidential.*

The advertisement made Valerie stop and think. "I had very easy pregnancies," she says, "and I didn't think it would be a problem for me to carry another child. I figured maybe I could help someone." And then there was the lure of the $10,000 fee. "The money could help pay for my children's education," she says, "or just generally to make their lives better."

The next day Valerie went for an interview at the Infertility Center of New York, a profit-making agency owned by Michigan Attorney Noel Keane, a pioneer in the controversial business of matching surrogate mothers with infertile parents. She was asked to fill out a five-page application, detailing her medical history and reasons for applying. Most applicants are "genuine, sincere, family-oriented women," says agency Administrator Donna Spiselman. The motives they list range from "I enjoy being pregnant" and an urge to "share maternal joy" to a need to alleviate guilt about a past abortion by bearing someone else's child. Valerie's application and her color photograph were added to 300 others kept in scrapbooks for prospective parents to peruse. Valerie was amazed when only one week later her application was selected, and she asked to return to the agency to meet the couple.

Like most people who find their way to surrogate agencies, "Aaron" and "Mandy" (not their real names) had undergone years of treatment for infertility. Aaron, 36, a Yale-educated lawyer, and his advertising-executive wife, 30, had planned to have children soon after marrying in 1980. They bought a two-bedroom town house in Hoboken, N.J., in a neighborhood that Aaron describes as being "full of babies." But after three years of tests, it became painfully clear that there was little hope of having the child they longed for. They considered adoption, but were discouraged by the long waiting lists at American agencies and the expense and complexity of foreign adoptions. Then, to Aaron's surprise, Mandy suggested that they try a surrogate.

Their first choice from the Manhattan agency failed her mandatory psychological test, which found her to be too emotionally unstable. Valerie, who was Aaron and Mandy's second choice, passed without a hitch. A vivacious woman who is an avid reader, she more than met the couple's demands for a surrogate who was "reasonably pretty," did not smoke or drink heavily and had no family history of genetic disease. Says Aaron: "We were particularly pleased that she asked us questions to find out whether we really want this child."

At first, Valerie's husband had some reservations about the arrangement, but, she says, he ultimately supported it "100%." Valerie is not concerned about what her neighbors might think because the family is planning to move after the birth. Nor does she believe that her children will be troubled by the arrangement because, she says, they are too young to understand. And although her parents are being deprived of another grandchild, they have raised no objections.

For their part, Aaron and Mandy have agreed to pay Valerie $10,000 to be kept in an escrow account until the child is in their legal custody. In addition, they have paid an agency fee of $7,500 and are responsible for up to $4,000 in doctors' fees, lab tests, legal costs, maternity clothes and other expenses. In April, Valerie became pregnant after just one insemination with Aaron's sperm. Mandy says she was speechless with joy when she heard the news.

Relationships between surrogate mothers and their employers vary widely. At the National Center for Surrogate Parenting, an agency in Chevy Chase, Md., the two parties never meet. At the opposite extreme is the case of Marilyn Johnston, 31, of Detroit. Johnston and the couple who hired her became so close during her pregnancy that they named their daughter after her. She continues to make occasional visits to see the child she bore and says, "I feel like a loving aunt to her."

Not all surrogate arrangements work so well. Some women have refused to give up the child they carried for nine months. As a lawyer, Aaron is aware that the contract he signed with Valerie would not hold up in court, should she decide to back out of it. "But I'm a romantic," he says. "I have always felt that the real binding force was not paper but human commitment." Valerie, whose pregnancy is just beginning to show, says she is "conditioning" herself not to become too attached to the baby. "It is not my husband's child," she says, "so I don't have the feeling behind it as if it were ours." She does not plan to see the infant after it is born, but, she admits, "I might like to see a picture once in a while." —*By Claudia Wallis.*
Reported by Ruth Mehrtens Galvin/New York

husband's sperm. However, if the wife is capable of producing a normal egg but not capable of carrying the child, the surrogate can be implanted with an embryo conceived by the couple. This technique has been attempted several times, so far without success.

The medical profession in general is apprehensive about the use of paid surrogates. "It is difficult to differentiate between payment for a child and payment for carrying the child," observes Dr. Ervin Nichols, director of practice activity for the American College of Obstetrics and Gynecology. The college has issued strict guidelines to doctors, urging them to screen carefully would-be surrogates and the couples who hire them for their medical and psychological fitness. "I would hate to say there is no place for surrogate motherhood," says Nichols, "but it should be kept to an absolute minimum."

In contrast, in-vitro fertilization has become a standard part of medical practice. The risks to the mother, even after repeated attempts at egg retrieval, are "minimal," points out Nichols. Nor has the much feared risk of birth defects materialized. Even frozen-embryo babies seem to suffer no increased risk of abnormalities. However, as Steptoe points out, "we need more research before we know for sure."

The need for research is almost an obsession among IVF doctors. They are eager to understand why so many of their patients miscarry; they long to discover ways of examining eggs to determine which ones are most likely to be fertilized, and they want to develop methods of testing an embryo to be certain that it is normal and viable. "Right now, all we know how to do is look at them under the microscope," says a frustrated Gary Hodgen, scientific director at the Norfolk clinic.

Many scientists see research with embryos as a way of finding answers to many problems in medicine. For instance, by learning more about the reproductive process, biologists may uncover better methods of contraception. Cancer research may also benefit, because tumor cells have many characteristics in common with embryonic tissues. Some doctors believe that these tissues, with their tremendous capacity for growth and differentiation, may ultimately prove useful in understanding and treating diseases such as childhood diabetes. Also in the future lies the possibility of identifying and then correcting genetic defects in embryos. Gene therapy, Hodgen says enthusiastically, "is the biggest idea since Pasteur learned to immunize an entire generation against disease." It is, however, at least a decade away.

American scientists have no trouble dreaming up these and other possibilities, but, for the moment, dreaming is all they can do. Because of the political sensitivity of experiments with human embryos, federal grant money, which fuels 85% of biomedical research in the U.S., has been denied to scientists in this field. So controversial is the issue that four successive Secretaries of Health and Human Services (formerly Health, Education and Welfare) have refused to deal with it. This summer, Norfolk's Hodgen resigned as chief of pregnancy research at the National Institutes of Health. He explained his frustration at a congressional hearing: "No mentor of young physicians and scientists beginning their academic careers in reproductive medicine can deny the central importance of IVF–embryo transfer research." In Hodgen's view the curb on research funds is also a breach of government responsibility toward "generations of unborn" and toward infertile couples who still desperately want help.

In an obstetrics waiting room at Norfolk's in-vitro clinic, a woman sits crying. Thirty-year-old Michel Jones and her husband Richard, 33, a welder at the Norfolk Navy yard, have been through the program four times, without success. Now their insurance company is refusing to pay for another attempt, and says Richard indignantly, "they even want their money back for the first three times." On a bulletin board in the room is a sign giving the schedule for blood tests, ultrasound and other medical exams. Beside it hangs a small picture of a soaring bird and the message: *"You never fail until you stop trying."* Michel Jones is not about to quit. Says she: "You have a dream to come here and get pregnant. It is the chance of a lifetime. I won't give up."

—By Claudia Wallis. Reported by Mary Cronin/London, Patricia Delaney/Washington and Ruth Mehrtens Galvin/Norfolk

ARE THE PROGENY PRODIGIES?

Glenn Garelik

"I want yummie!"

Dr. Afton Blake (IQ 130+), psychologist and mother, has barely had time to turn on the ignition of her beat-up Honda, before her three-year-old son Doron squirms out of his plastic car seat. "Gimme yummie!"

Before the age of two, Blake says, Doron (Greek for gift) learned to use all the proper parts of speech. He tested as a four-year-old at half that age, and, as a rule, he doesn't indulge in baby talk. But this is clearly a matter to which rules don't apply.

"He only uses baby talk for this one thing," says Blake, who specializes in child development. "Just this one thing."

And this one thing is?

She turns off the ignition. Doron is already in her lap, looking up at her with a smile that would melt the polar caps.

Dutifully, importantly, Blake unbuttons her blouse.

Doron William Blake was the second child to be born through the good offices of the so-called Nobel sperm bank, as the Repository for Germinal Choice has been known since its inception in 1980. The bank was founded by plastic-eyeglass-lens magnate and social-theory hobbyist Robert Graham (IQ 130+), then 73, in Escondido, Calif. for the express purpose of improving the human stock. It would collect and freeze the sperm of Nobel Prize-winning scientists and offer it, free of charge, to women of intelligence, and at least some means, whose husbands were infertile.

It used to be that to improve your children's lot you moved out of your crummy old neighborhood and headed for a suburb where the schools were top notch. But to many people the Repository seemed to be saying that, when it came to improving the kiddies' prospects, the old methods wouldn't do. Sure, fresh air and better teachers are nice, but if you really wanted to ensure a grander future for your children, you had to give them better genes.

Not surprisingly, this view drew heavy fire. Critics of a philosophical bent argued that it shouldn't be assumed that brighter meant better or that the Nobel Prize—or even IQ generally—was an accurate index of social usefulness or human worth. Scientists pointed out that there's no hard evidence that genes are the decisive component of intelligence, and that even if they are, intelligence can't be reliably reproduced. Sociologists invoked memories of the nightmare that was the German "eugenics" program of the 1930s and '40s. The Nazis not only glorified certain traits and attempted to breed children who had them, but murdered people who didn't.

The critical uproar was hardly allayed by the enthusiastic support given the Repository by *soi-disant* anthropologist William Shockley (IQ 129), a physicist who shared a Nobel Prize in 1956 for inventing the transistor—and who won the prize for chutzpah in the 1960s and '70s for preaching elitist genetics. Shockley was one of three Nobelists—the others have never been identified—who Graham says actually obliged with their sperm.

If the outcry helped keep down the number of Nobel contributors, it didn't at all inhibit the continuance of the Repository, which proceeded to pursue what it considered the next-best course—the participation of high-IQ,* high-achievement scientists who hadn't been honored by the Nobel committee.

Using the donations of less proven geniuses, the Repository—along with Heredity Choice, an offshoot run by Graham's former "research officer," Paul Smith (IQ 150)—has "sired" 20 children, with another 17 on the way. The essential question, with the brouhaha five years behind, is: Are the progeny prodigies? Do genes really make for genius?

Almost nothing about these children, except Doron Blake, has been revealed, and though we'll get to some of the other brave new children later, he remains the most accessible of the "Nobel" babies. What to make of him?

---

*According to Graham, although the Repository requires no minimum IQ, all its present donors have scores of more than 140. The highest is 182.

The Nobel sperm bank, founded to elevate the IQ of mankind, has produced 20 kids—none sired by laureates—and lots of foolishness. Are the children smart? It may not matter

Says Blake, annoyed because Doron's school toilet-trained him, "There's no containing him, and I don't try to. That's not appropriate to a three-year-old"

In a Los Angeles restaurant on a recent afternoon, Doron ignores his drink and takes his mother's iced tea instead. He adds sugar from the glass dispenser on the table. Then he starts to pour sugar into his mother's shrimp cocktail. "You have ca-ca," he says to no one in particular.

"Emotionally, he's his age, I'll grant," says Blake, according to whom Doron is nonetheless "intellectually advanced." That may be thanks in part, she says, to his anonymous father, who she says was described in the Repository catalogue as a handsome, blond, athletic science professor who scored 800 on his math SAT and won prizes for performing classical music. (His only drawback is a 30 per cent chance he'll pass on to his children a propensity for hemorrhoids.) Blake's genetic contribution, she says, goes back to the royal court of Norway on one side and to the poet William Blake on the other.

The results of Doron's heritage, she believes, are already in evidence. At the age of two, he was pictured in *Newsweek* at a piano, and she says he's considerate and cooperative. Indeed, almost unbidden, he releases the shrimp cocktail and pours the remaining contents of the sugar dispenser into his mother's tea. "You have ca-ca," he chirps again.

"For Mother's Day I took my mother to a very elegant restaurant," Blake says. "It had a fish pond. There was a statue by the pond, and Doron was leaning on the statue. He broke it and fell in." Ah, boys will be boys.

If only she could afford to have another. She's so pleased with the results of her pioneering venture that she has gotten an exclusive, as it were, on the sperm of Doron's father—catalogue designation 28 Red. (The colors in this identification system stand for nothing.) Daddy has discontinued his donations now, so Blake keeps vials of the original supply in her garage, frozen in a tank of liquid nitrogen.

The scene is another restaurant on another day: Doron is thirsty. "Do you have anything for a kid?" Blake asks a waitress.

"How about a couple of ropes?" mutters the woman before recommending something called Wild West punch.

Clearly, Doron is out of control. "There's no containing him," says Blake, who's less bothered by this than by the fact that Doron's day school recently toilet-trained him. He has mostly lost the interest in the piano he evinced at two, but he plays his mother like a violin. "There's no containing him, and I don't try to. That's not appropriate to a three-year-old."

That's an appealing line, one that reflects a philosophy of child rearing that in theory might work: if a kid can have maximum control of his world in his early years, he'll have no need to be controlling as an adult. But Doron isn't an adult, and the unbridled id is tyrannical. Blake is tense, nearly overwhelmed, but she's unyielding in her indulgence of her son.

Doron may some day be all she says he'll be, but who knows? Because intelligence is more than the ability to score well on tests, and because intelligence alone doesn't ensure success, even a towering IQ may bear little fruit if its possessor can't harness it to real tasks in the real world. Even Paul Smith—who, citing early tests of Doron, proudly describes him as an "unqualified success"—says, "Most child psychologists, like most dog breeders and trainers, recognize you've got to teach a child his limits."

If nothing else, the Blakes make it clear that the first generation of "Nobel" children and parents isn't all it might be. What's more, if they're any indication of what the Repository has wrought, it paradoxically may prove to be the laboratory from which definitive evidence finally comes that—at least as far as usable intelligence is concerned—nurture overwhelms nature.

There are some indications that this is the case. According to an ongoing study begun in 1921 by psychologist Lewis Terman of more than 1,500 California children with IQs over 135, the most reliable bench mark for adult success isn't a person's exact IQ but such qualities as persistence and the desire to excel.

The Terman study also suggests that the children of highly intelligent people are intelligent too—but not, on the average, as intelligent as their parents. This is true even though intelligent people tend to marry intellectual peers. (The failure of these children to attain the intelligence levels of their parents is a manifestation of a statistical phenomenon called regression toward the mean.) Because the Repository, for reasons that will be explained later, is less careful about screening its recipients than it is about choosing its donors, and because it's less careful than most men would be in selecting a mate, it would seem to follow that the average IQ of its offspring would be somewhat lower than that of donors' children by their wives.

Of course, the odds are extremely small that anyone—via a sperm bank or the conjugal bed—will reliably reproduce the special combination of factors that yields not just creative genius but significant accomplishment as well. (Even George Bernard Shaw, who advocated a kind of eugenics, recognized the whimsical nature of genetic inheritance. According to one story, a beautiful woman wrote him suggesting that, with her body and his brain, they could produce superior offspring. Shaw wrote back, "But, alas, what if the child inherits *my* body and *your* brain?")

However, says Graham, the production of laureates was never his aim; in fact, he says, "if we ever see a Nobelist out of this, I'll be astonished. We're happy just to produce competent, creative, interesting people who might otherwise not be born."

In theory, Graham says, that alone will improve the general gene pool and draw the species up by its bootstraps. And drawing the species up by its bootstraps is necessary, he thinks, because the comforts and advances of modern civilization have frustrated the process of natural selection by allowing the less fit to survive. "As a species," he says, "we've been experiencing an accelerating degeneration in recent centuries."

Besides, says Graham, a fringe benefit of humanity's general ascent through selective breeding might be the birth of a secular savior or two who will by some scientific insight or technical innovation lead mankind past its present impasses. These, anyway, were the ideas, as Graham put them forward in his self-published *The Future of Man* (1970), and as they were articulated before that by Hermann Muller, the late Nobel Prize — winning geneticist to whom Graham dedicated the Repository.

All this isn't as Hitlerian as some critics originally charged—the idea is merely that the more intelligent you are, the more children you should have. The breeding is voluntary; there's no sterilization of the less spectacularly endowed. And the application of the ideas is so limited, at least to date, that it represents a mere blip on the social horizon. In a way, it's almost charming—by turns innocent and inept, as noble in its way as it is nutty.

It's nearly 100 degrees in Escondido, but inside the Re-

pository, located at the rear of the Pacific Coast Savings Bank on South Escondido Boulevard, it's clean and cool, if a bit down at the heels. Julianna McKillop (who refused to disclose her IQ), the Repository's general factotum and principal recruiter of donors, picks up a plastic pitcher, as if she's about to pour a glass of lemonade. "You'll excuse me while I replenish the tank," she says, as she removes the top of a large canister of liquid nitrogen. Evaporating nitrogen billows from it, as it does from the pitcher, which she has filled from a larger container. She draws a "cane" of six ampules of frozen sperm from the tank, quickly examines it, and returns it. Then she pours some nitrogen from the pitcher into the tank.

"Life is an exercise in problem-solving," she says calmly, iterating one of the Repository's tenets. "Don't we all look for the best mate anyway, the best place to live? We want everything. Being smart may not guarantee happiness, but I'd rather be bright and miserable than dumb and miserable: it helps in solving problems. And after all, we're just trying to make it possible for infertile couples who want children to have them with the best germinal material."

As for what constitutes superior germinal material, Graham originally limited his definition to that of Nobel science laureates, and later to that of other notable scientists; among the virtues of using scientists, he says, is "there's less conjecture about their standing than there is about men in other fields.

"Anyway, we had to start recruiting *somewhere!* We just passed the buck and let the Nobel committee make the judgments."

But there was a problem. Even in samples from younger donors, as much as half the sperm fails to survive the freezing and thawing that are part of the Repository pro-

cess; the survival rate was wretched in men old enough to have distinguished records and then to have been recognized by an eminent body. However, it didn't really matter that there was a dearth of healthy Nobel sperm. Says Graham, "The women weren't choosing the Nobelists much, anyway." Few recipients asked for laureates' sperm (which, for the sake of anonymity, was never labeled as such), and none was impregnated by it. Whatever romantic stirrings other women might have experienced upon reading what were in fact Nobel donors' descriptions were apparently crushed by the ages that were listed: twice and three times their own. And as most recipients are also aware, older men's sperm is much more likely than that of younger men to harbor mutations.

In the face of all the problems with Nobelists, the Repository went from one extreme to another, and attempted to draw donors from a well known study of precocious adolescents. But the head of the program declined to offer up any of his callow subjects.

Clearly, whatever scientific rigor there was to have been in the program was quickly being lost; but then the Repository has always really been much more a social service than a social experiment. Besides, Graham says, his guiding principle has always been more complex than has been generally reported. "High IQ is far from enough for us," he says, "although it's a fairly good predictor of what we're after. What we really want there's no precise name for— although I suppose 'creativity' might do. We expect a person to have demonstrated achievement and an interest in contributing something to the world. The Repository won't settle for just smart."

In practice, not settling for just smart means not insisting

on smart when a potential donor has alternative attractions. This spring, for example, Graham recruited an Olympic gold medalist. Last fall he even signed up a student, a staple of less discriminating banks and, until that time, a practice that had been the butt of the Repository's scorn.

When donor Light Green 40 (IQ 150+), a graduate student in Wisconsin, read about the Repository, he says, he put himself "in the place of couples who're trying to have children and can't." With his wife's blessing, he decided it would be "personally gratifying" to help these unfortunate couples out.

But the Nobel sperm bank? Light Green 40 believes in its principle—"Parents concerned enough to go to the Repository are likely to be people who'll also provide the right environment for the child's intellectual growth," he says—and he believes, a lot, in himself. Though short a Nobel Prize—or any other comparable distinction—he called up Graham.

Graham invited Light Green 40 to Escondido and asked him to donate sperm—not, as Muller would have wanted, to be stored until he had proved himself in his career, but for immediate use. Says Dora Vaux (IQ unknown), one of Graham's recruiters, a little dreamily, "For him we made an exception—he's our only student. But he's not only smart, he's very well rounded, too, in physique and other things."

"Hmmm," says McKillop, now eyeing a visitor like a rancher surveying a slightly misshapen bull. If, for lack of distinguished scientists, the recruitment process is becoming more random, it has nonetheless not altogether lost its standards. After a minute or two of questioning, this unlikely prospect passes muster, and she's flush with excitement.

But now he flirts with failure by volunteering "I'm myopic. Nearsighted. Not terribly athletic, and nearsighted."

McKillop looks deflated. Her face drops. "Really? But you don't wear . . .?"

"Contacts. I wear contacts."

"Oh. I see. I mean, I couldn't see them. I mean, I didn't know. Do you really?"

"Yes. Nearsighted. I am. I do. But lots of your donors are nearsighted, aren't they?"

McKillop seems no longer to be listening.

No matter. Just then the proceedings are interrupted as Vaux bursts breathlessly into the inner office.

"Did you hear *that*?" she says. "Man came in was crazy—wanted to donate."

"What'd you say? Was he really? What'd he want? Was he crazy?"

"Man was crazy. *Said* he was crazy—not in his genes, but the result of an accident. Genes're still good, he says. He wants to donate while they still are."

"What was his IQ?" asks McKillop.

Theoretically, of course, Graham could offer his recipients a more liberal array of donors, even including self-proclaimed crazies, and still achieve his purposes. After all, he points out, "the final judges of whose genes will be perpetuated are the women."

To recruit women, Graham made his first public pitch in the bulletin of the San Diego County chapter of Mensa, an international organization, of which he's a member, for people who demonstrate intelligence in the top two per cent of the population. Two dozen women applied, but for a variety of reasons he found none of them suitable.

To a woman, the recipients of "genius" sperm say their bundles of joy are blessing enough; they won't demand little geniuses in the bargain. The world's a tough place, and they just want

S ays Smith, "We're not talking about producing geniuses but about shifting the mean toward the gifted"

to give their kids the best fighting chance they can. For example, Blake says that using the Repository's "smart" sperm was just a matter of "giving a child every possible advantage. But then he should go where his own path leads him. A child should have no obligation to the parent or the bank."

Whatever a parent claims to expect, a child who learns of his origins might expect an inordinate lot of himself. But more immediately dangerous is the parent who, despite protestations to the contrary, demands evidence of brilliance or perfect behavior.

That may have been what happened with the Repository's first recipient, Joyce Kowalski, now 42, of Scottsdale, Ariz. In 1980, Kowalski approached the Repository, claiming that her husband, Jack, had a sperm count too low to get her pregnant. In July 1982, three months after she gave birth to a baby girl she named Victoria, it was discovered that she and Jack, who's her second husband, had lost custody of her two children by her first marriage after their father charged the Kowalskis with abusing them. Apparently, the Kowalskis had expected the children to attain unreasonable standards of behavior and scholarship. In light of the extravagant expectations her parents expressed for Baby Vicki in an interview they sold to the *National Enquirer*, it's hard not to wonder what demands they will make of *her*. (Although the Kowalski affair "may have

been a social blunder," says Graham, "it wasn't a genetic one: Mrs. Kowalski's got an IQ of 130 or 140.")

But whatever "Nobel" mothers may ask of their children, and no matter how fussy they may be in their selection of donors (according to Smith, 18 of the 20 births to date are the offspring of only three men), few demands are put on recipients. "I never reject a woman because she's too 'umble," says Smith, who "doesn't quite make a living" running Heredity Choice and breeding border collies.

"There's a lot of sperm to go around, and if a woman wants to get pregnant she will," he goes on, employing the sort of logic usually reserved for arguments in favor of, say, legalizing marijuana. "If she doesn't get it from me, she'll get it from a less illustrious source. So my only concern becomes, will she make a good mother? Other than that—and unless she's mad as a hatter—I don't try to screen recipients at all."

According to Smith's calculations, a mother with an IQ of 100 and a father with a 200 will produce offspring whose IQ is halfway between their average (150) and the general mean (100)—that is, 125. "And that," he says, as if dusting chalk off his hands, "will become the populational average—still higher than any current group. We're not talking about producing geniuses but about shifting the norm toward the gifted."

Last year, after Smith had been working at the Reposi-

tory for several years, an off-hand remark he made in 1983 about a sperm bank he considered less selective led to a parting of the ways with Graham. The other bank—a fund of defectives, Smith called it—named not only Smith in its $3 million libel suit but also the Repository and Graham, whose fortune has been estimated at no less than $70 million and as high as twice that. The matter is scheduled to come to trial next year, and Graham decided he could ill afford any repetitions of Smith's indiscretion. Smith was sent packing.

It was hardly the first move he was obliged to make. In 1965, after a demonstration in Los Angeles against the Vietnam war, he was convicted of "willfully and maliciously disturbing the peace"; he fled the U.S. for Great Britain, where he lived for 15 years, got married, and fathered a son. Then, in 1980, British authorities found he didn't have a proper visa, and he was "removed," he says, to the U.S. "I've never had a lot of patience with authority," Smith says. His latest scrapes with authority involve a veritable Goliath: the Internal Revenue Service.

Upon being cast out of Escondido, Smith gathered up his effects and his collies, and—with promises of continued allegiance from two mathematicians and a scientist, all three of them popular with the ladies—wound up at Blake's house. There he set up operations until December 1984, when they had a falling out.

Smith then moved to the Chicago suburb of Oak Park, to collaborate with Richard Seed, an aptly named advocate of the more difficult technology of embryo transfer and the founder of an organization called Repro Clinic, Inc.

In England Smith had begun breeding border collies, whose shepherding skills are both legendary and clearly heritable. He brought several of what he describes as the best of the breed back to the U.S., but in Chicago no landlord was sufficiently impressed with them to rent him an apartment. Smith next made his way to a ramshackle farmhouse in the midst of a swampy New Hampshire wood. There he was recently encountered, conducting the business of selective breeding amid a great deal of clutter and the yap, pant, and clatter of his collies, which looked for all the world like five matted mutts.

Over to one side of the dining area are several tanks of liquid nitrogen. In them is the sperm of several donors, including the three he lured from the Repository—and of his prized collie stud, Kep. His old corduroy pants are grease-stained from working on his vehicle, a converted Air Force communications van. One of his shoes is untied, and his zipper is at half mast.

Smith acknowledges that of the several donors who so far account for all the "Nobel" breeding, one, conveniently enough, happens to have been a math student he knew while attending Berkeley. That man, Smith says, has fathered six of the children. A second donor, also a mathematician, has accounted for nine, and has another on the way. A third donor, the aforementioned 28 Red, produced three of the children, including Doron Blake, and is responsible for a pregnancy that began before Afton Blake got her "exclusive."

Smith believes that those with high IQs form an élite, but bristles at attempts to link specialized breeding to totalitarianism. On the contrary, he says, artificial insemination gives women more choices and, therefore, is more democratic. "The Nazis didn't invent breeding," he says. "It happens every time someone chooses a mate or

Says Stevens, "Even if I meet somebody, I'll probably go through with the insemination. What chance is there I'll meet a guy that could match this?"

has one chosen by her parents. If a woman's not very attractive, she'll have to settle for a man who's not so great phenotypically. With artificial insemination, there become available to her some very, very desirable men."

He hastens to excuse himself: "I've got to get a tank to the post office, to Express Mail, right away, before they close." A few days later he announces that this tank has arrived at the Los Angeles home of a 26-year-old lesbian.

"Of the single women who approach me," he says, "the heterosexuals are running out of time—most are in their late thirties and early forties, having wasted time imagining they were going to get married. Lesbians have no illusion about waiting for Mr. Right, so they tend to seek out sperm by their mid-twenties."

Smith's fascination with breeding for intelligence stems from his conviction that it's the one quality that's "uniquely human"—and that "the less bright find life boring and frustrating."

As for good looks, they're more than the window-dressing that Smith wishes they

were. "Recipients wouldn't be happy otherwise," he says. "I'd be delighted to have a Stravinsky, for example, but he'd not be popular with the ladies."

The phone rings. It's one of Smith's recipients, who's three months pregnant. "Hello," Smith says in his breathy, Britishy accent. He chats a bit, jokes. Then he turns silent and seems to blanch. The woman has miscarried. She says she's infected with *Chlamydia*—the most common cause of all venereal infections. The puzzle: She's married and monogamous, but she'd been artificially inseminated; could semen carry the disease? she asks Smith.

"Yes, theoretically," he says, shaken.

Could the semen he sent her have been carrying it, even though it was frozen? Could it have been carrying anything else—AIDS, for example? Could *Chlamydia* cause spontaneous abortion, and could it have done so at this stage of the pregnancy?

Smith offers tentative answers, tenders his condolences, and hangs up. He's in a hurry to consult his microbiologist. He dials, rattles off the questions, listens, hangs up again. He looks bleak. The answers: Yes. Yes. Yes. Yes. And yes. *

"I always expected to have a child late in life," says Irene Adkins, 41, a retired Naval officer with an IQ "between 145 and 155." Adkins is the mother of one-year-old Morgan—fathered by donor 27 Green, an eminent math professor who she says also fathered Vicki Kowalski and

---

*Although the connection is the subject of some uncertainty, a doctor at the Centers for Disease Control in Atlanta considers *Chlamydia* infection "very unlikely" as a cause of spontaneous abortion. As for the source of the woman's infection, Smith now claims she once used fresh sperm from another bank.

whose IQ Smith listed as 206 (nine points higher than the highest cited in the 1984 *Guinness Book of World Records*). Like Blake, she got around Graham's proscription against unmarried women by dealing with Smith.

Adkins lives in Portland, Ore. in a Victorian house she's renovating. There she devotes almost all her time to her son, because unlike most unmarried mothers, she doesn't have to work for a living; her Navy pension suffices. Although their relationship is platonic, it was her housemate and "younger brother," 38-year-old Bob Tybie, who helped inseminate her. Now he stands in as a father figure for Morgan, whom Adkins calls "the most beautiful, charming, gorgeous, brilliant baby that's ever been born."

It was perhaps inevitable that Smith—divorced from his English wife and thousands of miles from his son—would one day fall in love with one of his recipients. Adkins is the one; Smith has asked her to marry him—several times, he says.

Adkins laughs. "Sure, he proposed to me," she says. "But at first I didn't think he meant it. I told him, 'I'll bet you say that to all the recipients.' "

When it comes to childbearing, Serita Stevens (IQ 150+), *nom de plume* of a Los Angeles author of more than half a dozen paperback romances, is also getting old. Married in her twenties, divorced and childless in her early thirties, Stevens is now 36—and to her ears, anyway, the ticking of the biological clock sounds like tympani.

Stevens heard of Heredity Choice from Afton Blake last year, and now, with no prospective husband in sight, she has decided to go ahead. After Blake's recommendation, an hour's phone conversation with Smith, and a survey of his brochure of donors, she

1. DEVELOPMENT DURING THE PRENATAL PERIOD

felt ready to send away for sperm from the donor designated Blue-Brown.

It's fairly clear that to Stevens, living, breathing men are nearly superfluous in the New Age—mere catalysts in the process of reproduction. "Even if I met some guy now," she says, "it'd take several years to have a kid—getting to know each other first, and so on. By that time I'd be almost forty. And I'd like to have two children.

"To tell the truth, even if I meet somebody first—unless he's totally perfect—I'll probably go through with the artificial insemination. After all, what chance is there I'll meet a guy that could match *this*?"

She reads from the description of Blue-Brown:

"*Ancestry: Ashkenazic.* Me, too. *Eyes: hazel.* Mine, too. *Fair.* Me, too. *Brown hair.* Me, too.

"*Five-foot-eleven.* I'm five-seven.

"*Born in the 1950s, weighed 150 at 24. Normal face with strong, slightly dimpled chin. Extroverted, personable, very stable; enjoys swimming, skiing, photography, hiking, backpacking, bicycling, and animals.*

"*Significant film director, with law degree, produced scholarly published work in law. IQ: 155. Father, successful executive; mother, teacher. Music: evidence of good ability but untrained. Very competent swimmer; manual dexterity excellent. General health excellent. Grandfather lived to 82. Defects: dental malocclusion, -4 diopters myopia. Proved noncarrier of Tay-Sachs.*"

No fuss, no muss.

Stevens swears she cares nothing about the genius in her child's genes—but she's just as certain, thanks to several psychic friends, that the kid she and Blue-Brown might have would be a "leader in the New World"—a kind of spiritual and political Moses who'd conduct survivors of the Great American Quake out of the continental devastation and into exalted interactions in a sacred land.

"*There comes a time when we heed a certain call . . .*"

Ten-month-old Leandra is playing on the living room floor, cooing to herself, when suddenly the USA for Africa video comes on the television in the den. It's a siren call. She looks up and pauses for a moment. Then she stands—she has just learned to walk—and toddles off, nose in the air, like a puppy following a whiff of stew into a neighbor's kitchen. Grabbing the den's doorpost, she veers around the corner and barrels toward the TV screen, then sits mesmerized in front of it.

"*We are the world, we are the children . . .*"

She pumps her body up and down and flaps her arms to the chorus, then imitates the singers, verse by verse. There's Springsteen; Leandra concentrates, contorting her face. Now Tina Turner; she curls her lip and wrinkles her brow. Then what seems for her the culmination: "*Wowowo . . .*" wails Cyndi Lauper. Leandra flings her arms vehemently.

At ten months, Leandra is musical and empathic, and she has motor skills that some children haven't mastered until a year and a half. Her parents, Adrienne and David, are as happy to attribute her talents to the fact that Clear, her biological father, has an IQ "over 200" and "has won athletic competitions" as they are to credit her looks to his blond hair and blue eyes; it is, after all, what they bargained for when they sent for the tank from Escondido. Says Adrienne, "Go for the best! The genetic best! Try to create a person who'll contribute to society."

It's pleasing to wonder, as Leandra watches the televised appeal for a distant people, whether she won't be one of those Graham hopes for: someone whose intelligence may save a continent from famine or drought.

Then again, it's just as pleasant to imagine she won't; whether or not Leandra's life is given over to the public good, she's already an ineffable joy to her once childless parents. And that, in the New World, or any other world, is miracle enough.

LIFE
Before Birth

Geraldine Youcha

According to the traditional view, life before birth is the ultimate idyll. Clean, serene, protected from noxious substances by the complex filtering system of the placenta, shielded from the jolts, noise and stress of the world a few inches away, the fetus sleeps. Cradled in warm fluid, blissfully insensate, it floats cozily until it is thrust into the atmosphere and its brain, eyes and ears switch on for the first time. A pretty picture—but an inaccurate one. Scientists now know what the mother hears, feels and swallows *can* affect her baby. The placenta, that intricate web of blood vessels and membranes, is not a barrier but more like a sieve that lets many viruses, drugs and bacteria slip through and do their damage. One study investigated boys and girls who had been exposed to synthetic male sex hormones as fetuses when the drugs were given to their mothers to prevent miscarriage. Presented with hypothetical conflict situations, they reacted more aggressively than children who had not been exposed. The fetus spends some of its time sleeping and some of its time awake, and the mother's coughing, vomiting, twisting or turning can interrupt the serenity of the fetus's sleep.

Moreover, the noises it hears when awake are likely to be the rumblings of its mother's swallowing and digestion rather than the rhythmic pulsing of her heart. The fetus, it has recently become clear, also experiences muffled hints of the sounds that await it on the outside. As a matter of fact, researchers studying fetal hearing recently concluded that "the auditory experience of the fetal mammal may be considerably more extensive and . . . possibly of greater postnatal significance than has been believed."

REHEARSAL FOR LIFE

The fetus also sees the watery light that penetrates the thinned abdominal wall in late pregnancy and, in a rehearsal for later life, "breathes," sucks, grabs and sleeps. When it is born it is already like no other human being, having lived through unique experiences that its developing brain has recorded, reacted to—and perhaps even remembered.

What is life in this aquatic world really like? The fetus develops symmetrically, without flat areas from resting on one spot or another, because it is suspended in a fluid bath that exerts its pressure equally. In its sealed pool, the unborn baby swims and floats effortlessly, and after it emerges it can swim long before it can walk. In France and in Russia, in fact, there are unorthodox hospitals in which babies are delivered underwater as the mother relaxes in a pool. Still attached to the umbilical cord, the baby is able to stay afloat. Perhaps this should not surprise us. According to evolution theory, our ancestors crawled out of the primordial sea onto dry land, and we may still carry vestiges of our aquatic beginnings.

Another holdover from the prehistoric sea-to-land progression may be the newly discovered ability of the fetus to respond to sounds at frequencies its parents cannot hear. Adults usually detect sounds between 30 and 18,000 hertz. Does the fetus's sensitivity indicate a kinship with other water-borne mammals such as dolphins and whales, which can hear sounds at frequencies ranging from 16 to 180,000 hertz? What happens to this capacity? And is it located in the ears or someplace else? The mystery is still to be investigated. One theory suggests that it is the skin, the oldest and largest of our sense organs, that responds to the vibrations.

Normal hearing, too, develops early. Midway in pregnancy a fetus will jump in a startle reaction at a loud noise such as a slamming door. At six months, the fetus can be tested accurately for deafness by applying a tone-producing vibrator to the mother's abdomen and recording sudden changes in the baby's heartbeat as the tone goes on and off. As the time to be born approaches, the fetus can even distinguish different tones and rhythms.

When three researchers at the Institute of Animal Physiology in Cambridge, England, implanted hydrophones inside the amniotic sacs of pregnant ewes, they found to their amazement that even the sounds of ordinary conversation penetrated the uterus. One thing they were not able to hear was the sound they had expected—the maternal heartbeat. Perhaps, they speculate, other studies found this the dominant sound because they monitored noise with a microphone that was outside the amniotic fluid. Although they caution that what applies to sheep may not apply to human beings, they are convinced that a fetus hears more than has been assumed and that the influence thereof is only dimly understood.

In studies done after spontaneous abortions or abortions performed for medical reasons, scientists have found that very early in its development—when the embryo is only four inches long—stroking the skin in the area where the mouth will eventually develop makes the tiny crea-

1. DEVELOPMENT DURING THE PRENATAL PERIOD

These unique shots of the live human organism differ from classic photographs, which depict aborted fetuses at a later stage in development. These were taken through the window in the amniotic sac, a point at which a thinner wall allows a view inside. The means of approach? An endoscope that contains a series of lenses and a fiber-optic light source, inserted through the cervix.

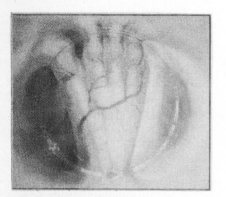

(Above) A human foot, nine weeks. (Above right) The sex organ at nine weeks. Sexual differentiation is not apparent until the 18th or 19th week; the organ above could still form either penis or clitoris.

ture pull away, indicating the existence of a rudimentary nervous system.

In a University of North Carolina study, newborn infants less than three days old learned to suck on a nonnutritive nipple in such a way as to elicit their mother's voice in preference to that of another female. In this ingenious experiment the mothers recorded part of a Dr. Seuss children's story shortly after delivery. Babies got their own mother's recording by sucking in one way and another voice if they sucked differently.

Though they were all in traditional nurseries with many female caretakers and were with their mothers only briefly for daytime feedings, they consistently chose their own mothers' voices. Perhaps, the researchers speculate, the babies

learned to recognize their mothers' voices very quickly, or—and the possibility is astonishing in its implications—they became familiar with her voice while still in the womb, even though the sound was filtered through flesh and fluid.

Now that we know sounds penetrate the uterus and that the fetus is a discriminating listener, shouldn't we ask what effect violent TV programs might have on it? How about loud arguments between husband and wife?

The notion that the mother's emotions affect the fetus is just beginning to gain

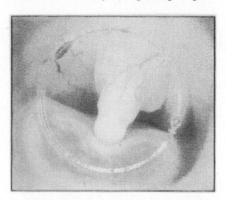

reluctant scientific respectability. For years, the idea of prenatal "impressions" was laughed at as fanciful folklore. But tests now show that the mother's anxiety increases the baby's heart rate because maternal anxiety causes an increased flow of hormones such as epinephrine, which constricts blood vessels and so interferes with uterine blood supply.

There are even hints that stress could predispose women to premature births or affect the rate of growth of the fetus, perhaps by causing muscle tension and changes in hormone levels. This sensitivity to its mother's feelings is evidently exquisitely tuned. In one experiment, pregnant women who smoked were deprived of cigarettes for 24 hours. The next day they were offered cigarettes again, and fetal hearts started beating faster even *before* their mothers could light up. Thus, although the fetus floats alone in its amniotic sac, it is intimately connected to another human being whose emotions and actions have profound effects.

Prenatal life as a glimpse of the future is evident in the development of the brain, and a hint of what the brain may be capable of comes earlier than had been thought. "Brain life," according to Dr. Dominick Purpura, dean of the Stanford

University School of Medicine, begins as early as the seventh month of gestation. And electroencephalograms of the fetus shortly before delivery show brain waves strikingly similar to those of an infant who has already been born, shaking the old assumption that entry into the world is necessary to turn on the brain.

All the phases of sleep, including Rapid Eye Movement, or REM, have been recorded in an unborn baby. In adults and children, REM often indicates dreaming. Does the fetus dream? Of what? If dreams are based on experience, does the fetus dream of a mother's indigestion or of the sensation of somersaulting in a warm, supportive solution?

The fetus may also "breathe" while floating in its watery first home. The regular rising and falling of the fetus's chest has been recorded by plotting the echoes of low-frequency sound waves bouncing off the fetus. Changes in these movements may be a better indicator of trouble than monitoring the fetal heartbeat.

Ultrasound pictures also show the fetus sucking its thumb, grabbing at the umbilical cord, hiccuping and smiling. It also kicks, rolls and stretches. All these movements are rehearsals for important later activities, and we know now that the fetus's day-to-day activity level in the womb is a good predictor of its activity level as late as the toddler age. The Chinese, who say that at birth a child is already one year old, could be closer to the truth than Westerners who think life begins at birth. The development of the brain, nervous system and endocrine system that began in the womb continues for many months after delivery. As the noted embryologist Keith L. Moore, of the University of Toronto, puts it, "Although it is customary to divide development into *prenatal* and *postnatal* periods, it is important to realize that birth is merely a dramatic . . . change in environment."

One of the most remarkable new discoveries of intrauterine research is that the placenta contains beta-endorphins, natural morphinelike substances produced by the brain. The mother's bloodstream, too, contains these natural painkillers, and their levels go up as delivery approaches. If the psychoanalysts are right and we all long to return to the snug safety of the womb, perhaps it is really because that was the time when we were blissfully high on a substance three times as powerful as morphine.

BEFORE THEIR TIME

Jack Fincher

Although a critical care nurse, Ann Starkey had no idea there was any serious problem with her pregnancy. Neither did her doctor, until she came to him complaining of a frequent and uncontrollable but futile urge to urinate.

A bout of bleeding early in her pregnancy suggested that Starkey's was a high risk case, but when she had not miscarried by the fifth month, her doctor felt the danger was probably past. No, this was only a bladder inflammation, the 35-year-old California woman thought.

But when analysis of a urine specimen revealed nothing out of the ordinary, the doctor decided to examine her cervix, something usually not done until labor starts unless there are cramps or further bleeding. "When he looked at me, he didn't say anything for a long time," she remembers. "I knew something was terribly wrong."

Ann Starkey's cervix may have been dilated for days—three months too early. She had been walking around with an open cervix, exposing membranes that could rupture at any minute and lead to massive contamination that could menace both her and her baby. Ann Starkey was the victim of premature labor, a distressing and dangerous condition for both mother and child that strikes up to one in 10 expectant mothers in the United States every year.

Seventy-five per cent of all premature labor can be traced to some underlying factor such as multiple pregnancy, toxemia, infection, irritation due to partial separation of the placenta or—as in Ann Starkey's case—a cervix unable to hold and carry the baby to term. An association has also been established between badly nourished mothers and early delivery. Many symptoms of high risk pregnancy can be detected in advance, and efforts are being made to delay delivery in these cases. A recent study at the University of California at San Francisco reports that special care of high risk mothers can reduce preterm delivery by 64 percent.

Still, fully one in four such deliveries remains as medically mysterious as the timing of labor itself, normal or otherwise. What's more, it can happen to any pregnant woman. Says Stanford Medical Center's Director of Maternal Fetal Medicine, Kent Ueland, "No one knows what starts labor. We know prostaglandins [natural body substances that trigger the smooth muscles of the uterus to contract] are involved. You can release them by stretching open the cervix with your finger. Just before labor, some mammals, such as sheep, have a rapid drop of the sex hormone progesterone and a concurrent surge of estrogen. But that hasn't been documented in humans, probably because it occurs so rapidly we just haven't been able to measure it as it happens."

Ann Starkey's doctor had her signed in to the labor and delivery unit at the Stanford Medical Center in Palo Alto. She was lucky because Stanford also has a top ranked neonatal intensive care unit where special care is provided after birth for very low weight infants. Starkey was put to bed and injected with a betamimetic, a powerful drug that often stops labor. Her blood, unfortunately, showed an elevated white cell count, a disturbing sign that her membranes might have ruptured, permitting bacteria to invade the egg-shaped amniotic sac which contained the fetus floating in an aseptic cushion of fluid. The presence of an infection there would dramatically increase the chances that the labor suppressant might induce a fatal heart attack or stroke. So much so that she had been asked to sign a paper saying she knew she was consenting to take a drug that threatened her life. Ann Starkey signed.

The injections were unavailing. Her heart pounding because of the drug, she watched anxiously as a stylus scratching away on graph paper mapped the progress of her contractions. They persisted, but remained weak in intensity, so about 8:00 P.M. sutures were put around the cervix and drawn tight. No use. At 1:00 A.M. her membranes started leaking amniotic fluid. Labor was accelerating.

Five hours later her white blood cell count had doubled. If there were to be any hope of a normal baby, Ann Starkey knew, it could no longer remain in the uterus. The sutures were cut and pitocin, another natural body substance that causes the uterus to contract, was given intravenously.

She lay in tears as Ron Cohen, an attending neonatologist at Stanford's neonatal intensive care unit, told her and her social worker husband Larry that their first child's chances of survival were dim. The Starkeys made it clear that, after Ann's safety, they wanted what was best for the baby. "We didn't want it to survive if it was going to be handicapped beyond a productive life," says Ann.

Gently, Ron Cohen informed them that babies weighing less that 500 grams—a little over one pound and common for 24 weeks' gestation—were usually too small to try to save. Cohen, they realized, was reaching the crucial decision for them. Their baby had to make the weight or it would be all over.

Ann Starkey never got to the delivery room. At about 1:00 P.M., lying flat on her back under an anesthetic that enabled her to sense pressure but not pain, she felt the baby passing through the birth canal. Sterile towels were thrown under her, a crew from the intensive care nursery was summoned with the push of a bedside button. Less than a minute later Matthew Starkey was born, one tiny wrinkled foot stirring the air as the doctor snipped the umbilical cord. "He opened his mouth, took a deep breath, and cried," says Ann, savoring that miraculous moment again with her husband. "He was pink, wasn't he pink? Usually kids that small come out blue, unable to breathe, with no muscle tone. They whisked him off on a portable ventilator and that was all I saw of him for the next nine hours."

The baby weighed 700 grams. Larry Starkey said, "When he popped out and cried, the thought that he might not make it just disappeared." That thought vanished too soon, the Starkeys would learn time and again to their frustration. But two hours later Ron Cohen roused Ann Starkey with the message she and her husband had been waiting for: "We've got a fighter."

Matthew Starkey had just joined the ranks of America's "Kilogram Kids," a legion of very premature infants weighing 2.3 pounds and under at birth. They are being saved by the advance of neonatal medicine and technology to face an uncertain future—sometimes normality if they are strong, well doc-

tored, and lucky, sometimes vegetating disaster if they are not. The fate awaiting can be blindness, deafness, chronic heart and lung disease, or severe brain damage. It can also be limited to minor handicaps such as minimal nerve "insults" that can only be detected by the minutest muscle stiffness, the most subtly discriminating neurological work-up or machine test or cardiopulmonary function. But at birth the outcome cannot be judged by even the most experienced neonatologists.

The term *premature* can be misleading. In the case of a 500-gram baby, the medical profession is given the task of turning a fetus into a human being. Many of these very premature babies weigh less than fetuses that can be legally aborted. And the task of saving one is no easy matter. Though 98 percent of Stanford's premature babies survive, 32 percent of those between 750 and 1,000 grams and 76 percent of those below 750 don't. About 242,000 premature babies were born in the United States in 1979, more than 18,000 of them weighing less than 1,000 grams. Although national survival rates are increasing rapidly, those rates are based on all babies weighing under 2,500 grams.

Still, neonatal intensive care units report not only increased survival rates but decreases in the severity of handicap of those babies that suffer handicaps. At the Children's Hospital of Los Angeles, for example, the survival rate for babies under 2,500 grams, who suffer not only from prematurity but from respiratory and congenital diseases as well, has increased from 40 percent in 1960 to 85 percent in 1980. Of the survivors only 15 percent suffer any sort of handicap, and only one percent are stricken with severe handicaps.

The economic question is major. For a baby requiring five to six months of intensive hospital care, the cost can be as high as $200,000, though usually it is within five figures. But that such babies by definition are not flawed but merely perilously ahead of schedule only confuses the moral dilemma. "After its size and sex," says Ron Cohen, "the question parents of a pre-

mature like Matthew always ask is, 'What's wrong with my baby?' I tell them, 'Your baby is normal. His form, his needs, how he's behaving—everything is exactly normal for that stage.'

"The baby is no more sick than I am. But he's as close to death as you would be if you were suddenly transported to the surface of the moon or the bottom of the ocean. Being on the moon or the ocean floor is not a disease, but it sure as hell is life threatening. We've taken him from an environment to which he is beautifully adapted to one that can be deadly.

"In other words, without an amniotic sac to protect him, without a placenta to feed him, breathe for him, oxygenate his blood, and eliminate his waste, he needs us to give him a space suit with all those tubes and wires and needles, because Captain Kirk can't beam him back into the uterus."

There are perhaps 50 to 75 top level neonatal intensive care units throughout North America, which like Stanford are laboratories outfitted to cope with the extreme highs and lows of this life-or-death confrontation. Banks of blipping lights, blinking numbers, and beeping alarms stand incessant sentry over water beds gently undulating in incubators controlled for temperature and humidity by the baby's own body. Indeed, so many electronic machines and computerized devices endlessly monitor, report, and sound warnings on such vital signs as brain waves, heartbeat, blood gases, and respiratory rate that a team of technicians must be retained to service them around the clock.

In 1980 Stanford admitted 339 prematures from throughout northern California. Forty-five were Kilogram Kids like Matthew Starkey, weighing less than 1,000 grams. Some were transported from dangerously less equipped rural areas by pre- and post-delivery mercy flights. The trend is to transfer high risk mothers before delivery so that babies like Matthew have the most sophisticated treatment at the outset. However, it is at the point of transfer to an intensive care unit when parents

Three "Kilo Kids"

Jeanette O'Kelly's fourth pregnancy was difficult from early on—a rupture of the amniotic sac, a swift infection that sent her temperature skyrocketing, an exhausting breech delivery. Still, the baby was alive and kicking. "About one pound, 11 ounces," says his mother, a former hairdresser. "I couldn't imagine anything so tiny. But he had everything: eyelashes, eyebrows, hair. The first time I saw him he was on his back with the respirator and three wires coming out of his umbilical cord. The IVs came later."

"The doctors were very gloomy," remembers her husband, Jerry, a structural engineer. "We didn't want to hear that." There were plenty of "red flag" complications. "But we couldn't visualize all those things," says Jeanette now.

The O'Kellys had to commute to the hospital at Stanford from a small coastal town 50 miles away. They made sure their three other children came with them whenever practical. "It was something happening to all of us, good or bad," Jerry says. "It was important for them to understand."

When one lung overexpanded and collapsed in a welter of cysts, the baby was put on a nerve-blocking drug that left him helpless, his tiny face swelling on the side where he lay, one eye bagging with gravity. "But then it got better," says Jeanette. "And we went on to something else." Michael's brittle bones kept breaking. "The nurses would put up

signs saying 'Don't touch this leg,' " says Jerry. "Then some technician would take blood out of it and there would be more damage. It must have been very painful. As an engineer you get used to trade-offs. But with this kind of trade-off, look at what the baby has to go through. Some of it didn't seem worth it."

"There were times I wanted them just to leave him alone," Jeanette says. "But I don't think I could have quit."

At one point the O'Kellys had to watch helplessly as their baby went blind. "The doctors told us he would," says Jeanette. "You could see his eyes getting whiter and whiter. I felt then I could deal with anything. We had come so far."

At another point the carbon dioxide in Michael's blood soared so high the doctors confessed that most people aren't able to remain conscious with that much in the system. Suddenly his oxygen intake improved; suddenly it got much worse. The O'Kellys were told that if he didn't get off the ventilator then, he might never do so. Removed, the tiny infant labored mightily and reached a precarious balance with a breathing hood only. But he never came completely off the oxygen.

His parents decided to take him home to the small hospital where he had been born. Doctors concurred, informing them pointedly that Michael O'Kelly was going to require prolonged hospitalization even if he made it. The transfer went well, but for the O'Kellys, the aftermath remains a nightmare. They began to doubt the hospital was equipped to meet their child's special needs.

(cont. on next page)

must make a major decision about their baby's future. When they move a child to a special facility they must be aware that the doctors and nurses there are trained to do everything possible to save the baby's life, whatever the handicaps.

There is something invincibly homey about Stanford's nursery despite the electronics. Shrewdly so, for the families'—and the staff's—sakes. Orange and yellow cats and dogs roam the fabrics of the curtains, and the woodwork and walls are painted in lollipop pastel. Tiny heads are covered with colorful wool caps to prevent heat loss. Crowns are fashioned from Styrofoam coffee cups to keep the wire leads of various telemetry carefully in place. Careful attention is paid to names. Gaily crayoned pink and blue signs proclaim, "My name is Juanita!" or "Happy Birthday, Jeffrey!" while platoons of green-gowned and blue-smocked

nurses, doctors, and technicians troop easily from crib to crib with their clipboards, stethoscopes, and treatment paraphernalia. Checking and probing, laughing and questioning, conferring and weighing, deciding and prescribing.

Parents arrive for a visit to be offered an Easter or Christmas card "from" their baby. And most of the homelike atmosphere in the nursery is provided by the nurses who not only serve as the babies' constant companions and day-to-day caretakers but as the major link between parents and doctors. Says Stanford's NICU head nurse Kathi Palange, "It's really hard to look at a 500-gram baby and think of it as a human being. We try to do everything we can to give parents and peers the sense of the baby as an individual." That includes hanging a profusion of snapshots amid the normal nursery mobiles. "After all," Palange continues, "many of

these babies are here for six months. This is home."

Yet there remains something supernaturally eerie about it, too, something otherworldly the visitor notices only at the sudden piercing wail of a baby just graduating from a life-saving hookup: An enforced peace alien to normal infancy. For everywhere around you the desperate struggle for life rages quietly on. The fight is not only against death and trauma but also against the instruments designed to forestall them: lung and nasogastric tubes, needles, warm metal disks that can irritate sensitive skin as they coax blood to the surface and register its oxygen content. The Kilogram Kids do not protest aloud as their bigger brothers and sisters might. They cannot. Their cries are blocked by fat tubes of soft rubber and plastic.

Presiding over all this with a buoyant humanity is a chunky neo-

(cont. from preceding page)

In the midst of one feeding, Michael O'Kelly went limp and his color faded. Looking on, his mother panicked. She yelled at the nurse to call the doctor. "Machines were going off, monitors buzzing," remembers Jeanette. "The doctor was on the phone shouting for suction. She did and in a split second he came to. All he had was a mucus plug."

A few days later Jerry O'Kelly was phoned that their baby was very sick. He arrived at the hospital to find the machines disconnected. No one had told him that Michael was dead.

Having gone through the experience, the O'Kellys now feel they can handle anything—even another prematurity if it happened again. Nor do they have any doubts that things would proceed pretty much as they did before.

"As soon as you walk into that hospital and the baby gets into trouble, you lose rights as parents," Jeanette says. "They will try to save that baby."

"It has rights and they will do whatever they can to accommodate those," Jerry says. "It makes sense. There's a lot of things they can't know, don't know."

"People take that for granted," Jeanette says, "until everything goes wrong. You can try to bargain—anything over 30 weeks and we'll chance it, anything under and we won't. Doctors say we as parents don't have that choice. They say they delivered a lady 25 weeks along and her baby did fine. After Michael was born, people kept telling us stories of uncles born under two pounds who were put in the oven to keep warm and now they're over six feet tall. They made it on their own, they tell us, without technology."

Jerry O'Kelly says, "I guess it's man's nature to try to achieve more than he understands. That's what a lot of these doctors are trying to do. There's no way around that."

From the beginning, David and Elsie Sobel—not their real names—had a premonition that the birth of their daughter Dawn would not go well. "I had bad cramping and a discharge, but I didn't know how to evaluate them," recalls Elsie. "I think women need to be better educated to the early signs of premature labor. If we had alerted our doctor a few days earlier he might have been able to stop labor and give me more steroid for Dawn's lungs.

"We were told she was going to be small, but we didn't know what that meant. Frankly my fantasy was that either she would die and we would be sad or that she would live and be okay. I had no concept there was this long, serious middle ground. When we took the tour of the labor and delivery rooms, they didn't want us to see the intensive care nursery. We did and we were shocked, never believing we would be there in three weeks."

But they were. Dawn was hurried into intensive care and placed on a ventilator a few hours later. She fought it and had to be given a curarelike drug to relax her muscles so the machine could do its work. Taken off after her breathing had stabilized, she fell unconscious five days later with a large bleeder. A spinal tap found blood. Her brain waves grew abnormal, her reflexes erratic. The Sobels →

natologist with the improbable name of Philip Sunshine. At 52, Sunshine patrols the halls of his unit day and night, like some genial northside Chicago janitor, a ring of brassy keys jutting from his hip pocket. Sunshine is one of a group of doctors in the relatively new field of premature medicine who have taken it over several watersheds in the last generation.

Having accomplished this, however, neonatologists are quick to caution that nothing about the care and treatment of the extremely premature can be certain. Constant, difficult trade-offs are the rule. If you increase liquid feedings to offset weight loss, for example, fluid may build up and strain the heart and lungs. Remove it with diuretics and you risk dehydration and disruption of vital sodium and potassium balance. Add a saline solution intravenously to restore it, and the whole heart-lung-fluid

strain is apt to start all over again.

At best, an infuriating paradox reigns. Robbed of their most rapid period of growth in utero (when body weight normally quadruples), Kilogram Kids have calcium-hungry bones that can be broken almost by a rough glance. Yet their powers of healing and recuperation are so phenomenal that fractured bones left unset will spontaneously realign, knit, and X ray as normal by age two or three. Lungs, so ravaged by acute respiratory distress that they resemble those of terminal emphysema sufferers, heal by age three or four.

Even as they are eager to grow and develop, the preemies below 1,000 grams face another maddening hazard. They are not yet equipped to coordinate simple sucking and swallowing, much less digest and move all the food they need through the gut. Nothing in the vast and growing armamen-

tarium of medicine can satisfy the preemie's need for calcium and other nutrients that would have been provided by the placenta. Put enough calcium in an IV and it precipitates out as "snow," minuscule stones that can clog the tubes. Deliver it directly to the digestive tract with a tube, and the baby throws it up or the intestinal lining sloughs off, admitting bacteria which can lead to gangrene of the bowel. Hold back on calcium and the breaking bones mend but do not grow adequately. "There's no way we can get enough calcium into these kids," laments Ron Cohen. "By age three they may be developmentally on par with their peers. But sizewise, it's beginning to look as if they may never catch up."

Finally, there is the hazard of apnea, or respiratory arrest. A 26-week-old fetus in the womb does not breathe. Why should it, with a mother and placenta so ready to do were told there was possibly serious brain damage. "Yet," says Elsie, "every time something else went wrong, and it did, they would try harder. More oxygen, more blood." The Sobels felt they were getting a mixed message: Dawn was bad off; Dawn might get better. "We didn't know if she would survive and, if she did, if she was so badly off that she wouldn't be any more than a vegetable. If we were going to end up with a severely damaged child, then maybe they shouldn't proceed with treatment so vigorously. I know it's a matter of personal values, but it comes down to a question of what quality of life is worth preserving. A lot of the nurses agreed with us."

"We were at our worst when we felt we were fighting the hospital," David adds. "I think they should say to parents, 'We really can't say if you're going to have a mentally disabled child or one in an institution. Think about this really carefully. This is what it might mean for the rest of your life.' And I think that's when parents should have input in which way to go."

The Sobels won what they thought was a firm, if, informal, commitment from the Stanford staff to let nature take its course. Told by a nurse the next night their daughter would probably not make it, the Sobels went to the hospital and held their first child for the first time. "It was a terribly ironic moment," remembers Elsie. "Knowing she wasn't going to make it, knowing that in many ways it was for the best." They left to sleep and make funeral arrangements, fully expecting to be awakened by a telephone call. It never came. The next morning they learned Philip Sunshine had intervened.

"They were all ready for the baby to die," Sunshine says. "I listened to her and there was no question what the problem was. A big pneumothorax—air pocket in the chest surrounding the lung. I put a chest tube in, and she immediately came out of her decline."

"We had very mixed feelings," David Sobel recalls. "We met with him afterwards and expressed some anger that we hadn't been consulted. I learned later some sixth sense told him Dawn was going to be all right."

"Parents don't have all the perspective, but neither, in truth, do doctors," says Elsie Sobel. "They have very little perspective on what it's going to mean if the child comes out less than perfect."

"They also have a vested interest in saving lives," David adds. "So they're not neutral and independent. They have their own emotional needs and desires and drives too. I think there has to be independent adjudication. Suppose they could be sued for wrongful life? Suppose a court decided agressive treatment leading to lasting damage was grounds for a suit? Doing what they should not do—malpractice—is one thing. But what if nobody knows all the consequences of what they're doing and they do it anyway?"

"In a situation like that, if real financial responsibility were attached, I think hospital administrators would be right in there," Elsie says. "You have to make your feelings known in any case. Because in the absence of pressure from you, it will just go on and on. Even with pressure, it may."

(cont. on next page)

it instead? "Why, then, should it know to breathe out here?" asks Cohen. "It's got a nervous system telling it it's 26 weeks old. Breathing is probably controlled by a reflex that's not fully developed at that gestational age. We've got kids out there who, if we turn off their ventilators, would stop breathing when they fall asleep. They don't know they're not supposed to."

Given such a mercurial prognosis, many troubled parents resent what they perceive as total hospital control over decision making. It is they after all who must live with the end results. It is a rare father or mother who doesn't at one point or another feel at least a twinge of angry resentment over this. At its most intense, according to Charlene Canger, one of the intensive care nursery's two social workers assigned full-time to such ongoing ordeals, the parental attitude can congeal into a mother's withering

blast: "You want to kill my child!"

At Stanford, typically, the tougher medical decisions are the product of what is candidly conceded to be a "benevolent despotism" operating on the venerable model of university administration—plenty of free speech for everyone, final authority coming from the top.

"The ultimate responsibility rests with one of the attending neonatologists," says David Stevenson, who divides that responsibility with two other doctors on a two-months-off, one-month-on basis. "You examine every patient, review all charts, several times a day. Ideally, you're not surprised by anything."

At Stanford the staff functions as a team. Doctors consult with each other, the social workers, and the nurses. Stevenson relies on the nursing staff for minute-to-minute guidance and information on the quick and subtle changes in the ba-

bies' conditions and for updates on the emotional status of the parents. Kathi Palange says that outsiders have difficulty thinking of their unit as intensive care because the patients are so small, but points to the tremendous strain of watching for each sign of change in these babies coupled with advising and comforting parents. "And it isn't just parents," she says. "The babies change from better to worse, worse to better, so frequently we often have to bolster each other as well."

The Stanford teams—doctors, social workers, and nurses—often consult with parents together. Stevenson admits that parents are "hostage to circumstance" in a way that makes "choice" a misnomer. "It would be naive to think that informed consent means you give them every detail of the information you have and let them decide on the best opinion," he says. "I prefer to see informed consent as

(cont. from preceding page)

Sunshine agreed that before making any more major decisions the Sobels would be consulted. As it turned out, none had to be made. Today Dawn Sobel is three years old. She has a limp in her left leg and a stiff left hand that require ongoing physical therapy, and a shunt that runs beneath the skin from her brain to her intestinal cavity with a pressure-activated valve behind her ear keeping natural pathways free of excess spinal fluid that could cause hydrocephaly. If the shunt fails, the symptoms—grogginess, out-of-it behavior—look like flu, and Dawn must be raced to the hospital. She will either outgrow the need for the shunt or face further surgery to have it lengthened.

"Dawn didn't get off scot-free, and neither did we," says Elsie Sobel. "But we feel we have a delightful child. Meeting her, you can see she's perceptive, extremely verbal, and socially oriented. We've been lucky. The problems have been tough. We can tolerate living with the ones she has."

The wonder of it is," marvels Cathy Doolittle, "we have a boy who is healthy." Michael weighed 900 grams at birth, and when he came home from the Stanford hospital two weeks before his originally projected birthdate, he weighed roughly 2,200 grams (four pounds, 12 ounces). A decade later he is at the bottom of the weight chart but about average in height and retains as the only medical residue of his hazardous experience the remnants of a "lazy eye."

"He was 20/20 in his good eye, 20/200 in the other," says his mother. "Now, thanks to exercises, he can go without glasses."

Adds his father, Chuck, "He's right in there. His stamina is good and he's not a shrimp. He's on the soccer team, on the swim team. Coming out of the pool he really freezes. But you don't know if that's his normal body makeup or the prematurity."

Compared with many his size, Mike Doolittle had a relatively uncomplicated stay at Stanford despite the lack of much of today's technology. His only real peril was persisting apnea. "They had all those probes on him," remembers his mother, "and every time he stopped breathing they would yell at him to get it going again."

A crack reader who has since recovered from early dyslexia, Mike is known as the family cutup. Though he was highly excitable in kindergarten and still has problems concentrating, Mike has grown into a talkative, active, outgoing 10-year-old. His only crisis in development came when his parents decided, at the doctor's recommendation, to hold him back a year in school because of his size. Mike at first felt he had flunked, but now his grades are fine. He's had more professional care than most children, say the Doolittles, but they've always been careful to treat him like the normal child he has grown to be. Their doctor had assured them nothing can happen now to Mike that wouldn't happen to any other child. "But even though we never coddle him," says his mother, "like all preemie parents, you never stop wondering, will he really be okay?"
—J.F.

words nearer to their etymological root—helping form or shape parent thinking so they come to understand, accept, and support the expert consensus of a staff that has more *knowledge* than they do, in consultation with a physician whose competence they assume.

"What happens, finally, is the parents often borrow that opinion. We manage the medical decision making, absolutely. They depend on me for that. They would be terribly upset if because of something I knew and they didn't, I let them decide on something they would ultimately regret."

Even then, adds Ron Cohen, such decisions are difficult enough. "Seventy-five percent of the babies below 1,000 grams that we put on the ventilator survive, 25 percent don't. And on day one, we can't tell you which will be which. I do think a lot of so-called ethical issues are really questions of good or bad

medicine. Not of letting a baby die or not, but of whether to treat or not treat."

Having a preemie, he goes on, can be an immense blow, "probably as much of an emotional sledgehammer as death. The parents have just had their dream of a big, bouncing baby killed, after all. They're in shock. Mourning, grieving, going through denial, anger. There are days you come in knowing that they hate your guts and can't take it any more. But how many change their minds? After you pull the plug, it's too late."

In the end, Cohen tells you, many of the parents with handicapped kids end up saying, "I love my kid. He's not perfect, he's not going to grow up to be president, but he's the joy of my life." "Is it up to me," Cohen asks, "to say, 'Sorry, mom. He's not good enough for you,' even though you think he is?"

Phil Sunshine agrees but re-

members the premature born to two Ph.D.s. The baby suffered no detectable permanent damage but his nervous system may have been affected by early oxygen deprivation. In any case, he had an average IQ—and a seven-year-old brother who tested as genius. The mother said, "I know you consider this a good outcome, but he can't do anything his brother can."

Sunshine didn't know what to say. The father said it for him: "Yeah, but he's pleasant to be around, happy, giving, and even-tempered. Maybe being able to do everything isn't all that important." Indeed, Phil Sunshine has his own fitting aphorism from the Indian savant Rabindranath Tagore: "Each newborn child brings the message that God is not yet discouraged with man."

Stevenson might agree. He says of the Starkey baby, "He is a mira-

Watersheds in premature medicine

Intermittent Positive Pressure Ventilation (IPPV)

A stroke of serendipity in the late 1960s rescued the respirator, or ventilator, from its grisly reputation as a tacit kiss of death for prematures. So unsophisticated were the early machines with their simple on-off rhythm that doctors dared put only dying infants on them. Philip Sunshine and his associates suspected a self-fulfilling prophecy. Sure enough, when two more complex adult-sized models were scaled down and a policy of aggressive early intervention initiated, the data clearly indicated that babies who were ventilated did better than those who were not. What's more, curiously, those ventilated on one of the two machines did statistically better than those on the other. Why? Sunshine had both dismantled and found that the better respirator had a broken valve that would not shut entirely off. There remained always a mother's kiss of pressure that prevented the immature lung's air sacs from totally deflating once they had been forcibly evacuated. This is caused by the lack in preemies' lungs of a chemical called surfactant (surface-active agent). Production of this chemical can be stimulated by giving the mother a steroid hormone when premature labor starts, in conjunction with a labor-suppressing drug to hold off delivery long enough for the steroid to work.

The transcutaneous oxygen monitor

No larger than a nickel, this platinum cathode, when taped to the skin and heated, can monitor the level of oxygen in the blood drawn to the surface. The alternative method of periodically drawing blood gas samples with a hypodermic needle depletes little bodies of too much blood and fails to reflect moment-to-moment changes in oxygen content that can be critical in determining proper ventilator settings. Explains Cohen, "When you draw a blood gas sample you have no way of knowing where on the fluctuating oxygen curve you are getting it. If your timing is unfortunate and every time you take one you are looking at the bottom of the curve, you will continue to turn up the oxygen." Too much oxygen, doctors have discovered, could play a part in retrolental fibroplasia, blindness due to retinal detachment from damaged blood vessels. The transcutaneous monitor now enables continuous fine tuning of blood oxygen.

Ultrasound scanning

Devised in wartime to detect enemy ships, sonar has since been refined to give neonatologists a means of plumbing premature heads and hearts in search of life-menacing anomalies. Echo is the nickname of a lightweight probe that bounces a beam of sound waves off deep body structures to form a picture on a portable TV screen. Echo can find hidden "bleeders," cranial hemorrhages that can spread and devastate the brain. Bleeders afflict almost half of all Kilogram Kids, usually in an atrophying area unique to the newborn known as the germinal matrix, a capillary-packed factory where brain cells are manufactured before they migrate elsewhere. Fortunately, most bleeders stop short of being extensive, and the preemie usually escapes grave neural damage.

PDA drugs and surgery

Echo can also determine the dangerous persistence of a vestigial cardiac shunt called the PDA, patent ductus arteriosis. Nature intends the PDA as a temporary bypass to spare the developing heart the unnecessary job of pumping large volumes of blood to fetal lungs, which are strictly unused machinery before birth, since the mother breathes for her unborn child through the placenta. Everybody is born with a PDA, but it usually closes a day or so after birth—provided birth is full term. In preemies, however, the shunt remains partially open, pouring three times too much blood back through the lungs instead of routing it to the waiting body. The result may be severe heart failure. Now, Echo can be employed to pinpoint a PDA problem without inserting a catheter into the heart, and it can be corrected with an antiarthritis medication that hastens the closing by interfering with prostaglandin production. If medication doesn't correct the problem, the duct can be tied off surgically.

—J.F.

cle. With his birth weight, no one could have guessed he'd do so well. He surprised us all."

After PDA surgery (see box above.) and complications that included brain hemorrhage, high bilirubin, fluid retention, dehydration, tube-caused infections, chronic lung disease, and distended bowel from forced feeding, Matthew Starkey was taken off the ventilator at seven and a half weeks and sent home a month later. At that time he weighed four pounds, four ounces (1,930 grams).

At just under six months old he weighed nine and a half pounds and has been gaining up to 12 ounces a week since he came home from the hospital. According to Ann Starkey, the only outward indication that he was a preemie is his size and a slight case of RLF disease in his eyes that may eventually lead to trouble with reading.

"He took incredible insults to his body," she says. "A preemie that small will usually be at death's door once he gets an infection. He will die before anyone figures out what's wrong. But he survived through a combination of natural resistance and the doctors' skill. For me, dealing with the doctors was one of the most positive experiences of my life. There was continual trust and caringness, and I always knew I was being well taken care of."

Development During Infancy

No age period in human development has received more attention during the past quarter century than that of infancy. Today researchers are more certain than ever before that the events of the first several years of life build a foundation for subsequent development. This does not imply, however, that the events of infancy determine later development. For example, high quality infant care lays the foundation for successful development during the preschool and school-age years, but optimal care during infancy does not make the preschool or school-age child any less vulnerable to inadequate caregiving. Child abuse first experienced during the school-age years can be as devastating to subsequent development as child abuse first experienced during infancy.

Although developmentalists now recognize that the events of one age period do not necessarily predetermine the events of a subsequent age period, it would be a mistake to underplay the importance of infancy. As indicated in "Making of a Mind," fundamental changes in the organization of the brain occur during infancy and early childhood, and these changes are intimately related to the quality, quantity, and timing of environmental stimulation to which the infant is exposed. In one sense the newborn's mind is like a blank slate; that is, no one yet has provided evidence to support the belief that the newborn's mind is filled with innate ideas. In another sense, however, the newborn enters the world with the full set of species characteristics that provide the basis for perceptual discrimination of colors and sounds, emotional expressiveness and responsiveness, learning and imitative behavior, communicative behavior and language acquisition, social interaction, and differentiation of personality. In short, the newborn begins life with all that is necessary to construct an active, adaptive fit to its environment. How the infant goes about constructing this adaptive fit and the importance of reciprocal exchanges between infant and caregiver are discussed in "What Do Babies Know?" "There's More to Crying Than Meets the Ear," and "How to Understand Your Baby Better."

The theme of parent-infant interaction continues in "Your Child's Self-Esteem." It focuses on the development of competence and mastery during infancy and early childhood, and on the possibility that poor parent-infant interaction may result in learned helplessness, poor self-esteem, and poor emotional adjustment. The last article, "The Roots of Morality," concludes that if it is true that moral behavior has its origins in the empathic behavior of infants and toddlers, then children with poor self-esteem and a sense of helplessness may also have difficulty responding to caregiving practices intended to socialize moral rules and behaviors.

When reading the selections in this section, students should keep in mind three important points. First, development during infancy is characterized by relatively great plasticity in brain-behavior organization; that is, the potential for change over time is relatively great. As a result, it is unlikely that any single time-limited event can have long-lasting consequences for subsequent development. For example, having immediate contact with a newly born infant can be a positive and rewarding experience for both mother and infant. However, constructing a positive, supportive, binding, loving relationship between mother and infant—generally referred to as attachment—requires considerably more time and interaction than the time-limited event of immediate post-natal contact allows. Second, the infant is an active agent in influencing the caregiving environment. The infant's cry is a communicative act that provides feedback about the infant's ability to influence events in the environment. The infant's vocalizations, smiles, laughter, and first unassisted steps, are behaviors destined to provoke responsiveness from caregivers. The nature of the caregiver's response is the basis for the third important point; namely, that both the quality and quantity of caregiving are important influences in structuring the match or fit between caregiver and infant, because it is the caregiver's responsiveness that provides feedback to the infant about the effectiveness of the infant's behavior. To one extent or another, each of the articles in this unit focuses on these three points: the plasticity of early development, the infant as an active agent, and the adaptive fit constructed by the infant and its caregivers.

Looking Ahead: Challenge Questions

Environmental stimulation is a key ingredient for brain organization. What additional factors must be taken into account when translating this prescriptive statement into

the practical world of infant care? Put another way, what are the relative advantages and disadvantages of quality, quantity, and timing of stimulation for brain growth and development?

Summarize the opinions of infant researchers with respect to the effectiveness of early infant stimulation programs. Explain why you agree or disagree with their opinions. Why do you suppose that it took so long for infant researchers to discover that infants possess a rich repertoire of abilities and competencies?

To what extent do sex differences in caregiving and sex differences in infants responses to caregivers reflect evolu-

tionary determinants of human behavior? To what extent do they reflect cultural stereotypes? Do you agree with the proposition that men can be as competent as women in providing primary care for infants?

Do you agree that competent social and emotional behavior seems to derive directly from early experiences provided by caregivers? Is it possible that some children respond appropriately to competent caregiving, whereas others do not because of a mismatch between infant temperament and caregiver temperament? How might peer interactions offset the negative consequences of this type of mismatch?

*The newborn's brain: registering
every flash of color, caress,
scent, and other stimuli vital to the*

MAKING OF A MIND

KATHLEEN McAULIFFE

"Give me a child for the first six years of life and he'll be a servant of God till his last breath."

—Jesuit maxim

A servant of God or an agent of the devil; a law-abiding citizen or a juvenile delinquent. What the Jesuits knew, scientists are now rapidly confirming—that the mind of the child, in the very first years, even months, of life, is the crucible in which many of his deepest values are formed. It is then that much of what he may become—his talents, his interests, his abilities—are developed and directed. The experiences of his infancy and childhood will profoundly shape everything from his visual acuity to his comprehension of language and social behavior.

What underlies the child's receptivity to new information? And why do adults seem to lose this capacity as they gain more knowledge of the world around them? Why is it that the more we know, the less we *can* know?

Like a Zen koan, this paradox has led scientists down many paths of discovery. Some researchers are studying development processes in infants and children; others search the convoluted passages of the cortex for clues to how memory records learning experiences. Still others are studying the degree to which learning is hardwired—soldered along strict pathways in the brains of animals and humans.

Another phenomenon recently discovered: Long after patterns of personality have solidified, adults may tap fresh learning centers in the brain, new nerve connections that allow intellectual growth far after fourscore years.

Although much research remains to be

done, two decades of investigation have yielded some dramatic—and in some instances unexpected—insights into the developing brain.

An infant's brain is not just a miniature replica of an adult's brain. Spanish neuroscientist José Delgado goes so far as to call the newborn "mindless." Although all the nerve cells a human may have are present at birth, the cerebral cortex, the gray matter that is the seat of higher intellect, barely functions. Surprisingly, the lower brain stem, the section that we have in common with reptiles and other primitive animals, dictates most of the newborn's actions.

This changes drastically in the days, weeks, and months after birth, when the cerebral cortex literally blossoms. During this burst of growth, individual brain cells send out shoots in all directions to produce a jungle of interconnecting nerve fibers. By the time a child is one year old, his brain is 50 percent of its adult weight; by the time he's six, it's 90 percent of its adult weight. And by puberty, when growth trails off, the brain will have quadrupled in size to the average adult weight of about three pounds.

How trillions of nerve cells manage to organize themselves into something as complex as the human brain remains a mystery. But this much is certain: As this integration and development proceeds, experiences can alter the brain's connections in a lasting, even irreversible way.

To demonstrate this, Colin Blakemore, professor of physiology at Oxford University, raised kittens in an environment that had no horizontal lines. Subsequently, they were able to "see" only vertical lines. Yet Blakemore had tested their vision just before the experiment began and found that the kittens had an equal number of cells that responded to each type of line.

Why had the cats become blind to horizontal lines? By the end of the experiment, Blakemore discovered that many more cells in the animals' brains responded to vertical lines than horizontal lines.

As the human brain develops, similar neurological processes probably occur. For example, during a test in which city-dwelling Eurocanadians were exposed to sets of all types of lines, they had the most difficulty seeing oblique lines. By comparison, the Cree Indians, from the east coast of James Bay, Quebec, perceived all orientations of lines equally well. The researchers Robert Annis and Barrie Frost, of Queens University, in Kingston, Ontario, attributed this difference in visual acuity to the subjects' environments. The Eurocanadians grow up in a world dominated by vertical and horizontal lines, whereas the Indians, who live in tepees in coniferous forests, are constantly exposed to surroundings with many different types of angles.

The sounds—as well as the sights—that an infant is exposed to can also influence his future abilities. The phonemes *rah* and *lah,* for instance, are absent from the Japanese language, and as might be expected, adults from that culture confuse English words containing *r* and *l.* (Hence the offering of steamed "lice" in sushi bars.) Tests reveal that Japanese adults are quite literally deaf to these sounds.

Infants, on the other hand, seem to readily distinguish between speech sounds. To test sensitivity to phonemes, researchers measure changes in the infants' heartbeats as different speech sounds are presented. If an infant grows familiar with one sound and then encounters a new sound, his heart rate increases. Although the evidence is still incomplete, tests of babies from linguistic backgrounds as varied as Guatemala's

Spanish culture, Kenya's Kikuyu-speaking area, and the United States all point to the same conclusion: Infants can clearly perceive phonemes present in any language.

The discovery that babies can make linguistic distinctions that adults cannot caused researchers to wonder at what age we lose this natural facility for language. To find out, Janet Werker, of Dalhousie University, in Nova Scotia, and Richard Tees, of Canada's University of British Columbia, began examining the language capabilities of English-speaking adolescents. Werker and Tees tested the subjects to see whether they could discriminate between two phonemes peculiar to the Hindi language.

"We anticipated that linguistic sensitivity declines at puberty, as psychologists have commonly assumed," Werker explains.

The results were surprising. Young adolescents could not make the distinction, nor could eight-year-olds, four-year-olds, or two-year-olds. Finally, Werker and Tees decided to test infants. They discovered that the ability to perceive foreign phonemes declines sharply by one year of age. "All the six-month-olds from English-speaking backgrounds could distinguish between the Hindi phonemes," Werker says. "But by ten to twelve months of age, the babies were unable to make this distinction."

The cutoff point, according to Werker, falls between eight and twelve months of age. If not exposed to Hindi by then people require a lot of learning to catch up. Werker found that English-speaking adults studying Hindi for the first time needed up to five years of training to learn the same phoneme distinctions any six-month-old baby can make. With further testing, Werker succeeded in tracking down one of the learning impairments that thwarted her older subjects. Although there is an audible difference, the adult mind cannot retain it long enough to remember it. "The auditory capabilities are there," Werker says. "It's the language-processing capabilities that have changed."

Even a brief introduction to language during the sensitive period can permanently alter our perception of speech. Werker and Tees tested English-speaking adults who could not speak or understand a word of Hindi, although they had been exposed to the language for the first year or two of life. They found that these adults had a major advantage in learning Hindi, compared with English-speaking adults who lacked such early exposure.

Werker and Tees's studies show that there is an advantage in learning language within the first year of life. But when it comes to learning a second tongue another study has revealed some startling findings: Adults actually master a second language more easily than school-age children do.

For four years Catherine Snow, of the Harvard Graduate School of Education, studied Americans who were learning Dutch for the first time while living in Holland. "When you control for such factors as access to native speakers and the daily exposure level to the language," Snow says, "adults acquire a large vocabulary and rules of grammar more quickly than children do. In my study, adults were found to be as good as children even in pronunciation, although many researchers contend that children have an advantage in speaking like natives."

Obviously not all learning stops when the sensitive period comes to a close. This observation has led some researchers to question the importance of early experiences. What would happen, for example, if a child did not hear a single word of any language until after one year of age? Would the propensity to speak be forfeited forever? Or could later exposure to language make up the deficit?

Because of the unethical nature of per-

❝It's probably fair to say that if you want bright kids, you should cuddle them a lot when they're babies because that increases the number of neural connections.❞

---

forming such an experiment on a child, we may never know the answer to that question. But some indications can be gleaned from animal studies of how early deprivation affects the development of social behavior.

In *An Outline of Psychoanalysis*, Sigmund Freud refers to "the common assertion that the child is psychologically the father of the man and that the events of his first years are of paramount importance for his whole subsequent development." At the University of Wisconsin Primate Laboratory, the pioneering studies of Harry and Margaret Harlow put this belief to the test on our closest living relative—the rhesus monkey.

"Our experiments indicate that there is a critical period somewhere between the third and sixth month of life," write the Harlows, "during which social deprivation, particularly deprivation of the company of [the monkey's] peers, irreversibly blights the animal's capacity for social adjustment."

When later returned to a colony in which there was ample opportunity for interacting with other animals, the experimental monkeys remained withdrawn, self-punishing,

and compulsive. Most significantly, they grew up to be inept both as sexual partners and parents. The females never became impregnated unless artificially inseminated. We don't know whether humans, like Harlow's monkeys, must establish close bonds by a certain age or be forever doomed to social failure. But an ongoing longitudinal study, the Minnesota Preschool Project, offers the encouraging finding that emotionally neglected four-year-olds can still be helped to lead normal, happy lives. To rehabilitate the children, the teachers in the project provide them with the kind of intimate attention that is lacking at home.

Perhaps one of the Harlows' observations sheds light on why the project was successful: During the critical period for social development, the Harlows found that even a little bit of attention goes a long way. During the first year of life, for example, only 20 minutes of playtime a day with other monkeys was apparently sufficient for the animals to grow into well-adjusted adults. L. Alan Sroufe, codirector of the Minnesota Project, tells the story of one four-year-old boy who was constantly defiant—the kind of child who would hit the other children with a toy fire truck. Instead of sending him to a corner, the teacher was instructed to remove him from the group and place him with another teacher. The message they hoped to impart: We are rejecting your behavior, but we're not rejecting you. Within a few months, the antisocial little boy learned to change his behavior.

If children aren't exposed to positive social situations until adolescence, however, the prognosis is poor. Like any complex behavior, human socialization requires an elaborate series of learning steps. So by adolescence, the teenager who missed out on many key social experiences as a child has a tremendous handicap to overcome.

Researchers are finding that each stage of life demands different kinds of competencies. This may be why sensitive learning periods exist. "When a baby is born it has to do two things at the same time," says biochemist Steven Rose, of England's Open University. "One is that it has to survive as a baby. The second is that it has to grow into that very different organism, which is a child and then finally an adult. And it is not simply the case that everything the baby does is a miniature version of what we see in the adult."

For example, the rooting reflex, which enables the baby to suckle, is not a preliminary form of chewing: There's a transitional period in which the child must begin eating solid foods. And then other sorts of skills become necessary—the child must learn to walk, talk, form friendships, and when adulthood is reached, find a sexual partner. "But the child does not have to know all that at the beginning," Rose says. "So sensitive periods are necessary because we have to know how to do certain things at certain times during development."

During the course of a sensory system's development, several sensitive periods occur. In the case of human vision, for example, depth perception usually emerges by two months of age and after that remains relatively stable. But it takes the first five years of life to acquire the adult level of visual acuity that allows us to see fine details. And during that prolonged period, we are vulnerable to many developmental problems that can cause this process to go awry. For example, a drooping lid or an eye covered by a cataract—virtually anything that obstructs vision in one of the child's eyes for as few as seven days—can lead to a permanent blurring of sight. This condition, known as amblyopia is one of the most common ophthalmological disorders. Treatment works only if carried out within the sensitive period, before the final organization of certain cells in the visual cortex becomes fixed. After five years of age, no amount of visual stimulation is likely to reorganize the connections laid down when the young nervous system was developing.

Like molten plastic, the nervous system is, at its inception, highly pliable. But it quickly settles into a rigid cast—one that has been shaped by experience. Just what neurological events set the mold is not known. Some suggestive findings, however, come from the research of John Cronly-Dillon, a professor of ophthalmic optics at the University of Manchester Institute of Science and Technology, in England.

Working with colleague Gary Perry, Cronly-Dillon studied growth activity in the visual cortex of rat pups reared under normal light conditions. To measure growth, researchers monitored the rate at which certain cells synthesized tubulin, a protein vital for forming and maintaining nerve connections. The researchers found that tubulin production in the visual cortex remained at a low level until day 13, which marks the onset of the sensitive period for visual learning. It coincides with the moment when the animal first opens its eyes. At that time, tubulin production soars, indicating a rise in growth activity.

Cronly-Dillon and Perry found that the rat's visual cortex continues to grow for the next week and then declines. By the end of the critical period, when the pup is roughly five weeks old, tubulin production drops to the level attained before the eyes open.

To Cronly-Dillon the surplus of tubulin at the beginning of the critical period and its subsequent cutback have profound implications. "It means that an uncommonly large number of nerve connections can exist at the peak of the critical period, but only a small fraction of them will be maintained at the end," he says. "So the question, of course, is which nerve connections will be kept?"

If Cronly-Dillon is correct, experience probably stabilizes those connections most often used during the sensitive period. "So

by definition," he says, "what remains is most critical for survival."

Cronly-Dillon's work elaborates on a theory Spanish neurophysiologist Ramon Y Cajal advanced at the turn of the century. According to this view, which has been gaining broader acceptance in recent times, brain development resembles natural selection. Just as the forces of natural selection ensure the survival of the fittest, so do similar forces preserve the most useful brain circuits.

The beauty of this model is that it could explain why the brain is as exquisitely adapted to its immediate surroundings, just as the mouthparts of insects are so perfectly matched to the sexual organs of the flowers they pollinate. The textures, shapes, sounds, and odors we perceive best may have left their imprint years ago in the neural circuitry of the developing mind.

There is also a certain economic appeal to this outlook: Why, for example, should Japanese adults keep active a neural circuit that permits the distinction between *r* and *l* sounds when neither of these linguistic components is present in their native tongue?

Yet another economic advantage of the theory is that it would explain how nature can forge something as intricate as the brain out of a relatively limited amount of genetic material. "It looks as though what genetics does is *sort of* make a brain," Blakemore says. "We only have about one hundred thousand genes—and that's to make an entire body. Yet the brain alone has trillions of nerve cells, each one forming as many as ten thousand connections with its neighbors. So imagine the difficulty of trying to encode every step of the wiring process in our DNA."

This vast discrepancy between genes and connections, according to Blakemore, can be overcome by encoding in the DNA the specifications for a "rough brain." "Everything gets roughly laid down in place," Blakemore says. "But the wiring of the young nervous system is far too rich and diffuse. So the brain overconnects and then uses a selection process to fine-tune the system."

The brain of an eight-month-old human fetus is actually estimated to have two to three times more nerve cells than an adult brain does. Just before birth, there is a massive death of unnecessary brain cells, a process that continues through early childhood and then levels off. Presumably many nerve connections that fall into disuse vanish. But that is only part of the selection process—and possibly a small part at that.

According to Blakemore, many neural circuits remain in place but cease to function after a certain age. "I would venture a guess," he says, "that as many as ninety percent of the connections you see in the adult brain are nonfunctional. The time when circuits can be switched on or off probably varies for different parts of the cerebral cortex—depending on what functions they control—and would coincide with the sensitive period of

learning. Once the on–off switch becomes frozen, the sensitive period is over."

This doesn't mean, however, that new circuits can't grow. There appears to be a fine-tuning of perception coinciding with these developmental events. And as the brain becomes a finer sieve, filtering out all but a limited amount of sensory input, its strategy for storing information appears to change.

"Studies indicate that as many as fifty percent of very young children recall things in pictures," says biochemist Rose. "And by the time we're about four or five, we tend to lose our eidetic [photographic] memory and develop sequential methods of recall."

To Rose, who is studying the neurological mechanisms that underlie learning, this shift in memory process may have an intriguing logic. "To be a highly adaptable organism like man, capable of living in a lot of different environments, one must start out with a brain that takes in everything," Rose explains. "And as you develop, you select what is important and what is not important to remember. If you went on remembering absolutely everything, it would be disastrous."

The Russian neurologist A. R. Luria had a patient cursed with such a memory—the man could describe rooms he'd been in years before, pieces of conversations he'd overheard. His memory became such an impediment that he could not hold even a clerk's job; while listening to instructions, so many associations for each word would arise that he couldn't focus on what was being said. The only position he could manage was as a memory man in a theatrical company.

"The crucial thing then," Rose says, "is that you must learn what to *forget*."

Some components of the brain, however, must retain their plasticity into adulthood—otherwise, no further learning would be possible, says neuroscientist Bill Greenough, of the University of Illinois, at Urbana-Champaign. While the adult brain cannot generate new brain cells, Greenough has uncovered evidence that it does continue to generate new *nerve connections*. But as the brain ages, the rate at which it produces these connections slows.

If the young brain can be likened to a sapling sprouting shoots in all directions, then the adult brain is more akin to a tree, whose growth is confined primarily to budding regions. "In the mature brain," Greenough says, "neural connections appear to pop up systematically, precisely where they're needed."

Early experience, then, provides the foundation on which all subsequent knowledge and skills build. "That's why it's extraordinarily difficult to change certain aspects of personality as an adult," says neuroscientist Jonathan Winson, of Rockefeller University. "Psychiatrists have an expression: 'Insight is wonderful, but the psyche fights back.' Unfortunately, one of the drawbacks of critical-period learning is that a lot of misconceptions and unreasonable fears can become frozen in our minds during this very vulnerable period in our development."

Greenough acknowledges that the system isn't perfect; nevertheless, it works to our advantage because you can't build on a wobbly nervous system. "You've got to know who your mother is, and you've got to have perceptual skills," he explains. "These and other types of learning have to jell quickly, or all further development would halt."

Can these insights into the developing brain help educators to devise new strategies for teaching?

"We're a very long way from being able to apply the work of neurobiologists to what chalk-faced teachers are trying to do," says Open University's Rose.

But he can see the rough outline of a new relationship between neurobiology and education, which excites him. "We can now say with considerable certainty that there are important advantages to growing up in an enriched environment," he says. "That does not mean that you should be teaching three-year-olds Einstein's theory of relativity on the grounds that you will be turning them into geniuses later on. But it's probably fair to say that if you want bright kids, you should cuddle them a lot as babies because that increases the number of neural connections produced in the brain."

Although early learning tends to overshadow the importance of later experience, mental development never ceases. Recent studies indicate that our intellectual abilities continue to expand well into our eighties, provided the brain has not been injured or diseased. Most crucial for maintaining mental vigor, according to Greenough, is staying active and taking on new challenges. In his rat studies, he found that lack of stimulation—much more than age—was the factor that limited the formation of new neural connections in the adult brain.

As long as we don't isolate ourselves as we grow older, one very important type of mental faculty may even improve. Called crystallized intelligence, this ability allows us to draw on the store of accumulated knowledge to provide alternate solutions to complicated problems. Analyzing complex political or military strategies, for example, would exploit crystallized intelligence.

There is a danger in believing that because the brain's anatomical boundaries are roughly established early in life, all mental capabilities are restricted, too. "Intelligence is not something static that can be pinned down with an I.Q. test like butterflies on a sheet of cardboard," says Rose. "It is a constant interplay between internal processes and external forces."

To be sure, many types of learning do favor youth. As violinist Isaac Stern says, "If you haven't begun playing violin by age eight, you'll never be great." But in the opinion of Cronly-Dillon, the best time for learning other types of skills may be much later in life. Although he will not elaborate on this until further studies are done, he believes we may even have sensitive periods with very late onsets. "There's a real need," Cronly-Dillon says, "to define all the different types of sensitive periods so that education can take advantage of biological optimums."

It is said that the ability to learn in later life depends on the retention of childlike innocence. "This old saw," insists Cronly-Dillon, "could have a neurological basis."

What Do Babies Know?

More than many realize, and much earlier, according to new research

"**F**antastic!" says Michael Lewis, a small, spry man with a gray-flecked beard. "This is great!"

What inspires such glee in Lewis is that two small and curly-haired sisters named Danielle and Stacy, ages twelve and 14 months, are starting to cry. The sound is heart-rending, but not to Lewis.

"Exactly what we expected," he says cheerily as the girls' parents arrive to comfort them. The wailing soon subsides. Lewis, 46, is not a sadistic Scrooge; on the contrary, he is an eminent and kind-hearted psychologist who presides over the Institute for the Study of Child Development at Rutgers Medical School in New Brunswick, N.J. His laboratory is a friendly place filled with dolls and Teddy bears and jigsaw puzzles; blue-red-and-yellow rainbows streak across the walls. Along one of those walls runs a ten-foot-long two-way mirror so that Lewis can study children unobserved and record their activities on two videotape cameras.

Danielle's parents had adopted Stacy. Lewis wanted to observe how two sisters of similar age and upbringing, but totally different genes, would interact with their parents. All four started by playing with toys and puzzles in front of Lewis' mirror. The parents left, first individually, then together. The girls resorted to playing with each other. Then a stranger entered, and that seemed to make the girls more sharply aware of their parents' absence and their own aloneness—hence the outburst of tears. But why is that "great" or "fantastic"? "We're trying to determine exactly what normal behavior is," says Lewis, who sees the large in the small, "in this case the child's developing sense of self, the sense that it is separate from other people."

In a 17th century brick building on Paris' Boulevard de Port-Royal, once the abbey where the Mathematician Blaise Pascal underwent religious conversion, a quite different kind of experiment is taking place. Into a small room of the Baude-locque Maternity Hospital marches a nurse bearing a tiny, wrinkled infant named Géry. He is four days old and weighs 6 lbs. 6 oz. The nurse carefully deposits Géry in a waist-high steel bassinet that stands next to a computer. The computer is attached to an empty nipple. The question to be tested: Exactly what sounds can young Géry recognize?

The nurse pops the nipple into Géry's mouth and then turns on a nearby loudspeaker. A recorded male voice begins to recite a random series of similar syllables: "Bee, see, lee, see, mee, lee, bee, see, lee, mee." Géry's infant fingers clutch at the orange base of the nipple. Whenever he hears a new sound he sucks harder, and his heart beats faster. When he gets used to these sounds, his attention fades, and his sucking slows down. The computer tirelessly counts the number of sucks per minute.

"Da," the loudspeaker suddenly says. Géry sucks harder, then begins to cry. He is hungry, and the empty nipple brings him no food. The nurse comforts him. Even at the age of four days, the lessons of life can be hard.

All across the U.S., all over the world, medical and behavioral experiments like these are under way. Each by itself is a small and seemingly inconsequential affair; the results are sometimes inconclusive, sometimes obvious. But taken all together, they represent an enormous research campaign aimed at solving one of the most fundamental and most fascinating riddles of human life: What do newborn children know when they emerge into this world? And how do they begin organizing and using that knowledge during the first years of life to make their way toward the mysterious future?

The basic answer, which is repeatedly being demonstrated in myriad new ways: babies know a lot more than most people used to think. They see more, hear more, understand more, and they are genetically prewired to make friends with any adult who cares for them. The implications of this research challenge some of the standard beliefs on how children should be reared, how they should be educated, and what they are capable of becoming as they grow up. Yale Psychology Professor William Kessen, who has been studying infants for more than 30 years, says in admiration of the newborn baby's zestful approach to life, "He's eating up the world." Harvard Psychology Professor Jerome Kagan, another pioneer, offers only one caveat about the new research: "Don't frighten parents! The baby is a *friendly* computer!"

Many parents do get frightened, of course, particularly when a flood of books and articles keeps telling them what to do and not to do—and above all not to get frightened. The current discoveries about how much a baby sees and hears and knows at the very moment of birth make the parental responsibility seem even more formidable. Most important, in a way, is that these findings are changing the way people actually see their own children, changing how they talk to them, what they expect of them. And these slow and almost imperceptible transformations can hardly help altering, in subtle and equally imperceptible ways, the babies themselves, and thus the adults they will some day become.

The traditional view of infancy was that of Shakespeare, who described the helpless newborn as "mewling and puking in the nurse's arms." Nearly a century later, John Locke proclaimed it as self-evident that the infant's mind was a *tabula rasa,* or blank tablet, waiting to be written upon. William James prided himself on more scientific observations but wrote in *The Principles of Psychology* (1891) that the infant is so "assailed by eyes, ears, nose, skin and entrails at once" that he views the surrounding world as "one great blooming, buzzing confusion." As recently as 1964, a medical textbook reported not only that the average newborn could not fix its eyes or respond to sound but that "consciousness, as we think of it, probably does not exist in the infant."

Such views have been increasingly re-examined and revised during the past two decades, and this research has now grown into a substantial industry. From the Infant Laboratory at M.I.T. to the University of Texas' new Children's Research Center to U.C.L.A.'s Child Study Laboratory, there is hardly a major university without teams of researchers poking and prodding babies. The number of studies of infant cognition has tripled in the past five years, according to Psychologist Richard Held of M.I.T. A conference of experts in Austin last year heard more than 200 research papers ranging from "Sleep-Wake Transitions and Infant Temperament" to "Right-Left Asymmetrics of Neurological Functions in the Newborn Infants." These multitudinous studies do not go unchallenged: researchers in various disciplines fight for their own specialties, psychiatrists differ sharply in their views from neurologists, judgments are often subjective, and babies themselves are as different as snowflakes.

The search for data is being steadily pushed back from childhood to earliest infancy and even before birth. One French obstetrician, for example, inserted a hydrophone into the uterus of a woman about to give birth and tape-recorded what the fetus could hear: the mother's loudly thumping heartbeat, a variety of whooshing sounds, the muffled but distinguishable voices of the mother and her male doctor, and, from a distance, the clearly identifiable strains of Beethoven's *Fifth Symphony*.

The obvious obstacle that long hindered scientific research on babies was that they could not talk,* could not tell what they saw or thought; the consequence was a widespread belief that they saw little and thought less. But that belief was based primarily on adults' dim recollections of their past. As early as the 1950s, a few psychologists were searching for laboratory methods to discover what babies could learn. Case Western Reserve Psychologist Robert Fantz made an important breakthrough in 1958 by demonstrating that babies' fascination with novelty could be turned into a form of silent speech. Specifically, Fantz watched infants move their eyes when he showed them two different objects; he carefully measured what they looked at and for how long. Given a choice, he showed, babies will look at a checkerboard surface rather than a plain one, a bulls-eye target rather than stripes, and in general they prefer the complex to the simple. Says Rutgers' Michael Lewis: "Out of such

*The very word infant derives from the Latin *infans*, meaning incapable of speech.

elementary observations, monstrously important consequences grew."

Once the basic approach was discovered, a whole world of previously untried research opened up; new technology made it possible to devise tests that would have been unimaginable a generation earlier. At the most rudimentary level, the videotape machine enables a psychologist to record a baby's wriggling and demonstrate that it often moves in rhythm with its mother's voice. At the most complex levels, surgeons at Prentice Women's Hospital in Chicago can diagnose prenatal hydrocephalus (a brain-damaging excess of cerebrospinal fluid) in a fetus, then introduce a plastic tube into the mother's uterus and into the fetus' head to drain off the surplus fluid inside its brain. Guiding many of these technological innovations is the ubiquitous computer, which can synthesize a mother's voice as easily as it can measure eye movements or count the times that young Géry sucks on his nipple.

The first area to attract a number of researchers was the newborn baby's senses, which were once thought to represent little more than hunger to be fed. Systematic testing soon showed that babies not only perceive a good deal but have distinct preferences in everything. An Israeli neurophysiologist, Jacob Steiner, found that a baby as young as twelve hours old, which has never tasted even its mother's milk, will gurgle with satisfaction when a drop of sugar-water is placed on its tongue and grimace at a drop of lemon juice. More mysteriously, a newborn will smile beatifically when a piece of cotton impregnated with banana essence is waved under its nose, and it will protest at the smell of rotten eggs. Other infant prejudices: vanilla (good), shrimp (bad).

The baby emerges from the darkness of the womb with a rudimentary sense of vision—it would be rated about 20/500, or "legally blind," as one expert puts it, but eyesight develops rapidly. Newborns start by looking at the edges of things, exploring. Even when the lights are turned out, as infra-red cameras show, an infant's eyes open wide to carry on its investigation of its surroundings. At eight weeks, it can differentiate between shapes of objects as well as colors (generally preferring red, then blue); at three months, it begins to develop stereoscopic vision.

Testing such perceptions can be complicated. At M.I.T.'s Infant Laboratory, for example, University of Tokyo Graduate Student Shinsuke Shimojo has programmed a computer to check whether seven-month-old Whitney Warren can differentiate between a straight bar and a slightly indented bar. The computer makes the indented portion of the second bar move slightly. If Whitney can see the indentation, he will see its movement, and Shimojo, crouching behind the computer

screen, can see his eyes move. Most babies spot the movement easily.

Despite their esoteric quality, such experiments can have an immediate practical value: some infants suffer from eye ailments, such as cataracts, severe astigmatism and strabismus, which benefit from treatment much earlier than would once have been possible. No less important, the new research has demonstrated that an impairment of infant vision can damage those parts of the rapidly growing brain that rely on visual information. That brain damage can be permanent unless the eye impairment is treated early.

Unlike the eyes, the baby's ears have been functioning even before birth, and the newborn arrives with a whole set of auditory reactions. As early as the 1960s, tests indicated that babies go to sleep faster to the recorded sound of a human heartbeat or any similarly rhythmic sound. More recent studies indicate that by the time they are born, babies already prefer female voices; within a few weeks, they recognize the sound of their mother's speech.

Many mothers believe they can understand different kinds of crying by their babies (a controlled experiment in 1973 showed they could not), and they believe even more strongly that their babies can understand a parent's murmurings. And perhaps they can. Though children do not ordinarily say anything very elaborate before the age of one year, Psychologist Peter Eimas of Brown University has demonstrated that infants as young as one month can differentiate between sounds in virtually any language. They also have a "very sophisticated" ability to organize sounds into various categories. "A baby already knows which sounds communicate," says Eimas. "I've never heard a baby imitate the sound of a refrigerator, for example. So a child can put all of his energy into learning how to use the rules of the language."

Pursuing the origins of language back into earliest babyhood is an interesting approach to understanding the infant intellect. No less so is the discovery that this intellect is at work long before any language is available as a tool. The key element in that discovery was the baby's desire to imitate its mother's facial movements. Jean Piaget, the celebrated Swiss psychologist who pioneered in this field with extended studies of his own three children, declared that such imitations began only at about eight to twelve months. Earlier than that, he reasoned, the baby could not understand that its own face was similar to that of its mother.

Olga Maratos, a Greek student who was testing seven-week-old infants for her doctorate, went to Piaget's house one snowy day early in 1973 to tell him of her progress. "Do you remember what I am doing?" she said. "I am sticking out my tongue at the babies, and do you know what they are doing?"

"You may tell me," Piaget murmured.

"They are sticking out their tongues right back at me! What do you think of that?"

The venerable professor puffed on his pipe for a moment as he contemplated the challenge to his theory. "I think that's very rude," he said.

Maratos' thesis was never published, so the credit for the discovery went mainly to two young psychologists who now teach at the University of Washington, Andrew Meltzoff and M. Keith Moore. Their study, published in 1977, showed that babies only twelve days old could imitate an adult sticking out a tongue. Meltzoff and Moore demonstrated that if a pacifier in the baby's mouth prevented the infant from imitating the adult, it would *remember* what it wanted to do until the pacifier was removed; then the baby would promptly stick out its tongue.

That first study by Meltzoff and Moore aroused considerable skepticism, so they repeated and elaborated it in 1981, eliminating all uncertainties and using still younger children. "We had one baby 42 minutes old, with blood still on its hair," recalls Meltzoff. "We washed it and tested it. We found that even newborns could imitate adults."

These experiments demonstrated the infant's very early capacity for what psychologists call "intermodal perception"—that is, to combine the brain's perceptions of two different activities, in this case vision and muscular action, which is virtually the first form of thinking. Says Yale's Kessen: "The past 15 to 20 years have demonstrated that the child has a mind. The next several years will be used to find out how it works."

Meltzoff pursued his exploration of intermodal perception by a different test of vision and touch. He gave ordinary pacifiers to a group of month-old babies and pacifiers with bumps on them to another group. He then had the babies look at models of the two kinds of nipples. The result, says Meltzoff, was that "they would look at the ones they had felt." Now, with Speech Professor Patricia Kuhl, he has extended those tests to language. The researchers showed infants two films of faces saying "ahh" and "eee," then placed between the two pictures a loudspeaker that could make either sound. The babies invariably looked toward the picture that fit the sound. "This means that babies can detect the relationship between mouth movements and the sounds they hear," says Meltzoff. "Essentially, babies are lip readers."

As they begin to develop this rudimentary capacity for thinking, babies develop an important ability to recognize categories. This was once thought to require language—how can the unnameable be identified?—but babies apparently can organize perceptions without a word. Psychologist Elizabeth Spelke of the University of Pennsylvania showed four-month-old babies a pair of films in which two toys bounced around on a surface in different ways, each with a corresponding sound track. She then played one sound track, and the babies were able to match the correct film to its sound. From the babies' "highly differentiated ability" to decide what goes with what, Spelke went on to deduce that children are born with an innate ability to divide their experiences into categories. Says she: "Obviously, in order to make sense of anything that you're confronted with, you have to bring to bear certain conceptions about the world. Our hope is that we'll learn something about what those initial conceptions are."

It is a puzzle, for babies repeatedly demonstrate a variety of skills and actions that seem to have no basis in their previous experience. Examples:

▶ Bradley Feige, age 11½ months, is sitting on a glass table at U.C.L.A.'s Child Study Laboratory. "Come here, Bradley, come here," his mother coaxes from the other side of the table, about six feet away. At her end, the cloth material under the glass top suddenly drops away to create the illusion that Bradley may plunge several feet if he does what his mother asks. At eight months, and again at ten months, Bradley ignored the illusion of peril and crawled across the table. Now he refuses to budge past the illusionary end of the table, not even when his mother holds out a toy as a lure. "We know that this response is not related to the experiences they've had," says Psychologist Nancy Rader, "but we've found that it relates to the age at which the baby starts crawling, and we're trying to find out why."

At Harvard's Center for Cognitive Studies, infants as young as two weeks were confronted with a cube (or sometimes only the shadow of a cube) that began moving slowly toward them. When it seemed about to hit them, they showed what psychologists call "a strong avoidance-reaction pattern." They turned aside and squirmed and tried to avoid being struck, though they had no previous experience that would make them think that the approaching object would hit them. When such a cube or its shadow approached the babies on an angled path that would miss them, however, the babies followed its motion with their eyes but showed no sign of anxiety. "The consummate skill of these infants in predicting the path of the moving object is astonishing," says Psychologist Jane Flannery Jackson, "and their evident wish to avoid objects on a collision course is even more so."

▶ At the University of Edinburgh, T.G.R. Bower and his associates have been conducting about 1,000 experiments a year on various infant abilities. One of their most startling claims is that babies can tell the gender of other infants they are looking at, and they prefer to look at those of their own sex. Bower made films of an infant boy and girl making various movements, and then deleted from the film all apparent signs of gender and even swapped their clothes. Some adult viewers had difficulty telling them apart, but something about the way the filmed infants moved enabled a group of 13-month-old children to distinguish the boy from the girl. Bower is still trying to figure out how they do that.

How babies do any of the things they do is a matter of considerable complexity. Some theorists, like Thomas Verny, a Canadian psychiatrist who wrote *The Secret Life of the Unborn Child,* believe the infant begins learning behavior patterns while it is still in the uterus. Most experts, however, assume that the genes still carry messages that primitive humans once needed for survival. The so-called Moro reflex,* for example, which makes a newborn infant reach out its arms in a desperate grasping motion whenever it feels itself falling, implies some monkey-like existence at the dawn of time. Says Lewis Lipsitt, director of the Child Study Center at Brown and a pioneer in research on babies: "The human infant is extremely well coordinated and put together for accomplishing the tasks of infancy. These are: sustenance, maintaining contact with other people, and defending itself against noxious stimulation."

One of the oddest elements in their development is that infants soon lose many of the skills they had at birth. A newborn baby that is held upright on a table is nearly able to walk while suspended; immersed in a tub of water, it makes a fairly impressive try at swimming. Those abilities deteriorate within a few months. The same process seems to occur with intellectual skills that are not used. Psychologists Janet Werker of Dalhousie University in Halifax, N.S., and Richard Tees of the University of British Columbia have shown that babies of six to eight months can distinguish sounds that are not used in their native language, but they have much greater difficulty by the age of twelve months. Japanese babies, for example, have no trouble with the "ell" sound that their parents find difficult.

Most experts now think a baby is born with a number of reflexes that are gradually replaced by the "cortical behavior" dictated from the cortex of its rapidly developing brain. Brown's Lipsitt believes that a period of "disarray" during the course of this transition may be an important element in the "crib deaths" that can mysteriously strike during the first year. The struggle to escape from accidental smothering in bedclothes, known as the "respiratory occlusion reflex," is automatic at birth but then needs to be learned. Says Lipsitt:

*Named for German Pediatrician Ernst Moro (1874-1951).

"The peak of 'disarray' is right at the point when crib death is most likely to occur, as if the baby doesn't know whether to be reflexive or cognitive. Suppose a child gets into a compromising situation where it has lost the reflex and has not acquired the learned behavior that has to come in to supplant the lost reflex." Lipsitt hopes to devise a specific test that will pinpoint those few children who may be in jeopardy.

Every test for every kind of trouble implies that there is a "normal" time for a baby to demonstrate various abilities. If it does not sit up by six or seven months or stand by nine or ten, a pediatrician may start neurological testing. The disciples of Yale's Arnold Gesell have applied this approach to all phases of childhood ("He wanders from home and gets lost at four," says the latest edition of the Gesell Institute's *Child Behavior*. "He demands to ride his bicycle in the street at eight").

Most current advice givers urge anxious parents not to take such standardization too seriously. Pediatrician T. Berry Brazelton *(see box)*, who is publishing a revision of his 1969 bestseller, *Infants and Mothers,* begins by declaring: "There are as many individual variations in new born patterns as there are in infants." Still, though a child's development during its first year is far slower than that of a monkey or even an elephant, it is nonetheless so dramatic—from lying flat on its back to the first creeping across the floor to the first faltering steps around the corner of the kitchen table—that scientists persist in trying to pinpoint when and how it learns each new accomplishment.

Two months, eight months and twelve months seem to mark major periods of change: in brain developments, in various skills and perceptions, in sociability. At about two months, for example, the baby is awake much longer than it was, it smiles a lot and stares with fascination at a new discovery: its own hand. At eight months, the infant is acquiring the important sense of its separate identity, and even an understanding of what Piaget called "object permanence," the realization that an object hidden from sight is still there. It begins to develop fears of strangers and of separation from its parents. At twelve months, the golden age, the baby has begun to walk and talk, and knows that the whole world awaits. Sometimes, clinging to a chair, waving a spoon in a fist, the one year old will throw back its head and crow in sheer delight.

These physical and social achievements have long been obvious: any mother can see them in her own children. What the new research demonstrates is that babies' mental growth can be as early and as striking as the rest of its development. Robert Cooper, a psychologist with Southwest Texas State University, is even testing a group of ten- to twelve-month-old children on their ability to recognize

The New Dr. Spock: "A Great Dad"

Dr. T. (for Thomas) Berry Brazelton, 65, says he is no scientist, which shows a becoming modesty, but he would have a hard time denying that he is the nation's pre-eminent baby doctor. A whole generation of pediatricians has studied and worked with him at Harvard Medical School and the Children's Hospital Medical Center in Boston. Tens of thousands of anxious parents have been reassured by his easygoing guidebooks (*Infants and Mothers, Toddlers and Parents, Doctor and Child, On Becoming a Family*). Millions of infants who never met him have been tested and evaluated by his Neonatal Behavioral Assessment Scale, generally known simply and naturally as the Brazelton.

It is a folksy sort of test, carried out with such implements as a pocket flashlight, an orange rubber ball, a paper clip, some popcorn kernels. The exam starts when the baby is asleep, and it gauges the infant's reactions to a series of stimuli, including light in the eyes, the sound of rattling, a scratch on the foot: 20 reflexes and 26 behavioral responses in all. After 20 minutes or so, a Brazeltonized baby is wide awake and none too happy about all the testing.

Brazelton began devising the exam some 30 years ago to solve a problem that bothered him: babies available for adoption were being kept in institutions until the age of four months because doctors were reluctant to certify that any younger infant was fully normal. "Four months is just too long to deprive anybody of a new baby," Brazelton recalls, with a trace of a Texas drawl that has survived his years in Boston. "That led me to say, 'Well, gosh, anybody can tell whether a new baby's O.K. or not. What is it we're going by?' Then I began to put together all these things that any good clinician uses. Very little about the scale was really new. It was a compilation of a lot of clinical observations that hadn't been documented."

That all sounds rather routine, but Brazelton's tone changes as he starts to talk about his test, about the way a baby's eyes jerkily follow a moving ball. "If you give him a human face to look at instead, his eyes will widen and he'll get more intense and he'll follow you," says Brazelton, "and as he follows, his face gets more and more alert and more and more involved, and you can feel yourself getting more and more involved back. This kind of visual involvement is more than just looking. You've got another component from the baby, which says to the person doing this, 'You're terribly important.' And the person is bound to feel important. What I'm getting at is that the baby's competence will call up competence from parents. We used to see the parents shaping the child, but now we see the child also helping to shape the parents."

Brazelton once wanted to be a veterinarian. At age eight, already an experienced baby sitter, he decided on pediatrics. He went to Princeton, starred in Triangle Club theatricals, even got an offer in 1940 to try out on Broadway for an Ethel Merman musical, *Panama Hattie,* but he held on to the goal of healing infants. His hero, he says, was Benjamin Spock, and although Brazelton is now regarded as the new Spock, he considers himself more a disciple than a rival of the older man.

Brazelton and his wife Christina have three daughters and a son, ages 19 to 32. He still worries about his high expectations and pushing his children too hard. His son, however, calls him "a great dad."

Like Spock, Brazelton makes it a cardinal rule to reassure anxious parents and to encourage them to trust their instincts. "Parents in our culture are so hungry for people to tell them what to do and so vulnerable as a result," he says. "I feel very strongly that telling them what to do is destructive. Supporting them for what they can do is constructive."

Brazelton is in the midst of a project of "intervention research" that involves studying 100 undersized babies and trying to see which of them will need special assistance. Babies that have been undernourished in the uterus are "very scrawny, very hypersensitive to any kind of stimulation, and they become very fussy and difficult for the parents," says Brazelton. "They need help to see their baby as a person. You have to help parents see that you're seeing the same baby they are. And that the baby doesn't need to be like them. And they don't need to be like it. It can be just as exciting to find another kind of person to learn about."

different numbers. They can master up to four, but he adds that "beyond four, there's some controversy." By showing his little subjects various groups of objects, Cooper demonstrates that they can tell the difference between three and five, he says, though the difference between four and five sometimes baffles them.

The idea that infants can start acquiring an education has tempted ambitious parents for centuries. At the age of three, John Stuart Mill learned Greek, and Mozart was playing the harpsichord. Both were taught by their hard-driving fathers. Today, New York City's fashionable nursery schools not only interview two year olds (and charge their anxious parents $1,200 a year for two mornings of schooling a week), but they also report applications outrunning openings by as much as 5 to 1.

The vogue is spreading. Gymboree, a franchise operation that started seven years ago in San Mateo, Calif., now has 61 outposts operating in 14 states that provide educational play for about 10,000 children. "Learning to read begins at birth," says one of Gymboree's brochures, but the $4 classes are mainly physical, ranging from "wee workouts" for beginners up to "gymgrad" for tots as old as four. "We've tried to create a 'yes' environment for the children, to place them in a setting they can master," says Gymboree's founder, Joan Barnes, a former dance teacher.

More strictly pedagogic is a Philadelphia organization called the Better Baby Institute, which offers a training course to enable mothers to "multiply their baby's intelligence." Specifically, the school claims that parents can learn in one week of intense instruction (for a fee of $500) how to teach their infants to swim, to read, to do math, to speak foreign languages and to play the violin at the age of two. You can't make it to Philadelphia? "Better Baby Video," a California-based spin-off, can provide the same lessons in a weeklong course offered primarily in West Coast cities. Some critics believe that all this mainly makes babies learn a few skills by rote, but it is difficult to obtain any scientific assessment of the five-year-old institute.

Many of these ventures in infant education are fueled by eager parents who will try anything to give their children a head start. Similar experiments are arousing interest in those who work among the poor. Dr. Joseph Sparling, for example, has developed and published a series of 100 educational games at the Frank Porter Graham Child Development Center at the University of North Carolina at Chapel Hill. These games, which range from specific subjects like language development to vague concerns like self-image, have been tried out with some success over the past five years in a federally funded program called Project Care. Researchers use the games both in day care centers and in weekly visits to children's homes. They report that the children get "significantly" higher intelligence-test scores at the age of one year than children in a control group who are not exposed to the games.

If nothing else, the push toward earlier education gives infants a valuable chance at making friends. Says Psychologist Colwyn Trevarthen of the University of Edinburgh: "They really have this intrinsic social capacity, and that's what human beings have evolved for, just as giraffes have evolved for eating high leaves."

But is early education itself really desirable? Does the discovery that a young child can absorb large quantities of knowledge require that it be stuffed like a Strasbourg goose? There were social reasons for launching Project Head Start in the 1960s to get poor children into preschool programs. Most psychologists engaged in the new research, however, are strongly opposed to any formal schooling before the age of three or four, even if the child is capable of it. "We know that babies are coming into the world with a lot more skills than we had previously thought, but I do not think reading, writing and arithmetic should be in their curriculum," says Psychologist Tiffany Field of the University of Miami School of Medicine. Warns Child Psychiatrist Robert Harmon, director of the Infant Psychiatry Clinic at the University of Colorado School of Medicine: "I think you're going to get children burned out on learning." And University of Denver Psychologist Kurt Fischer says of the baby's first year: "Don't worry about teaching as much as providing a rich and emotionally supportive atmosphere."

As Fischer's statement indicates, much of the new research emphasizes the extreme importance of the infant's relationship with its mother. (And/or its father, and/or what the linguistically liberated call the "caregiver.") She must not only feed it, and love it, but endlessly talk to it, play games with it, show it what is happening in the world. Rutgers' Lewis has tested 100 babies for mental development at three months and recorded their mothers' response to the infants' signs of distress. He was hardly surprised to find that those who had been more warmly cared for had learned more by the time they were retested at the age of one year. This kind of nurturing is essential to both emotional and intellectual growth; indeed, the two are inseparable. "The baby who doesn't smile may be giving us a more reliable indicator than cognitive tests," says Psychiatrist Eleanor Galenson of Manhattan's Mount Sinai Medical Center.

The baby's smile is also a kind of judgment on the care that its mother has been providing. "All these new data about how early the baby can distinguish things should upgrade motherhood, restore some prestige to it," says Dr. Benjamin Spock, 80, who taught a benign form of child rearing to a whole generation of Americans. "Motherhood has had an ever reduced amount of importance placed on it in our strange, overly intellectualized, overly scientific society."

According to traditional wisdom, all mothers know instinctively how to rear their children, but unfortunately that is not always true. Indeed, the instinct has been vehemently denied by Elisabeth Badinter, the French philosophy professor who wrote *Mother Love: Myth and Reality.* But even if a mother's nurturing is an instinct, it requires some experience as well, and if the ability is entirely a learned trait, it is sometimes none too well learned. To check on how consciously mothers interact with their babies, Psychiatrist Daniel Stern of the Cornell University Medical Center has been observing nearly 100 mothers playing with infants eight to twelve months old. "Whenever we notice that the baby has put on an emotional expression that the mother has seen, we look at how she responded to it," says Stern. "Then we ask her why she did it, what she thought the baby was feeling, what she expected to accomplish, and whether she knew what she was doing at the time." His preliminary findings: about one-third of the mothers were fully aware of what Stern calls the attunement with their infants, another third were quite unaware of it, and the rest were essentially unaware but could recall it when it was pointed out to them.

This extremely important emotional interplay, often described as "bonding," is a combination of love and play, but it is now seen as something else, a kind of wordless dialogue. The baby not only understands what the mother is communicating, or not communicating, but it is trying to tell her things, if she will only listen. Says Dr. Bennett Leventhal of the University of Chicago's Child Psychiatry Clinic: "We now know that babies send messages very early. In their first year of life, they are good students. They are also very good teachers, but they have to have someone to interact with them. There are sometimes very competent babies with very incompetent parents."

Many psychologists believe the new research enables them to anticipate future problems in even the youngest children. "We can now document where a baby may be unable to pick up sensory data; we can spot abnormalities in the emotional areas," says Stanley Greenspan, chief of the Clinical Infant Research Unit at the National Institute of Mental Health in Adelphi, Md. "There is no evidence that an infant's emotional problems are self-corrective. The environment that contributed to early damage will continue to contribute if one does not intervene."

One early and important symptom of trouble, says Greenspan, is the failure of mother or child to look at each other.

Greenspan makes videotapes of such cases. Here is Amanda, age four months, who turns her head away and generally shows what Greenspan calls "an active avoidance of the human world." Small wonder. Amanda's mother was raising her alone and suffered bouts of deep depression. Greenspan and his therapists spent four months playing with Amanda and engaging her interest; the videotape taken at eight months shows the baby cheering her mother along. Says Greenspan, with some satisfaction: "She developed coping facilities stronger than those of her mother."

Psychologists talking about "environment" often meant primarily the psychological structure of the family, but the social and economic environment is hardly less important to a child's development. Fully 13.5 million children in the U.S. live below the official poverty line. Nearly 7.5 million children are currently on welfare. More than half a million babies are born every year to American teen-agers.

The effects of such deprivation on infancy are hard to gauge scientifically, but Dr. Gerald Young of Manhattan's Mount Sinai Medical Center says flatly, "If you want to guess what a child will be like at age seven, look first to the socioeconomic background." This is not simply a matter of economic hardship or nutritional deficiency. Says Brown's Lipsitt: "The socioeconomic index is as powerful a predictor of later intellectual prowess as any variable we've got, but it doesn't operate in a **vacuum. It is a representation of the way** people live and relate toward each other, and the way they behave toward babies."

One interesting demonstration of this theory was undertaken more than a decade ago by a team of psychologists at the University of Wisconsin. Struck by the fact that many of the mentally retarded children in a Milwaukee slum had retarded mothers, they took 40 infants whose mothers had IQs of less than 75 and put 20 of them in special day care centers. From the age of three months on, the children began getting lessons in language and arithmetic as well as various other kinds of stimulation. By the time they reached school age, their average IQ was more than 100 (none was retarded); the 20 children who had received no special treatment had an average IQ of 85, and 60% were judged to be retarded.

The question of child rearing outside the home cuts across all classes. There are currently 4.1 million working women with children under the age of three, and one survey showed that nearly 70% of working women who have babies return to their job within four months. Overall, about 8 million of today's preschool children receive some form of day care (1 million in day care centers, 3.5 million in family day care homes and 3.5 million tended by relatives and baby sitters). If the nurturing mother is as important as psychiatrists say, hired substitutes may seem a poor alternative, but most psychological researchers reject any such conclusion. All a baby basically needs, they say, is at least someone who is consistently there and who really cares. All depends, obviously, on the quality of the day care—and of the home. In the case of the Milwaukee experiment with the potentially retarded, day care was a rescue service. But in one typical Maryland county, 788 regulated day care facilities have room for only 8,560 of the 65,000 children under 14 who have working mothers.

How good the average day care is remains something of a guess. Bernice Weissbourd, who founded the Chicago-based Family Focus groups to provide support and advice for new parents, argues that any day care service that has more than three infants per adult (and that includes most) is inadequate. "Too often," she warns, "the parents' main questions are simply how close to home is it and how much does it cost." But of day care as such, Cornell Psychologist Urie Bronfenbrenner says categorically, "There is no hard evidence that day care has a negative effect."

Whatever the difficulties, the overwhelming majority of parents want very much to do the best for their children, if only they can be sure what that best is, and that is anything but certain. Most experts say the need is great. "Not more than one child in ten gets off to as good a start as he could," says Burton White, author of *The First Three Years of Life.* Harvard's Kagan, on the other hand, urges parents to provide "a nurturant environment" and declares, "It's easy. Oh, it's easy. There's not a lot of witchcraft here."

Important changes come so slowly that they are taken for granted. Children's sheets used to be all white; now they are explosions of color. The mobile over the crib, which first seemed arty and pretentious, has become almost a basic piece of furniture. The backpacks that were once associated with Indian women carrying papooses are now sold everywhere, not only as a convenience for mothers but as an opportunity for the baby to get out of the house and see the world.

Thus the old keeps becoming the new. Much of what modern research is so elaborately documenting is what parents have always known—whether from instinct or from common sense or from the teachings of their own parents—that babies need and respond to love, attention, stimulation, education, in perhaps roughly that order. The research documents not only the importance of such needs but the damage that can occur when they go unanswered. Yet even these blessings of the latest orthodoxy can be overdone. "We are learning that everything will have an impact on an infant, but we still need to know exactly what happens," cautions Psychologist Rose Caron of George Washington University's Infant Research Laboratory in Silver Spring, Md. "It's conceivable that a child's competency might be diminished because of too much early stimulation."

"Do I contradict myself? Very well then I contradict myself," cried Walt Whitman. The creation of a baby is full of paradoxes and illogicalities. The cost of raising a child to 18, approaching $100,000 in the U.S., according to one estimate, would deter any sensible investor. So would the prospect of more than 20 years of anxiety and irritation. But having a baby remains, for most people, an act of faith. It represents a belief in better things to come, not just for themselves but for the world. That is a faith shared by the myriad baby researchers. Says Rutgers' Lewis: "Can we produce a better society with healthier children? The answer is yes." And in the very moment of birth, as a tiny, dark, wet head thrusts out into the world, every baby fulfills that belief. Then comes the first squall.

—By Otto Friedrich. Reported by Ruth Mehrtens Galvin/Boston and Melissa Ludtke/New York, with other bureaus

THERE'S MORE TO CRYING
THAN MEETS THE EAR

Barry M. Lester, Ph.D.

Assistant Professor of
   Pediatrics (psychology)
   Harvard Medical School

Director of Developmental
   Research
   Child Development Unit
   Children's Hospital
      Medical Center
   Boston, MA

When most of us think of crying in infancy, we think of the intractable cries of the baby with colic or temper tantrums in the toddler. But there is definitely more to crying than meets the ear! Parents have always known that their babies express different needs through the quality of the cry ("He just wants attention"); and the high-pitched "cerebral" cry of the brain-damaged infant has long been recognized in neonatal care. But the translation of these observations into objective measures opened the way for considering the cry as an acoustical signal that could reveal information about the developing nervous system, as well as a way for the baby to communicate basic needs.

A BIOSOCIAL PHENOMENON

Crying has direct and indirect effects on the subsequent developmental outcome of the infant. Direct effects are due to the cry as a measure of the integrity of the nervous system; indirect effects are due to the cry as a determinant of parent-infant interaction, which in turn affects the cognitive and socio-emotional development of the infant. These direct and indirect effects are related to the importance of crying as an acoustical signal for two reasons.

First, variations in acoustic features of the cry reveal properties of the developing nervous system and are important for medical diagnosis. Acoustic features such as frequency or pitch indicate physiological properties of the infant. Hence, physicians use the cry as a diagnostic tool to assess neurophysiological functioning. Second, the acoustic features of the cry affect parents; when they describe their aversive reaction to a "high-pitched, irritating" cry, they are responding directly to cry acoustics. Similarly, a pain cry is distinguished from a hunger cry by its rapid rise time, increase in pitch and longer phonation at the beginning of the cry, followed by longer breaths between each phonation. Crying is a biological sign that conveys a message of urgency.

As a behavior, crying is a social event that affects the development of the parent-infant relationship. It is now well established that how parents treat their infant is influenced by the infant's behavior. Crying can affect parents by its presence or absence ("He's such a pleasure, he hardly ever cries!"). Excessive crying or cries that sound aversive can cause disturbances in the developing parent-infant relationship and in extreme cases can lead to child abuse. It is important to acknowledge the strong feelings that crying generates in many of us. Feelings of anger and rejection are common and normal, even though, fortunately, they are rarely acted out against the baby.

This binocular view of infant crying as a biological and social event should help the clinician appreciate the importance of crying. As a regulatory system, it has a strong impact on parental caretaking behavior. For example, cold exposure leads to crying, which generates heat and improves thermoregulation, while at the same time it elicits physical contact from the caregiver. Cuddling the crying baby not only contributes to thermoregulation; it stimulates attention and visual scanning. In this social interaction the parent is doubly rewarded: the crying stops, and a mutually enjoyable experience occurs. Through this process, parents become attuned to the cues and signals of the infant; the rhythmic property of crying makes it easier for parents to understand and predict their infant's behavior. At the same time, the infant gains experience in effecting change in the environment.

DEVELOPMENTAL CHANGES IN CRYING

Infant crying varies greatly in cause, time of day, frequency and duration. Temperamental differences in infants are responsible for some variations; but for all infants, the functional significance of crying changes with age. Certain general developmental principles can be used by pediatricians as a guide to better understand the nature of crying.

In the first few weeks of life, crying has a reflexive-like quality and is mostly tied to the regulation of physiological homeostasis as the neonate is balancing internal with external demands. As physiological processes stabilize, periods of alertness increase, placing additional demands on regulatory functions. Crying is a normal mechanism for discharging energy or tension; it can occur in response to too much stimulation; and the need for tension reduction is especially acute at times of major develop-

 Reprinted from the *Child Care Newsletter,* Vol. 2, No. 2 (1983) pp. 1-5 with permission from the Johnson & Johnson Baby Products Company.

mental upheavals. Periods of so-called unexplained fussiness and sudden increases in crying occur in the first few months, probably related to maturational changes in brain structure and shifts in the organization of the CNS. Colic is probably a result of changes in the nervous system that accompany this early biobehavioral shift; but why some infants develop colic and others do not, or whether some infants are predisposed to colic, are important research areas.

Not only does the quantity of crying change over the first few months but so does the quality. As the infant gains more voluntary control over vocalization because of physiological and anatomical changes that occur around 1 to 2 months, crying becomes more differentiated. Mothers talk about the angry cry and the cry for attention as distinct from the hunger or pain cry. In the first few months, we see a change from crying as response to physiological demands, to crying as part of the development of affective expression. Toward the end of the first year (7 to 9 months), a second biobehavioral shift occurs, characterized by major cognitive and affective changes. Crying now may be caused by fear, most commonly fear of strangers and fear of separation from the caretaker. Crying also occurs in response to frustration as the infant becomes more mobile and explores the environment. Periods of fussiness often precede the acquisition of a new skill or attainment of a new developmental stage because of the tension and apprehension that are part of the natural process of growth. In the 2-year-old, tantrums are often due to frustration, and help reduce tension. The period between 17 and 24 months is an important time of maturational change in the nervous system. Emotions are strong at this time as infants want to remain infants but also to be separate. This struggle for autonomy as self-awareness increases is also a source of anxiety. We see crying in response to fear of failure as infants begin to internalize standards and develop a sense of right and wrong. The understanding of rules becomes an issue when parents set limits that the infant rejects but also wants. This is a difficult but exciting time for infants that makes them more volatile and cry outbursts more likely.

CRYING AND MEDICAL DIAGNOSIS

The original goal of the research on crying in newborns with medical disorders was to identify specific diseases or complications from the cry. The studies revealed, however, that though medical abnormalities and prenatal and perinatal trauma affect the character of the vocal signal, alterations in the cry may not be due to the same underlying physiological deficit. No single acoustic measure was found likely to distinguish normal from abnormal infants; instead, acoustic features were seen as a general statement about the status of the nervous system, not a specific disease indicator.

Clinical conditions do exist where the cry can be of use in medical diagnosis. Here the cry fulfills its function as a warning signal. First, it can be used to support a differential diagnosis as in, for example, asphyxia, hyperbilirubinemia, or cerebral insult. In these cases, infants usually present with other clinical signs, but it is often the abnormal-sounding cry that draws attention to the infant in the first place.

Second, in a group of at-risk infants, the unusual cry may alert the pediatrician or parents to the possibility of injury or damage due to known etiological factors such as prematurity or asphyxia. While many of these infants will go on to develop normally, others will develop mild to severe forms of mental or motor handicap in later infancy or early childhood. There is some evidence that acoustic cry features can differentiate in the newborn period which at-risk infants are headed for handicap. Figures 1 and 2 contrast the cries recorded at term of two preterm infants from one of our studies. Both were born at 33 weeks' gestational age with comparable birthweights. Although each suffered different medical complications associated with prematurity, their medical prognosis was the same; both were judged by pediatricians to be "moderately" at risk. At (corrected) age 18 months, their mental status was statistically predicted by the neonatal acoustical cry features and was unrelated to medical factors. The infant in Figure 1 whose cry looks like the cry of a term healthy infant showed above-average mental performance; the infant in Figure 2, whose cry appears abnormally high-pitched and variable, was mentally deficient, scoring two standard deviations below normal in mental performance. This is one illustration of how the cry can be used in the early detection of which at-risk infants are truly headed for handicap.

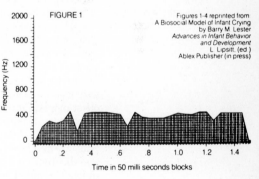

Figures 1-4 reprinted from: A Biosocial Model of Infant Crying by Barry M. Lester *Advances in Infant Behavior and Development* L. Lipsitt, (ed.) Ablex Publisher (in press)

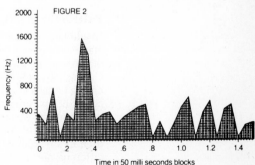

Sometimes the cry is the only presenting indicator. Other signs may be more subtle or not yet known. For example, there is preliminary research to suggest a relationship between un-

Crying is a normal mechanism for discharging energy and tension.

Some crying is a response to fear of failure.

usual crying patterns and the sudden infant death syndrome (SIDS), which could lead to identification of infants at risk for SIDS.

The pediatrician should view crying in part as a reflection of CNS integrity or stress that may indicate infants who are compromised or in jeopardy. During routine physical examination of the newborn and young infant, the pediatrician can "keep an ear out" for the following dimensions of acoustic features that could signify a compromised infant.

Abnormal crying may be an indicator of SIDS?

1. Extremes in pitch: high pitch that sounds shrill, piercing, squealing, or low pitch that is gutteral, throaty or harsh
2. Extreme variability in pitch: from little or no variability in which the cry sounds flat and monotonous, to rapid, extreme changes in pitch, as if the infant can't maintain a steady cry
3. Absence of harmonic quality of "richness": the cry sounds austere, stark, with little or no overtone
4. Lack of temporal patterning: the cry does not seem to have a rhythmic pattern and it is difficult to understand what the infant is trying to communicate
5. Presence of turbulence or non-harmonic distortion that masks the tonal quality of the cry: a harsh, rough, scratchy sound
6. Extreme variation in the length of each cry burst, from very short staccato cry phonations to unusually long cry expirations where the infant seems to have trouble coordinating crying and respiration
7. Short or long latency to cry, ranging from hypersensitivity where there is an extremely low threshold for stimulation that leads to crying, to hyposensitive, underaroused infants who even with a pain stimulus take a long time to cry or do not cry at all.

The presence of one or more of these factors does not necessarily mean that something is wrong with the infant. They are acoustic features which alone or in combination seem to vary with the status of the infant. There are other more qualitative features in the cry, more difficult to measure, but nonetheless real and important. Such cries, often reported by parents, are described as urgent, frantic, frenzied, and the infant is felt to be hysterical, beside himself, or out of control.

PARENTAL PERCEPTION OF THE CRY

In addition to listening to the cry, the pediatrician can use these acoustic dimensions when discussing the infant's crying with the parents. Crying is usually discussed in terms of how long crying periods last, how often and when they occur. But parents can also provide information about the quality of the cry that may be clinically useful. Mothers are often aware that their infant "sounds funny," that his or her cry is different, not "right." While some parents volunteer their observations, others do not because they cannot be specific. Some people are embarrassed to say to their pediatrician, "My baby sounds funny." By asking how the baby sounds along specific acoustic dimensions, the pediatri-

cian provides a framework for the parent to express these concerns. A list of rating scales of cry features that have been found useful in our research is shown in the table, "Characteristics of Infant Crying."

Characteristics of Infant Crying (rated on a 7-point scale)
Intense – not intense
Urgent – not urgent
Piercing – non-piercing
Irritating – non-irritating
Rough – smooth
Grating – not grating
Frantic – not frantic
Sick – healthy
Weak – strong
Disturbing – not disturbing
High-Pitched – low-pitched
Rhythmic – arrhythmic
Arousing – non-arousing
Discomforting – not discomforting
Loud – soft
Easy to soothe – hard to soothe
Cries for long periods – cries for short periods
Cries many times per day – cries few times per day
Easy to predict when he/she will cry – hard to predict
Easily gets upset – doesn't easily get upset
Easy to tell what's bothering him/her – hard to tell

For example, parents of difficult-temperament-infants rate their cries as more grating, piercing and aversive. The parents feel angry and irritated by these cries and were more physiologically aroused (skin potential) to these cries. These perceived qualities were confirmed by computer acoustic analysis as shown in Figures 3 and 4.

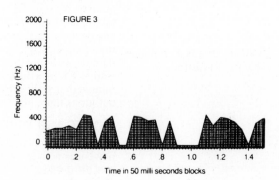

FIGURE 3

Time in 50 milli seconds blocks

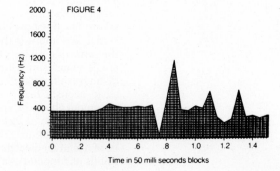

FIGURE 4

Time in 50 milli seconds blocks

**Parental perceptions:
an important source of
information**

Parents' perceptions of crying are also an important source of information because parents are most likely to be sensitive to changes in the cry. Sometimes a sudden change alerts parents that something is wrong. In discussing the cry with parents, the pediatrician should determine whether the baby always sounded this way, or if there has been a sudden change.

Once it is determined that the infant does have an unusual cry, the pediatrician should look for other "soft signs." What is the baby's behavior like in general? Are other aspects of the baby's behavioral (or physiological) functioning different or unusual? The cry may be a warning signal that says look more closely, and the pediatrician should be open to a range of possibilities — neurological problems, disease, colic, difficult temperament, or problems in the parent-infant relationship.

So far, we have talked about confirming parents' perceptions, but what if they are disconfirmed? The pediatrician may conclude that the crying is not unusual or excessive, that the issue is more parent- than infant-based. The clinician should be alert to this possibility, and perhaps treat it as a "cry" for help from the parents, who may be saying that *they* need help. Clinical management may consist of discussing their feelings, looking for related behavioral problems, or appropriate mental health referral.

CRYING AND PARENTING BEHAVIOR

Crying can have various negative or positive effects on parenting behavior. When the quality and/or quantity of crying pushes the parents beyond their tolerance limits, the cry can "turn off" parents, and the infant himself can be perceived as a negative, aversive stimulus. When this happens, it is difficult for parents to provide optimal caretaking; they may withdraw and be less sensitive to their infant's needs. Research shows that abusive mothers react more strongly to infant cries; and the tendency for preterm infant cries to be perceived as aversive could help explain the increased incidence of child abuse in low birthweight infants.

**Crying usually has
positive effects on
parental behavior**

However, most infants whose crying is perceived as aversive enjoy healthy relationships with their parents. In fact, crying usually has positive effects on parental behavior, because it ensures that the infant is well cared for. Unusual or aversive-sounding cries alert the parents that something is wrong with the baby, that he may need special handling. This increases their sensitivity to the infant and facilitates child care. The pediatrician can help the parents understand that the infant is communicating a need for special caretaking, and support the parents in that effort, while at the same time being aware of the potential for non-optimal parenting. This will require a talk with the parents, calling for the physician's best wisdom and tact.

CRYING AND PARENTAL FEELINGS

Discussing with parents how they feel about their baby's crying provides insight into the developing parent-infant relationship. Parents should be encouraged to distinguish between how the cry sounds (e.g., distressing, aversive, etc.) and how it makes them feel (e.g., angry), because these feelings may be overtly or covertly expressed toward the infant. The pediatrician can serve an important reality-testing function by confirming for parents that what they hear is "real," and inherent in the baby, rather than something they made up or caused. When parents understand that crying is part of the infant's constitutional make-up and not the result of the parents' failure, they can stop blaming themselves for what they have done or not done. It is also important to acknowledge the fears behind the feelings — fears that they might hurt or destroy their infant. By encouraging the acknowledgment of these fears, the pediatrician can let the parents know that having negative feelings toward their infant does not make them bad parents, nor does it mean that they will act out these feelings on the baby; in fact, acknowledgment reduces still further the possibility that the feelings will be acted out.

It is critical for parents to realize that the inadequacy, anger, guilt and fear elicited by crying are normal, natural and appropriate, and that they, as parents, are "okay." The energy consumed by coping with those difficult feelings can now be freed to facilitate the developing infant-parent interaction and maximize the baby's health and happiness.

MANAGEMENT OF CRYING

Infants have distinct preferences for specific soothing techniques, and these vary from infant to infant. Procedures from the Brazelton scale can be used by the pediatrician during the hospital discharge exam to show parents how their infants deal with crying. The most effective consoling techniques involve some combination of containment (holding, swaddling) or touch (massaging, caressing), rhythmic stimulation (rocking) and stimulating a response system that is incompatible with crying (sucking).

There are, of course, infants who will not be soothed, as in colic, where interventions are only effective for very brief periods, and parents feel helpless. In these cases, it is probably best to stop searching for the cause of the crying and changing caretaking routines and to help the parent get some time away from the infant. There are also "gimmicks," either commercially available, such as swings, or that parents discover or hear about, such as taking the baby for a car ride, that are effective in helping the baby calm down. However, there is a danger that the baby can become too dependent on them.

The immediate goal is to stop the crying, and at times that is all that really matters. But the longer-term goal is for the baby to learn to control his own crying behavior. Over time, we want the behavioral regulation to come more from the infant and less from environmental intervention. Infants who are quiet only when in a moving car or rocking in a swing create a new set of problems, both in terms of parental caretaking and the infant's development. Such infants have to be weaned from overdependence on these techniques, which are best used in moderation and in combination with strategies to teach the infant to control his own behavior by gradually withdrawing the intervention, while simultaneously praising the infant for staying calm and "doing it on his own." Exposing the infant to a variety of intervention techniques that vary in type, amount and intensity of stimulation helps the infant expand his repertoire of behavioral control and discourages the infant from exclusive reliance on one technique. It also allows parents to start substituting less stimulating interventions, such as music, for more massive interventions, such as holding and rocking, so that the infant learns he can manage more and more on his own.

A successful discussion with parents provides insight

FINAL COMMENT

Crying is really the first "test" for parental control of the infant's behavior. It is involved with early limit-setting and with the parents' issues of their own coping, their parenting abilities and their inadequacies. Crying is a microcosm of things to come, and the pediatrician who keeps an ear tuned to the infant's communication and the parents' communication can work toward preventative medicine, and maximize the opportunity for the development of the baby and the family.

It is hoped that this article will help the practicing pediatrician deal more effectively with the clinical management of crying. It is also hoped that it will help this psychologist who, when this article is published, will have a one-month-old infant of his own.

How To Understand Your Baby Better

New studies show infants do have complex emotions

LORI UBELL

HOW DO WE KNOW WHAT AN infant is feeling? Why a baby is crying? When is it fear or anger, when simply the need of a diaper change? Since he cannot tell us what is wrong, how can we help?

A 17-month-old baby wakes up, hysterical, night after night. Nobody knows why. Her mother tries to calm her by picking her up and rocking her to sleep. But nothing does any good. Another baby, 9 months old, continually spits up food, though there is no physical problem. In addition, he is sullen and defiant and seems to reject his mother. She can't understand what she is doing wrong.

In the past, such babies were classified as "difficult"; doctors said nothing could be done. Today, however, researchers say they have found the key to understanding what infants are really feeling and communicating to us. As a result, they have been able to devise simple ways to alleviate infants' distress.

LOVE
Baby's capacity for intimacy grows more complex between 2 and 7 months.

Until now, studies of infants were restricted to their physical and intellectual development. Babies were expected to master certain skills as they grew older—to crawl, speak, distinguish shapes at particular stages. But it was thought that infants did not experience complex *emotions* until later in life.

This changed in the early 1970s, when researchers studying emotionally dis-

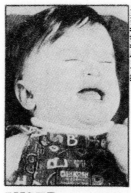

ANGER
At 2 to 7 months, anger and frustration allow baby to assert himself.

turbed 3- and 4-year-olds began to focus on even younger children. "We wondered how far back the problem began and if we could do more effective preventive medicine," says Dr. Stanley Greenspan, a child psychiatrist and clinical professor at George Washington University Medical School in Washington, D.C.

So he and his research team studied more than 200 babies. They determined that healthy children pass through six stages of emotional development from birth to age 4. Today, these milestones, which are described in the book *First Feelings*, by Dr. Greenspan and Nancy Thorndike Greenspan, are widely accepted by infant specialists.

Just as a series of small steps must be mastered in order to pass on to the next stage in physical development, so too in emotional development, the researchers found, a child must successfully learn a more simple emo-

ANXIETY
Between 3 and 10 months, baby sends out signals—and expects a response.

tion before he can progress to a more complex feeling. This process sets the foundation for every type of emotional involvement in life.

A newborn baby, in the first stage, faces two simultaneous challenges: learning to feel calm in a potentially overwhelming new environment and taking an interest in the world through the senses. A healthy baby, by 2 or 3 months, will respond with pleasure to voices, faces or touch and be able to focus his attention for longer and longer periods.

Knowing what to expect can help parents isolate what is wrong. Of course, every child is different, and some babies are naturally more extroverted than others. But if a physically healthy child is not looking around eagerly, something is probably wrong. Since the infant cannot tell us what that is, the parent must be a detective. Specialists stress that parents should observe their baby, themselves and their environment. It is important to notice *when* an infant seems particularly irritated. Perhaps he dislikes a certain way of being held. Perhaps she is sensitive to light. A parent's personality also can be a factor. A mother's depression can be linked to a baby's passivity. Dr. Greenspan urges parents to carefully examine their attitudes as well. Various subconscious fears—for example, fear of hurting the baby, of losing one's independence, even fear of closeness—can affect a baby's behavior.

Once parents understand how their behavior is affecting their baby, they can modify it. Sometimes a simple technique works. For example: "Some children overrespond to certain stimuli, like sound," says Elizabeth Curran, the children's program coordinator for the Women's Alcoholism Center in San Francisco, who has worked with infants from both alcoholic and nonalcoholic homes. "Parents can learn to use the other senses to interact with the baby. Instead of talking to the child, for instance, the parent can stroke him soothingly, and he will calm down."

Each of the next five stages presents the infant with a more complex challenge. During the second stage, from 2 to 7 months, the baby learns to relate to one or two special people. She gazes lovingly at her mother and begins to communicate feelings. With gestures, sounds and facial expressions, your child lets you know she is happy you are near.

In the third stage, 3 to 10 months, a baby learns about cause and effect. That is, he realizes that his joy can *cause* your joy and that he will be comforted if he signals pain. If a baby's signals are not correctly interpreted, however, this communication skill will be bypassed. If a baby smiles, for instance, and a mother consistently is too preoccupied to respond, the baby cannot learn the effect of showing happy feelings.

The 9-month-old boy who couldn't keep his food down was trying to communicate two things to his mother: physical discomfort and anger. The discomfort came from a hypersensitivity to touch. Every time his mother put a spoon in his mouth, he actually felt pain. Rather than rejecting her, he was protecting himself from unpleasant sensations by refusing to be fed. And since mealtimes had become a battle of wills, he grew sullen and angry at other times as well.

His mother learned to allow him to take the lead at feeding time; she would hold the spoon in front of him, and he would position his mouth so that it was comfortable for him. At playtime, too, she let him be in control; if he crawled away, she would crawl after him, making a game out of catching him.

Soon, the child was able to move on to the fourth stage. The 9- to 18-month-old toddler begins to develop a more complex sense of himself. He now is able to string together two or three behaviors in order to communicate emotions. At this stage, a child also learns to internalize the emotional messages he gets from his parents. He knows his parents love him even when they are at the other end of the room. This was the problem with the little girl who couldn't sleep. She had not learned to internalize her mother's comforting touch, to "take it with her" to bed. Consequently, when she woke up and didn't see her mother, she became hysterical. Once her mother understood this, the solution was relatively easy. During the day and at night, when her baby cried, she would reassure the child by means of words and gestures, even from across the room. Slowly, she put more physical distance between herself and her daughter.

During the last two stages, the child puts together everything he has learned. He shows that he understands the concept of love and nurturing. His pretend play becomes more complex, and he has emotionally appropriate responses. Outside of play, he is increasingly able to regulate the relationship between feelings, behavior and consequences.

Much of what follows depends on how well a child gets through this early period.

Today, doctors, parents and teachers are far more aware of the signs of emotional health in babies than they were 10 years ago. The American Academy of Pediatrics, for example, includes the latest information in its *Guidelines for Health Supervision*.

For parents, the new insights should not be cause for anxiety about "doing the right thing." Most babies develop normally; parents who are aware of their baby's emotional development will take extra pleasure in watching their newborn go from recognizing love to expressing it on his own. Indeed, understanding and responding to the "language" our babies speak can only bring us closer to them.

Your Child's Self-Esteem

Paul Chance

Paul Chance, Ph.D., is a free-lance writer who specializes in psychology.

Consider Alice, age five. Alice attends kindergarten, where she is making excellent progress. Her teacher thinks she is one of the brightest children in the class, though in fact she has no more natural ability than most. She is often the first to raise her hand when the teacher asks a question, waving it eagerly and sometimes calling out, "I know! I know!" If called on when she does not know an answer, she does not hesitate to make a guess. Sometimes these answers sound foolish to her classmates, but their laughter doesn't bother Alice—she justs laughs right along with them. Alice tackles almost every assignment with enthusiasm. If one approach fails, she tries another. If her persistence does not pay off, she asks the teacher for help. Alice is as comfortable with other children as she is with her schoolwork. She is a popular child, and in group activities she often takes the lead. At home, Alice is eager to do things for herself. She is proud, for instance, that she can already dress herself completely, buttons, shoelaces, and all, without help.

Now consider Zelda, age six. Zelda is in the first grade. She did not do very well in kindergarten, and her progress continues to be slow. Her teacher believes that she is one of the least intelligent children in the class, though in fact she has as much ability as most. She never volunteers, and if called on she usually refuses to say more than "I don't know." Zelda works on most assignments in a lackluster, mechanical manner and often abandons them at the first sign of difficulty. When her teacher asks if she needs help, she says merely, "I can't do it." Zelda is no more adept socially than she is academically. She has few friends, and in group activities she is usually the quiet one on the sidelines. At home, Zelda is more at ease and more loquacious, but not more self-reliant. She waits for others to do things for her because she "can't" do them for herself. Her mother still checks her each morning to be sure that she has dressed herself properly.

Competence has little to do with natural ability.

Alice and Zelda are as far apart as the letters *a* and *z*. The differences that separate them are not, however, due to differences in native ability. The differences are emotional and motivational. Alice is obviously self-assured and self-reliant. She likes herself and her world. Although she could not put her philosophy into words, she is an optimist. She believes that she has some degree of control over her destiny, that success and happiness are goals an individual can achieve through effort. Zelda, on the other hand, is as filled with self-doubt as Alice is with self-confidence. Her self-esteem is low and she thinks the world a harsh, unfriendly place. A philosopher would describe her as a fatalist, a person who believes that what happens is largely a matter of fate or chance. Zelda believes that she can do little to shape her future, that success and happiness are things that "just happen" to people who get lucky. Although Alice and Zelda have about the same amount of intelligence, it is clear that Alice is making far better use of her abilities. The result is that Alice is a highly competent child, while Zelda is best described as helpless.

Why do some children become Alices, while others become Zeldas? This question has received intensive study over the past decade. Most researchers seem convinced that experiences in infancy and early childhood play an especially important role in the development of competence, so their research efforts have focused on experiences in the first three years of life. This is not to say, of course, that whether a person becomes highly competent or utterly helpless is unalterably fixed by age three. People can change at any age. Nevertheless, the evidence suggests that the kinds of experiences that are important to the development of competence typically *begin* in early childhood. What are those experiences? The research on this subject is complex and not easily reduced to a few simple, clear-cut statements. But over and over again, the studies reveal four elements common to the backgrounds of the most competent children but conspicuously missing from the backgrounds of the least competent.

The importance of a strong parent-child bond.

It may come as no surprise to most parents that one common element in the backgrounds of very competent children is a strong bond between the child and the primary care-giver—usually, but not necessarily, the mother. Dr. Alan Sroufe, professor of child development, and his co-workers at the University of Minnesota, Twin Cities, have found, for instance, that infants judged "securely attached" at eighteen months of age were more successful at solving problems, such as getting an object out of a tube, at age two. They were also better able to elicit the help of their mothers to solve problems that were too difficult for them. They were, in other words, already more competent than children who lacked a strong bond with their mothers.

The signs of a secure child.

Dr. Sroufe notes that it is possible to predict which children will be successful preschoolers by studying the relationship a child has with his caretaker at twelve to eighteen months of age. "Even by two years," he says, "secure children will be more enthusiastic, persistent, and cooperative in solving problems than insecure children will be." Children with a good, secure relationship with an adult can function well in a nursery school at a younger age than can children without such a relationship. "Apparently," Dr. Sroufe concludes, "secure children

have learned early how to explore and master their environment and function within clear, firm limits."

Perhaps the most extensive work on the relationship between a close attachment and child development is that of Dr. Burton White, former director of the Preschool Project at Harvard University, and his colleagues. Their research followed the progress of 40 children, beginning at age one or two. The researchers went into the homes of these children every other week, 26 times a year, for one or two years, and then retested the children again at the ages of three and five. The researchers concentrated on the interactions of the infants with their mothers and others in the home. They concluded that a close social relationship "was a conspicuous feature in the lives of children who developed best."

Another way to study the benefits of a love bond, as it might be called, is by looking at the development of children for whom such a bond is notably lacking. One sometimes finds such children in poorly staffed institutions. Dr. Sally Provence, professor of pediatrics at Yale University's Child Study Center, who has made a special study of such children, observes that they often become "subdued and apathetic." Given a little tender loving care, however, these children often liven up dramatically.

Providing a stimulating environment.

Another common element in the backgrounds of competent children is a stimulating environment. Dr. K. Alison Clarke-Stewart, associate professor of education and human development at the University of Chicago, studied the interactions of mothers with their firstborn infants, ages nine to thirteen months. She found, among other things, that mothers of competent children talked to or made other sounds when interacting with their babies more often than did the mothers of less competent infants. Dr. White and his co-workers were so impressed by the role of verbal stimulation in the development of competence that they wrote that "live language directed to the child is the most consistently favorable kind of educational experience an infant can have during the eleven- to sixteen-month period." They go on to point out that language from a television or radio or speech directed elsewhere that the

child overhears has little if any beneficial effect.

Freedom to explore can make even an ordinary environment more stimulating than it is from afar. An environment that is full of interesting objects a child cannot get to is less stimulating than one with fewer objects that are within reach. Dr. White and his collaborators found, in fact, that freedom of movement was characteristic of the homes of competent children. The more freedom to explore about the house (within the limits prescribed by safety, of course), the more competent a child was likely to be.

The evidence for the benefits of a stimulating environment suggests that a dull, monotonous environment is a prescription for helplessness. This theory seems to be borne out by the classic research of renowned psychiatrist René Spitz, who studied the development of children living in the thoroughly monotonous world of a badly understaffed foundling home. These children had little to do all day but sleep or stare at the blank walls about them. Needless to say, such children do not develop normally, but the degree to which their development is retarded is striking. Dr. Spitz offers this description of the typical child reared in such an impoverished environment: "These children would lie or sit with wide-open, expressionless eyes, frozen immobile face, and a faraway expression as if in a daze, apparently not perceiving what went on in their environment."

Fortunately for such children, a little bit of stimulation can have substantial benefits. For instance, psychologists Wayne Dennis and Yvonne Sayegh gave institutionalized infants in an otherwise impoverished environment items such as flowers, bits of colored sponge, and a chain of colored plastic discs to play with for as little as an hour each day. It is hard to believe that so little improvement in their thoroughly monotonous environment would make a great difference, yet their developmental ages jumped dramatically.

Interactions—with parents . . .

It is, of course, possible to get too much of a good thing. Dr. White believes that too much stimulation, too many things going on around the child, may merely confuse him. This idea is supported by a study conducted by Dr. Jerome Kagan, professor of human development at Harvard University, who watched mothers as they interacted with their four-month-old infants. He noted when the mothers spoke to or cooed to their babies and what else they were doing at the time. He found that upper-middle-income mothers usually spoke to their children while facing them but did *not* simultaneously tickle them or provide other stimulation. Low-income mothers, on the other hand, were likely to talk to their infants while diapering, feeding, or burping them. It is probably not a coincidence, Dr. Kagan observes, that the children of upper-middle-income mothers typically show more precocious language development than do the children of low-income mothers.

Most researchers seem to agree that a varied environment is important to the development of competence, but the quality of the stimulation is more important than the amount.

A third characteristic of the backgrounds of competent children is frequent social interaction. Dr. White has found, for instance, that highly competent children have at least twice as many social experiences as their less competent peers. He says that "providing a rich social life for a twelve- to fifteen-month-old child is the best thing you can do to guarantee a good mind." He also notes that firstborn children have far more opportunities to interact with their parents than do later-born children. The rich social life of firstborns may have something to do with the fact that they are usually (though not always, of course) more competent than their siblings are. They are, for instance, more likely to obtain positions of leadership as adults than are later-born children.

. . . and with others.

Psychologist Michael Lewis, director of the Institute for the Study of Exceptional Children at the Educational Testing Service in Princeton, New Jersey, believes that the child's interactions with other people are more important than his interactions with any other part of the environment. "We learn about others through our interaction with them," he writes, "and at the same time we define ourselves." How does a child learn whether he is a boy or a girl, tall or short, strong or weak? Through his interactions with others who treat him as a boy rather than a girl, and who are taller or shorter, stronger or weaker, than he is. How does he learn whether

he is competent or helpless? Partly, argues Dr. Lewis, through his interactions with the people around him.

Studies of social isolation have shown that even when other forms of stimulation are available, a dearth of social experiences can have devastating effects. Dr. Harry Harlow and his colleagues at the University of Wisconsin at Madison found that monkeys reared alone grew up to be timid, wholly inadequate individuals. Monkeys reared by their mothers but separated from other youngsters of their kind fared better, but still developed abnormally. Thus it appears likely that the more opportunities for social encounters a child has, the better. It is even possible to see social competence emerge as a result of such experience. When, for example, California psychologist Jacqueline Becker gave pairs of nine-month-old babies the opportunity to play with one another, she found that they interacted more and more with each succeeding session. When these youngsters were introduced to a new baby, they were much more likely to approach him than were infants who had had less social experience.

A world that responds.

Probably the most important element in the environments of highly competent children is something that researchers call *responsivity*. A responsive environment is one that *responds to* the behavior of the child. There is, in other words, some correspondence between what the child does and what happens to him. Under ordinary circumstances there is at least a minimal amount of responsivity in the life of every child. Take, for example, the baby nursing at his mother's breast. As Dr. Martin Seligman, professor of psychology at the University of Pennsylvania, writes: "He sucks, the world responds with warm milk. He pats the breast, his mother tenderly squeezes him back. He takes a break and coos, his mother coos back. . . . Each step he takes is synchronized with a response from the world." When a child's behavior has clear, unequivocal consequences, he not only learns about those consequences but "over and above this," writes Dr. Seligman, "he learns that responding works, that in general there is *synchrony* between responses and outcomes." This means, in turn, that the child exerts some control over his environment, and many

researchers agree with Dr. Seligman that "how readily a person believes in his own helplessness or mastery is shaped by his experience with controllable and uncontrollable events."

A responsive environment, then, inclines a child toward competence, while an unresponsive environment inclines a child toward helplessness. Dr. Lewis illustrates the difference by describing the experiences of two infants, Sharon and Toby. One morning Sharon wakes up wet, hungry, or perhaps just lonely, and cries. Nothing happens. She cries again. Still no response. She continues crying for several minutes, but no one comes. Finally she falls asleep, exhausted. On the same morning another infant, Toby, awakes. She, too, is wet, hungry, or lonely, and cries. Within seconds she has the attention of a warm hand, a smiling face, and the food or dry diaper she needs. Sharon's world is unresponsive; her behavior has no effect. Toby's world is highly responsive; her behavior gets results almost immediately. Now, what is the lesson each child is taught by her respective experience? Sharon learns that making an effort to affect one's condition is useless. Things happen or they don't; what she does is unimportant. Toby learns that her efforts are worthwhile. What happens depends, in part, upon what she does.

Some readers may think at this point that responsivity is just another name for permissiveness. Give the child what he wants, pander to his every whim, deny him nothing. In other words, spoil him. Not so. A responsive environment is not one that gives a child everything and anything he wants. Responsivity means merely that an act produces clear, consistent consequences. Sometimes those consequences will be negative. For example, a four-year-old child who insists upon having cookies for breakfast may send his bowl of cereal flying. A parent might respond to this behavior by saying, "Since you've thrown away your breakfast, you'll have to go without." Another parent might insist that the child clean up the mess he has made. In each case the child's behavior has some effect, but the effect does not necessarily include getting a plate of cookies.

Handling a baby's cry.

But what about Toby's crying? Sure, Toby learns that she can master her environment, but doesn't she inevitably become a crybaby in the process? Doesn't she learn that the way to get what you want is by making a fuss? Interestingly enough, the answer is no. Psychologists Silvia Bell and Mary Ainsworth conducted a study of the effects of responsiveness on crying. They observed the interactions of mothers and their infants during the first year of the child's life. There were wide variations among the mothers in how often and how quickly they responded to their baby's cries. The most responsive mother, for instance, responded 96 percent of the time, while the least responsive mother answered only 3 percent of her baby's cries. Many parents would predict that the first infant would cry constantly. What actually happened, though, was that the children who could control their environment by crying soon learned to use more subtle tactics to exert control, and they also learned to do things for themselves. The researchers conclude that "an infant whose mother's responsiveness helps him to achieve his ends develops confidence in his own ability to control what happens to him." This means that he comes to do more things for himself, which means there are fewer occasions for calling upon Mom.

Other research supports the notion that a responsive environment leads to competence. In one study, for example, Dr. Lewis and psychologist Susan Goldberg watched mothers interact with their three-month-old infants. They noticed that the behavior of some mothers was likely to be a reaction to what the baby did, while the behavior of other mothers tended to be independent of the baby's activity. The researchers found that the first infants, those whose mothers were responsive, were more interested in the world around them and were more attentive.

Toys that foster competence.

Dr. John S. Watson of the University of California at Berkeley demonstrated that the responsiveness of the physical world also is beneficial. Dr. Watson designed a mobile that would rotate whenever a baby exerted pressure on a pillow. When an infant turned his head this way or that, the mobile would spin. Dr. Watson found that with just ten minutes of practice a day, the infants learned to control the mobile within a few days. They also smiled and cooed as the mobile spun to and fro, apparently enjoying the

control they exerted over the object. Other babies who saw the mobile spin but had no control over its movement did not show the same reaction.

Providing a child with responsive toys does not necessarily require anything so elaborate as Watson's motorized mobile. Psychologist Robert McCall notes that a mobile can be made responsive simply by tying a piece of soft cotton yarn loosely around a baby's wrist and tying the other end to the mobile. When the baby moves his hand, the mobile moves. Another inexpensive but highly responsive baby toy is the rattle, since it makes a noise only when and if the baby shakes it. Yet another example is a mirror, perhaps made of shiny metal so that there is no danger of broken glass. The child looks in the mirror and sees someone looking back. The person looking back does all sorts of things—smiling, laughing, frowning—but only if the child looking into the mirror does them first. As the child gets a bit older, a spoon and a pie pan provide responsive, if noisy, diversion. It may well be the case that the more responsive toys tend to be the least expensive. Toys that "do it all" rarely require much activity from the child. Thus, the *least* responsive toy available is probably the $500 color television set, while one of the most responsive toys around is the $1 rubber ball.

There is evidence that if a child's environment is thoroughly *un*responsive, he is almost certain to become helpless rather than competent. Dr. Seligman and his colleagues have conducted a number of studies that demonstrate just how devastating the lack of control over events can be. They have found that when a laboratory animal is unable to escape an unpleasant situation, it eventually quits trying. More important, when the animal is later put into another situation from which it could readily escape, it does not do so. In fact, it makes no effort to escape. Psychologist Donald Hiroto got similar results when he studied the effects of uncontrollable unpleasant events on college students. Some students heard an unpleasantly loud noise, which they could do nothing about. Other students heard the same noise but learned to control it by pushing a button. Afterward, the students were put into another situation in which they could turn off a noise simply by moving their hand from one part of a box

to another. Those who had learned to control the noise in the first situation did so in the second, but most of those who could do nothing about the first noise made no attempt to escape the second. They simply sat there and did nothing. They had been made helpless, at least momentarily.

Helping a child master his world.

Researchers have not, of course, deliberately set out to make children helpless by exposing them to unpleasant situations from which they cannot escape. They have, however, noted that children reared in unresponsive environments are not likely to become highly competent. Dr. Seligman points out that a lack of control was characteristic of the environment of Dr. Spitz's institutionalized children and may have been more important to their helplessness than the lack of stimulation they received.

All children are subjected to some unpleasant events that they cannot control. An infant's diaper rash is beyond his control, as is the misery of most childhood illnesses. And even the brightest child must eventually experience failure. But if the child is usually able to exert some control over the events in his life, this may give him some immunity against the adverse effects of unpleasant events he is powerless to control.

It appears, then, that the kind of experiences a child has in the first few years of life plays an important role in his development. The child who has a close, warm relationship with an adult; who lives in interesting surroundings; who has ample opportunity to interact with other people; and, most important, who lives in an environment that is responsive, has an excellent chance of becoming competent. The earlier these experiences begin, the better. "I believe," Dr. Seligman told me, "that motivation and emotion are more plastic than intelligence. I am no longer convinced that special kinds of experiences will raise a child's IQ by twenty points or induce him to write piano concertos at age five, as Mozart did. But I am convinced that certain kinds of experiences during childhood will produce a child who is helpless, while other experiences will produce a child who is competent."

There is in every child an Alice and a Zelda. The question is, which is to prevail?

THE ROOTS OF MORALITY

Does our moral sense
arise from emotions
or reason? And when
does it happen?

JOSEPH ALPER

*Joseph Alper is a free-lance writer living in
Washington, D.C.*

IN HER SUBURBAN HOME, 18-month-old Julie was excited when another baby, Brian, came for a visit. But Brian was less than pleased at being with strangers and soon began to shriek and pound his fists on the floor. Almost immediately Julie's delight vanished, her body stiffened, and she looked worried, startled, and anxious. Julie's mom put Brian in a high chair and gave him cookies, but he continued screaming and threw the cookies to the floor. Julie, who usually tried to eat everyone else's cookies, put them back on the high chair tray. Brian's crying continued, and Julie tried to stroke his hair. She then went to her mom, grabbed her by the hand, and brought her to Brian. Julie's mother, concerned over her daughter's discomfort, took Brian into another room.

Julie, whose behavior had been observed and recorded by her mother as part of a National Institute of Mental Health study, was obviously concerned over the distress of her young friend. But she did more than just express concern, she also tried to alleviate that distress even though it did not benefit her in any overt manner. Her actions, in fact, may represent early signs of what is known as prosocial behavior, or behavior for the benefit of others. Many psychologists believe this signals the start of moral development.

Signs of empathy also occur among infants: A nurse dropped a bottle of formula on the floor of a quiet hospital nursery, but aside from the crash of breaking glass, the room remained silent. A little later, one newborn started crying. An infant in a nearby crib joined in, then two more in the middle of the room. In minutes, the air was filled with the cries of dozens of babies, all demanding comfort.

Among older children the evidence is often verbal and therefore clearer: Freddy had asked his nursery school teacher why his friend Bonnie was not in class. The teacher replied that Bonnie's mother had died. The four-year-old stood quietly for a minute, then said, "You know, when Bonnie grows up people will ask her who her mother is and she will have to say 'I don't know.' It makes tears come to my eyes."

In each of these situations, young children apparently have sensed distress in others and, as a result, become distressed themselves. Feeling someone else's emotions is common in adults; we call it empathy. Only in the past several years, however, have psychologists grown interested in the development of emotion (as opposed to reason) in children. This new area of study reveals that children empathize at a very early age, and some of the investigators believe this behavior is innate, genetically programmed like sucking and crying.

Empathy may do more than trigger a sympathetic response. "The data strongly suggest that empathy is part of the developmental foundation for the child's future system of moral behavior, as well as for social behaviors such as altruism and sharing," says Martin L.

Hoffman, developmental psychologist at the University of Michigan.

Marian Radke-Yarrow, chief of the Laboratory of Developmental Psychology at the National Institute of Mental Health (NIMH) and her colleague Carolyn Zahn-Waxler have found that children as young as 15 months show complex empathic behaviors. "Over the next three to six years, depending on the individual, this empathy develops into what we consider moral or prosocial, behavior," says Radke-Yarrow. "Many psychologists still claim, in spite of extensive evidence from several labs, that young children are not capable of moral reasoning."

The NIMH researchers have found much individual difference in the quality of empathy—how sensitive a child is, a toddler's particular style of reacting to others' distress. Nancy Eisenberg, a developmental psychologist at Arizona State University in Tempe, reports similar findings. "Certain qualities of early empathic behavior appear to be associated with certain types of moral reasoning," she says.

Not everyone agrees that moral development starts early in life or that empathy plays a part in it. William Damon, a psychologist at Clark University, believes that the most important aspects of morality do not start forming until a child is adolescent. "It's true young children show empathic behaviors, even sharing, which you might say is a precursor of fairness. But I think a lot of those early signs are just playfulness," says Damon. "Caring is obviously important to any society, but the key question in moral behavior has to do with fairness." He claims that if there are any early indicators of future moral behavior in an individual, they would be the appearance and development of what is called distributive justice—how someone resolves competing claims for goods.

Most psychologists, however, would say that kindness and mercy, and respect for another person's feelings, must be included in a definition of moral behavior, in addition to honesty, justice, and fairness. Biblical writers, Confucius, Aristotle, Kant, Hume, Adam Smith, and Freud have probed the dimensions of moral decision-making and its development without much agreement. It is an issue that no single theory stands much chance of explaining fully.

The first modern-day psychologist to take a crack at explaining moral development in children was Jean Piaget. Piaget

directed most of his attention to determining stages of reasoning and intellectual capabilities in children. It was from this perspective of cognitive development that he addressed morality. He believed that children's social maturation, including the ability to make moral decisions, depended on two factors: the acquisition of the rules and patterns governing behavior, and the ability to reason and solve problems using these rules.

In the mid-1930s, Piaget postulated a two-stage development progression based on how children between ages six and 12 interpreted and reacted to various moral dilemmas. The following example is the most well-known:

John was in his room when his mother called him to dinner. John went down and opened the door to the dining room. But behind the door was a chair, and on the chair was a tray with fifteen cups on it. John did not know the cups were behind the door. He opened the door, the door hit the tray, bang went the 15 cups, and they all got broken.

One day when Henry's mother was out, Henry tried to get some cookies out of the cupboard. He climbed up on a chair, but the cookie jar was still too high, and he couldn't reach it. But while he was trying to get the cookie jar, he knocked over a cup. The cup fell down and broke.

Which boy was naughtier and why?

Usually, children younger than 10 replied that John was—he broke more cups. Children 10 and older tended to say that Henry was—he was up to no good when he broke the single cup.

Piaget postulated that younger children are self-centered and do not yet have the cognitive skills to understand the purpose of society's rules nor the ability to use them in a reasoned approach. Thus, they could only make their decisions on objective grounds: who broke the most cups. But around the age of 10, children's cognitive skills reach the point where they can interpret society's rules and become aware of the effects of violating them. They also start using their emerging intellectual capabilities to reach higher level moral judgments, which include motive and intent.

Though Piaget was devoted more to intellectual than to moral development, his work stimulated much of the more specific research on the subject and served as the starting point for Lawrence Kohlberg, a psychologist at Harvard University. Like Piaget, Kohlberg used stories containing moral dilemmas to learn about children's abilities to make moral

judgments and to identify styles of reasoning they used to make decisions. This is one of the Kohlbergian dilemmas:

In Europe, a woman was near death from a special kind of cancer. One drug that the doctors thought might save her was a form of radium that a druggist in the same town had recently discovered. The drug was expensive to make, but the druggist was charging 10 times what the drug cost him to make. He paid $200 for the radium and charged $2,000 for a small dose of the drug. The sick woman's husband, Heinz, went to everyone he knew to borrow money, but he could only get together about half of what it cost. He told the druggist that his wife was dying and asked him to sell it cheaper or let him pay later. But the druggist said, "No, I discovered the drug and I'm going to make money from it." So Heinz got desperate and broke into the man's store to steal the drug for his wife. Should the husband have done that? Why?

Based on the answers to these questions, Kohlberg posited six ordered stages of moral development. In the lowest, decisions are based solely on the likelihood of getting punished for a given action. In the sixth stage, moral decisions are based on one's principles. Two factors promote progression through the stages: exposure to levels of moral reasoning that are immediately higher than one's current level, and new experiences in role taking.

According to Kohlberg, most children up to about age 10 are at stage two. He claims that this is because they do not yet have the logical skills to engage in sophisticated reasoning. Other psychologists have criticized Kohlberg for ignoring the emotional and social roots of morality. In addition, the dilemmas he used to test for development are criticized as too intricate and remote for young children.

"To say that children can't reason, can't make moral judgments on the basis of anything but rules or for self-centered motives flies in the face of what we start seeing in preschoolers," says Nancy Eisenberg. For example, in one study she conducted, 35 four- and five-year-olds were told various stories in which the protagonist had to choose between an altruistic behavior and one that would benefit themselves. One of the stories was as follows:

One day a girl named Mary was going to a friend's birthday party. On her way she saw a girl who had fallen down and hurt her leg. The girl asked Mary to go to her house and get her parents so the parents could come and

To say children can't make
moral judgments flies in the face
of how preschoolers behave.

*take her to the doctor. But if Mary did run
and get the child's parents, she would be late
for the birthday party and miss the ice cream,
cake, and all the games. What should Mary
do? Why?*

Nearly all of the children in the study
gave answers based on empathic reasoning (as opposed to egocentric) some of
the time. Though few used empathic reasoning consistently, Eisenberg suggests
they all had the capacity. A sample answer demonstrating altruism might be
"Mary should go get the girl's parents because her leg hurts and she needs help."

Kohlberg's theory does not account
for this type of reasoning. "Cognitivists
don't accept the idea that emotion generated from inside the child plays any
significant role in developmental processes," says Hoffman. "And if you discount the idea of empathy, you can't
explain the type of behavior Nancy and
Marion have observed in young kids."

This is where Hoffman's theory differs
from past approaches. He, too, assumes
that cognitive development, especially
role taking and the ability to distinguish
self from others, is necessary for the
development of moral behavior. But he
also postulates that emotions from
within the child, especially empathy, provide the motivation for moral behavior.

Hoffman starts with the belief that
empathy is innate. He cites incidents like
the hospital nursery scene as support for
this idea. He has found in studies that in
a group of newborns, more will cry in
response to another child's cry than will
respond to another equally loud noise.
Because an infant cannot differentiate
between self and others, distress cues
from others are confounded with unpleasant feelings in himself. An 11-
month-old girl, on seeing a child fall and
cry, looked as if she herself were about to
cry, put her thumb in her mouth, and
buried her head in her mother's lap as if
she were hurt.

Around the end of the first year, children start recognizing themselves as being physically distinct from others. Children at this stage realize when others are
experiencing distress but assume that the
other's internal state is the same as their
own. When Julie fetched her mother in
response to Brian's distress, it was because she assumed Brian's emotional
needs were the same as hers in a similar
situation; most likely, she would have
gone for her own mother even if Brian's
had been present.

Role taking is part of a three- to four-year-old's experience and continues to be a major factor in the development of empathy. By seven or eight, children are better able to distinguish their own feelings from others' and to choose actions that are appropriate to help others. In addition, they begin to deal with feelings more abstractly. Freddy was extremely precocious, demonstrating this behavior at age four. He felt sad even though his own mother had not died; he had not yet witnessed Bonnie's sadness, but he imagined her future sorrow.

"In the early years of childhood we see expressions of empathy and sympathy and also serious attempts to help, share, protect, and comfort," says Radke-Yarrow. "If that's not moral behavior, then I guess I'm in the wrong field of study."

It is also during early childhood that people begin experiencing empathic guilt—feeling distress when they have been the cause of someone else's discomfort. "As the child learns that he can cause distress in others and that this can cause distress in himself, he will start behaving in ways that don't produce this self-caused distress," Hoffman says.

"Parents can use the child's sense of empathy and empathic guilt to reinforce good moral values," says Hoffman. The point is to discipline the child for wrong-doing, obtain compliance, and at the same time provide an explanation that will induce sympathy for the person being affected by his misbehavior: "You shouldn't call Johnny names because it hurts his feelings, and you wouldn't like it if your feelings were hurt." Radke-Yarrow and Zahn-Waxler found that children of parents who provide explanations for their own behavior have more highly developed empathy.

There are several ways empathy can be triggered, according to Hoffman. In a newborn, the cries of one child can trigger tears and unhappiness in another. This response is probably innate, but later in life whether or not one reacts empathically to a situation depends on one's experience and reasoning ability. One may automatically feel pain upon seeing someone else in distress because it brings back memories of one's own pain under similar circumstances. Or empathy may stem from the purposeful act of imagining how one would feel if in the shoes of the victim.

The NIMH researchers resorted to an unusual experimental technique that many say has generated the richest data on childhood behavior. "We felt that

past experimental methods had some serious shortcomings for studying certain developmental issues such as empathy. Situations that call for empathic behavior don't happen frequently or predictably," says Radke-Yarrow. "With young children, using hypothetical situations to represent real experience has doubtful validity. In real life, children experience events of consequence—real hurts and conflicts." Radke-Yarrow and Zahn-Waxler decided to address the problems by training mothers as research assistants and having each observe her child's behavior in its natural setting over periods of several months. By having day-to-day observations they could monitor continuing development rather than compare behaviors at selected times. This approach also avoids the necessity of getting children to respond to complex stories or questions that they either may not understand or not know how to answer with their limited verbal skills.

dren engaged in some form of helping interaction only rarely.

In succeeding months, the general distress crying waned, occurring very infrequently among the 20- to 29-month-old children. More controlled and positive actions toward the distressed person occurred. At first, these were simple behaviors, such as touching or patting. By age one, all the children in the study showed this reaction at some time. By the latter half of the second year the children were initiating contacts, embracing the victim, seeking help from a third party, inspecting the distress in a more active manner, and giving the victim gifts. "What the children chose to give was not random," Radke-Yarrow says. Sometimes the choice of gift was appropriate for the distressed child, and sometimes it was more appropriate for the comforter.

The studies also revealed that imitation plays an important role in the development of these early behaviors. For

In the earliest years of life,
children make serious efforts to protect and
comfort others who are unhappy.

In the first of several studies, for example, 24 mothers made systematic observations of distress—pain, physical distress, sadness, fear, anger, and fatigue—occurring in their children's immediate environments. The mothers used portable tape recorders to report their observations. Some of the children began the study at 10 months, some at 15, and some at 20. During the nine months of observation, a researcher came to each home at three-week intervals to check on the mother's reports and carry out some simulated distress experiments. This served as a way of comparing what the children did in simulations with what the mothers were reporting. When the children were six or seven years old the study was resumed for a three-month period.

From a series of these studies the researchers gleaned a basic understanding of patterns and individual differences in moral development. Children's earliest reactions in the presence of another person's distress were silent, tense standing, agitation, crying, and whimpering; these behaviors accounted for the majority of all responses between the ages of 10 and 14 months. In this early period, the chil-

example, one mother reported that after she had bumped her elbow, said "ouch," and then rubbed it, her child showed a pained expression, said "ow," rubbed his own elbow, and then rubbed his mother's. The child is still struggling with determining what is the self and what is the other person.

Certain children stood out as being more emotional than others, both in their level of arousal and the way they reacted toward other people. Others had very little apparent emotional reaction and took a more reasoned approach—inspecting, exploring, and asking questions about the situation. Some were aggressive in their interactions—hitting the person who caused a friend's distress, pushing one parent in the midst of a parental quarrel, tearing up the newspaper that had made mother cry. And still other children reacted adversely to the distress, trying to shut it out by turning or running away from the source.

Children who were most outgoing and emotional at age two continued to display this type of behavior, in a more developed manner, at age seven. Similarly, those who avoided distress by running away behaved in a similar way five years later. However, the data also revealed striking differences in the rate at which each child developed moral reasoning. "Our data showed the value of examining each child individually, because there wasn't much of a developmental pattern when we looked at the group as a whole," said Radke-Yarrow.

ing stood out from the rest. One group of children would give help only when they were asked. These children were generally nonassertive and lacked initiative. At the opposite end of the scale was a group of children who frequently engaged in spontaneous, or unsolicited, sharing. These children tended to be more outgoing, assertive, and social.

that behavior." She argues that it is high-cost behavior when a child gives up something for someone else's benefit. The decision to share spontaneously, because its cost is higher and requires more initiative, is therefore more likely to be a reflection of a child's stronger moral development.

Those researchers who support empathy's role in the development of moral behavior readily say that it is only one of many factors involved in this process. It is nonetheless attractive to speculate that an innate behavior, empathy, is central to behavior so crucial to human civilization—subjugating one's own needs, interests, and desires to those of society, for that is what moral behavior truly is. "That's one of the key questions of human civilization: 'Why do people go out and help one another?'" says Hoffman. "Empathy can provide that motive."

Why do people go out and help one another? That's one of the key questions of human civilization.

In studies at the Arizona State University Child Study Laboratory nursery school, Eisenberg obtained results similar to those of the NIMH group. While analyzing her data, she noticed two groups of children whose style of behav-

They also scored higher than average for their ages on moral reasoning tests.

"We see that spontaneous sharing correlates well with preschoolers' ability to make moral judgments," she says. "And I think this relates to the costs involved in

Development During Childhood

- Social and Emotional Development (Articles 12-18)
- Cognitive and Language Development (Articles 19-24)

Most of the changes that occur during the transition to childhood—as well as those that occur during childhood itself—involve social, cognitive, and language development. These changes are so dramatic that the preschool to school age transitional period is sometimes referred to as the "five-to-seven shift." During this transitional time significant changes occur in the child's attention span, memory, learning, and problem solving skills. Articulation improves, vocabulary size increases, and the child achieves a new understanding of the syntactical features of language. Entrance into the formal world of school is the most significant change in the child's environment during childhood. School not only expands the child's contacts with significant adults and peers, but also expands the child's concept of neighborhood.

Social and Emotional Development. During the school years the child's social network expands. School extends the child's peer group beyond the confines of the immediate neighborhood and exposes the child to a new set of authority figures. The quality of the child's interactions with available role models influence his or her sense of social and emotional competence, according to "Children's Winning Ways," and helps to structure attitudes and beliefs about others.

All life transitions create some degree of stress. Most children handle everyday stress fairly well; they are able to draw upon a variety of coping skills learned through past interactions with family members and peers. In "Stress and Coping in Children," Honig notes thata there are both positive and negative aspects of stress. Stress often serves as an effective motivator, enabling children to mobilize their cognitive, social, and emotional resources and direct their energy toward effective problem solving. On the other hand, social, cultural, familial, and interpersonal pressures can also create stressful situations that overwhelm the child's coping mechanisms. School can provide an environment where children are exposed to belief and value systems that severely stress their coping mechanisms. "Racists Are Made, Not Born" advances the hypothesis that prejudicial attitudes and beliefs can be tied directly to social and family influences on the child's cognitive expectations about other people. The extent to which conflict affects personality development during childhood is a topic of special interest to contemporary developmentalists, particularly with respect to the pervasiveness of aggression and violence throughout the world. This concept is reviewed in "Aggression: The Violence Within."

Children often resort to physical or verbal aggression in their attempts to resolve conflict or to deal with stress. Rarely are such tactics successful. Indeed, such aggression usually begets aggression. Other children may retreat from stress by drawing into themselves and becoming apathetic, despondant, and depressed. Although depression once was thought to be an affective disorder restricted to adulthood, it is now evident that it can occur as early as infancy. In "Depression at an Early Age," Alper reviews evidence pointing to childhood depression and suggests that the affective disorders may become the major health problems of the 1990s.

Although some children react to stress by striking out against the perceived source of the stress and others may withdraw and attempt to isolate themselves from stress, there are children who seem tc take stress-producing situations in stride. That is, however acquired, their coping mechanisms are more than adequate for dealing with the most distressing situations. These children have been referred to as "invulnerable" or "resilient." Resilient children have developmental histories that may include extreme poverty or chronic family stress. Most people would consider both factors to be potentially damaging to the development of personal and social competence skills. Yet, these resilient children do not become victims but rather develop effective interpersonal skills. Without question, studies of these children will contribute important knowledge to the understanding of personality development and child-rearing practices. There is some evidence to suggest that the development of a successful attachment relationship with a significant caregiver is a characteristic of such children. It would indicate that mastery of social competence skills in combination with a sense of self-control and personal worth may contribute to the resilient child's ability to cope with stress. If so, these characteristics may have their developmental origins in the initial emotional interactions of infants and their caregivers, a topic reviewed by McDermott in "Face to Face, It's the Expression That Bears the Message."

Cognitive and Language Development. During the past two decades the study of cognitive development was dominated by Piaget's theory. Although Piaget's theory provides a rich description of what the child can and cannot do during a particular stage of development, it is less adequate for explaining how the child acquires various cognitive skills. Thus, many developmentalists have turned to information processing models in an effort to integrate cognitive psychology with cognitive developmental theory. The articles in this section focus on developmental changes in cognitive processing, context effects on encoding of information into memory, cerebral specialization of function and memory processes, and challenges to IQ test definitions of intelligence.

Prior to the 1960s, theories of language development stressed its environmental determinants. Noam Chomsky's counterpoint shifted emphasis to genetic explanations. While the issue is far from being resolved, most contem-

and developmental differences between *Homo sapiens* and other species that led to the unique structure of the human speech apparatus. Interestingly enough, the evolutionary uniqueness underlying human speech may have negative consequences as well. "Why Children Talk to Themselves" addresses the subtle interactions between the child's emergent control of language and the various environmental influences that will play a role in structuring the social and regulatory aspects of language.

Looking Ahead: Challenge Questions

If prejudices arise from violations of expectancies (a cognitive dimension) that are reinforced by parental beliefs and values (a behavioral dimension), what strategies would be most effective in combating such prejudicial systems as racism or sexism?

In aggressive and violent cultures, what chance do parents have to suppress aggressiveness in their children? Why, do you suppose, males are generally found to be more physically aggressive than females in nearly every human culture? What adaptive function does aggressiveness play in the grand scheme of things?

Children who are described as resilient are an enigma. How would Erik Erikson's theory help to explain such strength of personality in the face of so many potentially disruptive influences in their lives? Could cognitive theories provide any better explanation? What coping mechanisms seem to provide resilient children with such strength of character?

The ability to solve problems ranks high among lay definitions of intelligence. Yet problem-solving ability is not exactly the same as the intelligence measured by IQ tests. Which do you believe is the better measure of intelligence? The fact that all human beings are not equally intelligent, anymore than they are equally tall, suggests biological variation in the distribution of intelligence. Skill-learning theories of intelligence also point to variation among individuals in specific aspects of intelligent behavior. What characteristics of an individual do you think of when you refer to someone as being intelligent? Is your definition closer to Gardner and Sternberg, or to Jensen?

Do you believe that the American education system is capable of embracing the implications of current research in cognitive processing and intelligence? For example, how would the schools have to change in order to place emphasis on lateral thinking, context effects on information processing, and skill-based approaches to intelligence? Would such changes lead to tracking of students at an early age so that, perhaps, by seventh grade one would be pegged for specific fields of specialization?

porary developmentalists stress language acquisition models which assume that environmental influences are built onto a biological foundation. In the "Origins of Speech," Robert Finn draws attention to the evolutionary

Children's Winning Ways

SOME CHILDREN ARE NATURALLY SKILLED AT SILENT PERSUASION. THEIR REWARD: AFFECTION AND POWER.

Maya Pines

Maya Pines, a contributing editor of Psychology Today, *is a science writer who specializes in the field of human behavior. She is the author of* The Brain Changers: Scientists and the New Mind Control *(Harcourt Brace Jovanovich).*

When Hubert Montagner, a French ethologist noted for his studies of wasp communication, moved in 1970 to the University of Besançon, near the Swiss border, psychologists at the university's Institute for School Psychology threw him a challenge he couldn't refuse: "They thought that existing psychological tests and psychoanalytic theory were not sufficient to understand children's behavioral disorders," Montagner recalls. "So they asked me to begin a systematic observation of children in school and to establish behavioral scales which could predict certain difficulties," he says. "They told me they'd be very interested in getting some new tools with which they could better understand why children behaved as they did in school."

This was an unusual request to make of an expert in social-insect behavior, but Montagner bit the bullet and devoted most of his time in the ensuing years to studying humans, not wasps. (Colleagues in his Laboratory of Psychophysiology study humans as well as bees, frogs, tadpoles and a variety of small rodents.)

The fruits of his labors, some 15 years and 200 miles of film later, are new ways to interpret the social behavior of young children, particularly the individual nonverbal communication styles that affect long-term social roles in groups, such as who will be attractive to others or who will become a leader.

Working in the tradition of Nicolas Blurton Jones, William MacGrew and Iranëus Eibl-Eibesfeldt, who began to apply the methods of ethology to humans in the 1960s, Montagner has developed a system for observing and classifying the gestural language of children. He has also tracked down some of its origins. In addition, his long-term observations of children have given him a sense of the normal and abnormal developmental timetable for the emergence of specific types of gestures and body language. By now he can calculate how many of a young child's gestures belong to certain categories of behavior and can then derive a series of significant ratios, such as the ratio of pacifying to aggressive behavior, which appear to predict some behavioral problems.

Montagner's other contributions include careful studies of how biological factors, particularly stress hormones, are related to behavioral characteristics. In his search for the roots of children's behavioral patterns, he has gone back to the earliest moments of interaction between mothers and newborns and has explored the role of scents in their communication.

In sum, this 45-year-old researcher, whose work is not yet well known in the United States, has provided new insights into critical aspects of social interaction and communication. And he has done it by studying children in their natural settings—not in a laboratory—just as his fellow ethologists study birds, apes or wasps.

When Montagner began his study of children, he chose to observe 2- and 3-year-olds at a local day-care center. "I thought I could use an ethological approach to study their nonverbal interaction," he says. "They do speak at that age, but only a few words. And later on I found out that even when children use language, they continue to show the same sequences of behav-

 From *Psychology Today*, December 1984, pp. 58-65. Copyright © Maya Pines. Reprinted by permission.

ior as before, in the same contexts, with the same functions."

Coming from outside the psychological tradition, Montagner paid no attention to how much these children understood about objects, a prime concern of the famed and influential Swiss psychologist Jean Piaget. Nor did he try to analyze the underlying emotions revealed by children as they played with dolls, worked on puzzles or engaged in make-believe play. Instead, hidden behind a screen, he and his team focused exclusively on how these children interacted with one another. The researchers recorded and classified with great precision all the children's gestures toward others. By now they have observed the interactions of 1,500 children between the ages of 6 months and 6 years, including 100 children whose behavior was studied very intensively for a period of three years, and some whom they followed up to their teens.

The observations yielded five general categories of social behavior (see "The Behavioral Building Blocks") and the realization that certain children used gestures from some categories more than from others. Even as early as age 2, some children regularly used more gestures from the aggressive category than from the pacifying-or-attaching category. Other children used pacifying-or-attaching gestures four times as frequently as aggressive gestures.

Those unwitting choices became increasingly consistent as the children grew up, and resulted, for many, in relatively stable ways of relating to others. Thus, each child spoke a gestural language that other children could understand clearly, even if adults could not necessarily decode it, and it affected how others treated the child. Surprisingly, tiny gestures seemed to spell the difference between success and failure in getting along with others, winning friends and becoming a "leader" whom others would willingly follow.

The building blocks of the gestural language Montagner has identified emerge early, "between the ages of 9 and 12 months," he says. And they appear to be universal. He notes that the very same gestural repertoire has been observed among children in other parts of France and in the Congo; most of them also have been described among American and English children by such researchers as Blurton Jones, MacGrew and by David Lewis. All children develop the ability to perform the full range of gestures, but the combinations they "choose" and how frequently they choose them differ widely from child to child.

According to Montagner, the secret of success with others—at least in the preschool set, though possibly beyond—is to use many gestures from the pacifying-or-attaching category.

---

*E*ACH CHILD SPOKE A GESTURAL LANGUAGE OTHER CHILDREN COULD READ CLEARLY, EVEN IF ADULTS COULDN'T.

Perhaps the most magical sequence within this category is a tilt of the head over one shoulder combined with a smile. Sometimes it includes extending one hand toward the other child, as if to shake hands. Either way, the sequence triggers a friendly response or an offering of some kind in 80 percent of the cases Montagner has observed. Children on the receiving end seem to melt. They will often calm down and show affection, he says, and some will even give up cherished objects of their own free will.

The power of this sequence is evident in many of the films Montagner's team shot at local day-care centers. In one film, a 2-year-old girl approaches two boys and tilts her head sideways while smiling at one. The boy immediately smiles back at her. Meanwhile, the second boy gets up, reaches for his toy car—his only one, according to Montagner—and offers it to her.

Children often use this sequence to get the things they want. However, its primary effect is to trigger warm feelings, Montagner says. Especially among younger children (ages 18 months to 2 1/2 years), this sequence produces signs of affection or love.

Adults can use it, too, Montagner points out. In fact, some parents do so instinctively. "It is one of the most efficient ways to appease a crying child," he notes. He has repeatedly observed that when young children see this tilt of the head and a smile on someone they like, they will reach toward this person, stroke her or rush into her arms.

He is aware that advertising agencies exploit this sequence very effectively. Just look at the pictures of young women with their heads tilted ingratiatingly in TV commercials or fashion magazines, he notes.

Despite its power, this sequence is not used by all children, according to Montagner. Some hardly ever use pacifying-or-attaching gestures and therefore never receive similarly affectionate or friendly gestures in return. This failure may spoil their relationships with others for the rest of their lives, Montagner believes.

By now, Montagner can read the behavior of 2-year-olds well enough to know pretty well what these children will be like, socially, at age 10. By age 2, certain children are already on their way to being extremely "attractive" to others, Montagner says, meaning that their behavior "provoke[s] approaches to the child by two or more children." This attractiveness is rooted in their extensive use of items from the pacifying-or-attaching category.

According to Montagner, the children who become real "leaders" (meaning that they engage in acts that provoke others to follow and/or imitate them) emerge from the group of "attractive" children, rather than from the more aggressive ones. Adults often make the mistake of thinking that children who rely on overt aggression, such as biting or hair-pulling, and who push others around to get what they want, are the leaders. But these "dominant-aggressive" children, as Montagner calls them, cannot make lasting coalitions. They are generally disliked by others, and any groups that form around them disintegrate rapidly in a chaos of aggressive acts.

The real leaders—the ones whom other children seek out and follow—are quite different, Montagner says. They use pacifying-or-attaching sequences as frequently as children who are merely attractive. But in addition, they participate in many competitions for desirable toys or preferred space, and they generally succeed. Their signals are clear: They do not seek or start fights but will fight and stand their ground if attacked. "These are the children who start new activities for the whole group more than 75 percent of the time," Montagner says. "We call them 'leaders,' even in French."

He emphasizes that children who are leaders never mix gestures from different categories. "When a child offers a toy with one hand and pulls hair with the other hand, his whole behavior is decoded as aggressive," he says. Children who mix their signals in this way are highly unpopular with others, according to Montagner. "When they approach other children, the others tend to run away or cry," he observes.

Other children who are highly attractive seldom take any leadership role (except in groups of two or three children) because, for some reason, they hardly ever participate in competitions. They don't even try to get the goodies that others seek.

Montagner has also identified another group of children who do not get involved in competition but alternate between periods of self-isolation and sudden, apparently pointless, acts of aggression. "They will bite another child without provocation and without trying to take any object away," says Montagner, who is still puzzled by this behavior.

What produces these different patterns? Montagner does not believe that genetic factors play a very important role. But he does see a correlation between the children's behavioral profiles at age 2 or 3 and "the kind of behavior usually expressed by their parents, especially their mothers, towards them."

"We have observed that the parents of children who become leaders communicate with them a great deal, using mimicry, gestures, words and squatting down at their level to talk to them," he says. "They often ask the child what he wants to do. Does he want to climb on their back? Does he want to play a particular game? They listen to what the child says and pay close attention to his spontaneous behavior. They don't threaten, except in potentially dangerous situations, and

they are neither aggressive towards him nor overprotective. Their behavior is very stable."

On the other hand, "When we looked at how parents of dominant-aggressive children treated them, we observed that the mother was either rather aggressive or totally permissive, paying no attention to what her child was doing," Montagner says. "Let me give you a typical case: At the end of the day, when his mother comes to get Nicolas, she opens the door of the playroom and calls for him. But when he comes running towards her, she doesn't interact with him. Instead, she turns toward the day-care worker and asks, 'Well, was he naughty today? You see, I have so many difficul-

---

*A*DULTS OFTEN MISTAKENLY THINK THAT AGGRESSIVE CHILDREN WHO PUSH, BITE OR PULL HAIR ARE LEADERS.

---

ties with him. He hits his little sister, etc.' So as the child approaches but gets no attention from his mother, he turns away and runs to another part of the room. Then the mother begins to shout at him, 'Come on, Nicolas, I'm in a hurry!' But by then the child is in the distance, smiling. So she becomes impassioned and aggressive, first in words, then in her behavior. It's not rare to observe overt aggression when she finally succeeds in catching her child, not rare at all!"

Montagner has not found any difference between the parents of the attractive children who become leaders and those who don't, however. "So we don't know why some children rarely participate in competitions."

As to the most fearful children, "we found that usually the mothers were overprotective," he says. "They appeared constantly anxious about their children's health, behavior or the possibility of accidents, and very rarely

let them go outside to play with the other children."

The youngest children's patterns sometimes changed when their home situations changed, Montagner says. Thus, one little boy who had been a nonaggressive leader with a 1.58 ratio of pacifying behavior to aggressive behavior suddenly became highly aggressive when his baby brother fell ill and his anxious mother no longer had any time or patience for him. Within one month, his ratio of pacifying to aggressive behavior fell to .42; five months later it was down to .25 and he had become the most aggressive child in his group. When the baby got well and the mother returned to normal, however, the boy's aggressiveness waned. His ratio of pacifying to aggressive behavior climbed back to 1.15 and he became a leader again. In one or two cases in which parents became more attentive to their children after talking to Montagner's team, the children began to use many more pacifying-or-attaching sequences.

"But in children older than 5, we did not observe any major changes," says Montagner, who notes that unless the feedback mechanism between parent and child is altered early, the behavior of both becomes more and more reinforced as the years go by.

"The children who had been most aggressive at age 2 or 3 were still very aggressive when we looked at them at age 10 or 11," he says, "And the most attractive ones remained among the most attractive. Only the fearful ones seemed changed: Some of them appeared to have overcome their fears—perhaps their parents stopped being overprotective. Yet some of the most fearful appeared to be still fearful."

Throughout his work, Montagner has tried to combine physiological measurements with his observations of behavior. For example, he monitored some of the children's stress hormones regularly (as measured in their urine) and found that the daily curves of the leaders' hormone levels were most stable over time while those of the most aggressive children varied greatly.

Even his studies of mother-child interaction have a biological side, for he has focused on how children and their mothers communicate through odors. Since Montagner is an expert on communication among insects, which is of-

THE BEHAVIORAL BUILDING BLOCKS

Five types of actions, when combined into certain sequences, can make or break social relations in nursery school and possibly beyond, claims French ethologist Hubert Montagner, based on his behavioral observations of young children. As he has categorized the actions, they are:

□ *Actions that pacify others or produce attachment:* offering another child toys or candy, lightly touching or caressing the other child, jumping in place, clapping one's hands, smiling, extending one's hand as if begging, taking the other child's chin in one's hand, cocking one's head over one shoulder, lean-ing sideways, rocking from left to right, pirouetting or vocalizing in a nonthreatening way.

□ *Threatening actions that generally produce fear, flight or tears in the target child:* loud vocalization, frowning, showing clenched teeth, opening one's mouth wide, pointing one's index finger toward the other child, clenching one's fist, raising one's arm, leaning one's head forward, leaning one's whole trunk forward or shadow boxing.

□ *Aggressive actions:* hitting with hands or feet, scratching, pinching, biting, pulling the other child's hair or clothes, shaking the other child, knocking the other child down, grab-bing something that belongs to the other child or throwing something at the other child.

□ *Gestures of fear and retreat:* widening one's eyes, blinking, protecting one's face with bent arms, moving one's head backward, moving one's trunk or one's whole body backward, running away or crying after an encounter with another child.

□ *Actions that produce isolation:* thumb-sucking, tugging at one's hair or ear, sucking on a toy or a blanket, standing or sitting somewhat apart from other children, lying down, lying curled into the fetal position or crying alone.

---

ten accomplished through chemical messages, this choice of subject matter is not so surprising.

Montagner showed, in 1974, that 2- and 3-year-olds can generally recognize the smell of a T-shirt worn by their mother, and that when they are given such a T-shirt, their behavior changes. "They may stop whatever they were doing and rub this shirt on their face, put it in their mouth or lick it," he wrote. "They become less aggressive than on other days and have fewer interactions with other children."

This clue led him to track down the origins of some of the feedback mechanisms between parent and child. So he went back to the very first days of life, studying mother-infant communication through smells, which had been demonstrated in the pioneering work of MacGrew and Aidan Macfarlane.

Infants react differently to pieces of gauze that their own mothers have worn against their necks and breasts than to gauze worn by other mothers, Montagner's team showed in a series of experiments with newborns that began in 1978. The researchers calculated that as the infants turned toward these pieces of gauze, the area swept by their noses and arms was narrower and their behavior more peaceful than when they were confronted with alien smells. This difference appeared on the third day of life, the team found, just when the blindfolded mothers began to recognize the odor of T-shirts worn by their own infants and to distinguish such shirts from those worn by other newborns. Olfactory signals of this sort can play a role in the attachment between mother and child, Montagner suggests.

---

*S*CENT
SIGNALS CAN
PLAY A ROLE IN
THE ATTACHMENT
OF MOTHERS AND
NEWBORNS.

Now Montagner and his colleagues are looking for early signs of trouble in the mother-child relationship, as well as in children's behavior toward other children. With Albert Restoin and Danilo Rodriguez, he has roughed out a timetable for the normal development of communication behavior.

They know, for instance, that it is normal for 12-month-old children to isolate themselves from others occasionally and to lie down flat on their stomachs. But the researchers also know that this particular posture usually disappears around 18 months of age, and that older children normally lie down on their sides.

Along the same lines, they have found that at 12 months, a child who seeks comfort in an adult's lap will generally sit facing the adult directly; by 15 months, this child will sit sideways. A certain kind of rocking that is normal for 1-year-olds also disappears before their second birthday. "When a child of 2 or 3 often lies down flat on his tummy, away from others, rocks himself or offers toys to a child who is facing in another direction, it is a sign of serious trouble," Montagner warns. Such children tend to become extremely isolated from the others as they grow up or to alternate between isolation and bouts of aggression.

The researchers also check whether young children know how to threaten others effectively, a very important skill, according to Montagner. When children begin to walk, they encounter more conflicts and competitions, and must develop a system to participate in these with a minimum of aggression. "Threats can prevent aggression," Montagner explains. "We have looked at many instances of conflict and calculated that children between the ages of 7 and 18 months who don't use threatening behavior such as loud vocalization when it is appropriate are the targets of other children's aggression 59 percent of the time. Those who do use threats receive aggression only 41 percent of the time."

In this way, Montagner's group is beginning to build a list of possible danger signals which, if heeded, might prevent many of the difficulties children encounter both in school and later in life. Their work fits in very nicely with that of other researchers who have studied the development of social competence in monkeys.

"In our animals, the first danger signals come from peer interaction. If a young monkey who had pretty smooth relationships with peers becomes very aggressive or stops participating in play, usually it's a sign of some disturbance in the mother-child relationship," says Stephen J. Suomi, an ethologist who conducts primate-behavior research for the National In-

PEER INTERACTION 'IS A NICE BAROMETER OF OVERALL PSYCHOLOGICAL WELL-BEING.'

stitute of Child Health and Human Development and the National Institute of Mental Health. Suomi observed hundreds of monkeys while collaborating with Harry Harlow at the University of Wisconsin.

Though monkeys normally try to play with whomever they can, "play behavior is very fragile," according to Suomi, and in case of trouble or stress, it is the first behavior to disappear. Thus, peer interaction is "a nice barometer of overall psychological well-being," he believes. It can serve to diagnose problems and also to predict future adjustment. "A young monkey who's hyper-aggressive towards others, or else unusually shy and reluctant to participate in peer interactions, will generally show more serious difficulties later in life," Suomi says.

Similarly, careful monitoring of children's gestural language and relationships with others could serve as a barometer of their mental health and as a possible predictor of how they will function in the future.

Stress and Coping in Children

Alice Sterling Honig

The article was written and edited by Alice Honig, Ph.D., Professor, Department of Child, Family, and Community Studies, Syracuse University, Syracuse, New York.

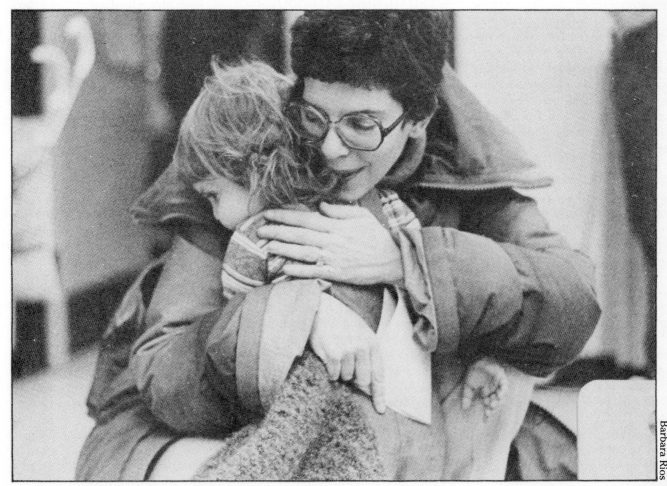

Barbara Rios

Teachers and parents can act as protectors to create environments low in stress and as facilitators to help young children cope better with stress.

Reprinted by permission from *Young Children*, May 1986, pp. 50-63. Copyright © 1984 by National Association for the Education of Young Children, 1834 Connecticut Avenue, N.W., Washington, DC 20009.

Stress is to be expected in the course of human development. The very act of being born includes a stressful and sometimes perilous journey for a baby, even an infant seemingly untouched by genetic, prenatal, or perinatal risk factors. Some newborns have neurological immaturities that lead to chin trembling and long bursts of crying. Some infants' digestive systems do not work well and colic stresses them for months (Honig, 1985a).

Inevitably, in the course of their first years in human society, children will be socialized, gently or more harshly, into acceptable toileting habits, civil (*please* and *thank you*) forms of getting needs met, and table manners that make public eating experiences less stressful for adults.

Stress continues to mark the achievement of developmental milestones. How often an infant, on the verge of toddling, stumbles, lurches, falls, crashes, and recommences bravely. Not all stresses are harmful. The struggle to learn to walk is a good example of how some stresses can be perceived as challenges that impel a child to strive toward more mature forms of behavior.

What is stress? Selye (1982), the father of stress research, defines stress as a *stimulus event of sufficient severity to produce disequilibrium in the homeostatic physiological systems.* Stress also has been conceptualized variously as a *nonspecific response of the body to any demand that exceeds the person's ability to cope,* as a *person-environment relationship that threatens or taxes personal resources,* and as a *mental state in response to strains* or daily hassles (Lazarus & Launier, 1978; Lazarus, DeLongis, Folkman, & Gruen, 1985; Rutter, 1983).

A variety of dissimilar situations that happen to children can produce stress, including physical illness; pain; concentration; overexertion; anxious anticipations of failure due to overly strict or high parental

expectations; fear and tension before a test or during a visit to the dentist; being a latchkey child after school; being teased about facial features, allergies, or asthma; humiliation; fear of abandonment; feelings of being unloved; racial slurs; living in a dangerous neighborhood; heavy doses of violent television; physical abuse; and even joyous overstimulation. Witness the temper tantrums that some small children burst into on a too-exciting Christmas morning (Grune & Brooks, 1985).

Different kinds of stress situations vary in their outcomes. In a study of behavioral stress re-

---

Adults need to be sensitive to the *meaning* of a particular stressor for a child.

---

sponses of public school children, Felner, Stolberg, and Cowen (1975) found that children who had lost a parent through death increased in timidity, shyness, and withdrawal. Children from separated or divorced families more typically increased aggressive, antisocial behaviors. Stress is difficult to research, partly because of the wide variety of stimuli that are potentially stressful, their differential intensity, duration, and the *interactions* of different stressors in a child's life.

Stress can arise from *internal* factors. A young baby with severe gas pains from colic cries miserably as she flexes her tiny legs. Stress can arise from illness or from the painful stomach aches of a young child lying in bed and listening each night to his parents' violent quarreling in the next room. Stress can arise from *external* factors, too. A kindergarten child, forced by a recent family move to attend a new school, finds walking the new route alone a terrifying experience. He

may arrive home with soiled pants because he loses bowel control.

Some stresses are *acute* in a child's life. They arise suddenly, are isolated instances, and their impact may not last long. An infant's sudden fever after a DPT shot or a preschooler's first days of adjustment to a high quality nursery school program are examples of acute stress. Some acute stresses, if isolated, such as a single hospitalization for a child, are associated with short-term emotional disturbance, but *not* with long-term upset years later.

Other stresses are *chronic*. Their impact may be cumulative even for the most well-adjusted child and can lead to long-term disturbances. An alcoholic, an unpredictably abusive parent, or bitter family recriminations long after parental divorce—these are chronic stresses that may impair even the psychologically sturdiest child's functioning.

This two-part review will first examine the components and stages of stress. Then, research findings will be used to illustrate varieties of intrapersonal, ecological, catastrophic, and interpersonal stress factors in the lives of children. Some measures of stress will also be noted. Research on children's coping skills and on "invulnerable" children will be cited. Finally, suggestions will be offered that can help caregivers recognize, prevent, and alleviate child stress as well as enhance children's coping skills to deal with stress.

Components and stages of stress

Researchers have focused on several components of stress: a *stressor;* how a child *perceives* that stressor; the *coping resources* a child has; the *support systems* available internally and externally for the child; and the child's *skill* in making coping or adjusting responses when stressed.

A *stressor* is "an acute life event or a chronic environmental situation that causes disequilibrium" (Blom, Cheney, & Snoddy, 1986, p. 9). The severity of stress consequences depends partly on how a child understands and feels about the stressor. Some children are born with temperamental and neurological vulnerabilities that impair their ability to think about and deal adaptively with even small stressors in their lives. Teachers and parents need to be sensitive to the *meaning* of a particular stressor for a particular child. Circumstances change meanings. What would seem to be a stressor for one child, such as father leaving home, may be a relief for another child who has been tyrannized and abused by that parent.

The stress response, seen as an "imbalance between requirements to make an adaptive response and the repertoire" of the stressed person (Zegans, 1982, p. 140) shows several stages:

1. Stage of alarm

Selye's early research (1936) into the initial alarm reaction of the body referred to a "general adaptation syndrome" of stress. Heart rate increases, hormones such as ACTH (adrenocorticotrophic hormone) are discharged, and the galvanic skin response changes. Adrenaline is secreted to make energy available to the body. High blood pressure, bleeding stomach ulcers, and, ultimately, feelings of exhaustion, are body alarm reactions to stress. Thus, life situations which threaten a child's security and evoke attempts at adaptive behavior also evoke significant psychosomatic alterations in the function of bodily tissues, organs, and systems. "These physiological changes, in their turn, will lead to a lowering of the body's resistance to disease" (Rahe, Meyer, Smith, Kjaerg, & Holmes, 1964, p. 42). As a result, stressed children may get sick more often.

2. Stage of appraisal

Many parents and teachers assume that stressors in a child's life can be easily identified. Yet differences in the *cognitive appraisal* of stressors lead to marked individual differences in children's reactions to potential stressors. Appraisal refers to the evaluative process that imbues a situational encounter with meaning for the person. Some children's psychological makeup or value system (such as beliefs, commitments, and goals) may predispose them to perceive particular events as highly threatening to personal security (Lazarus & Folkman, 1984). Other children may react with zest to unknown or unexpected life challenges.

Madeleine's father invited her friend to go sledding with them in the deep snow of a local park. On the way home, the car had a flat tire. The children had to get out of the car and wait in the fast-falling snow as dusk fell, while papa changed the tire. Madeleine held the metal nuts and handed them to papa as he needed them. She felt they were having an adventure and was proud of being a good helper. Her friend, however, became very worried. What if they could not fix the car? How would they ever get home? Her mother might be very angry with her. Such fears became overwhelming, and the little friend started to sob.

Darren was always ready to fight physically if another child at his center touched him. Even if the child just brushed past Darren while running excitedly toward a friend, Darren would whirl, put up his fists, and hit out at the perceived assaulter.

Arthur moved a great deal because of financial troubles in his single-parent family. Arthur, the youngest of four children, felt a sense of high adventure every time he woke in a new apartment. He felt special, for he got a chance to live in many different places compared to other children. Arthur felt special-privileged, rather than special-unfortunate.

3. Stage of searching for a coping strategy

Coping resources young children possess include tears, tantrums, thinking skills, and the ability to become absorbed in play with peers. Ability (1) to ignore unpleasant situations, (2) to find compromise solutions to social conflicts, and (3) to find and accept substitute satisfactions and comforts when stressed are *adaptive* coping resources.

If children have had inappro-

priate or ineffective caregiver models for coping, their particular strategies may only result in increased stress. In one family, where cursing and lashing out physically were typical parental responses to family frustrations, the child learned well the familiar strategies of cursing and kicking. But these brought only more grief and stress when applied to his first-grade teacher.

4. Stage of implementing coping responses

Coping responses are attempts to resolve life stresses and emotional pain (Billings & Moos, 1982). Some coping responses are *defensive* processes. These may distort or even deny a disturbing reality. Compulsive behaviors, such as insisting rigidly on certain ritualistic actions over and over or on certain room arrangements only, may be ways a child tries to ward off anxiety.

Adrien's parents fought violently at home. In the child care center, he pushed a toy car across a low table and watched it crash down as it tipped over the far edge. Then he picked up the car and rode it across the table again and again to let it fall precipitously over the edge of the table.

Eight-year-old Danny pulled the window shades up and down many, many times each evening before bedtime until he got them just so, to a particular level above the window sill. He opened and shut his closet door over and over in getting ready for bed.

Some children cope by *internalizing*. Children with internal controls are more likely to accept responsibility for their actions. If caught misbehaving, such children may honestly face up to their misbehavior, feel contrite, and realize the consequences of the unacceptable behavior.

Some children are *externalizers*. They attribute control to fate or to others. Garmezy (1981) compared children with these different behavioral processing styles. Externalizers are more likely to act surly or angry, fight with others, accuse others when they themselves misbehave, and show little empathy for children they may have hurt. Behavioral adjustments of children who cope with stress by externalizing are less favorable than for children who are internalizers.

George and Main (1979) observed that abused toddlers who were enabled in group care not only hit their peers more than nonabused toddlers did, but they showed no empathy for peers who were stressed by physical hurt or psychological upset. They did not react with concern, nor did they offer help. Too much stress in these toddlers' lives had left them emotionally unresponsive to others'

Male children are more vulnerable to stress than female children.

troubles. They ignored friendly adult overtures in the classroom.

Strong emotional responses to stress may or may not be negative. A strong emotional response may reflect a child's determination to deal with the stress as a challenge to be surmounted rather than a threat that leads to panic or emotional disintegration. Some coping strategies include *problem-solving* responses to deflect pain or lessen stress. In *instrumental coping,* a child uses skills and knowledge to make a stressful situation better.

Successful coping involves a pattern of behavioral responses to novel situations, obstacles, and conflicts in which there is an effortful searching for solutions, direct action, and shaping of events. Haan (1982) has identified five properties of successful coping:

• flexible and inventive creation of response options

• open consideration of options and choices

• orientation to reality and to the future implications of situations

• rational, conscious consideration and purposeful thinking

• governance and control over one's disturbing negative emotions

Every caregiver experienced with infants, toddlers, and preschoolers will note that very young children (as Piagetian theory would predict) will not yet have the sequential logical thinking skills nor the cognitive classification skills to permit such optimal coping processes in responding to stress. Does this mean that teachers should always expect young children to fall apart in the face of daily or long-term stressors? Indeed not. Successful coping also depends on internal personality strengths and external supports. Teachers and parents can effectively act as *protectors* to create environments low in stress and as *facilitators* to help young children cope better with stress.

What do we know about the variety of stressors in children's lives?

Personal child variables

Variables such as prematurity, sex, temperament and neurological sturdiness, age of child, and intellectual capacity are associated with different kinds of stress in children's lives.

Prematurity

Prematurity has been associated with severe stress in infant state management. Brazelton and others caution that when such babies are sent home from intensive care, parents must be alerted to the possibility that overstimulation in loving interactions can lead to acrocyanosis (a condition, associated with pain and numbness, that causes the extremities to turn bluish) and other dangers (1979). Lack of recognition of infants' active self-organization efforts may force premature babies to expend too much energy shutting out stimuli or using poorly modulated sensorimotor responses to cope (Als, 1985; Honig, 1984).

Additional stresses for preterms may arise from parental expectations. When mothers of 18-month-old infants, born preterm, were asked to record their expectations for achievement of developmental milestones, their expectations *lagged* for the first 2 years in comparison with mothers of normal babies. But after that, the mothers of preterms expected precocity in development. "Mothers of preterm infants may harbor unrealistic expectations for their young children" that can cause later dysfunctional family stresses (Leiderman, 1983, p. 151).

If children have inappropriate models, their coping strategies only result in increased stress.

Sex

Male children are more vulnerable than female children. Boys have higher rates of bed wetting, dyslexia, and delinquency. In a study of metropolitan child care centers serving low-income families, male toddlers made significantly more distress bids and sought help more than females from their caregivers (Honig & Whittmer, 1982). In an ecological study of multiple factors influencing child abuse, "the most salient characteristic of the abusive families was that 75% of the target children were male compared with 56% for nonabusive families" (Conger, McCarty, Yang, Lahey, & Kropp, 1984, p. 237).

In 1975, Seligman proposed the concept of "learned helplessness," a cognitive style which differentiates people who perceive coping outcomes as within or outside of their own control. Feelings of helplessness about their fate make children less able to deal with stress.

When children in elementary classrooms received feedback from adults that they were failing, boys tended to respond with greater efforts. Girls tended to give up and attributed their failing to their own lack of ability (Dweck, Davidson, Nelson, & Enna, 1978). Rutter has suggested that in classrooms there is a sex differentiated pattern of feedback from teachers. "Boys are given the message that their failure is a consequence of their misbehaving or not trying hard enough and hence that they *could* cope if they chose to do so" (1983, p. 27).

Certain stressors may differentially affect boys and girls. Significantly more negative behaviors were found for 24 nursery-school-age boys and 24 girls from white, middle-class families of divorce in comparison with children from intact families. Two years later, the *boys,* but not the girls, in divorced families were still having adjustment problems (Hetherington, Cox, & Cox, 1978).

Temperament

Three major child temperament patterns have been described: easy-to-adapt, slow-to-warm-up, and irritable/irregular (Thomas & Chess, 1977). Does temperament modify a child's ability to cope with stressful events? No research has found a correlation between security of attachment of infants with mothers and infant temperament classification. Yet Rutter (1978) has found that in families of mentally ill patients, children with adverse temperamental characteristics were twice as likely to be criticized by parents (Werner, 1986). Temperament vulnerabilities will be discussed later in relation to disturbances in mother-infant attachment.

Age and intellectual capacity

Different kinds of stressors affect children at different ages. For example, Grune & Brooks (1985) found that older children were distressed by anticipation of tests, by report cards, and by their personal appearance. Infants in the first year of life, particularly if they are born with medical problems, will be more stressed simply in trying to maintain physiological well being or homeostasis. Also, since stranger anxiety sets in at about age 8 months, infants who enter group care at about that time will be particularly stressed.

In Project Competence, interactive effects of life stressors and personality variables were studied from third to sixth grade in three groups of children: normal, heart-defect, and mainstreamed physically handicapped. Among the two stressed and one normal group, boys had significantly higher classroom disruption scores (Garmezy, Masten, & Tellegar, 1984). IQ functioned as a protective factor. High IQ children scored higher on an academic achievement test despite higher or lower levels of personal or family stress in their lives as measured by Coddington's Life Events Questionnaire.

Ecological stressors

Housing and neighborhood

The living environments of some children seriously increase the risk of stress because of increased neighborhood crime; criminal and antisocial role models; and unesthetic, dreary, or garbage-cluttered streets. Some apartments are very crowded, so that the number of persons per room (household density) does not permit the privacy and play space children need (Zuravin, 1985). In the United States, high household density seems to increase both the extent to which parents hit their children and the number of verbal quarrels (Booth & Edwards, 1976). In Hong Kong however, high density does not act as such a stressor. Therefore, cultural accommodations and attitudes must be taken into account in assessing the extent to which household factors may be stressors. Also, a child's *privacy* requirements increase from infancy to adolescence, so the age of the child may be a factor in whether or not high density is perceived as a stressor.

In a study of immigrant North African mothers and toddlers living in Paris, Honig & Gardner (1985) found that density *per se,* and even total number of persons in the apartment, did not differentiate families where mothers reported feeling overwhelmed with stress. However, significantly more overwhelmed mothers *lacked two or more facilities* such as toilet, kitchen, or bathing facility, in comparison with nonstressed mothers.

Childrearing stress is greater in high-rise apartment houses. Children's play is often restricted to the apartment interior, since parents fear accidents or crime if children are allowed outside on their own. This can create tensions, aggravate conflict among family members, and decrease neighborliness (Becker, 1974). Parents also restrain noisy play to decrease their own and neighbors' stress in high-rise apartments with poor sound-proofing. In low-rise housing, children often combine outdoor play with frequent trips indoors to see parents who can then get their own housework done.

Socioeconomic status as a stressor

Poverty

Poverty as a chronic family stressor may intrusively interfere with effective family functioning. When stress is severe in the lives of low-income mothers of infants, then attachment is more likely to change from secure to insecure between 12 and 18 months (Vaughn, Egeland, Sroufe, & Waters, 1979).

In the Oakland Growth study, Elder and colleagues (1985) found that when fathers lost jobs they became pronouncedly more severe, tense, explosive, and rejecting with their *less attractive* daughters. Mothers did not change in supportiveness. Attractive daughters, however, were not likely to be maltreated by fathers during adolescence, no matter how severe the economic pressures.

The Infant Accident Study included 24 infants (less than 12 months old) who had either been hospitalized for child abuse *or* accident (Elmer, 1978). Children were matched on age, race, sex, and socioeconomic status. All the infants were from low-socioeconomic status families. As infants, the problems suffered by the abused babies marked them as more stressed. More of the abused infants were below the third percentile in height and weight, had significant health problems, and were below norms developmentally. When followed up 8 years later, however, the effects of poverty proved almost as pernicious as earlier abuse or accident status in infancy. Aggression was pronounced in 6 abused and 8 accident children. Language and communication problems existed for 11 of the 12 abused and 6 of the 12 accident children. Eight of the abused and 3 of the accident children were not performing up to their ability in school. Half of each group was reported to have nervous mannerisms (by the teacher, pediatrician, or parent).

What was grievously apparent was that *all* of the children were damaged: psychologically, in language development, in the ability to achieve and learn in school. None seemed free to enjoy, learn, and grow in a reasonably healthy manner. They displayed a pervading sense of sadness, anxiety, and fear of attack by others.... We must inquire whether membership in the lower classes is itself dwarfing the potential of young children (p. 18).

During the 1960s and '70s, programs proliferated that offered enriched early child care to poverty families. Children who participated in such preschool enrichment, regardless of form of program delivery, showed better cognitive achievement and positive social behaviors (Honig, 1979, 1983). Ten or more years later, low-income children whose families had received these services had been held back in school less frequently, spent fewer years in special education classes, and had less delinquency

than children without such programs (Consortium, 1983).

Low-income single mothers with normally born infants were given medical and social services, including child care. Increased parental nurturance was the main goal of the program. Ten years later, the poverty children with intervention had better school attendance, and the boys were less likely to require costly special school services in comparison with the control group. Control boys were also rated much more negatively than intervention boys by their teachers

---

The stress of poverty can be mitigated by programs that nourish the social and motivational roots of early learning.

---

and, to a serious degree, they were described as more disobedient (Seitz, Rosenbaum, & Apfel, 1985).

Thus, poverty as a stressor in the lives of children can be mitigated by social programs that *nourish the social and motivational roots of early learning* and provide nourishing food and cognitive activities.

Catastrophes and terrors

Figley and McCubbin (1983) have addressed ways in which families face and resolve the effects of catastrophes like hospitalization, natural disasters, childhood illness, and war. Some of these stressors are discussed here.

Hospitalization

Whether parent or child is hospitalized, age of the child is strongly related to the degree of the child's stress. Stress from separation is more upsetting from about 6 months to 4 years (Schaffer & Callender, 1969). Interference with an

infant's development of a strong attachment to a parent leads to severe infant disturbance. The Robertsons (1971) found that stress was reduced when individualized, loving family care was provided for toddlers whose mothers were hospitalized. If parents are allowed unlimited access to hospitalized babies, separation stress may be much reduced.

Hospitalization experiences can be fearful in their own right. When researchers attempted to decrease such fears through prior home visits, peer modeling, films, and reading books to children about going to the hospital, children thus prepared before a tonsillectomy, for example, showed decreased emotional disturbance during and immediately after admission (Ferguson, 1979; Wolfer & Vistainer, 1979).

A single hospitalization does not seem to cause long-term noxious outcomes (Rutter, 1979). But two or more hospitalizations for a child who, in addition, comes from a deprived or disturbed family, have been found to result in more long-term, persistent child disturbance after discharge. Rutter comments, "It was not just that the adverse effects summated, but that they potentiated one another so that the combined effects of the two together was greater than the sum of the two considered separately" (p. 22).

Societal disasters

Disasters like the 1985 earthquake in Mexico can leave hundreds of infants and young children homeless and orphaned. Other potential disasters, such as the threat of nuclear warfare, are more amorphous but may increase anxiety and fearfulness in children. Clinical interviews reveal that specific trauma, such as the Chowchilla kidnapping in California, result in nightmares and emotional disturbances for several years. In that event, 26 children on a school bus were kidnapped and held hostage underground for 27 hours by three masked armed men. Several years later, stress symptoms remained moderate for 6 of the children and severe for 17. The children had kidnap-related dreams and felt guilty for their lack of awareness of preparedness for the trauma that had occurred (Terr, 1981).

Other disastrous traumas, like the assassination of President Kennedy or the explosion of the Space Shuttle Challenger, may preoccupy children with nightmares, worries, and gory imagery in their drawings. After the Challenger disaster, the New York Times conducted a poll with 1,120 families, of whom 175 had children ages 5 through 8 years of age. Older children, 9 through 17, 224 were also interviewed. According to the poll, girls were more likely than boys to say they had been upset a lot. Thirty percent of the parents of the younger children said the children were upset by the shuttle accident, and slightly more than half said that they thought it was because a teacher had been on the flight. After the tragedy, 40% of the younger children, as opposed to over 80% of the older children, watched a lot of television coverage. This difference may account for the lowered upset of the younger vs. older children polled (Clymer, 1986).

Nuclear war threat. Awareness and fear of the threat of nuclear war are infrequent among preschoolers but rise as children's cognitive sophistication increases. In a study by Beardslee & Mack (1986), about 40% of students between 5 and 12 years of age were concerned about or afraid of the nuclear threat. Older children, such as adolescents, "begin to have the capacity to understand the (nuclear) danger as well as the fact that people are doing something about it ..." (Yudkin, 1984, p. 25). During the World War II blitz in London, children showed *more* severe neuroses when separated from their parents and sent to safe foster homes in the

country than children left with their parents in the city. When adult caregivers are calm and provide security, young children will not pick up fears that cause deep concern in older children.

War and terrorism. Radio and television have brought global terror intimately into the lives of young children. Many a preschooler eats dinner with family, while on the TV screen all are watching someone in a far away land blow up or shoot at human beings. One can only wonder at either the stress *or* premature insensitivity to cruelty that is engendered when very young children see that their family

seems unperturbed by these bloody events on the TV screen.

In Israel, in border communities where terrorism has been a pervasive threat, children's nervousness, nightmares, and sleep difficulties have necessitated the active intervention of child psychologists in child care centers. This story was repeatedly told to me during 1985 visits to child care centers near the Lebanese border. Children in border kibbutzim (collective agricultural settlements) who lived under terrorist attacks and shelling showed more teeth grinding than children on kibbutzim farther inland, out of range of attacks (Ziv & Israeli, 1975).

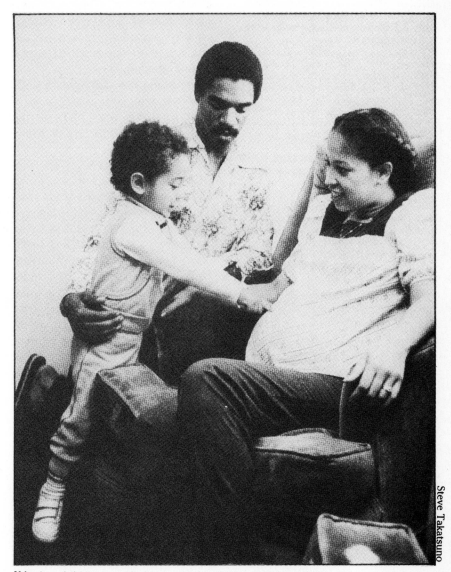

Steve Takatsuno

If loving child care workers, parents, and other family friends increase positive attention paid to older children, then the new baby may not turn out to be the severe stressor that it sometimes has been found to be for siblings.

Community cohesiveness and strong social support networks increase child stress resistance. Kibbutz methods of rearing children in group care with calm daily routines, beloved stable caregiving persons, and close relationships with peers and caregivers have been found to buffer children from war anxiety (Milgram, 1982). During a period of heavy artillery shelling of a particular kibbutz on the Lebanese border in 1970:

> The preschool children had become visibly distressed and their *metaplot* (caregivers) were uncertain as to how to handle the situation. The consultant held group discussions with these women and quickly identified the problem: the underlying anxiety of some *metaplot* and preschool charges. Once they were able to acknowledge their concerns for their husbands and their own children, they could begin to draw upon the resources of the kibbutz to bolster their flagging confidence. Thereafter, they could develop teaching and training strategies to enhance the sense of security of the children under their care. They set up telephone communication between their charges and their mothers and also communicated with their own children elsewhere on the kibbutz. They encouraged the children to decorate the shelters in which they had to sleep. They permitted expression of apprehension through class discussions and through drawing. In sum, group work with the outside consultant enabled the child care workers to cope more effectively with their own fears and to help their preschool charges cope more effectively with theirs. (p. 662)

Unfortunately, in many war-torn communities, such as sectors of Northern Ireland, psychological services for young children are not widely available. One can only guess at the stresses suffered by children in displacement camps or starving persons in drought countries where famine strikes.

War orphans invariably suffer a painful experience. A recurring problem in Israel revolved around how and when to tell children about their father's death (Milgram, 1982).

> It became clear that mothers who were evasive or who lied altogether only confused and upset their children. One four-year-old child was told that his father was in heaven. He subsequently became anxious whenever he heard an airplane overhead because he feared (it was learned subsequently) that the airplane might run his father over. He also got upset whenever it rained because the rain would make his father cold and wet. Another child of three years was told that his father had fallen. The word "fall" in Hebrew clearly implies death, but this child became disturbed whenever he went up or down stairs, fearing, as he later explained, that if he were to fall on the stairs, something terrible would happen to him as it had to his father who had fallen. (p. 663)

Family events as stressors

Birth of siblings

For some children, the birth of a new baby into the family can be a stressful occasion. Moore (1975), in a study of London children, found that 15% of the children he studied developed difficulties, predominantly through a disturbed mother-child relationship or acting out behaviors, after the birth of a new baby in the family.

Dunn and her colleagues (1981) reported that after the birth of a sibling, about one fourth of the children they studied developed sleep troubles. Of the 40 2-to-3-year-old children, almost one half showed new toileting problems, and more than one half of the children cried more easily. Dunn and Kendrick (1980) attributed these child stress responses to a changed mother-oldest child relationship. Following the new baby's birth, mothers were less likely to play with the oldest child, and they increased their prohibitions, confrontations, and negative verbal interactions with the older children. Thus, the birth of a sibling can be a stressor. But if loving child care workers, parents, and other family friends increase positive attention paid to older children, the appearance of a new baby may not become the severe stressor that it sometimes has been found to be for siblings.

Bereavement and loss

Death of a parent. Loss of a parent is perhaps the greatest stress a child must cope with. Death of a parent increases a child's sense of profound helplessness and loss, particularly when mental resources are not equipped for coping with such a tragedy (Furman, 1974). Researchers have reported that loss of a parent "creates profound reactions, including: denial, reversal of affect, identification with the parent, an intense attachment to the lost parent, fantasies of the parent's return, idealization of the parent, persisting demands to be cared for, a vindicative rage against the world, efforts to force the parent to return by suffering, and a self-inflicted repetition of the loss via other relationship" (Adams-Greenly & Moynihan, 1983).

The age and sex of the child at the time of a parent's death affect the intensity of stress. Research by the Institute of Medicine of the National Academy of Science indicates that girls under age 11 who lose their mothers and adolescent boys who lose their fathers are particularly at risk for enduring emotional problems all their lives (1984). For 47 children, both kibbutz and city reared, whose fathers had been killed in the Yom Kippur war, a significantly higher rate of emotional pathology was found among the young boys, aged 3 to 6 years, than among girls of the same age (70% vs. 20%) both at 1 year and 1½ years after the father's death (Elizur & Kaffman, 1983).

Rutter's research (1983) on children of sick, dying parents revealed that immediate grief reactions of young children were both milder and of shorter duration than those of adolescents. Yet the long-term consequences in terms of psychiatric disturbance were greater for the young children. Stress factors *consequent* upon the bereavement may, however, account for the long-term severity of loss for young children. For example, after a parent's death, a child may be forced to undergo a move to new housing, may be cared for by many others, live impoverished, and be forced into a supportive, prematurely responsible role by the grieving, surviving parent.

Bowlby's attachment theory suggests that early loss of a parent sensitizes a child, who then becomes more vulnerable to later stresses, particularly later losses and threats of loss. For children who lose a mother before they are 11 years old, later stress events are more likely to provoke depression (Brown & Harris, 1978).

Of 2,000 murders in this country each year, 10% involve children as witnesses. Poussaint studied 10 Black and Hispanic children where family death had occurred through murder. For these bereaved children, rage was overwhelming. "They felt violated, like rape victims—they wanted to seek revenge, hunt for the killer, and exhibited terror of their environment" (cited in Turkington, 1984, p. 16). Emotional pathology was likely when the killer was not caught, and the children became obsessively preoccupied with thoughts about the murder rather than able to go through the grief work that leads to more normal personality reorganization.

Psychiatric symptoms in death of a parent. Arthur Kemme (1964) looked at the symptoms of 23 emotionally disturbed children who had lost a parent. The children showed severe psychiatric symptoms that included:

● deep insecurity and inability to trust

● narcissistic self-involvement

● extreme impulsivity

● unprovoked violence toward peers, adults, and property

● overtly seductive invitations for sexual contact

● nightmares, night terrors, and phobic reactions

● guilt toward the dead parent and reactive depression episodes

A more recent study randomly selected bereaved spouses and control families with children at a mean age of 10 to 11 years. Child age, sex, number of children, and the socioeconomic status of the family were carefully controlled. One year after the death, according

to the bereaved parents who were interviewed, the children showed a greater frequency of sadness, crying, irritability, sleep and eating troubles, temper tantrums, and bedwetting. During the first year after parental death, bereaved children were reported to fight more with siblings and be more disinterested in school than control children. None of the control children, but almost one fifth of the bereaved children, reportedly were doing worse in school.

Death of a sibling. A dying sibling represents a severe shock for a

---

When adults are calm and provide security, young children will not pick up fears.

---

child (Rosen, 1985). Family members experience fear, guilt for resenting the disruption in their lives, and anger and sorrow at the impending loss. Bewilderment and isolation increase for the healthy sibling. In a study of therapeutic work done with children dying of cancer and their families, Sourkes has noted the typical effect of the stress of a sibling dying on younger children in the family:

> Parents learn to recognize and interpret stress reactions in the siblings. Susan's little brother began to wet the bed nightly during her prolonged hospitalization (for cancer). The mother thought that bedwetting was his means of delaying her daily departures to the hospital, since she would wash and change his linens before leaving. The mother thus understood the enuresis as a symptom of the child's stress, rather than as an annoying additional burden. (1977, p. 67)

Handicapped sibling: A stressor?

Research prior to the past decade tended to show that siblings of a handicapped child were more likely than other children to suffer from emotional or behavioral disorders, especially if the normal sib-

ling was close in age. Recent research has not found a difference in the proportion of emotionally disturbed children whose siblings are handicapped (Breslau et al., 1981). Where support services are low, the siblings of autistic children have been found to be more poorly adjusted and family stress higher (Bristol & Schopler, 1984). Male siblings may be more vulnerable. Breslau (1982) found that boys with handicapped brothers and sisters suffered greater psychological difficulties than did girls. Normal siblings coped best when their parents spoke openly about the child's handicap, included the normal child in the family decision-making process, and asked for the child's help in integrating the handicapped sibling into the community (Garland, Swanson, Stone, & Woodruff, 1981). Thus, open communication, support, and parental sensitivity can diminish the potentially stressful impact of a handicapped sibling on a normal child.

Spouse problems: A child stressor

Separation and divorce

Like bereavement or loss of community imposed by war and dislocation, separation and divorce are devastating stress situations for children. Schools have reported higher rates of disrupted learning, erratic attendance, increased tardiness, school dropout, and social misbehaviors among children from divorced families.

Wallerstein and Kelly (1980) carried out extensive clinical investigations of 131 children and parents from 60 middle class, mostly White California divorcing families. Approximately one third of the children continued to experience overt parental discord even 5 years after the divorce. Stress thus can be severe and even terrifying for children not only when parents fight prior to a divorce but long after a marriage is dissolved.

The patterning of divorce-as-stressor is different from the long-term chronic stressor poverty or from an acute stressor, such as a single hospitalization. Family instability generally lasted in this sample for 3 to 5 years—well over half the lifetime of a preschool child. Long-term effects may well be more serious than recovery from acute/immediate postdivorce griefs would suggest. Compared with loss through death, divorce stress "is greatly increased by the child's accurate perception that the parents are the agents of his distress, and that they have become such agents voluntarily" (Wallerstein, 1983, p. 272). "Anger at parents and guilt-restoration fantasies may even be more powerful and longer lasting" (p. 273) for children of divorce compared to bereaved children.

Wallerstein conceptualizes several different stages of the adjustment to divorce process:

1. *Acute Phase.* Precipitated by the decisive separation of the married couple, and (usually) the father's departure from the home. Verbal violence, rages, and evocations of hurt and sense of loss of internal control often characterize this phase.

2. *Transitional Phase.* Economic hardship, moving, and radically altered parent/child relationships may occur.

3. *Stabilizing Phase.* The postdivorce family is reestablished as a stable, functioning unit.

In the acute phase, the custodial parent tends to be less competent. Hetherington and his colleagues, (1978), in a longitudinal study of children after divorce, noted that household disorder increases. New patterns of mother-work often result, with young children having to put themselves to bed and to spend a lot of time with sitters as the now single parent tries to create a new social/sexual life. Wallerstein has also observed "greater disorder, poorly enforced discipline, and diminished regularity in enforcing household routines. The root causes underlying the deteriorating household's order (were), in addi-

tion to fatigue and overloading, the mother's fear of rejection by her children" (1983, p. 277). During the first postdivorce year, children were more irritable, edgy, accusatory, and rebellious.

Age was an important variable in delineating the effects of divorce stress. The preschool child is

> more likely to regress behaviorally; more likely to worry about being abandoned by both parents; more likely than the older child to feel responsible for causing the divorce; is likely to become intensely aware of all separations and to be very frightened at routine separations during daytime and especially at bedtime; is likely to be tearful, irritable, and more aggressive; and is likely to suffer an inhibition of play.
>
> Children in the early school-age group are likely to show moderate depression; to be preoccupied with the father's departure from the home; to grieve openly and to long intensely for his return; to fear replacement ("Will my daddy get another mommy, another dog, another little boy?") and to experience the father's departure as a rejection ("If he loved me, he would not have left"). Approximately half of the children in this age group experienced a disruption in their learning at school as well as deteriorated relationships with their peers during the year following the decisive marital separation. (p. 280)

Sex of child. Both of the investigators cited have found that compared with sons, girls in mother-custody homes adapt significantly better after the divorce for the first couple of years, in their overall adjustment and in cooperation with their mother.

Wallerstein notes six tasks children have in coping with divorce. The first two are: *acknowledge the marital disruption* and *regain a sense of direction and freedom to pursue customary activities.* The next four tasks become more salient for adolescents than for younger children: *dealing with loss and feelings of rejection; forgiving the parents; accepting the permanence of the divorce;* and *resolving issues of relationship.*

The first task is very difficult for small children, partly because fantasy and reality are so hard for preschoolers to separate. Fears were

rampant. One half of the children in Wallerstein's study feared that their father would totally abandon them; one third expected the mother would do so; a few feared they would be placed outside the home. Fears and anxieties "evoked by visions of a cataclysmic disaster" (p. 285) decreased over the first postdivorce year particularly as children were assured of continued love and care. These absurd fears are typical when children have experienced loss:

> Kim, a Korean 6-year-old child who had been adopted 2 years before, seemed very worried and sad as his adoptive family excitedly prepared to move into the new house they had recently purchased. When questioned, the child explained, "But I don't know how to cook and I won't have anything to eat." He assumed that when the adoptive family moved to the new house, he would be left behind in the old house and would starve alone, since he could not cook for himself.

Caregivers need to realize the full extent of such child terrors in order to become more sensitive and attuned to the dimensions and intensity of divorce and other disturbances, as they may be acted out in group care. Teachers can be particularly helpful with Task 2. Master teachers can refocus and sustain children's interest in play activities, help children commit energy and persistent interest to appropriate learning tasks. *Empathy* rather than exasperation (with children's mood swings, agitated tensions, and disruptions in concentration) can energize caregivers. Adults need to galvanize their energies to help provide nurturant, stable, secure, curiosity-challenging play opportunities and interactions at a level that encourages stressed children to cope.

Wallerstein notes that the most difficult task for many children to master is "assimilating the grief over the departure of one parent from the home, and coming to terms with the partial or total loss of that parent" (p. 285). Where visitation patterns did not live up to the child's yearnings, an intense sense of suffering loss, of dejection, and of being unlovable or unworthy continued. Low self-esteem then interfered with effective learning and with friendships. Teachers may want to give priority in program planning to building secure self-esteem, and devising cooperative games to enhance peer pleasures for stressed children of divorce.

Some children in this study failed to achieve the special coping tasks for dealing with divorce. Developmental arrest was evident in the continuation of intense sorrow, anger, feelings of rejection, and the never-ending search for the absent parent combined with poor self-esteem. Divorce stressors added heavy burdens to the children's regular developmental tasks. Successful resolution *did* occur for many children 5 to 10 years after the divorce, as revealed in increased children's pride in independence and in their mastery of divorce stress.

Stepfamilies

The Vishers (1983) have noted that "many stepfamilies seem to start with the assumption that the new parent and the stepchildren will come together and love one another at once, as though they had grown up together. This unrealistic expectation can lead to deep disappointment and guilt in the family and stress for the parents and children" (p. 137). Based on their work with stepfamilies, the Vishers counsel that settling basic custody and visitation issues, developing more courteous relationships with ex-spouses, and finding social support networks can ease stressful tensions for children in stepfamilies.

Working mothers: A source of stress for children?

Research on the effects of other-than-mother care for children with working parents has proved inconclusive. Some studies show no negative effects, especially where the mother's attitude toward her work is positive (Yarrow, Scott, de Leeuw, & Heinig, 1978). Others have shown that boys are differentially stressed when their mother works; their school achievement suffers, and their fathers have less positive regard for them (Gold & Andres, 1978).

In a study of toddlers of working and nonworking mothers, the toddlers of nonworking mothers possessed significantly higher IQs (Schachter, 1981), but children of employed mothers were more peer-oriented and self-sufficient.

Vaughn & colleagues (1985) studied 90 mothers with 24-month-old children, who represented these three groups, respectively: one where the mother returned to work/school early and placed the child in group care prior to 12 months; one "later work" group where infants were placed between 12 and 18 months; and one "no-work" group of mothers. All the infants were observed at both 12 and 18 months in the Ainsworth Strange situation. [This is a procedure that consists of 8 three-minute episodes involving the presence and absence of mother and stranger for a baby left in a room with toys. Attachment is coded as secure, avoidant, or ambivalent/anxious depending on how positively the baby accepts comfort upon reunion with mother (after two episodes of her departure) or whether the baby ignores mother or squirms away from her attempts

Empathy for, rather than exasperation with, children's mood swings, agitated tensions, and disruptions in concentration can energize caregivers.

Betty C. Ford

to comfort baby upon rejoining him or her.]

At 2 years of age, the children were asked first to participate in toy play, then cleanup, and finally in solving difficult tool-using tasks without adult help. The results of this study were dramatic and suggest that early return to work was stressful for these young children.

Babies who had been identified as securely attached at 18 months and whose mothers had returned to work early showed more oppositional behaviors when mothers gave directives than did the children previously identified as insecure. They also had "a less optimal experience during the problem-solving tasks than the insecure children" (p. 131). Securely attached children in the late-work group ignored fewer of their mothers' directives and were rated as better able to cope with the stresses of the difficult problem-solving tasks.

Children in the "no-work" group showed the most optimal responses. The no-work group yielded numerous significant differences between children identified as securely versus insecurely attached. In this group, secure children were less likely than insecure children to behave in an oppositional way, to say no, to whine, or to display frustration behaviors in the face of maternal directives. In addition, the secure children were less likely than their insecure counterparts to ask for mother's help. These securely attached children were rated as more persistent, as better able to cope with the stress of the problem-solving procedure, as showing more positive affect (and less negative affect) during the procedures, and as being more enthusiastic in approaching the problems. Further, the secure children were rated as less noncompliant, frustrated, and angry than the insecure children were. Finally, mothers of securely attached children (in the no-work group) were rated as being better able to structure and control the situation for the child. (pp. 131–132)

Thus, in this research, when mothers went to work early (before 12 months), there was a deterioration in the quality of adaptation over the period of 18 to 24 months for toddlers rated as securely attached, so that by 24 months, the formerly secure children were no longer distinguishable from insecure children in their adaptations. It should be noted that all groups of anxiously attached children were rated as somewhat maladaptive in the 24 month tasks. Thus, the *timing* of mother's return to work, if too early, may serve as a stressor to interrupt continuity in positively adaptive organization for infants who have been up to then securely attached. Much more research is needed to give us a more complete understanding of the effects of maternal employment on child development. [YC]

References

Adams-Greenly, M., & Moynihan, R. T. (1983). Helping Children of fatally ill parents. *American Journal of Orthopsychiatry, 53*(2), 219–229.

Als, H. (1985). Patterns of infant behavior: Analogues of later organizational difficulties? In F. H. Duffy & N. Geschwind (Eds.), *Dyslexia.* Boston: Little, Brown.

Arthur, B., & Keeme, M. L. (1964). Bereavement in childhood. *Journal of Child Psychology and Psychiatry, 5,* 37–49.

Beardslee, W. R., & Mack, J. E. (1986, Winter). Youth and children and the nuclear threat. *Newsletter of the Society for Research in Child Development, Inc.* 1–2.

Becker, F. D. (1974). *Design for living: The residents' view of multifamily housing.* Ithaca, NY: Center for Urban Development Research.

Billings, A. G., & Moos, R. H. (1982). Stressful life events and symptoms: A longitudinal model. *Health Psychology, 1,* 99–117.

Blom, G. E., Cheney, B. D., & Snoddy, J. E. (1986). *Stress in childhood: An intervention model for teachers and other professionals.* New York: Teachers College Press.

Booth, A., & Edwards, J. (1976). Crowding and family relations. *American Sociological Review, 41,* 308–321.

Brazelton, T. B. (1979, December). *Assessment techniques for enhancing infant development.* Paper presented at the meeting of the National Center for Clinical & Infant Programs, Washington, DC.

Breslau, N. (1982). Siblings of disabled children: Birth order and spacing effects. *Journal of Abnormal Child Psychology, 10*(1), 85–96.

Bristol, M., & Schopler, E. (1984). A developmental perspective on stress and coping in families of autistic children. In Blacher (Ed.), *Severely handicapped young children and their families: Research in review.* New York: Academic.

Brown, G. W., & Harris, T. (1978). *Social origins of depression: A study of psychiatric disorder in women.* New York: The Free Press.

Clymer, A. (1986, February 2). Poll finds children remain enthusiastic on spaceflight. *The New York Times,* pp. 1, 16.

Conger, R. D., McCarty, J. A., Yang, R. K., Lahey, B. B., & Kropp, J. P. (1984). Perception of child, child-rearing values, and emotional distress as mediating links between environmental stressors and observed maternal behavior. *Child Development, 55,* 2234–2247.

Consortium for Longitudinal Studies. (1983). *As the twig is bent: Lasting effects of preschool programs.* Hillsdale, NJ: Erlbaum.

Dunn, J., & Kendrick, C. (1980). The arrival of a sibling: Changes in the pattern of interaction between mother and first-born child. *Journal of Child Psychology and Psychiatry, 21,* 119–132.

Dunn, J., & Kendrick, C. (1980). The arrival of a sibling: Changes in the pattern of interaction between mother and first-born child. *Journal of Child Psychology and Psychiatry, 22,* 1–18.

Dweck, C. S., Davidson, W., Nelson, S., & Enna, B. (1978). Sex differences in learned helplessness: II. The contingencies of evaluative feedback in the classroom, and III. An experimental analysis. *Developmental Psychology, 14,* 268–276.

Elder, G. H., Van Nguyen, T., & Avshalom, C. (1985). Linking family hardship to children's lives. *Child Development, 56,* 361–375.

Elizur, E., & Kaffman, M. (1983). Factors influencing the severity of childhood bereavement reactions. *American Journal of Orthopsychiatry, 53*(4), 668–676.

Elmer, A. (1978). Effects of early neglect and abuse on latency age children. *Journal of Pediatric Psychology, 3*(1), 14–19.

Felner, R. D., Stolberg, A., & Cowen, E. L. (1975). Crisis events and school mental health referral patterns of young children. *Journal of Consulting and Clinical Psychology, 43,* 305–310.

Ferguson, B. F. (1979). Preparing young children for hospitalization: A comparison of two methods. *Pediatrics, 64,* 656–664.

Figley, C. R., McCubbin, H. I. (Eds.). (1983). *Stress and the family. Vol. 2: Coping with catastrophe.* New York: Brunner/Mazel.

Furman, E. (1974). *A child's parent dies.* New Haven, CT: Yale Univ. Press.

Garland, C., Swanson, J., Stone, N., & Woodruff, G. (1981). *Early intervention for children with special needs and their families.* WESTAR Series Paper #11, Westar States Technical Assistance Resource. Monmouth, OR (ED 207–278).

Garmezy, N. (1981). Children under stress: Perspectives on antecedents and correlates of vulnerability and resistance to psychopathology. In I. A. Rabin, J. Aronoff, A. M. Barclay, & R. A. Zucker (Eds.), *Further explorations in personality.* New York: Wiley.

Garmezy, N., Masten, A. S., & Tellegen, A. (1984). The Study of Stress and Competence in Children: A Building Block for Developmental Psychopathology *Child Development, 55,* 97–111.

George, G., & Main, M. (1979). Social interactions of young abused children: Approach, avoidance and aggression. *Child Development, 50*(2), 306–318.

Gold, D., & Andres, D. (1978). Developmental comparisons between 10-year-old children with employed and non-employed mothers. *Child Development, 50*(2), 306–318.

Grune, A. L., & Brooks, J. (1985, April). *Children's perceptions of stressful life events.* Paper presented at the meeting of the Society for Research in Child Development, Toronto, Canada.

Haan, N. (1982). The assessment of coping, defense, and stress. In L. Goldberger, & S. Brezner (Eds.), *Handbook of stress: Theoretical and clinical aspects.* New York: The Free Press.

Hetherington, E. M., Cox, M., & Cox, R. (1978). The aftermath of divorce. In J. H. Stevens, Jr., & M. Mathews (Eds.), *Mother-child relations.* Washington, DC: NAEYC.

Honig, A. S. (1979). *Parent involvement in early childhood education.* Washington, DC: NAEYC.

Honig, A. S. (1983). Evaluation of infant/toddler intervention programs. In B. Spodek (Ed.), *Studies in educational evaluation* (Vol. 8) (pp. 305–316). London: Pergamon.

Honig, A. S. (Ed.). (1984). Risk factors in infancy [Special issue]. *Early Child Development and Care, 16*(1 & 2).

Honig, A. S. (1985a). High quality infant/toddler care: Issues and dilemmas. *Young Children, 41*(1), 40–46.

Honig, A. S., & Gardner, C. (1985, April). *Overwhelmed mothers of toddlers in immigrant families: Stress factors.* Paper presented at the biennial meeting of the Society for Research in Child Development. Toronto, Canada.

Honig, A. S., Wittmer, D. S., & Gibralter, J. (in press). *Discipline, cooperation, and compliance: An annotated bibliography.* Urbana, IL: ERIC Clearinghouse on Elementary and Early Childhood Education.

Institute of Medicine. (1984). *Report on bereavement.* Washington, DC: National Academy of Sciences.

Lazarus, R., & Folkman, S. (1984). *Stress, appraisal and coping.* New York: Springer.

Lazarus, R. S., DeLongis, A., Folkman, S., & Gruen, R. (1985). Stress and adaptational outcomes: The problem of confounded measures. *American Psychologist, 40*(7), 770–779.

Lazarus, R. S., & Launier, R. (1978). Stress-related transactions between person and environment. In L. A. Pervin & M. Lewis (Eds.), *Perspectives in interactional psychology.* New York: Plenum.

Leiderman, P. H. (1983). Social ecology and childbirth: The newborn nursery as environmental stressor. In N. Garmezy & M. Rutter (Eds.). *Stress, coping and development in children.* New York: McGraw-Hill.

Milgram, N. A. (1982). War related stress in Israeli children and youth. In N. Garmezy & M. Rutter (Eds.), *Stress, coping, and development in children.* New York: McGraw-Hill.

Moore, T. (1975). Stress in normal childhood. In L. Levi (Ed.), *Society, stress and disease: Childhood and adolescence* (Vol. 2). London: Oxford Univ. Press.

Rahe, R. H., Meyer, M., Smith, M., Kjaerg, G., & Holmes, T. H. (1964). Social stress and illness. *Journal of Psychosomatic Research, 8,* 35–44.

Robertson, J., & Roberston, J. (1971). Young children in brief separation: A fresh look. *Psychoanalytic Study of the Child, 26,* 264–515.

Rosen, H. (1985). *Unspoken grief: Coping with childhood sibling loss.* Lexington, MA: Lexington Books.

Rutter, M. (1978). Early sources of security and competence. In J. Bruner & A. Garton (Eds.), *Human growth and development.* New York: Oxford Univ. Press.

Rutter, M. (1979). Protective factors in children's responses to stress and disadvantage. In M. W. Kent & J. A. Rolf (Eds.), *Primary prevention of psychopathology. Vol. III. Social competence in children.* Hanover, NH: Univ. Press of New England.

Rutter, M. (1983). Stress, coping, and development: Some issues and questions. In N. Garmezy & M. Rutter (Eds.), *Stress, coping, and development in children.* New York: McGraw-Hill.

Schaffer, H. R., Callender, W. M. (1969). Psychologic effects of hospitalization in infancy. *Pediatrics, 24,* 528–539.

Seitz, V., Rosenbaum, L. K., & Apfel, N. H. (1985). Effects of family support intervention: A ten year follow-up. *Child Development, 56*(2), 376–391.

Selye, H. (1936). A syndrome produced by diverse nocuous agents. *Nature, 138,* 32.

Selye, H. (1982). History and present status of the stress concept. In L. Goldberger & S. Breznitz (Eds.), *Handbook of stress: Theoretical and clinical aspects.* New York: The Free Press.

Schachter, F. F. (1981). Toddlers with employed mothers. *Child Development, 52,* 958–964.

Sourkes, B. (1977). Facilitating family coping with childhood cancer. *Journal of Pediatric Psychology, 2*(2), 65–67.

Terr, I. C. (1981). Trauma: Aftermath: The young hostages of Chowchilla. *Psychology Today, 15*(4), 29–30.

Turkington, C. (1984, December). Support urged for children in mourning. *APA Monitor,* pp. 16–17.

Vaughn, B., Egeland, B., Sroufe, L. A., & Waters, E. (1979). Individual differences in infant-mother attachment at twelve and eighteen months: Stability and change in families under stress. *Child Development, 50,* 971–975.

Vaughn, B. E., Dean, K. E., & Waters, E. (1985). The impact of out-of-home care on child-mother attachment quality: Another look at some enduring questions. In I. Bretherton & E. Waters (Eds.), Growing points of attachment theory and research. *Monographs of the Society for Research in Child Development, 50*(1–2). Serial No. 209.

Visher, E., & Visher, J. (1983). Stepparenting: Blending families. In H. I. McCubbin & C. R. Figley (Eds.), *Stress and the family (Vol. I)): Coping with normative transitions.* New York: Brunner/Mazel.

Wallerstein, J. S., & Kelly, J. (1980). *Surviving the breakup: How children and parents cope with divorce.* New York: Basic Books.

Wallerstein, J. S. (1983). Children of divorce: Stress and developmental tasks. In N. Garmezy & M. Rutter (Eds.), *Stress, coping and development in children.* New York: McGraw-Hill.

Werner, E. E. (1986). Resilient children. In H. E. Fitzgerald & M. G. Walraven (Eds.), *Annual editions: Human development.* Sluice Dock, CT: The Dushkin Publishing Group.

Wolfer, J. A., & Vistainer, M. A. (1979). Prehospital psychological preparation for tonsillectomy patients: Effects of on children's and parent's adjustment. *Pediatrics, 64,* 646–655.

Yarrow, M. R., Scott, P., deLeeuw, L., & Heinig, C. (1978). Child-rearing families of working and non-working mothers. In H. Bee (Ed.), *Social issues in developmental psychology* (2nd ed.). New York: Harper & Row.

Yudkin, M. (1984). When kids think the unthinkable. *Psychology Today, 18*(4), 25.

Zegens, L. S. (1982). Stress and the development of somatic disorders. In L. Goldberger & S. Breznitz (Eds.), *Handbook of stress: Theoretical and clinical aspects.* New York: The Free Press.

Ziv, A., & Israeli, R. (1973). Effects of bombardment on the manifest anxiety levels of children living in the *kibbutz. Journal of Consulting and Clinical Psychology, 40,* 287–291.

Zuravin, S. (1985). Housing and child maltreatment: Is there a connection? *Children Today, 14*(6), 8–13.

Racists Are Made, Not Born

Kim Brown

Kim Brown is a New York–based free-lance writer who specializes in psychological topics.

What did she do? Prejudice results from people's fears and misconceptions. Its effects are frightening—and destructive.

The year was 1960. Ruby was about to enter first grade. Her parents signed a permission slip for her to attend an all-white school, thinking that she would be one of several hundred black children. They had no idea what kind of ordeal awaited their daughter. What happened to Ruby that year in New Orleans, recounted by Robert Coles in *Children of Crisis,* makes an unforgettable chapter of American history.

Ruby was all alone, the sole black child in a tempest of rage and hate. A mob congregated each morning to greet her, shouting and cursing and holding up signs. They screamed things like "You little nigger, we'll get you and kill you" and "We're going to poison you until you choke to death." For many weeks Ruby had nearly the whole school and its teachers to herself; the whites boycotted.

Ruby endured. Just six years old,

she called upon courage she never knew she possessed. The first white woman to break the boycott by sneaking her children past the mob to a back entrance of the school was also courageous. In an interview with Dr. Coles, she said this about the experience: "After a while they didn't scare me one bit. I wouldn't call it brave; it was becoming *determined.* That's what happened, really. We all of us— my children, my husband and me— became determined."

The mob persisted, and day by day her ideals grew stronger and clearer. Her house was battered, windows broken, her husband's office picketed. Her children had to be protected and escorted by the police, along with Ruby. But she didn't waiver. And her children went to school.

The prejudice and racism Ruby and others in New Orleans had to face in 1960 were, indeed, extreme. We

would like to think that events such as these are confined to the past, but they are not. For the last few years, the opposition to busing (our society's latest attempt to achieve integration) has erupted into violent and abusive behavior in Boston and other American cities. However, even though such outbreaks still occur, they are probably less acceptable to larger numbers of people today than they were a generation ago. Since Brown *v.* Board of Education, the landmark Supreme Court decision in 1954 that made segregation in public schools illegal, we have made strides toward creating a more egalitarian society. But even children who are raised by parents concerned about human rights may harbor biases and prejudices, as do the parents themselves. Prejudice is, in fact, as old as civilization and exists in every society. And it is not restricted to differences in race—all that's needed is the presence of a minority group in a larger community.

What is a "minority group"?

Years ago, social psychologist Kurt Lewin defined the key characteristic of a minority group as an "interdependence of fate." Individuals within the group may have little in common except a label. Yet others perceive them through this label and consequently will treat them according to whatever preconceptions they already have. We call such opinions bias or prejudice.

There are hundreds of categories of minorities that are discriminated against. Outside of race, ethnicity, and religion, groups are stigmatized because of their sex, age, sexual preference, language, occupation, educational level, economic class, physical handicaps, deformities or disfigurements, and mental handicaps. Probably most Americans can be included within at least one minority group—which means that others may prejudge them without grounds.

Why is a light-brown person classified as a black when someone with darker skin may be called white? Clearly, our designations of race are arbitrary and have much more to do with the prevailing ideas of the times (which today are "genetic"), rather than actual color or physical characteristics. The United Nations currently recognizes three predominant races: Caucasoid, Negroid, and Mongoloid; 60 years ago, studies used as the basis of United States immigration policy designated five racial divisions plus numerous subdivisions. These studies were used to prove in a pseudoscientific way that various immigrant groups were fixed races, some less desirable than others. Thus, whether there are five races, three races, or one race depends on how we want to see it rather than on innately objective subdivisions of humanity.

Effects of bias.

Many studies have documented the consequences of bias for minority individuals. For example, Dr. Stephen Richardson, a psychologist at the Albert Einstein College of Medicine in the Bronx, New York, found in his investigations of children's values with respect to disabilities that children with disabilities had the same preferences and biases as the nonhandicapped. "The children with disabilities had assimilated the values of the majority culture," he explains. "This was a sad finding, for it means that they tend to denigrate themselves."

Perhaps people wouldn't hold on to their biases if they didn't benefit from them in some way. Sociologist Jack Levin observes that for the group in power, prejudice allows individuals to blame the things that go wrong on other people. Minority groups in America and elsewhere have traditionally been scapegoat targets, held accountable for diverse social and economic ills. Some psychologists argue that prejudice helps those with a poor self-image to restore their feelings of worthiness by comparing themselves favorably with "inferiors." By dividing the world into teams of good guys and bad guys, majority-group members can protect themselves from facing disturbing and difficult choices.

Kids and categories.

At what age do children learn about belonging to a minority group? If they have a visible distinguishing characteristic, such as skin color or a handicap, they will probably be aware of it by the time they enter nursery school. Depending on the importance of other defining characteristics in their lives, by age three or four they may also know about several groups with whom their families identify. But even before then, children begin to classify others—it's simply part of their process of orientation in the world. To categorize and generalize is a natural tendency of all human beings. Unfortunately, it can easily lead to bias and unwarranted prejudgments.

A toddler delights in naming things. "Mommy, what's this?" he asks over and over again. Each and every parcel of the data must be categorized. The toddler relates every new thing to something he has already experienced, although sometimes this connection is made unconsciously. Thus, as they are naming, children are simultaneously building mental categories of like objects and events.

"The human mind must think with the aid of categories," writes Gordon Allport in *The Nature of Prejudice*. "Once formed, categories are the basis for normal prejudgment. We cannot possibly avoid this process." Allport points out that these categories are often associated with emotional responses. If a young girl has a dog and enjoys playing with him, then she is likely to enjoy her friend's dog as well. But if she has been frightened by dogs and has never had a chance to get accustomed to one, she might avoid opportunities to do so in the future, thinking of all dogs as scary.

Though some categories or generalizations may have an element of truth or be based on firsthand experience, others may be completely irrational and built from hearsay. These illogical and false categories can also be powerfuly tinged with emotion—sometimes even more so, since what we imagine can be more threatening and terrifying than the real thing.

Negative influence of groups.

When individuals have unfair biases or fears based on false categories, groups function in a way that can help perpetuate them. This is because members of groups may engage in "group-think," a term coined by Yale University psychologist Irving Janis. This means that they are likely to gear their behavior toward gaining the group's approval, often losing sight of good judgment. Have you ever heard of a *lone* Ku Klux Klansman causing a public disturbance? Members of groups can feel less personally responsible for the group's actions when the group acts maliciously. This has been proved in the laboratory: when part of a group, people will apply electrical shock to a subject more frequently than when alone.

Dr. Richardson observes that biased groups often try to rationalize their immoral behavior. For example, at one time the mentally ill were thought to be inhabited by devils. Treatment often took the form of cruel and prolonged punishment—justified by the "rightness" of the religious beliefs of the wider society. Another example is that many of the white slave owners in the seventeenth, eighteenth, and nineteenth centuries did not consider the Africans to be fully human and believed that the slaves could not look after themselves and needed owners.

Children may behave in a group in a manner they might later be ashamed of. It often happens that a group of youngsters will pick on an unpopular child. Afterward, individuals may regret their behavior. It's important for parents to acknowledge the pressure that children feel from their peers. Yet we must also emphasize how miserable it feels to be the one who is excluded or humiliated.

"Just an exception."

In most instances, even when confronted with experiences that contradict their preconceptions, people resist changing their ideas about certain categories. They are more prone to think: "This evidence conflicts with my ideas—but it's probably just an exception." Take someone who believes that most Italian-American business owners are connected with organized crime. If he meets a kind, honest businessman from an Italian family who obviously has no such ties, he will be inclined to think of this person as "odd."

Why is this so? Most people find comfort in the familiar. Our own favorite chair, a worn quilt, the stew we've been making the same way for years, and our preconceived ideas about other people—all of these things make us feel secure. We seek to spend time with those who have things in common with us, who are more familiar than strange in their lifestyle and values. Sociologist Robert Jackson observes that "all people have an investment in finding themselves justified and validated. We all seek people who reinforce our conceptions of ourselves and the world."

Violation of expectation.

Almost as natural as our love of the familiar is our fear of the different. In her book *Oneness & Separateness: From Infant to Individual,* Louise Kaplan notes that all babies are wary of strangers. "As soon as a baby has a special attachment to a special mother, the sight of a strange face will sober him," she notes. "By the time he's seven or eight months old, the baby's sobriety is heightened with at least a tinge of worry."

When a group of Harvard researchers showed four-month-old babies models of human faces with scrambled features, the infants cried. The images upset the babies because they were so unlike the faces the babies had grown used to expecting. Dr. Richardson calls this a "violation of expectation."

"From an early age we begin to incorporate a schema of what we expect to see," Dr. Richardson explains. "As we get older, this schema includes a person's height, his weight for his height, his facial appearance, his skin color, the way he moves, dresses, speaks, his gestures." When a person's appearance is outside our nor-

mal range of expectation, we react with fear and anxiety.

What is it about such a shock that frightens an infant? Is he afraid of being hurt? Not exactly. In fact, the surprise reminds him of one of life's earliest lessons—that he is an individual in an uncertain world. As Louise Kaplan notes, "The stranger reminds him of terrible things: of his vulnerability, of being done to, of his separateness from mother, and most of all, of unpredictability." A strange face transports the baby from the known to the unknown, and raises the fear of losing sight of—and therefore of losing—the mother. Yet the stranger is also compelling to the baby; many an infant will take stock of the new person before returning to the familiar parent's lap. Indeed, the stranger represents all kinds of possibilities, alternatives, and new feelings.

Whether or not a baby becomes accustomed to a new person depends on his parent's show of trust. For instance, a mother has a friend whose presence at first may send her baby into a fit of howls. But if the baby hears his mother's relaxed, normal tone of voice, feels the steady beat of her heart, listens to her chatting in a friendly and easy way with the new person, he will be reassured. And if he sees this friend fairly often, he will adjust to being with her. If, on the other hand, the mother reacts anxiously to this other person, her child will probably mirror these feelings.

Expressions of bias.

Thus, bias is taught when the child's fears or anxieties are reinforced, even unconsciously, by adults. It is the raised eyebrow, the clenched jaw, the strained tone of voice, the quickened heartbeat, and the abruptly grasped hand that teach prejudice. It can also be the false overtures of friendliness or overemphatic gestures that tell the child that this person is different, the relationship is unnatural, and he or she must be treated in a special condescending way.

To avoid this kind of unconscious teaching, parents must first find out what causes their own expectations to be violated. "Most people will experience violation of expectation, though what causes this violation will differ depending upon the person's experience," Dr. Richardson observes. Do you find yourself feeling nervous or tense in the presence of people with disabilities? Are you in the habit of

treating people of other races with special attention? Such questions will help you identify your own boundaries of "normalcy."

Once you can see beyond the characteristic that shocked or disturbed you, it becomes obvious that the individual in question has many unique attributes, and the exceptional trait is no longer the only crucial factor. For instance, a person with cerebral palsy, in addition to the problem of controlling his muscles, may have an outstanding sense of humor. Upon getting to know this man better, you may find that his humor has become the quality that you remember most about him. Fred Davis has called this "breaking through." Breaking through has consequences for both parties—it is the moment that an honest relationship is possible.

Encouraging openness.

The broader the range of experiences that we and our children have, the more accustomed we will be to expect variety rather than similarity from our fellow man. This is not easy to do, given the vast number of ways that our society fosters segregation. Few of our neighborhoods and schools have a real cultural, age, or racial mix. Movies, plays, books, advertisements, television—all of these just reinforce these divisions. Have you ever seen a person with handicaps on television drinking soda or changing a diaper? "Parents must make the effort to integrate their lifestyles," says child psychiatrist Dr. Phyllis Harrison-Ross. "It's the most important thing they can do."

From time to time children are bound to make inappropriate comments about other people. Rather than act alarmed or angry, use the incident as an opportunity to talk about individual differences. Yes, there are differences—your child may have been observant enough to notice them. But different doesn't mean better or worse, and differences may be minor compared to the overwhelming similarities. Children are usually curious about these matters and welcome the chance to explore them.

It takes courage to be the kind of family that routinely faces the unknown, fear, vulnerability, hostility, and the reality of human interdependence. But the prize won for eradicating prejudice is a less hostile, more peaceful world to live in.

AGGRESSION

THE VIOLENCE WITHIN

MAYA PINES

Maya Pines has been writing on the brain since research in the field really took off in the 1970s. Her book The Brain Changers: Scientists and the New Mind Control, *which described research on the two hemispheres and on mind-altering drugs, won the National Media Award of the American Psychological Association in 1974. She has also written a book on retarded children and one on the enormous potential of the very young child.*

When his supervisor made a sarcastic comment about the latest production delay, the young engineer said nothing. But he could feel his blood boil. On the way home he had a few drinks to calm down.

That didn't help much. He was not bearing the blame well. At home, he took some Valium. Then he got into an argument with his wife. Suddenly he exploded and punched her. He then smashed a chair, stormed out of the house and got into his car, taking off at top speed. Going out of control, he crashed into another car and wrecked it. The collision broke the other driver's neck.

As the incident illustrates, aggression is not confined to the New York City subways (see box, next page). It lurks in suburban streets and sometimes invades our homes. The so-called civilized world is riddled with violence, both sanctioned and unsanctioned. At a time when more and more people have access to atomic weapons, a single person's aggressive impulses—or perhaps a nation's—could be the most dangerous force on Earth.

Today, we are all at risk of becoming the targets or accidental victims of some kind of violence. Cars driven by aggressive people can be the instruments of suicide or murder. Stabbings, shootings and rapes are now so commonplace that most newspapers don't bother to report them. Law-enforcement agencies reported a total of 1.2 million violent crimes in 1983; probably at least as many other episodes went unrecorded. More than 50,000 Americans are murdered or commit suicide annually.

In the hope of finding ways to prevent or reduce aggression, psychiatrists, brain researchers and behavioral scientists are now making a determined effort to understand its causes. They are not concerned with the kind of drive that fuels ambition and makes people stand up for their rights, nor with aggressive thoughts, but with physical aggression—outright attacks that result in injuries or death. Among some recent findings:

■ Harsh punishment produces aggressive behavior in children. So does the example of violence, at home or on television.

■ Extremely impulsive and aggressive people of both sexes have unusually low levels of a brain chemical that inhibits the firing of nerve cells.

■ Men who are highly aggressive have higher levels of the male hormone testosterone.

■ Treatment with lithium reduces aggressive behavior in highly impulsive and violent people.

While much of the research on the biology of aggression has been carried out on animals, it is known that human beings can resist their biological drives more efficiently than other creatures, because their neocortex—the thinking part of the brain—is more highly developed.

"By the time the human brain matures, the normal individual is controlled very largely by social norms," says Estelle Ramey, a professor of physiology and biophysics at the Georgetown University School of Medicine. "Men don't urinate on the living-room floor, even when they're in agony; they wait until they get to the bathroom. Although urination is a normal, instinctive biological drive, it's very quickly brought under control. Similarly, young men learn quickly who it's safe to be aggressive toward. A man may be a meek and mild Caspar Milquetoast in the office, and kowtow to his boss, and then go home and beat up his wife. Is he an aggressive male?" She answers her own question: "He's aggressive when it's safe for him to be aggressive."

Much depends, therefore, on the level of aggression that is acceptable in a particular society at a particular time. Cultural change is possible, argued John Paul Scott of

EXTREME AGGRESSION TENDS TO RUN IN FAMILIES—A FINDING THAT HAS SET OFF A DEBATE OVER THE CAUSE.

Bowling Green State University at a recent session on evolutionary theory and warfare. Warlike habits can give way to very peace-loving ones. Although Scandinavian culture spawned the Vikings, for instance, it is now represented by people who are "among the most pacific in the world."

Much depends on one's early training. Extreme aggression tends to run in families—a finding that has set off a furious debate over the cause. In the nature-versus-nurture controversy, strong proponents cite genetic causes while others claim children learn from their parents' example.

Last winter, a team of researchers reported on an extraordinary study of aggressive behavior that was made by tracking three generations. It revealed how much children are influenced by their parents' aggression. Led by L. Rowell Huesmann and Leonard Eron of the University of Illinois at Chicago, the team began in 1960 by testing 870

third-graders, who were asked to rate one another on such questions as "who pushes and shoves children?" Their parents were interviewed as well. The researchers then followed up more than 600 of these children and their parents for 22 years—by which time many of the original children had children of their own, who then also became subjects of the study.

The researchers found that the children who most frequently pushed, shoved, started fights and were considered more aggressive by their classmates at age 8 turned into the more aggressive adults. These men were very likely to have criminal records by the age of 30. If their behavior did not land them in jail, they were apt to get into fights, smash things when angry, drive while drunk and abuse their wives. Many of their own children already showed signs of the same type of aggressive behavior.

At the beginning of their study, the team had learned that the more aggressive children had parents who pun-

VIGILANTISM

When Bernhard Goetz drew his gun and shot four youths who, he claims, were harassing him on a New York City subway, he captured the nation's imagination. Not since the days of the Wild West and lynch mobs in the South had people been so preoccupied with a so-called vigilante: A *Washington Post*-ABC News poll found that nearly half of those interviewed supported Goetz. Was his personal crime-fighting crusade an isolated incident, or did it presage a national trend?

"There's a lot of rage and frustration in our society about the ineffectiveness of law enforcement," says Ralph Slovenko, a professor of law and psychiatry at Wayne State University, in Detroit. "Crime is a very safe profession. Only two percent of the people who commit serious crimes are actually sentenced."

A prior mugging victim who had seen his assailants get off easy, Goetz was hardly a vigilante in the traditional sense: Rather than pursuing a specific target in order to right perceived wrongs, he was ready to lash out at the next person who went after him.

"I believe strongly that Mr. Goetz was a man whose emotional state deteriorated in the time after his previous victimization," says psychologist Morton Bard, of the Center for Social Research at the City University of New York. "He was acting out a revenge fantasy that virtually all crime victims have."

What is particularly disturbing is that Bernhard Goetz may be a harbinger of things to come. Everyone agrees that conventional law enforcement is becoming less and less effective; the court system is slow, trial costs are high, and overcrowded prisons mean an increasing reliance on what's known as turnstile justice.

As yet, vigilantism is a predominantly urban phenomenon, fueled by the crowded, alienating environment. The self-styled Army of God is systematically bombing abortion clin-

ics nationwide. A Massachusetts man wounded a teenager as revenge for hitting his car. Abroad, crime-fighting citizens patrol Amsterdam's streets at night, and squads in Dublin combat a growing drug problem. But rural areas can be affected, too: Two men in backwoods Arkansas recently castrated a man charged with rape.

"The American public is fed up," notes James Turner, a clinical psychologist at the University of Tennessee Center for the Health Sciences, in Memphis. "The system no longer protects them. It protects the guilty and punishes the innocent—the victims." He is now writing a book, called *Victims to Vigilantes*, that will document this progression in public opinion.

Researchers have only just begun to consider the effects of crime on the victim. Bard, who recently chaired an American Psychological Association task force on the victims of crime and violence, believes these studies are long overdue. "Victims should have services available to help them deal with the emotional and physical trauma of crime," he says. "Until now, there has been little sensitivity to victims and their 'invisible wounds.'"

Sociologist Emilio Viano, at American University's School of Justice, thinks better treatment of victims will forestall future Bernhard Goetzes. "Victims have been ignored and abused, used only as sources of information. The failure of the court system has weakened the fabric of society. People have nowhere to turn for help. If they want anything done, they feel they must do it themselves."

James Turner is investigating various nonlethal means of self-defense. "We must teach people how to deal with violence, or they will become victims," he says. "Basic police classes are moderately effective; karate needs to be practiced in order to be useful. Mace can make an attacker even angrier, and it's illegal in many major cities." Turner is now reviewing a "stun gun" that uses electronic pulses to temporarily override an attacker's neuromuscular system, causing him to collapse in a daze. ∎

–Andrea Dorfman

CHILDREN ARE COPYCATS, AND WHEN THEY SEE REPEATED VIOLENCE ON TELEVISION, THEY TEND TO IMITATE IT.

ished them far more severely than the less aggressive children. Now the pattern was repeating itself as these aggressive adults severely punished their own children. The most likely explanation for this repetition, the researchers concluded, was that children learn to be aggressive by copying what their parents do to them and to others.

In their report, they said it was most impressive that "the children who are nominated as more aggressive by their third-grade classmates on the average commit more serious crimes as adults." They also found that the degree of aggressiveness in the third-graders was, 22 years later, even more strongly related to the aggressiveness of their children than it was to their own aggressiveness as adults. In short, the study concluded that the aggressive child is father to the aggressive child.

Once aggression is established as a child's "characteristic way of solving social problems," it becomes a relatively stable and self-perpetuating behavior, the researchers emphasized. And by the time this behavior comes to the attention of society, "it is not readily amenable to change."

The idea, then, is to prevent such behavior from becoming fixed before adolescence. One approach is to limit the amount of violence to which children are exposed. Children are copycats, and when they see repeated violence on television, they tend to imitate it. The evidence for this is so strong that the American Academy of Pediatrics is now warning parents about the effects of TV violence and urging them to limit—as well as monitor—what their children watch.

Another approach is to put clear limits on children's aggressive behavior at an early age. Psychologist Gerald Patterson, of the Oregon Social Learning Center in Eugene, and others have found that one of the most effective means of stopping excessive aggression is to isolate children in a room for about five minutes before the child's behavior becomes extreme. Patterson calls this a time-out procedure—a nonpunitive way of saying "that's not acceptable behavior."

If the pattern of violence is not broken before adulthood, little can be done by psychological means. Highly aggressive adults generally resist psychotherapy, according to Gerald Brown, a psychiatrist at the National Institute of Mental Health (NIMH), although behavior therapy is sometimes helpful. Anybody who thinks he or she can reform one of these volatile and violent men through love "is in for a lot of trouble," Brown says. "That idea has brought a lot of grief into marriages. These people can idealize you one minute and attack you the next."

Generally these highly aggressive people are men. Men, in fact, commit about 90 percent of all violent crimes in the United States. This is in part the result of social conditioning. As children grow up, they learn that "it's socially unacceptable for women to become aggressive, but if a man throws his weight around, that's *manly!*" points out Estelle Ramey. But there are also biological reasons for the higher level of aggression in males, as indicated by studies involving certain brain chemicals and male sex hormones.

One chemical that seems to play a key role in preventing or releasing aggression is serotonin, which carries inhibitory messages from cell to cell in the brain. Serotonin is difficult to measure directly, but a substance called 5-HIAA, which is a breakdown product of serotonin, can now be measured in spinal fluid. Several experiments have shown that highly aggressive people have lower levels. This holds true for both men and women, but on the average, men have less 5-HIAA.

Because research involving 5-HIAA requires sometimes painful spinal taps, few studies have been conducted. At the National Naval Medical Center in Bethesda, Maryland, 26 marines and sailors, aged 17 to 32, who had come to the attention of psychiatrists because of their histories of repeated assault, agreed to undergo taps.

"They were the kind of people who'd had temper tantrums as small children and lots of fights in grade school and who would go into bars and tear up the place," says Brown, the NIMH psychiatrist who directed the spinal-fluid study. "They had a very short fuse—they'd be provoked by things others would not find provoking."

Predictable Aggression

Although aggression is highly valued in the military, Brown points out that "it must be controlled and predictable aggression. If people are too unpredictable and keep getting into trouble, they're not suitable for the service." In fact, the young men under study were being examined by a board of officers who were to decide whether they should be discharged from the Navy.

The laboratory that analyzed their spinal fluid found that 14 of the young men had low levels of 5-HIAA, while the rest had nearly double the amount. The first group included the men who had the worst records for impulsive acts of aggression, and 12 were discharged.

"We could have looked at their spinal fluid and predicted with eighty-five percent accuracy which people would be removed," declares Brown.

The results of this study complement earlier research in which scientists lowered the level of serotonin in animals' brains through chemicals or brain surgery and saw a dramatic increase in the animals' aggressive behavior—at least in the kind of aggression characterized by explosive attacks.

Brain researchers have known for decades that there are at least two unrelated kinds of aggression. In the 1920s, the Swiss physiologist Walter Hess, who later won a Nobel prize, described the cat's characteristic "bad-tempered aggression": With dilated pupils and bristling hair, the cat hisses, spits and growls. By contrast, "predatory aggression" is more cold-blooded and deliberate, as when the cat stalks a mouse, kills it quietly and eats it. Only the first variety seems to involve low levels of serotonin.

MOST OF THE OUTRIGHT VIOLENCE IN THE UNITED STATES COMES FROM PEOPLE WHO ARE CHRONICALLY AGGRESSIVE.

Testosterone has long been associated with aggression. A time-honored method of making male animals less aggressive is to castrate them—a procedure that eliminates the source of testosterone. When male mice are castrated at birth, for instance, they do not begin to fight each other at one or two months of age, as normal mice do upon reaching sexual maturity, but giving these castrated mice injections of testosterone makes them fight as if they had not been altered.

"Testosterone increases the biological intensity of stress, so that more adrenaline is released," explains Ramey. This produces a state of anxiety and irritability but also damages the lining of blood vessels and may lead to heart disease. That's why male animals generally die earlier than females, she says—unless they have been castrated, in which case they tend to live as long as females.

In one study, blood from hockey players rated by coaches and teammates as particularly aggressive was found to contain relatively high levels of testosterone. In another study, involving prisoners jailed for violent crimes, the men who were most aggressive toward other prisoners had twice as much testosterone in their blood as those who were not.

While the effects of such biological differences should not be underestimated, researchers emphasize that biology can only set the stage for aggressive acts or make such acts more likely. Biological forces can produce rage or the urge to attack, but they cannot dictate whether an attack will actually take place.

Even dogs react differently when annoyed by their masters, whom they seldom bite, and when annoyed by strangers, whom they will attack with much less provocation. They are particularly lenient toward young children.

In certain circumstances, such as self-defense, any of us may become violent. But when previously unaggressive people suddenly become violent for no apparent reason, they may have had too much alcohol or taken drugs such as PCP (angel dust), both of which often trigger aggression. Psychologist Claude Steele, of the University of Washington, recently found that alcohol accelerates aggressive behavior to an extreme level because, when in a state of conflict, a person using alcohol tends to lessen inhibitions by blocking thoughts of negative consequences.

Most of the violence in the United States comes from people who are chronically aggressive. Some of them simply earn their living from crime. Others have such a short fuse that aggression has become a way of life. For these impulsive people, a new kind of treatment now appears possible in some cases: the use of lithium, a drug that is generally prescribed to treat manic-depressives.

Lithium seems to affect the activity of several brain chemicals, including serotonin, preventing both highs and lows. Scientists don't yet understand exactly how it works, but it has been shown to reduce aggression in rats, mice and fish. On this basis, several psychiatrists have tried it on prison inmates with histories of repeated impulsive assaults. Joe Tupin, a professor of psychiatry at the University of California, Davis, gave lithium to 27 particularly aggressive male prisoners in a maximum-security prison in California. "More than two-thirds of them responded to the lithium," he reports. "It removed the explosive quality of their violence."

The men who responded best were those who would become extremely angry after trivial provocations. Their violence was not only inappropriate but very rapid, Tupin explains. "Between the provocation and the violence, these people didn't think—as if they didn't have the capacity to," he says. "They didn't stop for half a second of internal review during which they could think 'Gee, it was just an accident'; they didn't look at the possible consequences of their acts; they went totally out of control, with no in-between stages."

During the nine months that they were treated with lithium, these men became more reflective and had fewer violent episodes, Tupin says. The drug did not stop them from attacking their fellow prisoners deliberately from time to time, but it did prevent many hair-trigger explosions.

A number of psychiatrists are now giving lithium to patients of this impulsive type. According to Tupin, lithium is "a good choice" for this kind of patient; however, it would be ineffective with people who commit calculated, predatory violence or with psychotics, who are aggressive because of their delusions.

Controlling Behavior

Ideally, psychiatrists agree, highly aggressive behavior should be dealt with in childhood, by psychological means and by improving the economic and social conditions in which children grow up. But inevitably—either because of their brain chemistry or the way they were raised—a small percentage of people will go on being extremely aggressive and dangerous throughout their adult lives. If further research shows that lithium or other drugs are truly effective in such cases, does society have the right to prescribe them to control these people's behavior?

"I certainly wouldn't try to answer that question," says Gerald Brown. "But it is something the public will have to start thinking about, just as it is now thinking about the wisdom of putting mechanical hearts into people." Sooner or later, he says, society will have to decide what is the most humane and rational way to deal with people who keep on hurting others through uncontrolled aggression.

DEPRESSION
AT AN
EARLY AGE

It strikes in childhood,
and it's on the rise.

JOSEPH ALPER

Joseph Alper is a contributing editor to Science 86.

JANET WAS A BRIGHT GIRL, well liked by her teachers and friends. When she entered junior high and started having trouble paying attention in class, no one made much of it. After all, kids can get pretty restless at that age. In high school she did rather poorly, but since she was quiet and didn't make trouble, her teachers left her alone and assumed she just wasn't much interested in school. Her parents were a bit worried, but there didn't seem to be anything they could do. And the school counselor reassured them that Janet's attitude was just a normal part of adolescence, that she would eventually outgrow her problems.

Instead they got worse. The summer before her senior year, Janet started having trouble getting up in the morning, and when she did make it out of bed, she couldn't get motivated to do anything; it was as if she were stuck in low gear. She lost her appetite, slept poorly, and couldn't stop crying. Her parents decided this was no longer just part of growing up and took her to their family physician. He referred them to the affective disor-

ders clinic of their local hospital, where a psychiatrist who had treated dozens of kids like Janet talked to her and her parents. He told them that Janet's lack of energy, motivation, and appetite were signs of depression and prescribed amitriptyline, an antidepressant.

Within a month, Janet was feeling better than she had for as long as she could remember. She started making new friends, and her schoolwork improved dramatically. The following year she made the dean's list at college. And today, several years later, she continues taking her medication, does well in school, and has a good outlook on life.

"Until maybe 10 years ago, we believed that severe depression was solely an illness of adults," says psychiatrist Frederick K. Goodwin, scientific director of the National Institute of Mental Health in Bethesda, Maryland. "Adolescents didn't develop 'real' depression—they just had 'adolescent adjustment problems,' so most psychiatrists didn't and still don't think to look for it in kids. Now, however, we know that idea is dead wrong. Adolescents, even children, suffer from major depression as much as adults do."

Research is only just beginning, so estimates vary of how many young people suffer from major affective disorders, which include depression and manic-depression, an illness characterized by elation, hyperactivity, or irritability, alternating with depression. Depending on the age of the youngster and how his illness is defined, the estimates can range from one to six percent. There is no question, however, that the problem is serious and that it is growing. One study indicates that the percentage of older teenagers with major affective disorders has increased more than fivefold over the past 40 years.

"The chilling fact is that we may be on the verge of an epidemiclike increase of mania, depression, and suicide," says Elliot S. Gershon, chief of the clinical psychogenetics branch at NIMH. "The trend is rising almost exponentially and shows no signs of letting up. I would go so far as to say this is going to be *the* public health problem of the 1990s and beyond if the trend continues."

Buttressing Gershon's concern are studies revealing that many adults with affective disorders showed the first signs of their illness when they were teenagers or children and that the earlier the onset, the more severe the disease. "So if we are seeing more depression in kids today," he says, "we could be in for real trouble when these kids hit their 30s," the prime time for showing the classic swings of depression.

Not everyone agrees with Gershon's gloomy outlook. Some claim the dramatic rise is due to better reporting and diagnosis, not to rising incidence of disease. But they don't disagree with the trend.

"Depression is a crucial problem," says G. Robert DeLong, a pediatric neurologist at Massachusetts General Hospital, "because this illness clouds a child's or adolescent's perceptions at such a critical time in his social and psychological development. As a result, young people often develop long-lasting problems aside from the original depression or mania. So the earlier we diagnose these kids and treat them, the better chance they have of developing into adults who can enjoy a more normal life."

These findings have important clinical implications. Some psychiatrists believe that if young people can control their affective disorders with drugs during the particularly stressful adolescent years, their biochemistry may stabilize so they need not continue taking drugs for a lifetime as many adults now must.

Furthermore, there is mounting evidence that a constellation of harmful behaviors that accompany depression—suicide attempts, drug abuse, anorexia, bulimia, and juvenile delinquency—may be methods that young people use to try to cope with the anguish they feel. So if depression can be curbed, many of these disorders might disappear as well.

"We're talking about a whole spectrum of problems to which affective disorders seem to be linked," says Joseph T. Coyle, head of child psychiatry at Johns Hopkins. "This is not to say that every kid with these problems has a major affective disorder. But the odds are good he has."

Ben, for example, was eight when his parents brought him to Robert DeLong in 1974. Ben was smart, with an IQ of 128, a nice kid who had become aggressive and nasty. His teachers complained that the boy was extremely disruptive, and neighbors kept calling his folks to report that he had broken a window or beaten up their kid.

DeLong, one of the few nonpsychiatrist physicians to take an interest in childhood mental illness, talked extensively with Ben, who reported feeling

When depression starts in childhood or adolescence, it is likely to have a genetic link.

very sad and confused at times. "He was a classic manic-depressive—wild, aggressive, manic behavior followed by profound sadness," DeLong says. He put the boy on lithium, which is often used to treat manic-depression in adults but rarely in children. "Everyone who knew the child was stunned by his change for the better." Today, thanks to early diagnosis, lithium, and supportive psychotherapy, Ben is a well-adjusted college student who no longer needs lithium.

Ben was fortunate that he was sent to a physician who hadn't swallowed the orthodox view that children can't be manic-depressive. Grace, who is now in her mid-20s, was not as lucky. When she was about 13, she began feeling a bit down in the dumps. Then, at 15, though a promising musician, she started doing poorly in her musical studies, and she began vomiting after meals. It improved her mood, she said.

When her parents started finding jars of vomit hidden around the house, they took her to a psychiatrist, who made a diagnosis of borderline personality disorder, a serious illness involving self-destructive, impulsive behavior that usually does not respond to medication and has a poor prognosis. Grace had a lot of trouble over the next nine months, but then she spontaneously got better. That lasted until she was 18, when she again became depressed. This time she started cutting her wrists, explaining that it made her feel better. She was hospitalized. Again she recovered nine months later. She went to a prestigious college on a music scholarship and did well for another two years but once again became depressed and wound up at the Johns Hopkins Hospital. This time, Raymond DePaulo Jr., director of the affective disorders clinic, diagnosed Grace as manic-depressive and treated her with an antidepressant and lithium. Within a month she was feeling good.

But Grace can't cope with the idea of being seriously ill, so she has stopped taking her medicine, believing that she can control her moods with vomiting and wrist slitting. The years of turmoil have torn her family apart and seriously damaged her musical career. "This is a good example of how early diagnosis and

treatment of an affective disorder would have probably alleviated many of the psychosocial problems this girl and her family now have," DePaulo says. "The odds are she'd be in much better shape today if she'd been diagnosed correctly at 15."

But diagnosing affective disorders is not easy, especially since young people show a more diverse set of symptoms than do adults. And it can be hard to tell which kids are showing typical signs of rebellion and which are depressed. Before diagnosis can become routine, psychiatrists may need a fundamental change in their understanding of mental illness.

Such a change took place in the 1950s, when drugs were discovered that alleviated many of the symptoms of severe psychiatric illnesses, raising the possibility that affective disorders and other mental diseases could be understood in terms of biochemistry and genetics. As researchers learned more about biochemical abnormalities, they hoped to develop better tests to detect psychiatric problems and differentiate among them.

But there still is no perfectly reliable diagnostic test for any psychiatric illness, and just as they've always done, doctors must rely heavily on subjective criteria. "It's a probability game," says Ray DePaulo. "We talk to the patient and to their relatives."

"To gauge depression," says Coyle, longtime friend and collaborator of DePaulo, "we determine if people are sleeping poorly or if they've had a big weight change recently. We try to find out if they have a poor self-image, if they believe they are deficient and feel guilty about their shortcomings, if they feel empty and sad."

To judge manic-depression, DePaulo says, "we look to see if there have been any sudden mood changes. If so, do they take place for a while and then disappear, and then reappear later?"

"A big problem," Coyle says, "is getting this information from kids, because, unlike adults, they have a hard time connecting their unusual behavior to the way they feel. And even when we do get the kids to talk about their feelings, it's rare to find one who satisfies all these criteria."

One important factor in making a diagnosis is a family history of affective disorders. The evidence is compelling that when depression starts in childhood or adolescence, there is likely to be a genetic link.

Mary, for example, was diagnosed last

Between 1960 and 1980, the suicide rate among 15- to 19-year-olds increased by 136 percent.

year as a manic-depressive. She was 26 years old. As early as age eight, she had showed episodic symptoms—times when she just wasn't interested in her classes alternating with periods when she wanted to answer every question the teacher asked. Alarmed, her parents took her to a psychiatrist who blamed Mary's troubles on a too-close relationship with her mother.

In a way, he was right. Last year, Mary's mother, who was being treated for manic-depression, realized that Mary's problems were probably related to her own, and she and her husband brought Mary to Johns Hopkins. Looking into the family history, DePaulo learned that Mary's sister had a mood disorder and that Mary's father had been suffering from serious depression for years. The father's brother and an uncle had symptoms that sounded manic-depressive, and two of his relatives had committed suicide. "Mary's symptoms were very mild, and they could be overlooked," DePaulo says. "But that family history was too strong to overlook." Mary is now taking medication, and her prognosis is good.

Family histories are playing an increasingly important role in diagnosing affective disorders in young people. Studies have shown that a youngster with one parent who suffers from manic-depression stands at least a 25 percent chance of developing an affective disorder as an adult. If both parents are ill, the odds rise to between 50 and 75 percent. But genes merely contribute to the vulnerability to depression and manic-depression, and some psychological, social, or biological factor is needed to bring the illness to the fore.

"I think the rise in depression, suicide, and eating disorders is mostly environmental in nature," says Fred Goodwin. "Sure, genetics provides the vulnerability, but over the past 20 years there has been an erosion of sources of an external sense of esteem—religious identity, family cohesiveness, patriotism, etc. The many traumas that have rocked our society since the 1960s have forced people to turn inward, analogous to the grief process, so if there is a genetic vulnerability, it could touch off the problems of affec-

tive disorder. That is why it is so important to find some predictive genetic marker."

At Yale University School of Medicine, epidemiologist Myrna Weissman and geneticist Ken Kidd are looking for such markers by rounding up large families with many members who have affective disorders. Their goal is to use DNA samples from family members to identify genetic markers for affective disorders, in much the same way that researchers two years ago isolated a genetic marker for Huntington's disease. "It will be more difficult than the Huntington's marker because the affective disorders probably involve more than one gene," says Weissman. "We're a long way from the success of the Huntington's marker, but at least it's a beginning."

Such a diagnostic test would, among other things, save lives. "Suicide is the third leading killer in adolescents, and one of the most important risk factors for suicide in this age group is untreated affective disorder," says psychiatrist Susan Blumenthal, chief of behavioral medicine at NIMH. She estimates that at least a third of the adolescents who commit suicide have untreated or undiagnosed affective disorder.

Like depression, suicide seems to be on the rise among adolescents. Between 1960 and 1980, the suicide rate among 15- to 19-year-olds increased by 136 percent. Among boys, who commit three-fourths of the suicides, the suicide rate soared 154 percent. Suicide attempts, three-fourths of which are made by girls, are estimated to be 100 to 150 times more common than actual suicides. And although most psychiatrists now believe that the majority of people who attempt suicide do not mean to succeed, as many as 20 percent of those who make an attempt later complete the act.

The best lead so far in predicting which young people are at high risk for ending their lives is a low level of the neurotransmitter serotonin. Yet many girls with low serotonin levels don't try to commit suicide. They tend to develop bulimia instead.

Of all the behavioral problems that strike during adolescence, bulimia is perhaps the most bizarre. The disease is

characterized by an uncontrollable urge to binge—as many as 100 calories a minute or 5,000 calories per 30- to 60-minute binge up to five times a day—and then purge by vomiting or taking laxatives. Like suicide, bulimia is on the rise, and recent studies suggest that as many as 15 percent of adolescent girls may have the disorder. While between 70 and 80 percent of these young women maintain their normal weight and thus are difficult to spot, the remainder lose so much weight they become dangerously malnourished. Between 50 and 70 percent of bulimics have a major affective disorder, and there is a strong link with depression that runs in families.

There is good reason to believe that bulimia is associated with reduced serotonin in the hypothalamus, a small structure in the brain that serves as the master control center, regulating such functions as appetite, weight, body temperature, the endocrine system, and response to stress. The hypothalamus also serves as the major link between the limbic system, the part of the brain that controls mood, and the rest of the body. "We think the key may be found by studying the hypothalamus, the focal point for these illnesses, that something messes up serotonin function there, and that may be related to the symptoms of bulimia," says Harry E. Gwirtsman, assistant professor of psychiatry at University of California, Los Angeles, school of medicine. "We don't know for sure that bulimia is a brain disease, but certain hypothalamic tumors produce a disease indistinguishable from it." Furthermore, some antidepressants that raise brain serotonin levels reduce or eliminate bulimic binges.

This does not explain, though, why girls with low serotonin levels develop bulimia while boys take their own lives. "We find that as kids develop, boys tend to act out their depression as aggressive behavior, while girls seem to direct their feelings into themselves," says Elizabeth Susman, a research psychologist at the NIMH Laboratory of Developmental Psychology. "We believe this has something to do with the different cues that are important in forming a boy's self-image versus a girl's self-image during puberty."

Susman is examining how various hor-

"We're now taking monkeys with a high risk of depression and placing them with nurturing mothers."

monal and physical changes of adolescence combine with social events to affect a child's psychological development. Her work may offer new insight into the factors that set off the profound mood changes that accompany depression, eating disorders, and suicide.

Some researchers believe they have already identified one system in the body that ties all these seemingly disparate behaviors together. "The evidence strongly suggests that an abnormal biochemical response to stress, mediated through the hypothalamus, eventually produces the variety of biochemical and behavioral changes we find in people with these disorders," says Fred Goodwin.

It is known that hormones produced by the hypothalamus in response to stress cause mood changes in animals. Recently receptors for some of these chemicals have been found in the mood-controlling limbic system of humans, suggesting that stress hormones can affect our moods as well. In addition, nerves have been found leading from the limbic system into those parts of the hypothalamus that control hormone release. "What we could be seeing is a reduced ability of the body's stress-handling system to adapt, so that the levels of neurotransmitters, such as serotonin, do not change the way they should," says NIMH psychopharmacologist William Potter. He theorizes that genetic defects somehow diminish the system's resilience, so that repeated stresses wear it

down further, eventually provoking depression or manic-depression.

An intriguing study of primates tends to support Potter's theory. For more than 15 years now, psychologist Steven J. Suomi, who recently moved from the University of Wisconsin, Madison, to a joint position at the National Institute of Child Health and Human Development and NIMH, has been looking at the behavior of infant rhesus monkeys. While at Wisconsin, he noticed that certain offspring, when separated from their mothers for brief periods, showed symptoms that if seen in humans would be diagnosed as depressive: agitation, withdrawal, and certain biochemical abnormalities. They also responded well to antidepressants.

But what makes this a good model for human depression, Suomi says, "is that most of the monkeys in our colony eventually start coping with the separation, just as most humans adjust to stressful situations." Monkeys that become despondent when separated from their mothers also seem to get depressed if they are faced with stressful situations as they get older. When there is no stress, however, these monkeys behave in a perfectly normal way.

Suomi points out other parallels with human depression. "In infancy and childhood, the depressed animals are withdrawn and shy. During adolescence they are more agitated and delinquent, and in fact, we think we may be seeing some suicidal behavior. Then in adulthood they become withdrawn again. This

really mirrors the different appearance of human depression throughout development."

From the start Suomi noticed that depressed infants behaved differently from normal babies. They showed greater muscle tension and were more responsive to stimuli. By taking all these factors into consideration, "we were able to develop a profile that was very good at predicting which infants would be susceptible to later behavioral problems."

Suomi also found a hereditary risk factor. "I was very skeptical at first, but one look at the pedigree of the monkey colony, and you have to be convinced there is a genetic component to this disorder." But these unlucky monkeys are not necessarily destined for a sad life. "We're now taking monkeys with a high risk of depression and placing them with extremely nurturing mothers in the colony," says Suomi. "While these experiments have only started recently, the preliminary results seem to indicate that the nurturing environment can ameliorate the genetic susceptibility. The data are very exciting."

Elizabeth Susman thinks so too. "We are finding many of the same signs in children that Steve has seen in his monkeys," she says. "If we can learn enough about the genetics, the biochemistry, and the social factors that combine to produce behavioral problems in kids, we can help reduce the incidence of depression and make life better for them now and in the future."

Resilient Children

Emmy E. Werner

Emmy E. Werner, Ph.D., is Professor of Human Development and Research Child Psychologist, University of California at Davis, Davis, California.

Research has identified numerous risk factors that increase the probability of developmental problems in infants and young children. Among them are biological risks, such as pre- and perinatal complications, congenital defects, and low birth weight; as well as intense stress in the caregiving environment, such as chronic poverty, family discord, or parental mental illness (Honig 1984).

In a 1979 review of the literature of children's responses to such stress and risks, British child psychiatrist Michael Rutter wrote:

> There is a regrettable tendency to focus gloomily on the ills of mankind and on all that can and does go wrong.... The potential for prevention surely lies in increasing our knowledge and understanding of the reasons why some children are *not* damaged by deprivation.... (p. 49)

For even in the most terrible homes, and beset with physical handicaps, some children appear to develop stable, healthy personalities and to display a remarkable degree of resilience, i.e., the ability to recover from or adjust easily to misfortune or sustained life stress. Such children have recently become the focus of attention of a few researchers who have asked *What is right with these children?* and, by implication, *How can we help others to become less vulnerable in the face of life's adversities?*

The search for protective factors

As in any detective story, a number of overlapping sets of observations have begun to yield clues to the roots of resiliency in children. Significant findings have come from the few longitudinal studies which have followed the same groups of children from infancy or the preschool years through adolescence (Block and Block 1980; Block 1981; Murphy and Moriarty 1976; Werner and Smith 1982). Some researchers have studied the lives of minority children who did well in school in spite of chronic poverty and discrimination (Clark 1983; Gandara 1982; Garmezy 1981; 1983; Kellam et al. 1975; Shipman 1976). A few psychiatrists and psychologists have focused their attention on the resilient offspring of psychotic patients (Anthony 1974; Bleuler 1978; Garmezy 1974; Kauffman et al. 1979; Watt et al. 1984; Werner and Smith 1982) and on the coping patterns of children of divorce (Wallerstein and Kelly 1980). Others have uncovered hidden sources of strength and gentleness among the uprooted children of contemporary wars in El Salvador, Ireland, Israel, Lebanon, and Southeast Asia (Ayala-Canales 1984; Fraser 1974; Heskin 1980; Rosenblatt 1983). Perhaps some of the most moving testimonials to the resiliency of children are the life stories of the child survivors of the Holocaust (Moskovitz 1983).

All of these children have demonstrated unusual psychological strengths despite a history of severe and/or prolonged psychological stress. Their personal competencies and some unexpected sources of support in their caregiving environment either compensated for, challenged, or protected them against the adverse effects of stressful life events (Garmezy, Masten, and Tellegren 1984). Some researchers have called these children *invulnerable* (Anthony 1974); others consider them to be *stress resistant* (Garmezy and Tellegren 1984); still others refer to them as *superkids* (Kauffman et al. 1979). In our own longitudinal study on the Hawaiian island of Kauai, we have found them to be *vulnerable, but invincible* (Werner and Smith 1982).

These were children like Michael for whom the odds, on paper, did not seem very promising. The son of teen-age parents, Michael was born prematurely and spent his first three weeks of life in the hospital, separated from his mother. Immediately after his birth, his father was sent with the Army to Southeast Asia for almost two years. By the time Michael was eight, he had three younger siblings and his parents were divorced. His mother left the area and had no further contact with the children.

And there was Mary, born to an overweight, nervous, and erratic mother who had experienced several miscarriages, and a father who was an unskilled farm laborer with only four years of education. Between Mary's fifth and tenth birthdays, her mother had several hospitalizations for repeated bouts with mental illness, after having inflicted both physical and emotional abuse on her daughter.

Yet both Michael and Mary, by age 18, were individuals with high self-esteem and sound values, caring for others and liked by their peers, successful in school, and looking forward to their adult futures.

We have learned that such resilient children have four central characteristics in common:

- an active, evocative approach toward solving life's problems, enabling them to negotiate successfully an abundance of emotionally hazardous experiences;
- a tendency to perceive their experiences constructively, even if they caused pain or suffering;
- the ability, from infancy on, to gain other people's positive attention;
- a strong ability to use faith in order to maintain a positive vision of a meaningful life (O'Connell-Higgins 1983).

Protective factors within the child

Resilient children like Mary and Michael tend to have temperamental characteristics that elicit positive responses from family members as well as strangers (Garmezy 1983; Rutter 1978). They both suffered from birth complications and grew up in homes marred by poverty, family discord, or parental mental illness, but even as babies they were described as active, affectionate, cuddly, good natured, and easy to deal with. These same children already met the world on their own terms by the time they were toddlers (Werner and Smith 1982).

Resilient children tend to have temperamental characteristics that elicit positive responses from family members as well as strangers.

Several investigators have noted *both* a pronounced autonomy and a strong social orientation in resilient preschool children (Block 1981; Murphy and Moriarty 1976). They tend to play vigorously, seek out novel experiences, lack fear, and are quite self-reliant. But they

are able to ask for help from adults or peers when they need it.

Sociability coupled with a remarkable sense of independence are characteristics also found among the resilient school-age children of psychotic parents. Anthony (1974) describes his meeting with a nine-year-old girl, whose father was an alcoholic and abused her and whose mother was chronically depressed. The girl suffered from a congenital dislocation of the hip which had produced a permanent limp, yet he was struck by her friendliness and the way she approached him in a comfortable, trustful way.

The same researcher tells of another nine-year-old, the son of a schizophrenic father and an emotionally disturbed mother, who found a refuge from his parents' outbursts in a basement room he had stocked with books, records, and food. There the boy had created an oasis of normalcy in a chaotic household.

Resilient children often find a refuge and a source of self-esteem in hobbies and creative interests. Kauffman et al. (1979) describes the pasttimes of two children who were the offspring of a schizophrenic mother and a depressed father:

> When David (age 8) comes home from school, he and his best friend often go up to the attic to play. This area ... is filled with model towns, railroads, airports and castles. ... He knows the detailed history of most of his models, particularly the airplanes. ... David's older sister, now 15, is extraordinarily well-read. Her other interests include swimming, her boyfriend, computers and space exploration. She is currently working on a computer program to predict planetary orbits. (pp. 138, 139)

The resilient children on the island of Kauai, whom we studied for nearly two decades, were not unusually talented, but they displayed a healthy androgyny in their interests and engaged in hobbies that were not narrowly sex-typed. Such activities, whether it was fishing, swimming, horseback riding, or hula dancing, gave them a reason to feel proud. Their hobbies, and their lively sense of humor, became a solace when things fell apart in their lives (Masten 1982; Werner and Smith 1982).

In middle childhood and adolescence, resilient children are often engaged in acts of "required helpfulness" (Garmezy, in press). On Kauai, many adolescents took care of their younger siblings. Some managed the household

when a parent was ill or hospitalized; others worked part-time after school to support their family. Such acts of caring have also been noted by Anthony (1974) and Bleuler (1978) in their studies of the

---

Most resilient children establish a close bond with at least one caregiver from whom they received lots of attention during the first year of life.

---

resilient offspring of psychotic parents, and by Ayala-Canales (1984) and Moskovitz (1983) among the resilient orphans of wars and concentration camps.

Protective factors within the family

Despite chronic poverty, family discord, or parental mental illness, most resilient children have had the opportunity to establish a close bond with at least one caregiver from whom they received lots of attention during the first year of life. The stress-resistant children in the Kauai Longitudinal Study as well as the resilient offspring of psychotic parents studied by Anthony (1974) had enough good nuturing to establish a basic sense of trust.

Some of this nuturing came from substitute caregivers within the family, such as older siblings, grandparents, aunts, and uncles. Such alternate caregivers play an important role as positive models of identification in the lives of resilient children, whether they are reared in poverty (Kellam et al. 1975), or in a family where a parent is mentally ill (Kauffman et al. 1979), or coping with the aftermath of divorce (Wallerstein and Kelly 1980).

Resilient children seem to be especially adept at actively recruiting surrogate parents. The latter can come from the ranks of babysitters, nannies, or student roomers (Kauffman et al. 1979); they can be parents of friends (Werner and Smith 1982), or even a housemother in an orphanage (Ayala-Canales 1984; Moskovitz 1983).

The example of a mother who is gainfully and steadily employed appears to be an especially powerful model of identification for resilient girls reared in poverty, whether they are Black (Clark 1983), Chicana (Gandara 1982), or Asian-American (Werner and Smith 1982). Maternal employment and the need for sibling caregiving seems to contribute to the pronounced autonomy and sense of responsibility noted among these girls, especially in households where the father is permanently absent.

Structure and rules in the household and assigned chores enabled many resilient children to cope well in spite of poverty and discrimination, whether they lived on the rural island of Kauai, or in the inner cities of the American Midwest, or in a London borough (Clark 1983; Garmezy 1983; Rutter 1979).

Resilient children find a great deal of emotional support outside of their immediate family.

Resilient children also seem to have been imbued by their families with a sense of coherence (Antonovsky 1979). They manage to believe that life makes sense, that they have some control over their fate, and that God helps those who help themselves (Murphy and Moriarty 1976). This sense of meaning persists among resilient children, even if they are uprooted by wars or scattered as refugees to the four corners of the earth. It enables them to love despite hate, and to maintain the ability to behave compassionately toward other people (Ayala-Canales 1984; Moskovitz 1983).

Protective factors outside the family

Resilient children find a great deal of

emotional support outside of their immediate family. They tend to be well-liked by their classmates and have at least one, and usually several, close friends and confidants (Garmezy 1983; Kauffman et al. 1979; Wallerstein and Kelly 1980; Werner and Smith 1982). In addition, they tend to rely on informal networks of neighbors, peers, and elders for counsel and advice in times of crisis and life transitions.

Resilient children are apt to like school and to do well in school, not exclusively in academics, but also in sports, drama, or music. Even if they are not unusually talented, they put whatever abilities they have to good use. Often they make school a home away from home, a refuge from a disordered household. A favorite teacher can become an important model of identification for a resilient child whose own home is beset by family conflict or dissolution (Wallerstein and Kelly 1980).

In their studies of London schools, Rutter and his colleagues (1979) found that good experiences in the classroom could mitigate the effects of considerable stress at home. Among the qualities that characterized the more successful schools were the setting of appropriately high standards, effective feedback by the teacher to the students wtih ample use of praise, the setting of good models of behavior by teachers, and giving students positions of trust and responsibility. Children who attended such schools developed few if any emotional or behavioral problems despite considerable deprivation and discord at home (Pines 1984).

Early childhood programs and a favorite teacher can act as an important buffer against adversity in the lives of resilient young children. Moskovitz (1983), in her follow-up study in adulthood of the childhood survivors of concentration camps, noted the pervasive influence of such a warm, caring teacher.

Participation in extracurricular activities or clubs can be another important informal source of support for resilient children. Many youngsters on Kauai were poor by material standards, but they participated in activities that allowed them to be part of a cooperative enterprise, whether being cheerleader for the home team or raising an animal in the 4-H Club. Some resilient older youth were members of the Big Brothers and Big Sisters Associations

which enabled them to help other children less fortunate than themselves. For still others emotional support came from a church group, a youth leader in the YMCA or YWCA, or from a favorite minister, priest, or rabbi.

There is a shifting balance between stressful life events which heighten children's vulnerability and the protective factors in their lives which enhance their resiliency.

The shifting balance between vulnerability and resiliency

For some children some stress appears to have a steeling rather than a scarring effect (Anthony 1974). But we need to keep in mind that there is a shifting balance between stressful life events which heighten children's vulnerability and the protective factors in their lives which enhance their resiliency. This balance can change with each stage of the life cycle and also with the sex of the child. Most studies in the United States and in Europe, for example, have shown that boys appear to be more vulnerable than girls when exposed to chronic and intense family discord in childhood, but this trend appears to be reversed by the end of adolescence.

As long as the balance between stressful life events and protective factors is manageable for children they can cope. But when the stressful life events outweigh the protective factors, even the most resilient child can develop problems. Those who care for children, whether their own or others, can help restore this balance, either by *decreasing* the child's exposure to intense or chronic life stresses, or by *increasing* the number of protective factors, i.e., competencies and sources of support.

Children who are left to fend for themselves because of a difficult family structure must find a great deal of emotional support outside that structure. This support allows them to deal with life situations. They can then develop into productive adults.

Implications

What then are some of the implications of the still tentative findings from studies of resilient children? Most of all, they provide a more hopeful perspective than can be derived from reading the extensive literature on problem children which predominates in clinical psychology, child psychiatry, special education, and social work. Research on resilient children provides us with a focus on the self-righting tendencies that appear to move some children toward normal development under all but the most persistent adverse circumstances.

Faith that things will work out can be sustained if children encounter people who give meaning to their lives and a reason for commitment and caring.

Those of us who care for young children, who work with or on behalf of them, can help tilt the balance from vulnerability to resiliency if we

• accept children's temperamental idiosyncrasies and allow them some experiences that challenge, but do not overwhelm, their coping abilities;

• convey to children a sense of responsibility and caring, and, in turn, reward them for helpfulness and cooperation;

• encourage a child to develop a special interest, hobby, or activity that can serve as a source of gratification and self-esteem;

• model, by example, a conviction that life makes sense despite the inevitable adversities that each of us encounters;

• encourage children to reach out beyond their nuclear family to a beloved relative or friend.

Research on resilient children has taught us a lot about the special importance of surrogate parents in the lives of children exposed to chronic or intense distress. A comprehensive assessment of the impact on siblings, grandparents, foster parents, nannies, and babysitters on the development of high risk children is elaborated upon in Werner (1984).

Outside the family circle there are other powerful role models that give emotional support to a vulnerable child. The three most frequently encountered in studies of resilient children are: a favorite teacher, a good neighbor, or a member of the clergy.

There is a special need to strengthen such informal support for those children and their families in our communities which appear most vulnerable because they lack—temporarily or permanently—some of the essential social bonds that appear to buffer stress: working mothers of young children with no provisions for stable child care; single, divorced, or teen-age parents; hospitalized and handicapped children in need of special care who are separated from their families for extended periods of time; and migrant or refugee children without permanent roots in a community.

Two other findings from the studies of resilient children have implications for the well-being of all children and for those who care for them.

(1) At some point in their young lives, resilient children were required to carry out a socially desirable task to prevent others in their family, neighborhood, or community from experiencing distress or discomfort. Such acts of *required helpfulness* led to enduring and positive changes in the young helpers.

(2) The central component in the lives of the resilient children that contributed to their effective coping appeared to be a feeling of confidence or faith that things *will work out* as well as can be reasonably expected, and that the odds *can* be surmounted.

The stories of resilient children teach us that such a faith can develop and be sustained, even under adverse circumstances, if children encounter people who give meaning to their lives and a reason for commitment and caring. Each of us can impart this gift to a child—in the classroom, on the playground, in the neighborhood, in the family—*if* we care enough.

Bibliography

Anthony, E. J. "The Syndrome of the Psychologically Invulnerable Child." In *The Child in His Family 3: Children at Psychiatric Risk,* ed. E. J. Anthony and C. Koupernik. New York: Wiley, 1974.

Antonovsky, A. *Health, Stress and Coping: New Perspectives on Mental and Physical Well-being.* San Francisco: Jossey-Bass, 1979.

Ayala-Canales, C. E. "The Impact of El Salvador's Civil War on Orphan and Refugee Children." M.S. Thesis in Child Development, University of California at Davis, 1984.

Bleuler, M. *The Schizophrenic Disorders: Long-term Patient and Family Studies.* New Haven: Yale University Press, 1978.

Block, J. H. and Block, J. "The Role of Ego-Control and Ego-Resiliency in the Organization of Behavior." In *The Minnesota Symposia on Child Psychology 13: Development of Cognition, Affect and Social Relations,* ed. W. A. Collins. Hillsdale, N.J.: Erlbaum, 1980.

Block, J. "Growing Up Vulnerable and Growing Up Resistant: Preschool Personality, Pre-Adolescent Personality and Intervening Family Stresses." In *Adolescence and Stress,* ed. C. D. Moore. Washington, D.C.: U.S. Government Printing Office, 1981.

Clark, R. M. *Family Life and School Achievement: Why Poor Black Children Succeed or Fail.* Chicago: University of Chicago Press, 1983.

Fraser, M. *Children in Conflict.* Harmondsworth, England: Penguin Books, 1974.

Gandara, P. "Passing Through the Eye of the Needle: High Achieving Chicanas." *Hispanic Journal of Behavioral Sciences* 4, no. 2 (1982): 167–180.

Garmezy, N. "The Study of Competence in Children at Risk for Severe Psychopathology." In *The Child in His Family 3: Children at Psychiatric Risk,* ed. E. J. Anthony and C. Koupernik. New York: Wiley, 1974.

Garmezy, N. "Children Under Stress: Perspectives on Antecedents and Correlates of Vulnerability and Resistance to Psychopathology." In *Further Explorations in Personality,* ed. A. I. Rabin, J. Aronoff, A. M. Barclay, and R. A. Zucker. New York: Wiley, 1981.

Garmezy, N. "Stressors of Childhood." In *Stress, Coping and Development,* ed. N. Garmezy and M. Rutter. New York: McGraw-Hill, 1983.

Garmezy, N. "Stress Resistant Children: The Search for Protective Factors." In *Aspects of Current Child Psychiatry Research,* ed. J. E. Stevenson. *Journal of Child Psychology and Psychiatry,* Book Supplement 4. Oxford, England: Pergamon, in press.

Garmezy, N.; Masten, A. S.; and Tellegren, A. "The Study of Stress and Competence in Children: Building Blocks for Developmental Psychopathology." *Child Development* 55, no. 1 (1984): 97–111.

Garmezy, N. and Tellegren, A. "Studies of Stress-Resistant Children: Methods, Variables and Preliminary Findings." In *Advances in Applied Developmental Psychology,* ed. F. Morrison, C. Lord, and D. Keating. New York: Academic Press, 1984.

3. DEVELOPMENT DURING CHILDHOOD: Social and Emotional Development

Heskin, K. *Northern Ireland: A Psychological Analysis.* New York: Columbia University Press, 1980.

Honig, A. "Research in Review: Risk Factors in Infants and Young Children." *Young Children* 38, no. 4 (May 1984): 60–73.

Kauffman, C.; Grunebaum, H.; Cohler, B.; and Gamer, E. "Superkids: Competent Children of Psychotic Mothers." *American Journal of Psychiatry* 136, no. 11 (1979): 1398–1402.

Kellam, S. G.; Branch, J. D.; Agrawal; K. C.; and Ensminger, M. E. *Mental Health and Going to School.* Chicago: University of Chicago Press, 1975.

Masten, A. "Humor and Creative Thinking in Stress-Resistant Children." Unpublished Ph.D. dissertation, University of Minnesota, 1982.

Moskovitz, S. *Love Despite Hate: Child Survivors of the Holocaust and Their Adult Lives.* New York: Schocken Books, 1983.

Murphy, L. and Moriarty, A. *Vulnerability, Coping and Growth from Infancy to Adolescence.* New Haven: Yale University Press, 1976.

O'Connell-Higgins, R. "Psychological Resilience and the Capacity for Intimacy." Qualifying paper, Harvard Graduate School of Education, 1983.

Pines, M. "PT Conversation: Michael Rutter: Resilient Children." *Psychology Today* 18, no. 3 (March 1984): 60, 62, 64–65.

Rosenblatt, R. *Children of War.* Garden City, N.Y.: Anchor Press, 1983.

Rutter, M. "Early Sources of Security and Competence." In *Human Growth and Development,* ed. J. Bruner and A. Garton. New York: Oxford University Press, 1978.

Rutter, M. "Protective Factors in Children's Responses to Stress and Disadvantage." In *Primary Prevention of Psychopathology 3: Social Competence in Children,* ed. M. W. Kent and J. E. Rolf. Hanover, N.H.: University Press of New England, 1979.

Rutter, M.; Maughan, B.; Mortimore, P.; and Ouston, J; with Smith, A. *Fifteen Thousand Hours: Secondary Schools and Their Effects on Children.* Cambridge, Mass.: Harvard University Press, 1979.

Shipman, V. C. *Notable Early Characteristics of High and Low Achieving Low SES Children.* Princeton, N.J.: Educational Testing Service, 1976.

Wallerstein, J. S. and Kelly, J. B. *Surviving the Breakup: How Children and Parents Cope with Divorce.* New York: Basic Books, 1980.

Watt, N. S.; Anthony, E. J.; Wynne, L. C.; and Rolf, J. E., eds. *Children at Risk for Schizophrenia: A Longitudinal Perspective.* London and New York: Cambridge University Press, 1984.

Werner, E. E. *Child Care: Kith, Kin and Hired Hands.* Baltimore: University Park Press, 1984.

Werner, E. E. and Smith, R. S. *Vulnerable, but Invincible: A Longitudinal Study of Resilient Children and Youth.* New York: McGraw-Hill, 1982.

This is one of a regular series of Research in Review columns. The column in this issue was edited by Elizabeth H. Brady, M.A., Professor and Chair, Department of Educational Psychology, California State University, Northridge, Northridge, California.

Face to face, it's the expression that bears the message

Scientist Paul Ekman demonstrated that all people make the same basic faces, and his studies launched new research into emotions

Jeanne McDermott
Jeanne McDermott is a frequent contributor to the magazine. Her article on vision appeared in April.

Paul Ekman has a funny face. It is not the long mouth, the black eyebrows or the broad furrowed forehead, but the way his face tumbles and contorts into an unabashed exhibition of expressions. After practicing before the mirror for several years, Ekman has learned how to control each individual muscle in his face, a dexterity few mimes and actors will claim. "Most people can't do this," he says, snapping into the odd grin that became Charlie Chaplin's trademark. "When I was a kid, my mother said 'Stop making all those crazy expressions on your face,'" he recalls. The odd expression softens into one of bemused contentment. "Now I'm making a living at it."

As an experimental psychologist, Ekman has devoted his entire career to discovering what is in a face. His quest and curiosity are hardly novel. In Helen of Troy's time, there was enough mystique to launch a thousand ships. Even in faces without memorable beauty (which studies of beauty-contest winners suggest is only the idealization of the species' juvenile characteristics), there is plenty. Nothing that we see, no other object in the world, possesses such hypnotic fascination, such wealth of meaning or such power in its confluence of expression and identity.

So rich is the face that each generation of artists has revealed something new about the way we see it. According to Gertrude Stein, Picasso painted faces the way they might appear to a newborn baby: flat, without dimension, features straying off course. With calligraphic brushstrokes, Henri Matisse made quick sketches of his friend, the poet Louis Aragon, capturing how much the face changes from one moment to the next. And with the technology of the 20th century, New York City artist Nancy Burson creates faces that belong to the horizon of dream and imagination.

The woman in one of Burson's photographs looks like a *Vogue* model, today's image of a natural beauty (p. 115). But look again. There is nothing natural about this face. It belongs to no one who has ever existed or ever will. Rather, it is a chimera, a composite created by camera and computer. For more than a decade, Burson has been developing the software that allows her to merge two or more faces into one believable and utterly impossible composite. When a plastic surgeon told Burson that people came to him with a list of ten things they wanted—"like Audrey Hepburn's nose"—it stimulated her interest in the notion of what we call beautiful. Rather than making a patchwork of the desired features, she superimposed the faces of five icons from the 1950s: Grace Kelly, Sophia Loren, Bette Davis, Audrey Hepburn and Marilyn Monroe. Then she blended the faces of today's icons: Brooke Shields, Jacqueline Bisset, Diane Keaton, Jane Fonda and Meryl Streep. Alone, each composite face is a ghost of beauty. Together, they remind us that our ideal face, and Hollywood's, changes with time.

Darwin's overlooked theory of faces

Like artists, scientists have also been drawn to faces as a subject of study. Some contemporary scientists see the face as a key to understanding how we communicate, how we see and how we remember emotional experience. For the first time, and with the usual loose ends and debates, they are tendering some precise answers to many long-standing questions. Charles Darwin was the first to subject the nature of facial expressions to modern scientific analysis. In *The Expression of the Emotions in Man and Animals*, a best-seller in 1872, he argued that the expressions of the face are, in large measure, universal and innate. But despite Darwin's clout and the common sense in his argument, the book had little impact in the late 19th century's flux of ideas. The public still believed in phrenology, divining a person's character from the bumps on his head, and physiognomy, divining a person's character from the features on his face (SMITHSONIAN, November 1980). The infant science of psychology, the study of

the mind, was breaking off from philosophy and from physiology, the study of the body. Out of this turmoil came the conviction that universal expressions of the face did not exist, a belief later reinforced by the behaviorists and cultural anthropologists in the 20th century.

Only in the mid-1960s did Paul Ekman, along with Carroll Izard, also a psychologist, resurrect Darwin's ideas, which they had both encountered in Silvan Tomkins' book, *Affect, Imagery, Consciousness*. In appearance and style, the two offer a contrast. Ekman is broad-shouldered and works in a mask-filled clutter on the fog-cool San Francisco campus of the University of California. Izard is thin to the point of gauntness with a superclean desk at the University of Delaware. But you cannot help noticing that both possess very expressive faces.

Although individuals can read nearly infinite shades of meaning into a face, Ekman and Izard wanted to know if any expressions carry the same meaning, regardless of the observer, the word we attach to it, the culture or the context. Ekman, who was initially skeptical and Izard, who was not, used photographs of faces that a large number of people in this country agreed depicted one and only one simple emotion. On what Ekman now calls the "universals expedition," he and Izard found the same responses among literate people in Europe, South America, Africa and Japan. Ekman then showed these photos to tribespeople in Borneo and New Guinea who did not read or write and who had never been exposed to modern media. When asked which face illustrated a story, such as "she is angry and about to fight," the tribespeople judged the faces in the same way we do. While each culture used different words to label the emotions, all associated the same expressions with the same feelings. Around the world, brows lowered and drawn together,

Computer artist Nancy Burson turns digitized images into composites.

Catwoman composite was made by blending photo of Burson's face with a picture of her feline pet.

"Ideal" of 1950s is mix of Audrey Hepburn, Grace Kelly, Sophia Loren, Bette Davis, Marilyn Monroe.

Today's "ideal" blends faces of Brooke Shields, Jane Fonda, Jacqueline Bisset, Diane Keaton, Meryl Streep.

tightened lower eyelids and pressed lips mean anger; a wrinkled nose signals disgust. For fear, happiness, sadness and surprise, they also discovered universal expressions of the face. Izard in his travels in literate areas identified three more—contempt, interest and shame.

Once identified, Ekman and Izard spent the 1970s describing the six universals and their myriad variations by videotaping and analyzing hundreds, if not thousands, of faces. Slow-motion videotaping led Ekman to discover what he calls microexpressions. While an average expression lasts from one to one-and-a-half seconds, these flash across the face in a fraction of a second, too fast for anyone to register consciously. They are fully formed expressions at distinct odds with whatever shows on the person's face. Ekman found them when a patient had attempted suicide only shortly after doctors had granted her leave from a psychiatric hospital. Perplexed, the doctors said she had given them no clue to her despair. Only by replaying a videotape of her last session in slow motion did Ekman catch the fleeting microexpressions of sadness mingled with her otherwise happy face.

Ekman's technique for analyzing the face is based on the underlying muscles. "Unlike other mammals, humans have only facial muscles attached to the skin," he says. "While a horse can flip a fly from its flank, you can only flip a fly off your forehead." Ekman remapped the face's muscles, uncovering ones that anatomists had overlooked. In total, he found 44, assigned each a number and then spent many years with a mirror learning which muscles contract to form what expressions. He has found more than 10,000 possible anatomical combinations and now speaks of faces in terms of numbers. "That's a 13," he says of the Chaplin smile. Only one expression has no number. "A neutral face is one without any muscular contraction evident. But you

can read anything into a neutral face, which is what we do," he says, deadpan.

Only a short time after Ekman and Izard gave the scientific community powerful tools for analyzing the face, researchers trooped off to the nursery to find out if the universals are also innate, present from birth. Just how and when and why these expressions emerge is being hotly debated now. Newborns lack the use of only one facial muscle that adults possess, Number 13, in fact. We are born to communicate with the face.

No one believes this as strongly as Tiffany Field, psychologist at the University of Miami Medical School. "We work a lot with newborns. Just playing. The tendency is to make funny faces at them. And their tendency is to make funny faces back," she says. When Field did a systematic study of this playing, she found that babies only 36 hours old can and do imitate happy, sad and surprised expressions on the face. Even prematurely born babies were able to do it.

Newborns were once thought to be blind at birth and are now believed to have limited perception of depth and visual acuity; how can they see a face well enough to mimic it? Many scientists are highly skeptical, and Field herself admits that "the imitative behavior is surprising." Is it a reflex? How much voluntary control does the infant have? Field believes that newborns have an innate ability to compare what they can perceive with what they can do, a built-in resonance to faces and their expressions. Half-jokingly, she says this may explain why smiles are contagious.

The earlier work of Andrew Meltzoff, psychologist at the University of Washington in Seattle, inspired Field's research (p. 118). He looked at actions like sticking the tongue out and opening the mouth. Under conditions that optimized a newborn's ability to see, he found it would readily imitate the adult. But like many scientists, Meltzoff gets wary when it comes to making claims about emotion. "My work is about the imitation of basic facial actions," he says, "not about complex facial patterns like emotional displays."

What babies may be doing in the first weeks of life is the facial equivalent of babbling, testing the "equipment," so to speak. Since infants seem to be sensitive to faces almost from birth, it is a fair assumption that the "equipment" includes nerve cells in the brain that are attuned to faces. Robert Emde, psychiatrist at the University of Colorado Medical School in Denver, has been studying what newborns can do with this basic equipment. Although Emde has a sleek, Scandinavian-style analyst's couch in his office, he has been interested since graduate-student days in what infants and children communicate without words. From the day of birth, perhaps even before, babies smile. Family folklore says it is a gas pain, or more poetically that the child is smiling at the angels. Emde says that the first smiles are associated with REM, the rapid eye movement stage of brain activity that occurs in adults during the dreaming stage of sleep. In infants, **REM** occurs 50 percent of the time, when the infant is both awake and asleep. The mother has no control over the REM smiles. "That so violates our intuitive understanding of the principles of communication that some mothers say the smile must be caused by gas or the deities," he says. As the infant ages, the smiling associated with the REM stage changes, paralleling a growth spurt in the brain. By three weeks infants smile, irregularly, to changes in the environment. By two months, infants smile to faces, anybody's face, and soon after, to the caregiver's face more than any other.

Long before children speak or understand language,
they speak and understand a powerful language of the face. While working on her doctoral thesis several years ago, Denver psychologist Mary Klinnert discovered a process now known as "social referencing." Watching a session in a camera-lined laboratory-playroom, she noticed how strongly the mother's facial expressions influenced the baby's actions. So she invited toddlers, 12 and 18 months old, to play with scary toys like Incredible Hulk and a remote-controlled spiderlike robot. Before making a beeline for the toy, some children hesitated and checked the mother's face for guidance. Klinnert had carefully trained the mothers in the expressions of approval or apprehension. When the mother smiled serenely, the child went ahead and played. But when the mother grimaced or showed fear, most children backed off and some cried. Not a word was required to keep the dubious infant away from uncertainty. All it took was a convincing expression on the mother's face. Klinnert explains: "The child learns ways of seeing the world from other people's expressions."

In the early days of Klinnert's study, however, the children's response was not always as clear-cut. Sometimes, instead of backing off in fear, they began to laugh. When Klinnert played the videotapes in slow motion, she saw what the children had seen: the mothers of laughing babies had changed their posed expression of fear ever so slightly. Rather than lifting the eyebrows, they let them relax. Rather than dropping the corners of the mouth, they began to raise them. The expression of surprise was creeping into that of their fear and the babies, who were attuned to peekaboo games, sensed the very subtle difference. The problem was solved by placing a wireless microphone in the

In his search for universal expressions, Paul Ekman asked New Guinea natives how they would feel if: (left) a friend visited, (center) your child died, or (right) you saw a dead pig, lying there a long time.

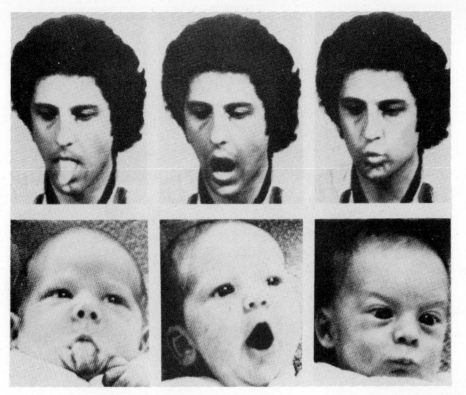

Psychologist Andrew Meltzoff, in top row of pictures, makes a series of faces at infants, whose responses are shown in corresponding pictures in bottom row. The infants in this study were two to three weeks old. Prior to Meltzoff's work, scientists thought that infants could not imitate facial gestures until eight months of age.

mother's ear and coaching her to keep a fearful look.

If the infant's face is initially a pure medium of expression, then only with time does it become a mask, molded by what the family believes are proper ways to show feeling. "By the age of one, kids are learning enormous amounts about rules for the expression of emotion," says Emde who is studying faces for clues to how different families teach these rules. At this age, children first learn not to show emotion, particularly anger, and they learn to put on expressions, particularly smiles. The tight coupling between a facial expression and the feeling itself begins to stretch.

While the bond between what the face reveals and what the person feels may loosen, it is never truly severed. Ekman studied actors trained in the Stanislavsky method, which teaches them to physically become the characters they play. When asked to assume certain facial expressions, these actors reported feeling the emotion that accompanies the expression—and by such objective physical indicators as heart rate and skin temperature, they did. Izard had earlier obtained similar results when, on a sabbatical in Moscow, he tested actors trained at several institutes following Stanislavsky. If the physical action triggers the emotional sensation, does that mean you will feel happy if you put on a happy face? Not with a smile, says Ekman, but he has found a close connection with the emotions of anger, depression, fear and sadness. Feigning those expressions can trigger corresponding internal sensations and physiological changes, he says.

By the age of two, the face is not only a medium and mask for communication but it also becomes the locus

of identity. Children learn how to recognize their own faces as reflections of themselves at this age, and then all through childhood they become increasingly adept at recognizing other people's faces. Curiously, this ability momentarily falters around ages 11 through 14, which MIT psychologist Susan Carey attributes to the hormonal and cognitive reshuffling of puberty. But in the midteens, it snaps back and by adulthood, the number of faces that we are able to recognize is as boundless as the universe of possibilities. "We are all Einsteinian physicists when it comes to recognizing faces," Carey says.

For scientists who study how we see and remember, the process of recognizing faces is an enigma. "It's impossible to say in detail how we do it," says Alvin Goldstein, psychologist at the University of Missouri, Columbia. As an object of perception, the face is in a class by itself. Nothing in the environment gives so much information essential to survival. Is the other friend or foe? Kin or outsider? The answer is in the face. And despite the transformation brought by aging, no other object is probably recognized so automatically and on the basis of such slender clues.

When the Defense Advanced Research Projects Agency (DARPA) decided to fund research in teleconferencing, MIT's Media Technology Laboratory initiated an investigation called Transmission of Presence. The question came up: What is needed to carry on a face-to-face conversation between people at remote places? Susan Brennan, now a researcher at Hewlett-Packard Labs, proposed the caricature as one answer. For her master's thesis, she developed software

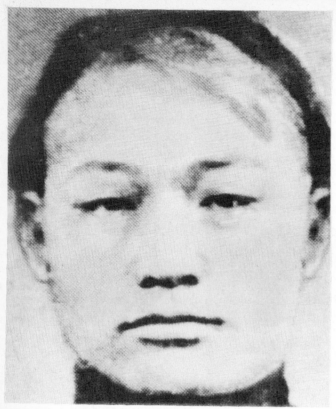

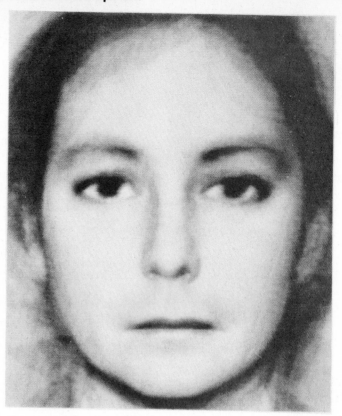

Face of Mankind (from world population statistics) is 57 percent Asian, 26 percent white, 7 percent black.

Androgyny is represented in this face composed by combining the photographs of six men and six women.

that transforms faces into caricatures. Her system starts with a realistic line drawing, which it compares with an "average" face composed of all the other faces stored in the computer's memory. An operator exaggerates the differences between the line drawing and the average until an acceptable likeness is found. Then the caricature is automatically animated and driven by speech over ordinary telephone lines.

Caricatures proved more acceptable to teleconferencers than another approach to the MIT project called Lipsync, which borrowed from the traditions of puppetry. Looking vaguely like an advertisement for the Rocky Horror Picture Show, it animated a digitized photograph of the face with only the lips moving in synchrony to the person's words. One reason for the caricature's relative success (it has yet to be used commercially) is that it may be a visual shortcut, a potent distillation of the eye and brain's own strategy for recognizing faces. From the exaggerated lines of a caricature emerge a likeness that just isn't there in the equally simple but more faithful lines of a silhouette. The caricaturist knows that the face is not merely the sum of its features, but something whole—an arrangement of three-dimensional relationships. Unlike a silhouette artist, the caricaturist takes information from several points of view and compresses and distorts them onto a flat page. A caricature does not simply distort the face's features but makes that distortion rela-

tive to a tacit understanding of what is unusual about the face and what is not. While Richard Nixon and Alfred Hitchcock made easy targets for caricature, Gerald Ford did not. His face looked too "average."

With the exception of such famous faces, the "average" faces that people carry in their minds for reference vary with experience. Cognitive psychologists believe that we may recognize people by learning the range of variations possible within a group of faces and by unconsciously flagging the ways each individual falls within the range. The well-known fact that people of one race have trouble distinguishing people of another reflects the segregation of society more than any biological predisposition. The white child without black friends will not learn about the range of variations in black faces, and thus what makes one face distinctive relative to others.

While no one knows how to describe just what we recognize, the memory for faces is both durable and flexible. A famous study done at Ohio Wesleyan University in the 1970s found that we can recognize the faces of high-school classmates (without necessarily remembering the names) even 50 years after graduation. And they are hardly the same faces. After the bones and cartilage of the face stop growing, time and gravity take over, until the skin begins to sag around the jaw, neck and eyes, and wrinkles, creases and crow's-feet make their appearance.

Researcher Susan Brennan's computer program created caricatures based on realistic line drawings.

The caricatures include (clockwise from top left) JFK, Faye Dunaway, Richard Nixon, Dianne Feinstein.

Although the face does not age in easily predictable ways, computer artist Nancy Burson has developed the means to electronically simulate the aging process. In 1969, she first imagined a computer-screen version of *The Picture of Dorian Gray*—Oscar Wilde's tale of a portrait that grows old and ugly on canvas. Computer scientists told her at that time that it could not be done. But Burson was tenacious. By the late 1970s, the technology had caught up with her and she was able to apply a realistic mask of wrinkles to anyone's face. After using it initially to see what celebrities like the Royal Family and Brooke Shields would look like as senior citizens, she turned her attention to the aging of miss-

ing children. For the case of Etan Patz, a New York City child who disappeared seven years ago on his way to the school bus, Burson interviewed the family and asked them to decide whom Etan most resembled. She blended one of his last photographs with a recent one of his older sibling. While the child has not yet been found, the FBI has expressed interest in using Burson's approach with other cases.

A person who can't recognize her own face

The answer to why most people never forget a face may eventually come from people who do. Nowhere is the process of facial recognition so disturbed as in people with a very strange disorder called prosopagnosia. Antonio Damasio, neurologist at the University of Iowa's College of Medicine and this country's expert, sees only about one case a year. It develops only after a stroke or accident injures the tiny section of the brain involved in processing information about the face. Once-familiar faces no longer appear familiar. In one case, according to Damasio, "a 60-year-old woman suddenly noted that she could no longer recognize the faces of her husband and daughter. To her amazement, she could not even recognize herself in the mirror, nor could she recognize the faces of her neighbors. Yet she always knew that a face was a face." By showing patients photographs of people they knew and monitoring their automatic (or unconscious) responses with skin electrodes, Damasio found that his patients actually did recognize the faces—unconsciously. But this recognition had somehow slipped away from their conscious control.

As Damasio has pointed out, prosopagnosics fail to recognize more than faces, an observation that may lead to a more global understanding of the structure of the brain's memory. Boston neurologist Michael Alexander has one patient, an electrician, who finds wiring panels as confusing now as the faces of his three daughters. Another patient is a dermatologist who is no longer able to distinguish skin diseases. Alexander speculates that "at a very young age, some basic neurological mechanism for recognizing faces develops and we may use this feature to distinguish other meaningful objects."

Alexander's speculation raises an intriguing possibility. Does this innate framework for facial recognition also serve as a foundation for an esthetic sense? "Our notion of beauty may well originate in the face," says Joseph Campos, director of the Infant Development Laboratory of the University of Denver. "A baby may first scrutinize the face and then generalize it to other objects." The face contains all the qualities artists traditionally celebrate—symmetry, proportion, contrast. And these qualities, in turn, may also explain why the face proves to be such an enduring and irresistible subject for artists.

Like the artists before them, scientists with many different perspectives are converging on the face. It took Ekman's and Izard's "universals expeditions" and their tools for analysis to make the face a respectable thing to study. By and large, Ekman shrugs off his pioneering role. Credit belongs to many. But then he says playfully, "I've never been able to understand why everybody doesn't study the face." Actually, most of us do. We just don't get paid for it.

Really and Truly

*UNTIL THEY ARE 4 OR 5, CHILDREN
DON'T UNDERSTAND THE DISTINCTION BETWEEN
APPEARANCE AND REALITY;
WHAT YOU SEE IS NOT ALWAYS WHAT YOU GET.*

John H. Flavell

John H. Flavell is in the department of psychology at Stanford University. This article is adapted from a Distinguished Scientific Contributions Award address presented at the recent meeting of the American Psychological Association.

It looks like a nice, solid piece of granite, but as soon as you squeeze it you know it's really a joke-store sponge made to look like a rock. If I ask what it appears to be, you say, "It looks just like a rock." If I ask what it really is, you say, "It's a sponge, of course." A 3-year-old probably wouldn't be so sure. Children at this age often aren't quite able to grasp the idea that what you see is not always what you get.

By the time they are 6 or 7 years old, however, most children have a fair grasp of the appearance-reality distinction that assumes so many forms in our everyday lives. Misperceptions, misexpectations, misunderstandings, false beliefs, deception, play and fantasy—these and other examples of that distinction are a preoccupation of philosophers, scientists, artists, politicians and other public performers and of the rest of us who try to evaluate what they all say and do.

For the past half dozen years, my colleagues and I have been asking children questions about sponge rocks and using other methods to find out what children of different ages know about the difference between appearances and reality. First we give the children a brief lesson on the meaning of the appearance-reality distinction by showing them, for example, a Charlie Brown puppet inside a ghost costume. We explain and demonstrate that Charlie Brown "looks like a ghost to your eyes right now" but is "really and truly Charlie Brown," and that "sometimes things look like one thing to your eyes when they are really and truly something else."

We then show the children a variety of illusory objects, such as sponge rocks, in a straightforward fashion and ask questions, in random order, about the reality and appearance of the objects: "What is this really and truly; is it really and truly a sponge or is it really and truly a rock?" "When you look at this with your eyes right now, does it look like a rock or does it look like a sponge?"

Or we show a 3-year-old and a 6-year-old a red toy car covered by a green filter that makes the car look black, hand the car to the child to inspect, put it behind the filter again and then ask: "What color is this car? Is it red or is it black?" The 3-year-old is likely to say "black," the 6-year-old "red." We use similar procedures to investigate the children's awareness of the distinction between real and apparent size, shape, events and the presence or absence of a hidden object.

In all these tests, most 3-year-old children have difficulty making the distinction between appearance and reality. They often err by giving the same answer (appearance or reality) to both questions. However, they rarely answer both questions incorrectly, suggesting that the mistakes are not random; the children are simply having conceptual problems with the distinction. By the time they are 6 or 7, however, most children get almost all the questions right.

Among 3-year-olds, certain types of illusory objects tend to elicit appearance answers to both questions (what we call a phenomenism error pattern), while others usually produce reality answers to both (an intellectual realism error pattern). The latter is the more surprising pattern because it contradicts the widely held view that

*I*F WE MAKE AN OBJECT THAT IS REALLY RED
OR SMALL OR STRAIGHT APPEAR BLACK OR BIG
OR BENT, MOST 3-YEAR-OLDS WILL SAY IT
REALLY IS BLACK OR BIG OR BENT.

young children respond only to what is currently most noticeable to them.

When we ask children to distinguish between the real and apparent properties of color, size and shape, they are most likely to make phenomenism (appearance) errors. If, for example, we use lenses or filters to make an object that is really red or small or straight look black or big or bent, most 3-year-olds will say the object really is black or big or bent.

But if we ask them what object or event is really present or has really occurred, most make intellectual realism errors. For example, children say the fake rock looks like a sponge. When they are shown a display consisting of a small object blocked from view by a large one, they say the display looks like it contains both objects rather than only the one they see. When they are shown someone who appears from the child's viewing position to be reading a large book but who the child knows is really drawing a picture inside the book, most children say it looks like the person is drawing rather than reading.

We also find that children make more phenomenism errors when we describe the same illusory stimuli in terms of their properties ("white" versus "orange" liquid) rather than in terms of their identities ("milk" versus "Kool-Aid"). Exactly why the appearance usually seems to be more important to young children in the first case and the reality more important in the second case remains a mystery.

Understanding of the appearance-reality distinction seems so necessary to everyday social life that it is hard to imagine a society in which normal people would not acquire it. To see if our findings applied in other cultures, we repeated one of our early experiments with 3-to-5-year-olds from Stanford

University's laboratory preschool with Chinese children of the same age at Beijing Normal University's laboratory preschool. Error patterns, age changes and even absolute levels of performance at each age level proved to be remarkably similar, suggesting that our results were not due to a simple misunderstanding of the English expressions "really and truly" and "looks like to your eyes right now." Instead, it seems that 3 or 4 years of age is the time when children of both cultures begin to acquire some understanding of the appearance-reality distinction.

We have not yet found effective ways to test for possible precursors of the appearance-reality distinction in children younger than 3, but we have tried to find out whether 3-year-olds really and truly lack competence in this area or only appear to. If there is one lesson to be learned from the recent history of developmental psychology, it is that the mental abilities of young children are often seriously underestimated simply because researchers at first fail to come up with accurate ways to measure those abilities.

To avoid this mistake, we devised a number of what we thought were easy appearance-reality tasks to be administered to groups of 3-year-olds. We used the same object-identity (fake objects) and color (objects placed behind colored filters) tasks as in our previous investigations, but we tried to make them easier for very young children. The tasks still demanded some genuine, if minimal, knowledge of the appearance-reality distinction but came closer than the standard tasks to demanding only that knowledge.

In one easy color task, for example, we left a small part of an object uncovered by the filter so its real color was still visible to the children when the

appearance and reality questions were asked. In another one, we took milk, whose real color is well known to young children, and used a filter to change it to a different color that they never see it have in reality. We thought this might help the children both keep the real color in mind and recognize the bizarre apparent color as mere appearance.

And since the repeated linking of questions about appearance and reality might confuse 3-year-olds, we further simplified matters by avoiding appearance and reality questions on some tasks. For example, at the beginning of the testing session, prior to any talk about appearances and realities, we asked the single "is" question about the toy car's color. Is it red, or is it black?

Similar strategies were used to create what seemed to be easier object-identity tasks: After a brief conversation about dressing up for Halloween in masks and costumes, the children were questioned about the real and apparent identity of one of the experimenters who had conspicuously put on a mask. We assumed that young children would be more knowledgeable about this sort of appearance-reality discrepancy through Halloween and play experiences than with those created by the fake objects and filters we had used in previous experiments.

Our use of easier-looking, less demanding tasks to study appearance-reality competence was surprisingly unsuccessful. Some of the children did perform slightly better on the easy tasks, but as a group their level of performance was almost the same as on the standard tasks. The results suggest that the typical young preschooler cannot think effectively about appearances and realities even when the

OUR USE OF EASIER-LOOKING, LESS DEMANDING TASKS WAS SURPRISINGLY UNSUCCESSFUL WITH 3-YEAR-OLDS.

tasks are deliberately made "child-friendly."

In a final test for hidden competence on appearance-reality tasks, we selected 16 3-year-olds who performed very poorly on such tasks and trained them intensively for five to seven minutes on the meaning of real versus apparent color.

We demonstrated, defined terms and repeatedly explained that the real, true color of an object remains the same despite repeated, temporary changes in its apparent color due to the use of a filter. We fully expected that this training would help the children, but when we retested them, only one showed any improvement, and that was slight. The difficulties 3-year-olds have with the appearance-reality distinction are apparently very real indeed.

Things soon begin to change, however. In both the United States and the People's Republic of China we found that performance on our appearance-reality tasks improves greatly between 3 and 5 years of age. This is consistent with the reported increase, at around 4 years of age, in the ability (probably related) to talk with playmates about pretend play.

While 6- and 7-year-olds are almost consistently error-free on these simple appearance-reality tasks, their development of knowledge about the distinction is not yet complete. We found, for example, that even though 6- and 7-year-olds answer the questions correctly, they continue to have trouble with the concept and find it difficult to talk about appearances, realities and appearance-reality distinctions.

Most don't show a well-developed, abstract understanding of the appearance-reality distinction until about the age of 11 or 12. In fact, it may not be until the time they reach adulthood that most people have a sufficiently rich and creative understanding of the concept that they can not only identify appearance-reality discrepancies but also reproduce them, change them or create new ones.

Our studies of how the appearance-reality distinction develops may shed light on a larger development—the child's understanding that mental representations of objects and events can differ both within the same person and between persons. I can be simultaneously aware, for example, that something appears to be a rock and that it really is a sponge. I can also be aware that it might appear to be some-

THE DISTINCTION IS WORTH STUDYING AS PART OF OUR KNOWLEDGE OF OUR OWN AND OTHER MINDS.

thing different under special viewing conditions, or that yesterday I pretended or fantasized that it was something else. I know that these are all possible ways that I can "represent" the same thing. In addition, I may be aware that you might represent the same thing differently than I do, because our perspectives on it might differ.

Knowledge about the appearance-reality distinction is but one instance of our more general knowledge that an object or event can be represented in different ways by the same person and by different people. The development of our understanding of the appearance-reality distinction, therefore, is worth studying because it is part of the larger development of our conscious knowledge about our own and other minds. And that's an area of development worth investigating—really and truly.

Voices, Glances, Flashbacks: Our First Memories

*CONSIDERING THE NOVELTY AND RICHNESS
OF THE FIRST FEW YEARS OF LIFE, WHY ARE
OUR EARLY MEMORIES SO FRAGMENTED?*

PATRICK HUYGHE

Patrick Huyghe is a science broadcasting fellow at WGBH in Boston, Massachusetts, and author of Glowing Birds: Stories from the Edge of Science.

"The earliest memory that I have is of waking up one morning with blood on my pillow and being extremely frightened."

"I don't remember how old I was, but I distinctly remember the joy of digging both hands into the dirt and stuffing it into my mouth."

"My first memory is of a chocolate birthday cake with white frosting and pink trim, and a little wooden train chugging around it."

Think back, for a moment, to your earliest memory. It is probably not your doctor's hands in the delivery room, or even your precarious first step. More likely, the tantalizing event occurred several years later. Perhaps you recall the birth of a sibling or the death of a family member or pet. Many first memories are of mundane events or images: sitting on the stairs, having a picture taken, eating a bowl of cereal.

That first childhood memory is notoriously hard to pin down. To be certain that you are in fact remembering, you must avoid the influence of family photographs and stories, which you may unintentionally substitute for true memories. While the distinction between memory and memento is easy to understand, in practice it is often hard to make. And even when you do manage to summon up an early experience, you may find it difficult to date accurately.

Or it may not have happened at all. Child psychologist Jean Piaget used to tell of a memory at the age of 2 in which he was nearly kidnapped as his nurse was wheeling him down the street in Paris. His recollection included the fact that the nurse's face was scratched by the kidnappers during the fracas. But when he was in his teens, the nurse confessed that she had fabricated the entire story.

Our earliest childhood memories have a magical quality about them, if for no other reason than their being the apparent beginnings of our conscious lives. These "islands in the sea of oblivion," as the novelist Esther Salaman called them, have fascinated psychologists for more than a century, and their studies of the phenomenon indicate that most people's early memories are remarkably similar on the surface.

"Almost all of our earliest memories are located in the fourth year of life, between the third and fourth birthdays," psychologist John Kihlstrom of the University of Wisconsin says. His survey of 314 high school and college students, conducted with Columbia University psychologist Judith Harackiewicz, found that most early recollections are visual, many in color. Their content, however, varies widely, and seems to fall into three broad categories: trauma, transition and trivia. Other studies have shown that the first memories of women appear to date back somewhat further than those of men, but the difference, which is no more than a few months, may be due to earlier brain development among girls.

The pioneers of psychoanalysis attached great significance to first memories. Freud believed that they could open the secret chambers of a person's inner life. Alfred Adler, originator of the Individual Psychology school, said, "The first memory will show the individual's fundamental view of life." Adler believed that childhood memories have a great diagnostic value, regardless of whether they are real or imaginary, because of their unique capacity for revealing a person's attitude toward self, others and life in general. Interestingly, Adler, who conceived of the inferiority complex, had a vivid first memory of sitting on a bench, sidelined by disease, watching his brother play.

In their autobiographies, various public figures show their tendency to fasten onto early experiences that are important to them. Golda Meir's earliest recollection, which she thought might have been a dream, was of a group of Jews being trampled by cossack horses in czarist Russia. The earliest memories of Seymour Papert, the

children, during the Oedipal phase, we repress anxiety-evoking memories of sexuality and aggression. All that remains, he noted, are "screen memories," memories that are totally lacking in feeling.

"Childhood amnesia does exist, but it's not necessarily Oedipal," contends Emory University psychology professor Ulric Neisser. "The child forgets everything about the self, not just sexual or aggressive memories. We begin

not remember anything before about age 3, not because they have forgotten, but because they were incapable of storing memories in the first place. The evidence for this is conflicting. "If you look at a 3- or 4-year-old in action," Neisser says, "you will see a person who remembers quite a lot, in the sense that you can ask a 4-year-old about things that happened the year before and get very intelligent answers. It's not that they have no memory, but when they become 10 or 12 or 20, they don't remember those things much anymore."

Psychologist Marion Perlmutter and her associates at the University of Michigan have been assessing the mnemonic abilities of preschool children in both experimental and naturalistic settings since the mid 1970s. In one study, she and psychologist Christine Todd examined the conversations between young children and adults to determine, among other things, the length of time children could retain information. They found that children between 35 months and 38 months old could remember events that had occurred more than seven-and-a-half months previously. The older children, those between 45 months and 54 months old, recalled episodes that had occurred as much as 14½ months ago.

Perlmutter was particularly impressed by the fact that in some cases the children "demonstrated a verbal recall for events that occurred prior to the time that they were speaking extensively." Her findings contradict some psychologists' long-held notion that children's autobiographical memory develops with language ability.

The study of memory in children who aren't yet speaking relies on the evidence of habituation and other forms of conditioned learning. In a study with psychologist Daniel Ashmead, Perlmutter asked parents to keep a diary recording the actions of their 7-, 9- and 11-month-old infants that revealed the use of memory. "Albert eating lunch," one typical entry reads. "Handed Dorine [the babysitter] his glass. Dorine saw it was empty and filled it. He did the same for me several days ago. Twice during one meal he handed me his glass. Each time it was empty. Each time I filled it."

While all the infants in the study showed some spontaneous memory,

PERSONALITY LEADS TO SELECTIVITY OF MEMORY. PEOPLE REMEMBER THINGS THAT ARE CONSISTENT WITH THE CONCEPT THEY HAVE OF THEMSELVES.

creator of the LOGO computer language for children, center on wheels, mechanical devices and figuring out what things do and how they work. Albert Einstein remembered receiving a magnetic compass at about the age of 4 or 5 and being awed by the needle's urge to point north.

"Some people think that these early experiences may somehow form personality, and that's why they get remembered," Kihlstrom says, "but I don't think that's right." Do our memories make us, or do we make our memories? "I think that personality leads to selectivity of memory," Kihlstrom says. "People remember things that are consistent with the concept they have of themselves."

Considering the novelty and richness of the first several years of life, it is perhaps surprising that adults have so few early recollections. This apparent amnesia has been a puzzle to psychologists ever since Freud observed it in his patients at the turn of the century.

The phenomenon, which he labeled infantile or childhood amnesia, applies only to our memories about the self, not to our memory for words or recognizable objects and people. Freud believed that we lose contact with most of our autobiographical memories from the first six years because as

to remember our life pretty well only from about the age of 5 or 6 because that's when we go to school and develop an organized structure for our lives."

But if schooling does allow children to better encode episodes for later retrieval, it would seem to follow that children who attend nursery school or other prekindergarten schools should have more early memories than those who did not. So far there haven't been any studies, however, to confirm this intriguing hypothesis.

"The evidence that the phenomenon of childhood amnesia even exists is mostly anecdotal," Kihlstrom says. Some people claim to have memories that date back before the age of 3, and most surveys of childhood memories indicate that there is no age when continuous, uninterrupted memories consistently begin. We are simply less likely to retain a memory as more time elapses from the event.

So perhaps what we call childhood amnesia is really no different from normal forgetting. "After all," says David Rubin, a researcher in human memory at Duke University, "childhood was a long time ago, and perhaps the reason we don't remember much of it is because we have just normally forgotten it."

People often assume that they can-such episodes were less frequent among the youngest infants. Perlmutter also found that older infants were more likely to reveal actual memory, rather than to just respond to a familiar environmental cue. "We think that this is evidence of something like recall memory beginning to appear in the older infants," she says.

When a child is between 8 and 12 months old, a change does seem to occur in memory abilities. Some psychologists see the change as a transition from conditioned recognition and response to recall memory, but others, such as Daniel Schacter and Morris Moscovitch of the University of Toronto, suggest that two different memory systems are at work. They refer to these systems as early and late memories.

Schacter and Moscovitch compared the performance of two groups: amnesiacs, who have an impaired memory system, and very young infants, who have not yet fully developed a memory system. They used a simple task that Piaget made famous in 1954. Piaget had observed that 7- to 8-month-old infants can easily find an object when it is hidden at the same location all the time. But after several successful searches at that location, many infants continue to search there even after seeing the object hidden someplace else.

O LDER INFANTS WERE MORE LIKELY TO REVEAL ACTUAL MEMORY, RATHER THAN JUST RESPONDING TO A FAMILIAR CUE.

"The amnesiac remembers where you put the object in the first place," Schacter says, "but then gets tripped up when you switch locations, just like the infant." But while amnesiacs continue to make the error, infants stop making it as they approach their first birthday. Schacter thinks that the ap-

pearance of a late memory system may explain this improvement. "The neural machinery that underlies the ability to remember the past may be in place within a year of birth. Any further developments in memory are probably the result of building up the knowledge base and integrating this machinery with other cognitive functions."

Studies of visual and auditory memory have shown that even the youngest of infants are consistently more responsive to novel stimuli than to familiar stimuli. These observations have led many psychologists to conclude that infants have a memory capacity from birth.

"Infants are able to encode and retain some information about their visual world from the first hours of life," Perlmutter says. Other researchers have shown that premature infants, with an average gestational age of 35 weeks, can discriminate between novel and familiar stimuli.

Despite evidence for early infant memory, psychologists have been reluctant to date the origins of memory before birth. Prebirth memory remains largely uncharted territory, and even those who willingly concede the possibility that the fetus has the rudimentary capacity to encode experience will cry foul at the claims for prebirth memories.

"On one level the subject is very, very controversial," Rubin explains, "but on another level it's totally dull. Why should the act of birth increase your learning abilities?"

Research by Anthony DeCasper at the University of North Carolina at Greensboro strongly implies the existence of prebirth memory. In his study, a newborn infant could choose

to hear a recording of its mother's voice or that of another woman by sucking on a nipple in a particular way. Infants as young as 30 hours consistently chose their mothers' voices.

It is very likely that they recognized their mothers' voices from what they

W HILE 3- AND 4-YEAR-OLD CHILDREN CAN PRODUCE REASONABLY GOOD GENERAL ACCOUNTS OF DINNER AT HOME, THEY HAVE DIFFICULTIES PRODUCING AN ACCOUNT OF A SPECIFIC DINNER.

heard in the womb. As Rubin points out, "The acoustics are there. There are studies in which microphones have been placed in the uterus of sheep and the sound is not muffled as much as you might think."

But in general, claims for birth and prebirth memories are regarded with suspicion by psychologists, most of whom tread more traditional ground in trying to answer the question, "When does autobiographical memory begin?"

Katherine Nelson, a developmental psychologist at the Graduate School of the City University of New York, has studied the question of early memories in connection with "scripts" that children have for familiar events. These scripts refer to the way children have organized their acquired knowledge in terms of general events. According to her theory, children have scripts for such familiar routines as eating dinner at home and going to the supermarket. This script-building appears by age 1, or earlier.

"The drive to build up these scripts seems to come out of a biological need to understand what is going on," Nelson says. "So the child doesn't need language or anything else; all that is needed is the background of experience."

While these scripts help children remember general events, they can also block or override memories of specific experiences. Nelson found that while 3- and 4-year-old children can produce reasonably good general accounts of dinner at home, they have difficulties producing an account of a specific dinner. They will speak about "what happens" rather than "what happened." Perhaps this explains why children insist upon routines, she says.

Nelson speculates that certain memories are lost as children enter specific experiences into their more general scripts. Unique events, like going to the circus, may be more memorable because they haven't been repeated or overridden by other similar experiences. "But after a while, if you don't go to the circus, you will forget about it, because it's not adaptive to hold onto that memory if it's not going to tell you anything about the future," Nelson says.

By the time children are 2 or 3, speech has developed and memory begins to show signs of social construction. "Children are also taught to remember by their parents," Nelson says, "when they say such things as 'Do you remember when we went to the store last week and you said such and such?' So their memory becomes at least partially formulated in terms of language. At about the age of 3, significant variations of emotionally involving events begin to create a memory string that is uniquely human and social. That's when autobiographical memory begins."

Although Nelson's view of early memory development does not pretend to explain the multitude of phenomena involved in memory, it certainly ties up a lot of loose ends regarding autobiographical memory. If, as she says, memory proceeds from the accumulation of single novel experiences, repeated and built into scripts, and then to unique events capable of being shared, it should be easy to see why we, as adults, cannot remember specific autobiographical memories before about the fourth year of life. Such memories could not form until a significant general base of event knowledge had been established. And this, of course, would take a number of years to build up. The reason that early memories are so elusive may be that the process of gaining autobiographical memories is like other developmental processes, something learned with time.

INSIGHTS INTO SELF-DECEPTION

Denial masks uncomfortable truths. This is one reason, say cognitive psychologists, why individuals and whole societies find a compelling need to lie to themselves. But there are risks to burying secrets.

Daniel Goleman

Daniel Goleman writes on psychology for The New York Times. This article is adapted from his book, "Vital Lies, Simple Truths: The Psychology of Self-Deception," published by Simon and Schuster.

THE WOMAN spoke about her father only after being reassured she would not be identified, because, in his time, the man had been famous. Even now, some 25 years after his death, the woman's voice was halting as she talked about her father's alcoholism and about how her family had somehow managed not to know how troubled he was:

"After my father died, we would find bottles of liquor hidden around the house, behind books, in the backs of closets. And, looking back, I can remember how poppa was always 'taking a nap,' as my mother would say. Sometimes he would get very loud and angry with my mother and push her around. She'd tell my little sister and me that he was 'in a mood,' and, without another word, she'd take us by the hand outdoors for a walk.

"He was an alcoholic, but somehow we stayed oblivious to it all. Once, after I was grown and married myself, I got up the nerve to ask my mother about it all. She denied it out-and-out." To this day, her mother has refused to admit the truth about her husband's drinking.

Family therapists have described this sort of denial and cover-up as "the game of happy family." It is just one aspect of the larger phenomenon of human self-deception, the nature of which is only now beginning to be understood by cognitive psychologists. The scientists' work explains how and why people lie to themselves. And patterns emerge from the scientific evidence that would seem to indicate that, just as individuals and families deceive themselves, so do larger groups of people, so do whole societies. The new research reveals a natural bent toward self-deception so great that the need for counterbalancing forces within the mind and society as a whole — forces such as insight and respect for truth — becomes more apparent than ever.

THE THEME OF BURIED SE-crets is so familiar and ancient in literature that it attests to the universality of the experience. The story of Oedipus revolves around such secrets. Willy Loman's tragic fall in "Death of a Salesman," testifies to the explosive potential of family secrets unmasked. Ibsen called this sort of secret a "vital lie," a myth that stands in place of a disturbing reality.

To acknowledge that it is commonplace for people to lie to themselves is not to understand why or how unpleasant truths can be buried so effectively. Freud explained it by proposing a range of psychological defenses, but his speculations came long before the detailed mapping of the mind's mechanics by cognitive psychologists, researchers who study how the mind perceives, processes and remembers information. Working in the laboratory with new techniques for measuring perception and memory, researchers have been able to sketch a scientific model of the mind, one that shows how and why self-deception can operate with such ease.

Among the major discoveries that have contributed to the modern understanding of the mind's architecture, and the place of self-deception in that design, are the following:

■ There is now firm scientific evidence that the unconscious mind plays an immensely potent role in mental life. The evidence includes the startling phenomenon known as "unconscious reading," in which, as psychologists at Cambridge University in England have shown, a person unconsciously registers the meaning of words that are presented to him in such a way that he has no conscious awareness of having seen them at all. The premise that *most* mental processes go on prior to awareness — and may never reach awareness at all — has now come into widespread acceptance among cognitive scientists.

■ Recently, psychologists at the University of Wisconsin have obtained evidence — for the first time ever — that suggests there is a specific mechanism in brain function associated with the psychological defense of repression. The transfer of information from one half of the brain to the other, they have found, is the point at which upsetting emotional experience may be blocked from awareness.

■ Self-deception itself is coming to be seen in a more positive light by psychologists, who find that it can serve people well as a psychological basis for self-confidence and hope. Researchers at the University of California at Berkeley have found that, in certain medical situations, those patients who deny the seriousness of medical risk fare better than those who dwell on it. This is not to say that self-deception is always to the good. But it may be that people fall prey to self-deception with such ease precisely because it has an appropriate, even essential, place in the ecology of mind.

ALTHOUGH RESEARCHERS are exploring self-deception by probing deep into the mechanics of the mind, the phenomenon itself can easily be observed in everyday life and at several levels of human activity. The roots of self-deception seem to lie in the mind's ability to allay anxiety by distorting awareness. Denial soothes. Freud saw that the mind, with remarkable alacrity, can deny a range of facts it would rather avoid and then not seem to know that it has done so.

At a dinner party, for example, a young woman commented on how close she was to her family, how loving family members had always been. She then went on to report, as evidence of their closeness, "When I disagreed with my mother she threw whatever was nearest at me. Once it happened to be a knife and I needed 10 stitches in my leg. A few years later my father tried to choke me when I began dating a boy he didn't like. They really are very concerned about me," she added, in all seriousness.

While the self-deception here is obvious, it often takes much more subtle forms, such as those that psychoanalysts track — defense mechanisms like denial and repression. All such mental maneuvers are part of a psychological calculus in which painful truths and soothing denials are the main variables. In the game of happy family, for instance, the rules call for twists of attention to bolster the pretense that nothing is wrong. Such psychological charades require that family members orchestrate their attention in an exquisitely coordinated self-deception.

As the Scottish psychiatrist R. D. Laing put it, "I have never come across a family that does not draw a line somewhere as to what may be put into words, and what words it may be put into." The line directs attention *here* and away from *there*. The rule works best when family members are not aware it exists at all but simply respect it automatically. In Dr. Laing's words, "If you obey these rules, you will not know that they exist."

Synchronized denial can take place in groups of all kinds. We slip so easily into group membership, as Freud saw, because we have learned the art of belonging as children in our families. The unspoken pact in the family is repeated in every other group we will join in life: Part of the price of membership, of being valued as part of a group, is to honor the implicit rules of shared attention and shared denial.

Such orchestrated self-deceptions were at work, for example, among the group that planned the Bay of Pigs invasion. Irving L. Janis, a psychologist at Yale University, studied in detail how the plans were laid for that fiasco. It was a textbook case of the collective defenses that Janis has called "groupthink."

Essentially, when groupthink is at work, group members hobble their seeking of information in order to preserve a cozy unanimity. Loyalty to the group requires that no one raise embarrassing questions, nor attack weak arguments, nor counter softheaded thinking with hard facts. "The more amiable the esprit de corps among the members of a policy-making group," Janis has observed, "the greater is the danger that independent critical thinking will be replaced by groupthink."

Looking back, Arthur Schlesinger Jr., who was then on the White House staff, observed how the meetings in which the Bay of Pigs plan took shape went on "in a curious atmosphere of assumed consensus." Yet, he suspects that had a single person voiced a strong objection, President Kennedy would have canceled the plan. No one spoke up. In a post-mortem, Theodore Sorenson, who had been special counsel to President Kennedy, concluded that "doubts were entertained but never pressed, partly out of a fear of being labeled 'soft' or undaring in the eyes of their colleagues." The rationalization, erroneous, as it turned out, that there would be a mass uprising against Castro once the invasion began, kept the group from contemplating such devastating information as the fact that Castro's army outnumbered the invading force by more than 140 to one.

The same dynamics that shunt discomforting facts from attention in groups operate in society at large. When some aspects of the shared reality are troubling, a semblance of cozy calm can be maintained by an unspoken agreement to deny the pertinent facts, to ignore key questions.

Take the case of Argentina in the late 1970's. While the military junta was in control there, the unaskable question within the society was: "What happened to the 10,000 or so political dissenters who mysteriously disappeared?" When the democratic regime took over in 1983, the unaskable question was the first to be asked. The answer, of course, pointed the finger of guilt at the junta itself.

To understand such self-deceit, whether individual or shared, cognitive psychologists focus on the mechanisms of the mind. A key element in the mind's architecture is rather dramatically represented by the phenomenon known as "blindsight." Certain functionally blind people — sightless as the result of stroke or brain injury rather than damage to the eye — have the uncanny ability to reach with accuracy for an object placed in front of them, even though, before they reach, they can not say where it is, or whether it is there at all. If asked to reach for the object, they will say it is impossible, since they cannot see it. But if they can be persuaded to try, they will find it with a sureness that amazes even themselves.

Blindsight is such a startling ability that some experts refuse to believe it can happen. Its authenticity is still hotly debated among cognitive scientists, some of whom are uncomfortable with the implication that only part of the mind can be aware of something. They argue that blindsight must be due to some form of cheating or sloppy research. One of those who defends blindsight is Anthony Marcel, a psychologist at Cambridge University. Marcel is more comfortable with blindsight than are some of his colleagues partly because he has done other experimental work that shows in normal people the mental capacity that seems most jarring in blindsight: That one part of the mind can know something, while the part that supposedly knows what is going on—awareness—remains oblivious.

Marcel had been doing studies of how people read when he chanced upon a strange effect. In his experiments, he would rapidly flash words on a screen, displaying them in a visual context so confusing they could not be read. When he asked his subjects to guess at the words that they thought they hadn't read, he was struck by a pattern of "clever mistakes." Often, the subjects would guess a word with a closely related meaning: "Day," for instance, might have been the word on the screen, and "night," the subject's guess.

Intrigued, Marcel began to flash words in such a way that observers did not even know that any describable image had been presented. Then he would project a pair of words and ask his subjects to guess which of the words meant or looked the same as the one they had not been able to perceive. He found that people guessed right more often than could be predicted by chance.

The results of these and subsequent studies involving the perception of words from strings of letters — Marcel calls these "unconscious reading" — make sense only if we adopt a rather radical premise in terms of how we normally think about the mind: Much consequential mental activity goes on outside awareness.

The whole process of recognition, sorting and selection takes a fraction of a second. Emanuel Donchin of the University of Illinois, a leading researcher in the field of cognitive psychophysiology, has done a great many studies using the evoked potential, a sophisticated brain-wave measure, to track the timing of the mind's operations. "In our research, we find that the mind recognizes a word within the first 150 milliseconds of seeing it," says Donchin. "But nothing shows up in awareness, as the subject reports it, for another 100 milliseconds or so, if it shows up at all."

At any given moment, then, most of what impinges on the senses, and most of the thoughts or memories that might come to mind as a result, never do come to mind. A huge amount of mental effort goes into sorting through and selecting a slim thread of

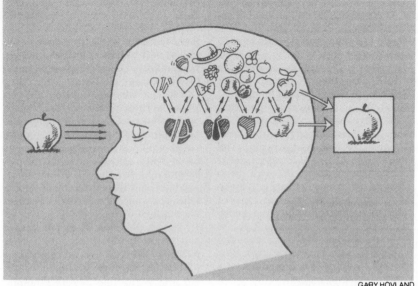

GARY HOVLAND

COMING INTO AWARENESS *According to many cognitive psychologists, the eye first registers an impression of an object. Then, in the unconscious, totally outside awareness, the mind scans and analyzes the object's basic elements, such as color, size, shape, and makes comparisons to what it knows — hats, fruits, bow ties, for instance. All this is done within a matter of milliseconds. Only the accurate perception that the object is an apple pops into awareness.*

consciousness from an immense array of mental candidates for awareness. The evidence is that the vast majority of possible thoughts and perceptions that might enter awareness are blocked from consciousness. There is a filter at work, and an intelligent one at that.

There are compelling reasons for this arrangement in the design of the mind. Awareness would be far too cluttered were the flow of information not vastly reduced by the time it arrived. If too much gets through, awareness is swamped by irrelevant information, as happens, in different forms, during anxiety attacks and in schizophrenia.

"Awareness is a limited capacity system," Donchin explains. "We don't know — and don't need to know — about most of the stuff the mind does. I have no idea how I search memory or get grammatically correct sentences out of my mouth. It's hard enough to handle the little that reaches awareness. We'd be in terrible shape if everything were conscious."

THE EXISTENCE OF AN INtelligent filter raises the question of just what intelligence guides the filter's operations. The answer seems to be that what enters through the senses gets a thorough, automatic scan by memory itself. There are several kinds of memory,

and this crucial gatekeeper's task seems to be performed, in part, by "semantic" memory, the repository of meanings and knowledge about the world. Semantic memory filters experience so that those messages that reach awareness are primarily those that have pertinence to current mental activity.

Donald A. Norman, a cognitive psychologist at the University of California at San Diego, who was one of the first to propose this design of the mind, argues that perception is a matter of degree. The judgment of relevancy is orchestrated by "schemas," the term psychologists use for the packets in which the mind organizes and stores information. All the contents of the mind are sorted into schemas; a train of association in thought is a road map through loosely connected schemas.

Schemas and attention interact in an intricate dance. Attention to one facet of experience — it is lunchtime, say, and you are hungry — activates other relevant schemas — thoughts of nearby restaurants, say, or of what is at hand in the refrigerator. The schemas, in turn, guide attention. If you walk down the street with these schemas active, your focus will be on the restaurants, not the other kinds of shops on the street; if you go to the refrigerator, your attention will fix on the cold cuts, not on the roast for the evening meal. Schemas choose this and not that; they determine the scope of attention. The interplay between attention and schemas puts them at the heart of the matter of self-deception. Schemas not only determine what we will notice: They also can determine what we do *not* notice.

Ulric Neisser, a psychologist at Emory University who wrote "Cognitive Psychology," the volume that put the discipline on the intellectual map, makes the point with an elegant, straightforward demonstration. He made a videotape of four young men playing basketball. The tape lasts just one minute. About midway, an attractive young woman carrying a large white umbrella saunters through the game. She is on the screen for four seconds.

Neisser showed the tape to visitors to his laboratory, who were asked to press a key whenever the basketball was passed between players. When Neisser asked afterward if they had seen anything unusual, very few mentioned the woman with the white umbrella. They had not noticed her; the schema guiding their viewing fixed attention on the ball. When Neisser then replayed the tape, they were astonished to see the woman.

I once asked Neisser whether there might be schemas that, in effect, say "do not notice that."

"Yes," said he, "I'm sure there are, at several levels. It probably starts from cases like the woman with the umbrella. People don't shift their attention from the task at hand. But the mechanism would be much the same when you have a pretty good suspicion of what's over there if you were to look, and you'd rather not deal with it. And you don't look; you don't shift your attention. You have a diversionary schema that keeps you looking at something else instead."

This kind of schema has a special potency in the mind: It operates on attention like a magician misdirecting his audience. Just such a mechanism seems to have been at work in a classic study conducted by Lester Luborsky, a psychologist at the University of Pennsylvania School of Medicine. Luborsky used a special camera to track people's eye movements while they looked at pictures. His apparatus allowed him to tell precisely where their gaze fell at each moment.

Some people gave a remarkable performance. When he had them look at pictures that were partly sexual in content, they were able to avoid letting their gaze stray even once to the sexual part of those pictures, though, presumably, their peripheral vision could detect it. Thus, when they looked at a drawing of the outline of a woman's breast, beyond which there was a man reading a paper, their eye did not fix on the woman at all, but focused only the man and his paper. Later, when asked to describe the picture, they had no recall of the sexual aspects; as it turned out, these people were particularly anxious about sexual matters.

"I think there's a lot of this kind of repression in everyday life," says Neisser, "lots of limits and avoidance in thinking about or looking at things. We all do that. There may be some painful experience in your life which, when you start to think about, you simply decide at some level not to pursue. So you avoid using your recall strategies. You could probably get pretty skilled at it, at not remembering what's painful."

In what may be the most telling results to date on the roots of self-deception, a team of researchers have pinpointed a brain mechanism associated with at least one defensive maneuver, a prospect Freud himself envisioned and then abandoned because of the primitive state of the brain sciences of his time.

The first step in this breakthrough was accomplished by Daniel A. Weinberger, now a psychologist at Stanford University, while he was still a graduate student at Yale. Weinberger was able to show that certain people, whom he called "repressors," consistently denied being anxious. In research on stress, he contended, they were being misclassified as being very low in anxiety, when, in fact, they displayed all the physical and behavioral signs of tension.

Weinberger presented college students identified as repressors with sexual or aggressive phrases. He would confront them, for instance, with "the prostitute slept with the student," or "his roommate kicked him in the stomach." He then asked them to free-associate from the phrases and found that their repression was obvious. Unlike other students who did the same task, the repressors offered associations that downplayed or avoided altogether the sexual or hostile tone of the phrases. At the same time, measurements of their heart rate, perspiration and forehead muscle tension revealed that they were, in fact, agitated.

There is, of course, the question of how conscious the repressors were of their self-deception: Were they lying about their feelings, or actually unaware of them?

An answer to that question has been suggested by very recent research.

Richard J. Davidson, a psychologist at the University of Wisconsin who had been a collaborator of Weinberger, carried the investigation one crucial step further. Davidson, working with Jonathan Perl and Clifford Saron of the State University of New York at Purchase, and using an ingenious technique, has been able to show that repressors suffer from a faulty transfer of information from one half of the brain to the other.

Davidson's experiments employed a device that, by means of a precise arrangement of lenses, projects a word so that it is seen by only that part of the retina that sends signals to the right hemisphere. Then the brain passes the information to the left. In a right-handed person, this means that the right hemisphere, which can register the meaning of words, must transfer the information to the speech center in the left before the person can speak that meaning.

Davidson had repressors free-associate to negative emotional words, many of which were sexual or hostile in meaning. When he presented these words to the right hemisphere, he found that a significant time elapsed before the subjects could utter their responses. Among those who study brain response, this slower reaction time is interpreted to mean that there is a deficiency in the transfer of information, in this case from the right to the left hemisphere. Of most significance was the specificity of the lag: It was for the negative words — which presumably posed a psychological threat — not for neutral or positive words. And the lag showed up only when the words were presented to the right hemisphere, not when shown to the left.

These findings take on special significance in light of the fact that the right hemisphere is strongly believed to be a center for emotions, such as fear and anxiety. Thus, in theory, when repressors experience anxiety, their emotional center in the right brain sends that information to the verbal center in the left over the same faulty circuits. In short, the entire pattern suggests that the repressor's denial of his anxiety is associated with deficient brain function centering on the transfer of information from the right to the left hemisphere. The findings suggest that the repressor is not lying about his lack of agitation, but is actually less aware of it than are most people. The same mechanism, Davidson believes, may operate whenever people repress threatening information.

A RANGE OF RESEARCH suggests a decidedly positive role for certain kinds of self-deceit. For example, in research at a hospital near San Francisco, Richard S. Lazarus of the University of California at Berkeley found that patients who avoided thinking about the surgery they were facing fared better afterward. Lazarus's colleague, Frances Cohen, interviewed patients about to undergo elective surgery, such as for gall bladder problems. Some patients, they found, were extremely vigilant about what would happen — and what might go wrong — during surgery, even reading medical texts to discover fine details of the procedure. Others completely ignored such facts, relying instead on faith that things would go right.

The avoiders, the researchers found, recovered more quickly after the surgery, and with fewer complications. In a similar study, researchers at the University of North Carolina have found that those patients who similarly avoided thinking about forthcoming dental surgery showed more rapid healing afterward.

Avoiding what is painful, to a great extent, seems to serve a positive function. There is a growing body of research evidence that shows there to be a pervasive mental tendency for people to ignore or forget unpleasant facts about themselves and to highlight and remember more easily the pleasant ones. The result is an illusory glow of positivity. When people become depressed, the illusion that things are better than a neutral weighing of facts might suggest disappears. Hope, the crucial mainstay in the face of all adversity, depends to a great extent on the same illusion. In short, self-deception, to a point, has a decidedly positive place in the human psyche.

Nevertheless, Lazarus is quick to point out that the context makes all the difference. "You shouldn't assume denial is *necessarily* good," he observes. "The presurgical patients had nothing to gain from their vigilance. Take, by contrast, the case of a diabetic; he's got to monitor his sugar levels constantly. If he denies his problem, he's in great trouble."

IF THERE IS A LESSON TO be drawn from the new research, it is the urgent need for compelling antidotes to self-deception. The more we understand how natural a part self-deceit plays in mental life, the more we can admit the almost gravitational pull toward putting out of mind unpleasant facts. And yet, as in the case of the diabetic cited by Richard Lazarus, there is often danger in giving in to denial, whether that denial is individual or collective.

Psychotherapy seeks to heal by exposing, not suppressing, hidden truths, and the therapist's stance is no different from that of the investigative reporter, the ombudsman, the grand jury or the whistle-blower. Each bespeaks a willingness to rock the boat, to bring into the open those facts that have been hidden in the service of keeping things comfortable.

We live in an age, we say, when information has taken on an import and urgency unparalleled in history. A mark of democracy, we maintain, is that information flows freely. It is totalitarian authority that must choke off alternative views and suppress contrary facts: Censorship seems the social equivalent of a defense mechanism.

Now that cognitive psychology is showing how easily our civilization can be put at risk by burying our awareness of painful truths, we may come to cherish truth and insight, more than ever before, as the purest of goods.

INTELLIGENCE
NEW WAYS TO MEASURE
THE WISDOM OF MAN

KEVIN McKEAN

Mr. McKean is a senior editor of Discover.

On the weekend of July 4th, 1942, when psychologist Seymour Sarason reported for work at the Southbury Training School for the mentally retarded, the place was in an uproar: one of the students had escaped. Southbury was a model new institution set in a lovely Connecticut valley. But the students, who were still more or less prisoners in those days, would occasionally evade supervision long enough to slip into the woods and strike out for home, obliging the school to send out a search party.

Sarason, who had been hired to set up a psychological service, paid little attention to the escapes at first. But as the months wore on, he noticed a curious thing. He was giving the students the Porteus Mazes Test, an IQ exam often used for retarded people since it required no language, simply challenging them to trace their way out of printed mazes. To his astonishment, many of the escapees couldn't work so much as the first and simplest puzzle. "These kids couldn't get from point A to point B on paper, so how did they plan a successful runaway?" says Sarason, now a professor at Yale. "That was when I realized that what these kids could plan on their own was in no way reflected by how they did on tests."

The lesson that there's more to intelligence than IQ is one that most people learn the hard way at one time or another. Everyone has known people with low IQs who get along in the world famously, and others with high IQs who never amount to much. Indeed, the venerable Intelligence Quotient has such an imperfect relationship to intelligence that many psychologists have dropped the term from their lexicon.

FASCINATION WITH IQs

Nevertheless, the subject of intelligence—and how to measure it—continues to engross psychologists and laymen, probably because intelligence is the principal ability that separates man from other creatures. People who readily agree that the results of IQ tests are meaningless shy away from revealing their own scores. Those who did well on such tests during grade or high school tend to recall the results with smug satisfaction. Those who did poorly remember in shame or forget.

A similar division prevails among scientists. Defenders of IQ, like Hans Eysenck of the University of London, points to its eight-decade record of service: "There's an indisputable body of scientific evidence showing that IQ tests do reflect actual cognitive abilities." Says Earl Hunt of the University of Washington, "The intelligence test is probably psychology's biggest technological contribution." Critics counter that IQ's many flaws render it useless. "The assumption that intelligence can be measured as a single number is just a twentieth-century version of craniometry," says biologist and author Stephen Jay Gould, referring to the nineteenth-century "science" that claimed a man's intelligence could be determined by measuring his head. Norman Geschwind, the noted Harvard neurologist, was fond of pointing out that some people with massive frontal-lobe brain damage, whose personality, motivation, and insight had been irreversibly damaged, could still attain near-genius IQ scores—a fact, he said, that showed the bankruptcy of IQ.

IQ fell on hard times through a combination of bad luck and misuse. Its bad luck was to have been invented at the turn of the century, when racial and nationalistic prejudices were more prevalent—or, at least, more apparent—than they are today. Some of the creators of IQ misused it as a justification for repressive measures against foreigners, blacks, and other "undesirables." And scientists are still hotly debating whether the differences in IQ among various races and nationalities mean anything. These controversies leave some researchers doubtful about whether IQ is worth rescuing. "The problem with IQ is that it's been marred for decades by the smell of political issues, race, and so on," says psychologist Robert Sternberg of Yale. "Intelligence research doesn't have to be that way. There's legitimate scientific inquiry to be pursued."

Sternberg and other young theoreticians are striving to render IQ obsolete by forging new and more realistic definitions of what it means to be intelligent. It would be wrong to characterize this group, which includes, among others, psychologists Howard Gardner at Harvard and Jon Baron at the University of Pennsylvania, as a "school"; their ideas are too diverse. Gardner identifies seven "intelligences," including social grace and athletic skill. Sternberg posits three, one of which, practical intelligence, resembles common sense. Baron stresses the need for rational thinking. Still other theoreticians argue that mental development doesn't cease at adolescence, as many defenders of IQ maintain, but continues throughout life.

HUMANISTIC PERSPECTIVE What these new theories share is an almost humanistic perspective. Their creators, mindful of the pitfalls of IQ, borrow from cognitive psychology and neuroscience to define smartness as a complex web of abilities; they construct new intelligence exams using realistic problems; they explicitly allow for national and cultural differences in the definition of intelligence; they argue that smartness results from an interaction of genes and environment, making the bitter "nature-nuture" argument pointless; and, contending that much of intelligence consists of learned skills, many of the theorists are devising programs to teach it.

No one expects the new theories to end complaints about intelligence testing, or to settle once and for all what that elusive quality we call intelligence really is. Yet they've already begun to change the scientific establishment's view of intelligence. Sternberg is devising a new exam for the Psychological Corporation, a major test marketer; Gardner is discussing a joint project with the Educational Testing Service, creator of the Scholastic Aptitude Test (SAT). While the direct influence of these researchers may be small, changes in the way such scientists view intelligence have important consequences when they filter into the world at large (as the fact that many people still equate intelligence with IQ shows). To understand what we will mean by intelligence in the next century, we need to understand what psychologists think of intelligence today. So the work of this diverse group of researchers is likely to have far-reaching effects, for better or worse, on our social and public policies, and on our view of ourselves, in the future.

ORIGINS OF THE IQ CONCEPT

The modern conception of intelligence has its roots around the turn of the century, when a number of scientists sought definitions for the term. Sir Francis Galton tried to measure intelligence using simple reaction-time tests. French psychologist Alfred Binet in 1905 published the first modern IQ-like test to help the government identify schoolchildren in need of remedial education.

STANFORD-BINET TEST Binet's test was taken up enthusiastically by American psychologists, chief among them Stanford's Lewis Terman, who, in 1916, produced an expanded version designed for subjects of any age. The Stanford-Binet test, as it was called, consisted of problems that would be familiar to modern IQ test-takers: vocabulary questions, tests of reasoning and logic, questions that involved completing a series of numbers. While Binet's scale had yielded a score expressed in terms of "mental age," Terman called his score an Intelligence Quotient—calculated by dividing a subject's mental age by his physical age and multiplying by 100. (Thus a six-year-old performing at the six-year-old level would have an IQ of 100; if he performed at the nine-year-old level, his IQ would be 150.) And while Binet meant his test simply as an educational tool, Terman had broad ambitions for wide-spread testing of adults. "Intelligence tests," he wrote, "will bring tens of thousands of . . . high-grade defectives under the surveillance and protection of society. This will ultimately result in curtailing the reproduction of feeble-mindedness, and in the elimination of an enormous amount of crime, pauperism, and industrial inefficiency."

Terman's dream of mass testing was quickly realized. A Harvard psychologist named Robert Yerkes persuaded the Army to examine some 1.75 million recruits during World War I. Whether because of poor test conditions—many test halls were so crowded that recruits seated in the back could scarcely hear the instruction—or the soldiers' lack of ability, the average able-bodied white recruit scored a mental age of about 13, a tad smarter than "moron." The test's authors were dismayed by the low scores, but encouraged that the racial and national break-down suited the prejudices of the day. Yerkes' disciple Carl Campbell Brigham calculated that Americans of "Nordic" descent, which meant, more or less, northern Europeans, had an average mental age of 13.28; "Alpines," or middle Europeans, 11.67; "Mediterraneans," or southern Europeans, 11.43; and blacks, 10.41.

With the Army exams as impetus, intelligence testing took off after the war. Brigham and others devised the first SAT; Terman issued a revision of the Stanford-Binet in 1937 (there was third revision in 1960); in the 1940s and '50s, psychologist David Wechsler of Bellevue Psychiatric Hospital in New York devised intelligence tests for adults and children that today rival the Stanford-Binet in popularity.

MODERN IQ TESTS

The modern versions of these tests are fairly straight-forward. The Wechsler Adult Intelligence Scale exam, for example, is given in a personal interview. It consists of eleven subsections that ask examinees to do such things as define words, solve math problems, recall strings of digits in forward or reverse order, and arrange blocks according to a specified design. IQ tests are used to identify children who are slow or speedy learners, to evaluate job candidates, as part of psychological or psychiatric exams, and, when coupled with other tests that assess interests or experience, to help draw a picture of a person's mental strengths and weaknesses.

Gardner well remembers his first exposure to such an IQ test battery. His parents, Jewish refugees from Nazi Germany who had settled in Scranton, Pa., knew they had a bright child and wanted to find out what to do with him. So, in the mid-1950s, when Howard was thirteen, they took him to Hoboken, N.J. for a week of testing. "It cost three hundred dollars—the equivalent of thousands of dollars today," says Gardner, now 42. "At the end of the week they told my parents, 'Your son tests well in everything, but he seems to be best at clerical matters.' Maybe that was when my skepticism about tests got its start."

With true clerical attention to detail, Gardner made sure that nearly every grade he earned at Harvard was an A (he got one B+). Then he headed to England for graduate study in philosophy and sociology. After returning to Harvard in 1966, he worked with gifted children in Project Zero, an innovative task group trying to understand artistic creativity, and also with brain-damaged patients at Boston University and the Boston V.A. Hospital.

This dual exposure, to talented children and grievously ill adults, helped mold Gardner's eclectic view of the mind. He and David Feldman of Tufts were impressed by the tender age at which some children manifest special abilities. Musically inclined pre-schoolers, for example, easily learned to play simple instruments "not only because they found music patterns easy to learn, but because they found them almost impossible to forget!" The observation reminded Gardner of the story of Stravinsky, who, as an adult, could still remember the tuba, drums, and piccolos of the fife-and-drum band that had marched outside his nursery.

At the V.A. hospital, Gardner was struck by the cruel and exquisite selectivity with which disease and injury can damage the mind. Patients with a left-hemisphere lesion might lose the power to speak but still be able to sing the lyrics to songs because the musical right hemisphere was intact. Right-hemisphere patients might read flawlessly but be unable to interpret what they read. (Gardner's findings suggested that the ability to read between the lines—that is, to get the point of something, including a joke—is largely a right-hemisphere function.)

These experiences persuaded Gardner that intelligence, far from being a unitary power of mind, consists of a set of mental abilities that not only manifest themselves independently but probably spring from different areas of the brain. In his book *Frames of Mind* (Basic Books, 1983), he hypothesizes that there are at least seven broad categories of intelligence. Three are conventional: verbal, mathematical, and spacial. But the other four—musical ability, bodily skills, adroitness in dealing with others, and self-knowledge—have sparked controversy because they're far afield of what's usually called intelligence. "If I'd talked about seven talents, nobody would've complained." Gardner explains. "But by calling them intelligences, I hope to shake up the community that wants to reserve this name for the results of a Wechsler exam or the SAT."

In defense of the label "intelligence," Gardner argues that each of the seven abilities can be destroyed by particular brain damage, each shows up in highlighted form in the talents of gifted people or idiots savants, and each involves unique cognitive skills. "Take an athlete like Larry Bird, who has a sixth sense of where to throw a basketball," says Gardner. "He has to know where his teammates and opponents are, judge where they are likely to go, and use analysis, inference, planning, and problem-

CATEGORIES OF INTELLIGENCE

solving to decide what to do. A number of different intelligences get involved in these decisions. But it's clear that, even in the bodily movements alone, there's a reasoning process.''

Gardner's theory makes room not only for Western definitions of intelligence, but also for other cultures. Intelligence among the Iatmul people of Papua New Guinea, for example, may consist of the ability to remember the names of some 10,000 to 20,000 clans, as adult males of that group do. In the Puluwat culture of the sprawling Caroline Islands, intelligence can be the ability to navigate by the stars.

CULTURAL VARIATION Because he emphasizes cultural variation, Gardner refuses to define a single IQ-like scale. ''When you measure people on only one measure, you cheat them out of recognition for other things,'' he says. But in a project with Feldman and Janet Stork of Tufts, and Ulla Malkus and Mara Krechevsky of Project Zero, he's seeking ways to characterize the skill patterns of children in a Tufts pre-school. To gauge musical ability, for example, the group may ask a child to listen to a melody and then re-create it on tuned bells. ''When they are three or four, we should, without subjecting children to anything intrusive, be able to profile their abilities,'' Gardner says. ''And that's the time when feedback to parents and teachers can make a difference. . . .''

Not content to be a mere taker of tests, Sternberg, the son of a Maplewood, N.J. dressmaker's supply salesman, decided to create his own. At age thirteen he dug up a description of the Stanford-Binet in the library and used it to devise what he called the Sternberg Test of Mental Abilities. He then set about assessing the mental abilities of his classmates until the school psychologist called him in for a scolding. ''The psychologist didn't think it was appropriate for seventh graders to be testing each other,'' Sternberg says.

This early setback didn't prevent Sternberg, now 35, from going on to become one of the foremost figures in the field. He's at once very similar to and strikingly different from Gardner. Both are children of immigrant Jews who fled the Nazis (in Sternberg's case, only his mother came to the U.S. as a refugee—his father had arrived earlier). Both graduated from college summa cum laude, Gardner from Harvard, Sternberg from Yale, and returned to their alma maters to teach. And both aim to broaden the definition of intelligence. . . .

Sternberg's ''triarchic theory,'' laid out in *Beyond IQ* (Cambridge University Press, 1985), breaks intelligence into three parts. The first deals with the mental mechanisms people use to plan and carry out tasks, with special emphasis on what Sternberg calls the ''meta-components'' of intelligence—the skills by which people plan and evaluate problem-solving. To Sternberg, planning is often more important than sheer mental speed: good test-takers, for instance, spend more time than poor ones on studying and digesting questions before trying to work them out.

Sternberg's studies also highlight the role of planning in reading comprehension, often a part of standard IQ tests. He and a colleague, Richard Wagner, asked volunteers to read four passages, one for gist only, one for main ideas, one to learn details, and one for analysis. The best readers, as measured by a standard Nelson-Denny reading test, devoted most of their time to the passages that had to be read with the greatest care, whereas poorer readers spent the same amount of time on all four selections. When Sternberg added the results of his time-allocation test to the Nelson-Denny profile, the combined score turned out to be a better predictor of how well people understood the passages than the Nelson-Denny score alone.

The study shows how Sternberg tries to make intelligence tests more realistic. ''Think of the reading comprehension tests on the SAT,'' Sternberg says, ''They tell you to read each passage carefully. If you did that in real life, you'd be reading and reading and reading. You'd never get done.''

The triarchic theory's second part deals with the effect of experience: the intelligent person not only solves new problems quickly, Sternberg argues, but also trains himself to solve familiar problems by rote in order to free his mind for other work.

The third part, which focuses on practical intelligence, asserts that common sense depends largely on what Sternberg calls tacit knowledge—which might loosely be defined as all the extremely important things they never teach you in school. He says success in life often depends more on tacit knowledge than on explicit information. ''A lot of people in a field don't know it [the tacit knowledge],'' says Sternberg. ''Or don't know that they know it. Or don't know what it is they know, even if they do know that they know it.''

Sternberg is convinced that, if tacit knowledge could be made explicit, it could be taught, and he tries to teach it to his graduate students. ''I sit them down and explain the strategy of scientific publishing, how it's better to publish in a journal with a circulation of 40,000 than 4,000, and how to tailor a paper to a journal''—things that many professors don't teach, in the belief that they're ''sleazy.'' Sternberg has also devised a number of tests of practical intelligence. Some pose typical quandaries in business, politics, or science. Others seek the testee's sensitivity to nonverbal cues: one type of question presents a picture of two people and asks which is the boss and which the employee (the boss is usually older and better dressed, of course, but also tends to look directly at the employee, who tends to look away). The aim of such questions is not to train people to spot bigwigs, though that alone might be useful, but to make intelligence tests more realistic. "Standard IQ tests are fairly good for predicting how people will do in school, but they have a very low correlation with job performance," says Sternberg, who has a $750,000 grant from the Army to develop practical intelligence tests for military job placement. "I want to be able to say who's going to be a good officer, a good business executive, or a good scientist." He's also worked up a program for teaching intelligence that is being published as *Intelligence Applied* (Harcourt, Brace, Jovanovich) this winter. . . .

THE NATURE-NURTURE DEBATE

To Arthur Jensen, programs that aim to raise intelligence, like Sternberg's and Baron's, are just wishful thinking. He believes that it's genes, not culture or environment, that do the most to determine intelligence. Specifically, Jensen argues that intelligence is a physical property of the brain; that IQ is a pretty fair measure of that property; that braininess is mostly inherited; that, as a result, there are sharp biological limits set at birth on an individual's intellectual capacity; and that there may also be clear-cut differences in average intellectual potential among races and nationalities.

It's not a position that has won many friends. When Jensen first suggested, in a 1969 article in the *Harvard Educational Review*, that lower intelligence might account for the fact that blacks don't do as well as whites in school, friendly critics called him naïve and hostile ones a racist. Pickets appeared at his University of California at Berkeley office; he had to move classes to a new location each day to avoid demonstrations. Hate mail arrived in such quantities that, for a year, the police bomb squad opened all packages, and campus cops accompanied him to and from work. "It was sort of like having a bear by the tail," recalls Jensen, whose lectures at other campuses are still occasionally disrupted by protests. "I could either run and get out of it, or I could stick to my research and see where it would lead."

As his words imply, Jensen stuck it out. And his research has made him the biggest thorn in the side of people like Gardner and Sternberg who would like to broaden the meaning of intelligence. In intelligence testing Jensen-style, the examinee sits before a control panel featuring an array of buttons. He holds his finger on the center button until one of the eight surrounding buttons lights up. Then he jabs his finger at the lighted button as quickly as he can, turning it off. This procedure is repeated with slight variations about 60 times. Although this apparently mindless exercise doesn't seem to involve much intelligence, performance on a battery of such tasks correlates quite well with performance on standard IQ exams—which implies that Jensen's button-pushing device is nearly as good at measuring IQ as the tests themselves.

THE "g" FORCE

Critics say this only confirms the meaninglessness of IQ, but Jensen, 52, sees more fundamental forces at work—in particular, a force that Charles Spearman called g. Spearman observed that there was a high degree of correlation among a wide variety of mental tests, even those that seemed wildly dissimilar. In a classic paper published in 1904, he argued that this correlation reflected a *general* mental ability—hence g—involved in all cognitive work, and presented a mathematical method called factor analysis for determining the extent to which g was involved in any given task.

Spearman's paper set off a debate that still rages over whether g is a real property of the brain (Spearman thought it was a free-floating mental energy, a notion that has since been abandoned). Jensen argues that his button-pushing task measures IQ mainly because the task is highly g-loaded, as the jargon goes. And a person's level of g, Jensen adds, may be controlled by genes. Identical twins score more nearly alike than fraternal twins on g-loaded subsections of the Wechsler exam. Japanese children whose parents are first or second cousins score lower on the g-loaded subsections (compared to non-inbred children) than on the portions that are not so g-loaded—suggesting that g is more vulnerable than other mental abilities to the harmful genetic effects of inbreeding.

In a June article in *Behavioral and Brain Sciences*, Jensen analyzed a number of standard IQ tests and concluded that those showing the largest difference between the average scores of blacks and whites were also the ones with the highest g-loading. The gap between the races was small—only about 15 points, less than the average variation between children in a family. Nevertheless, Jensen argues that a small average difference could markedly affect the black-white ratio at the highest and lowest levels. Only 16 percent of test takers score higher than one standard deviation above average (in IQ, one standard deviation—a conventional statistical measure of scatter—equals about 15 points). Fewer than three percent exceed two standard deviations above average. Thus, a group with an average IQ of 115 would have five times as many people over 130 as an equal-sized group with an average IQ 15 points lower—simply because 130 is two standard deviations above the lower group's average but only one standard deviation above the higher group's. "It doesn't take a big shift in average," says Jensen, "to make an enormous difference in the portion of the curve that's above a cut-off like 130."

Jensen thinks this may explain the fact that 39 percent of Asian-American high-school graduates in California meet Berkeley's entrance requirements, versus only 15 percent of other students; his studies of children from San Francisco's Chinatown suggest that Americans of Asian descent may be about 15 IQ points smarter than children of European descent. At the other end of the intelligence distribution, Jensen blames a lower average IQ for the disproportionate numbers of blacks in remedial education. School systems that try to restore racial balance by putting more whites in these programs, he says, only deny help to some who need it while retarding the progress of others who don't.

GENETIC DIFFERENCES In Jensen's view, as much as 70 percent of the differences in intelligence—or at least *g*—may be genetic. "There's no doubt that you could breed for intelligence in humans the way you breed for milk in cows or eggs in chickens," he says. Would it be worthwhile? Jensen says yes. "If you were to raise the average IQ just one standard deviation, you wouldn't recognize things. Magazines, newspapers, books, and television would have to become more sophisticated. Schools would have to teach differently." Of course, the transition wouldn't be smooth: "You would go from having three percent of the population over IQ 130 to something like 15 percent, meaning that you would have five times as many people wanting to become doctors, lawyers, professors, and so forth. If the change occurred suddenly, you might have to hold lotteries to decide who gets what job."

CULTURAL INFLUENCES

Jensen's critics regard this kind of talk as meaningless. "To say that because black children *don't* do well on IQ tests they *can't* do well is extraordinarily simple-minded," says Sandra Scarr, chairman of the psychology department at the University of Virginia. "Our studies show that blacks reared by whites have an IQ of about 110, the same as white adoptees raised in the same environment." And Berkeley anthropologist John Ogbu, a Nigerian,

points to studies showing that Third World schoolchildren who attend Western-style schools score as high as Westerners on IQ tests.

Ogbu and others add that tests designed for one culture are notoriously faulty when applied to another. A classic example is the study by Joseph Glick of Liberia's Kpelle tribesmen. Glick, of the City University of New York, asked the tribesmen to sort a series of objects in a sensible order. To his consternation they insisted on grouping them by function (placing a potato with a hoe, for example) rather than by taxonomy (which would place the potato with other foods). By Western standards, it was an inferior style of sorting. But when Glick demonstrated the "right" answer, one of the tribesmen remarked that only a stupid person would sort things that way. Thereafter, when Glick asked tribesmen to sort the items the way a *stupid* person would, they sorted them taxonomically without difficulty.

Stephen Jay Gould maintains that even if intelligence were 70 percent heritable, which he doubts, it wouldn't prove that racial or cultural differences were genetic. Imagine, he says, a group of malnourished Africans whose average height is a few inches less than that of North Americans. Height is highly heritable—about 95 percent. But that fact gives no assurance that these Africans would stay shorter if they were properly fed. The average height in Japan has gone up several inches since World War II, but no one argues that the Japanese gene pool has changed. As for Spearman's *g*, Gould believes it has no real existence but is simply a mathematical artifact expressing subtle but pervasive advantages of schooling, parental attention, expectations, and motivation. "The chimerical nature of *g* is the rotten core of Jensen's edifice, and of the entire hereditarian school [of IQ]," he wrote in *The Mismeasure of Man* (Norton, 1981).

Critics fear that Jensen's ideas will do psychological damage to the supposedly inferior groups, and could even be used to justify social or political oppression. His defenders counter that this is no excuse for limiting scientific inquiry: "Would you proscribe Charles Darwin because the Social Darwinists used his theory to justify *laissez-faire* capitalism?" asks Bernard Davis, an emeritus bacteriologist and geneticist at Harvard Medical School. To that, critics respond that anyone working in the intelligence field has to be sensitive to the political consequences of his ideas. "If you aren't," says Gardner, "you're just an ostrich."

Moreover, Gardner and the other new theorists believe the whole nature-nurture debate is a red herring. "Suppose we improved our social and educational systems so much that all schools and homes were absolutely equal," says Sternberg. "Then, regardless of how genetically influenced intelligence was, the heritability would go up to 100 percent—since there would be no other differences besides genes. That shows how meaningless heritability is." Utopian visions of a superintelligent future are naïve he thinks, because "some of the brightest people in history have been the biggest bastards."

BIOLOGICAL INFLUENCES ON INTELLIGENCE

Whether or not heritability is meaningless, it's certainly true that intelligence is sometimes biologically influenced, as the work of Julian Stanley and Camilla Benbow dramatically shows. Since 1971, Stanley, of Johns Hopkins University, has collected data on children who, before the age of 13, score 700 or better (of a maximum 800) on the math SAT. These children are culled from annual talent searches like the one Stanley founded at Johns Hopkins, and represents roughly one child out of every 10,000 in that age group nationally.

The startling discovery of Stanley and Benbow, who joined him in 1977, is that among the 292 high scorers on the math test, boys outnumber girls about twelve to one. The effect seems limited to math, since the sex ratio is roughly fifty-fifty among children who score high on the verbal SAT. "We were shocked at first," says Stanley, chief of the Study of Mathematically Precocious Youth, as the program is called. "We knew that by age eighteen, when most kids take the SAT, boys do better than girls. But by that age, boys have usually taken a lot more math. We assumed that at age twelve, when both sexes have taken the same amount of math, there would be little or no difference." And questionnaires aimed at finding a cultural explanation failed: boys and girls in the screening program answered similarly when asked whether they studied math, enjoyed math, or felt math was important to their careers.

Not all scientists have accepted the implication that bright thirteen-year-old boys may be innately better at math than girls. The journal *Science*, in which the report appeared, ran seven letters of rebuttal arguing that Benbow and Stanley had overlooked important social factors—such as the possibility that twelve-year-old girls might not even enter a "talent search" if they thought it would make them brainy or unattractive to boys. But Benbow and Stanley felt some of the criticism was unfair. "People were furious with us for finding what we found, and even more furious that, having found it, we had the ill grace to publish," says Stanley. "We're not saying bi-

ology is the cause," adds Benbow. "We're just saying it's premature to rule biology out."

It would be nice to say that the two scientists could at least rule out the stereotype of the bright student squinting through thick glasses and sniveling into a handkerchief. But here, too, their data lend curious support to an old prejudice. The high scorers are about four times as likely as other children the same age to be myopic, and twice as likely to have allergies (or other auto-immune disorders) and to be left-handed. This collection of traits is not as whimsical as it sounds. Geschwind, who died last year, believed that hormonal influences in the womb often caused male fetuses to experience greater development of their brains' right hemispheres at the expense of the left. This, he argued, could account for the fact that males are more likely to be left-handed (the right hemisphere controls the left side of the body), and also for the fact that left-handers are somewhat more prone to auto-immune disorders and reading disabilities.

The fact that the Johns Hopkins kids are anything but reading-disabled doesn't cut them out of this picture. Benbow notes that mathematical reasoning—as opposed to simple calculation—seems associated with the right hemisphere. Thus, if Geschwind's theory is correct, a right-hemisphere hormonal boost may be a double-edged sword, encouraging mathematical genius in some, sending brain development awry in others. Benbow, who moves to Iowa State next year, will pursue this biological puzzle by tracking a number of the children into adult life—a modern variation of Lewis Terman's massive study of high-IQ schoolchildren.

Whatever the cause, it's clear there's a qualitative, not merely a quantitative, difference between prodigies and ordinary bright kids. When these twelve-year-old mathematicians sit down to take the SAT, many of them have never been exposed to the algebra, geometry, and rudimentary calculus that a high school senior might know. Thus, they more or less have to re-invent these disciplines on the spot—deducing things like the Pythagorean theorem, or how to factor algebraic expressions.

What if such a child had been born in classical times? Would he have upstaged Pythagoras? Stanley thinks not. "He would've manifested his ability early and become a great surveyor or engineer," Stanley says. "But remember, if he'd been born in Rome in the year zero, he wouldn't even have had a zero to work with!" It was the decimal system, derived from the Arabs and Hindus, that introduced the mathematically important zero; the citizens of Imperial Rome were hampered by their cumbersome—and zero-less—numeral system.

QUALITATIVE DIFFERENCES

Stanley's point is that intelligence is meaningless when separated from its cultural roots. That's why Newton, for all his genius, wasn't being unduly modest when he said he stood on the shoulders of giants. Thanks to cultural storage and transmission of knowledge, every person stands on the shoulders of countless giants to whom he owes concepts, like inertia, that are so familiar they seem intuitive, but which were in fact brilliant insights.

IS INTELLIGENCE INCREASING WITH TIME?

That fact makes it extremely difficult to say whether intelligence is increasing with time. If one considers cultural advances, the answer is certainly yes: there's little doubt that at least in scientific and technical matters, today's culture is vastly advanced over that of centuries or even decades ago. And a randomly selected sample of modern Americans would easily outscore thirteenth-century Frenchmen in standard IQ. But would the contest be fair?

Gardner adds that different kinds of intelligence have been valued at different times in history. In the pre-literate era, the mark of wisdom was a prodigious verbal memory. Today, that sort of skill is more associated with *idiots savants*. And, while one can only speculate how modern geniuses would fare were they transported to the past, there have been examples of the reverse situation, in which a "primitive" scientist has been thrust into the present. One such case was that of Srinivasa Ramanujan, the renowned Indian mathematician, who was brought up in an isolated village. Ramanujan had an extraordinary talent for seeing the hidden properties of numbers: once, when a friend visited him as he lay ill in England, Ramanujan observed that the number of the visitor's cab—1,729—was the smallest that could be expressed as the sum of two cubes in two different ways. "This was amazingly rapid mathematical insight," says Gardner, "but not [one] that was at a premium, or even especially appreciated, in twentieth-century Britain. Beyond natural gifts, aspiring mathematicians need to be in the right place at the proper time."

Just as each type of intelligence has its time, so does each stage of its development in individuals. This was the insight of the renowned Swiss psychologist Jean Piaget, who charted the intellectual discoveries that make up the mental growth of children. For Piaget, who died in 1980, the last stage of mental development was reached during adolescence with the attainment of what he called "formal operations"—more or less the ability to pose and solve abstract logical problems. But some of Piaget's intellectual descendants think mental

development continues in adulthood, and a number of them have devised tests to try to prove it.

Harvard psychologist Michael Lamport Commons, for example, charts three stages of mental development beyond formal operations, each involving the ability to reason abstractly about the achievements of the previous stage. His colleague Lawrence Kohlberg has devised tests of an adult's ethical development. To assess the relationship of such measures to IQ, Commons recently gave a battery of his tests to 150 volunteers from Mensa—the worldwide organization of people with high IQs. The Mensa group didn't score any better than adults of average intelligence. On the contrary, those who claimed to have the highest IQs tended to have lower degrees of ethical judgment on Kohlberg's scale. Says Commons, "The notion that geniuses will solve our problems is a hoax."

Although IQ, by some measures, declines three to four points per decade after age 20, University of Denver psychologist John Horn finds that other aspects of intellectual performance actually increase with time. Horn's chart of mental abilities over a life-time shows most trailing dismally downward with age. But at least two of the lines climb hopefully upward. One represents an ability that Horn and others have called crystallized intelligence, amounting to the sum of a person's knowledge and experience. The other is what Horn calls long-term storage and retrieval, that is, the ability to call to mind experiences of long ago. Taken together, Horn argues, these two might well be called wisdom. "Their effect shows up on tests like the one in which you ask people to think of all the things they could do with a brick. The older person comes up with more ideas because he has been around longer and had more experiences. And the ideas are good ones, not silly ones."

Perhaps the most troublesome puzzle of intelligence is why some obviously intelligent people make such messes of their lives—as was the sad case with Leonard Ross. A child prodigy who won $164,000 on TV quiz shows in the 1950s, Ross graduated from Yale Law School—where he was editor in chief of the law journal—and seemed destined for a brilliant future. But he drifted from one job to another, never able to settle on a career or find personal happiness. Last May, at age 39, he was found floating in a motel pool in Santa Clara, Calif., apparently a suicide.

Sternberg, who has given this matter much thought, thinks problems like Ross' lie outside the realm of intelligence. He has drawn up a list of factors that undermine intelligence performance—lack of motivation, inability to persevere, uncontrolled impulsiveness, and failure to know one's own limitations. "You have to concentrate on those things you're best at," Sternberg says. "The trick of intelligence is playing to your strengths."

Gardner sees our inability to comprehend why the intelligent often live unintelligently as a consequence of the nature of intelligence testing. Most intelligence tests, even newer ones, last no more than a few hours, and it may simply be impossible to draw a realistic picture of a human being in that time. Gardner contends that testers should dump the short-answer quiz in favor of a system that would assign grades, or even intelligence scores, on the basis of performance on long-term projects. "There are too many people who are brilliant at short-answer quizzes and yet failures at life for that talent to be important," Gardner says. "What was it William F. Buckley said about beating the SAT—that all you had to do was figure out what the guy who made up the test wanted you to say? But creative people look beyond the superficial. It's the ones who don't accept the 'right' answer who come up with the really important ideas."

ORIGINS OF SPEECH

TALKING MEANS POWER, AND WE'RE THE ONLY ANIMALS WHO CAN DO IT.

ROBERT FINN

Robert Finn, a science writer with an M.S. in psychobiology, appears in Contributors.

Humans are the only animals that possess the biological machinery needed for speech. Chimpanzees may learn sign language, and honeybees may dance out a message to the hive, but only people speak to each other in words. And some researchers are beginning to believe that the evolution of speech may have been every bit as crucial to our mastery of the world as the evolution of our brains.

Until recently, however, there has been no way to discover how and when the organs evolved that make speech possible. No fossil remains have survived; the flesh, muscle and cartilage of which they are composed decay soon after death. Even so, painstaking detective work by anatomist Jeffrey Laitman, of Mount Sinai School of Medicine in New York City, and his colleagues has turned up fossil clues to the structure of speech organs in the earliest prehumans, or hominids. And some careful work in comparative anatomy is revealing just how special the human vocal apparatus is.

"Human speech is a two-part system," Laitman explains. "You must have the brain for it and you must have the vocal tract for it. Apes are very intelligent primates. They certainly have an advanced communication system. But do they have the ability to produce speech? The answer is no."

The hominid and ape lines began to diverge about 8 million years ago, and predecessors of *Homo sapiens* arose some 4 million years later. At some point since then, the organs that produce sounds in all mammals shifted, subtly but decisively, in a way that probably permitted man or his ancestors to produce rapid, articulate speech. Even today, one of the most striking things about our speech organs is the fact that they are remarkably similar to the "standard plan" vocal apparatus found in all mammals, apes included.

Sound starts at the vocal cords, which are located in the larynx, or voice box. This is the structure in the neck that creates the bulge known as the Adam's apple. The sound is then modified by the supralaryngeal vocal tract, which consists of three resonant chambers—the mouth, the nose and the pharynx (the area at the back of the throat where the nose and mouth meet the esophagus). Complex neural mechanisms direct the distribution of sound within these chambers by coordinating the interplay of the soft palate (the continuation of the roof of the mouth), the tongue and the lips.

When viewed from above, the vocal cords, more properly called the vocal folds, look like two lips constricting the larynx. As the lungs push air upward, the vocal cords flap open and closed repeatedly, releasing tiny puffs of air to the speech structures farther up the line. The vocal cords work in much the same way your lips do when you sound a raspberry or a Bronx cheer. And just as you can alter the raspberry's pitch by adjusting the tension on your lips, you can alter the "fundamental frequency of phonation" by changing the tension of the vocal cords. You can feel your vocal cords vibrate by touching your fingers lightly to your throat as you hum.

Other primates have vocal cords much like ours. According to otolar-

Our throats are like trumpets—versatile enough
to modulate sound in many ways. Other animals
can't do this; their throats are like bugles.

yngologist Jan Wind, of the Institute of Human Genetics at the Free University, Amsterdam, "The chimpanzee larynx produces sounds that have about the same qualities as the human voice." He maintains that if it were possible to graft a chimpanzee larynx onto a human throat, speech would probably be unaffected. Vocal cords aren't even entirely necessary for speech. Many people who have lost their larynx through surgery for a malignancy swallow air into the stomach and use a series of controlled burps to actuate the speech sound.

The uniqueness of the human vocal apparatus lies less in the organs themselves than in their location. The larynx is lower than in the standard plan, making the pharynx larger; and the tongue, which in the standard plan lies entirely within the mouth, extends

down into the throat. To characterize these differences precisely, Yale University anatomist Edmund Crelin, working with Jeffrey Laitman, then his graduate student, used cineradiography—X-ray movies—to study the vocal tracts of adults, children and monkeys in the act of vocalizing.

Since most of the speech organs are transparent to X-rays, the subjects were given barium sulfate to drink. "With barium you get an outline of the throat, the oral cavity and the esophagus," says Crelin. "We also put little metal clips on the tip of the uvula and on the tip of the epiglottis. [The uvula is the teardrop-shaped structure hanging from the back of the roof of the mouth; the epiglottis is a flap that's open when you breathe and covers the larynx during swallowing.] This way we knew the exact position of these structures. Otherwise, on the X-ray, they're practically invisible."

More recently, Crelin used casts taken from human and animal cadavers to make rubber replicas of the vocal tract. He attached each of these to a plastic larynx, and then, he recalls, "we altered the shape of the rubber tract to produce the sounds of speech, especially the vowels."

☐

These studies showed that the main departure of human speech organs from those of other mammals is in the location of the larynx, whose lower position in the neck results in a greatly expanded pharynx. Crelin and Laitman compare the standard-plan vocal tract to a bugle and the human one to a trumpet. Both instruments are made from the same materials and make more or less the same initial sounds. But the trumpet is more versatile because it has three valves that modify sound by directing it down a number of passageways. The

How Sounds Are Made

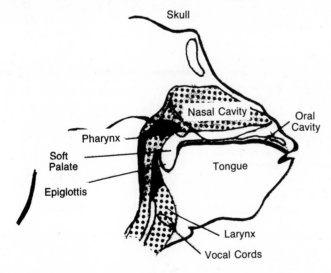

Sound starts at the vocal cords in the larynx. The soft palate, tongue and lips distribute sound among three resonant chambers: the pharynx and the oral and nasal cavities. But the "standard" vocal plan of chimpanzees (above) and other mammals does not permit speech.

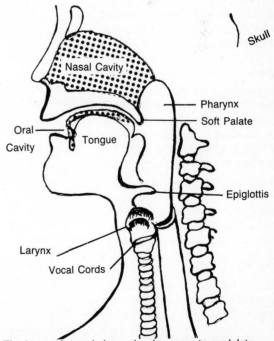

The human larynx is lower, leaving room to modulate speech. Humans are born with the standard plan, but as the larynx descends, the distance between the roof of the mouth and the rear of the skull base decreases. The center of the base buckles into an arch (arrow).

bugle's tubing is a fixed length and cannot be modified, whereas the trumpet's three valves allow sound waves to travel an additional length of tubing. This permits sound coming from the mouthpiece to be modified, something that isn't possible with a bugle. The tubing of the trumpet is similar to an adult human's throat.

"Having the larynx lower in the neck increases the area available to modify sound," Laitman explains. "We've gained the ability to produce a greater array of sounds than any other mammal." Among the sounds unique to humans are the vowels in the words *team, tomb* and *tom* and the initial consonants in the words *gut* and *cut.*

We pay a high price for this versatility, however. The opening to the trachea, or windpipe, is situated fairly low in the neck; food passes by on its way to the esophagus, so we have a much greater chance of choking than do animals with a standard-plan vocal tract. It's virtually impossible for most other mammals to choke, because their windpipes make direct contact with the nasal passages. An ape can breathe through its nose while simultaneously drinking liquids or swallowing small bits of food; the material simply passes to the left and right of the larynx.

Human infants retain the advantages of the standard-plan vocal tract. "The airway is identical in the monkey and the newborn human," Crelin says. "The tongue is all in the mouth, and the larynx locks into the pharynx.

But beginning at six months, there is a gradual unlocking and descent. At six years of age, it has reached its final point. And that is the time, linguists have found, at which the child can make all the sounds of speech."

Before they could make any firm statements about the evolution of speech, Crelin and Laitman needed to find some aspect of the vocal apparatus that doesn't decay after death. Since the upper end of the pharynx abuts the base of the skull, they reasoned that the anatomy of the skull might tell them something about the anatomy of the speech organs. And so it did: Animals with standard-plan vocal tracts, including dogs, cats, apes and human infants, have a skull base that's relatively flat. But human adults and children older than six years have a deep arch in the center of the skull base.

"The distance between the hard palate [the roof of the mouth] and the foramen magnum [the opening through which the spinal column passes to join the brain] gradually decreases so that the skull base buckles into an arch," Crelin explains. The deeper the arch, the farther the larynx has descended.

Since the base of the skull is often preserved in fossilized remains, the researchers could now reconstruct the speech organs of our early ancestors. These reconstructions indicate that the australopithecines, who are thought to have lived between 4 million and 1.5 million years ago, possessed the standard-plan vocal tract. (Thus, the famous fossil known as Lucy would not have been able to speak as we can today.) "The first evidence that the vocal anatomy had begun to change appears in forms such as *Homo erectus*, about one and a half million years ago," Laitman says. "The anatomical structures necessary for the types of sounds that are made by present-day humans probably did not become fully in evidence until the rise of archaic *Homo sapiens*, some three to four hundred thousand years ago."

The details of the tangled net of evolutionary pressures that caused this change remain a major bone of contention among scientists in the field. Jan Wind has isolated at least 97 factors that influenced the evolution of speech either directly or indirectly. Among these he includes the increasing size of the brain (and the addition of the neural mechanisms necessary for speech perception and production), the flattening of the muzzle (which pushed the tongue and larynx down the throat) and the need for communication during cooperative hunting.

Which Factors Were Key?

It's difficult to say which factors were most significant. Crelin cites the change in vocal-tract anatomy. "The brain didn't initiate speech. You first have to have an end organ," he says. "The peripheral structures evolve, and they evolve by accident. Once you've got them, the brain will adjust so that you can get maximum function from them. The vocal tract had to evolve first—and then the brain followed suit and used it to the maximum."

Wind disagrees. "I think that the brain is more important than the vocal tract. Both are needed, of course, but I think that a human equipped with a chimpanzee's vocal tract would be able to produce vocalizations that we would call speech. Maybe that human being would speak with a funny sort of voice—an accent— but I maintain that such a being would be able to speak."

According to Philip Lieberman, professor of linguistics and cognitive science at Brown University, the most important point is not *whether* an animal with a stan-

dard-plan vocal tract could make certain sounds, but *how rapid* the rate of speech would be. He cites as an example Neanderthal man, who, though anatomically modern in many respects, retained the standard-plan vocal tract long after archaic *Homo sapiens* began changing over to the modern configuration.

During their time, the Neanderthals were quite successful. Yet they were outclassed in the struggle for existence by archaic *Homo sapiens,* who possessed brains no larger in total volume than theirs, who were physically much weaker and whose only obvious advantages were somewhat larger frontal lobes in the brain and the modern vocal tract.

Language itself probably did not set archaic *Homo sapiens* apart from the Neanderthals, who may well have had a language of their own. The crucial difference, Lieberman claims, was our ancestors' ability for rapid, syllabic speech. Modern humans can speak at a rate of 20 to 30 sounds per second—about 150 words per minute—and can talk and listen for long periods of time. We achieve this breakneck rate by melding sounds together. When we say the word *bet,* for example, we do not enunciate the *b,* then the *e* and only then the *t.* Instead, as soon as we start forming the *b* sound, our mouths begin moving toward the configuration for the *e.* And before we get all the way there, the transition to the *t* begins.

In contrast, the Neanderthals' standard-plan vocal tract and the corresponding lack of neural speech mechanisms would probably have limited them to conveying one sound at a time—and to transmitting their thoughts only one-tenth as fast as we can. Neanderthal man's language would have been restricted to very short sentences and very simple thoughts.

Neanderthal man's language would have been restricted to very short sentences and simple thoughts.

But fast speech poses something of a problem. The acoustical characteristics of even a simple word like *bet* vary greatly according to the length of the speaker's supralaryngeal vocal tract. One speaker's *bet* might be acoustically identical with another's *bit.* In order to perceive speech properly, we must resolve such ambiguities automatically, instantaneously and unconsciously. We do this, according to Lieberman, by subconsciously listening for acoustic cues. One of the most accurate is the "supervowel" [i]—the vowel sound in the words *be, beat* and *beef.*

Like the other vowels, the exact acoustics of [i] will vary from person to person according to his own vocal tract. But [i] is so different acoustically from other vowels that it is easily distinguished. Once [i] is pronounced during a conversation, the listener's brain can instantly resolve all subsequent ambiguities in pronunciation.

Adapted for Rapid Speech

Since only the human vocal tract can produce the supervowel, Lieberman believes humans are the only living animals adapted for rapid, syllabic speech. But others remain skeptical of his conclusions. Jan Wind says, "If you could deduce the physical qualities of the vocalizations of ancestral hominids, as Lieberman has done, there's still the problem of whether or not you should call these vocalizations speech. The definition of speech depends not on the purely physical qualities but, rather, on the communicational and informational qualities."

What was ultimately responsible for the Neanderthals' extinction? Says Crelin, "I would love to say that it was a virus or a bacterium or cold weather, but when our *Homo sapiens* ancestors were really able to talk, they became the most vicious big-game hunters the world had ever seen. I'm sure a Neanderthal would have been easy game for them. They just killed them all off, either directly or indirectly, by being so good at getting the goodies out of the environment without leaving anything for the poor Neanderthal."

The end result was that in a mere 40,000 years—a blink of the eye in evolutionary terms—mankind took control of the Earth. We owe all this to our articulate ancestors, who seem to have fast-talked their way onto the world stage.

For Further Reading

The Biology and Evolution of Language, by Philip Lieberman, Harvard University Press, Cambridge, Massachusetts, 1984.

"The Anatomy of Human Speech," by Jeffrey T. Laitman, *Natural History,* August 1984.

"Primate Evolution and the Emergence of Speech," by Jan Wind, in *Glossogenetics: The Origin and Evolution of Language,* edited by Eric de Grolier, Harwood Academic Publishers, 1981.

Why Children Talk to Themselves

Laura E. Berk

Laura E. Berk, Ph.D., is Professor of Psychology at Illinois State University, Normal, Illinois.

Among the hundreds of daily verbal exchanges experienced by children in a typical group setting, an astute observer will notice an intriguing but puzzling form of language behavior which psychologists call *private speech*. It refers to speech uttered aloud by children which appears to be addressed to either themselves or to no one in particular. Unlike adults, who self-consciously talk to themselves only in solitary moments, young children frequently use private speech in public. A closer look at children's private speech reveals that it assumes a variety of forms.

"Where's another green piece? Here's one! It doesn't fit in. . . ." says Sarah to herself as she works on a puzzle.

Omariah experiments with words as he sings to himself, *"Put the mushroom on your head. Put the mushroom in your pocket. Put the mushroom on your nose."*

Rico and Joan are working with rolling pins and clay. *"My mommy made chocolate chip cookies,"* says Rico as he flattens out his clay. *"I'm going to the park after school,"* Joan responds irrelevantly.

When observed in school environments, children between the ages of 4 and 10 use private speech in about 20% (Berk & Garvin, 1984; Kohlberg, Yaeger, & Hjertholm, 1968) to 50% (Piaget, 1926) of their language. If children talk to themselves so frequently, what role does such language play in their psychological growth?

Early views

Egocentric speech

Piaget's earliest work, *The Language and Thought of the Child* (1926), first called attention to children's "egocentric speech"—speech that was either not directly addressed to another or not expressed in such a way that others could easily understand its meaning. Piaget observed three types of private speech in children between the ages of 4 and 7:

repetition—the child imitates syllables and sounds or echoes the verbalizations of another. "Lev [after hearing the clock strike 'coucou']: *'Coucou . . . coucou.'*" (p. 35).

monologue—the child carries on a verbal soliloquy during an activity or while walking about. "Lev sits down at his table alone: *'I want to do that drawing, there. . . . I want to draw*

something, I do. I shall need a big piece of paper to do that.'" (p. 37).

collective monologue—the speech of one child seems to stimulate speech in another, but the remarks of the second child are not a meaningful and reciprocal response to those of the first. "Pie: *'Where could we make another tunnel? Ah, here Eun?'* Eun (responding egocentrically): *'Look at my pretty frock.'*" (p. 76).

By calling children's private speech *egocentric*, Piaget expressed his view that such language is an indication of the young child's cognitive immaturity. Young children, Piaget interpreted, engage in egocentric speech because they cannot take into account the perspectives of others. For this reason, their talk is often "talk for self," much of which is not meaningful or understandable to a listener.

From this early work emerged a major Piagetian concept—*egocentrism*. Piaget frequently referred to egocentrism as responsible for the limitations of early childhood thinking. According to this concept, young children assume that their own viewpoint is the same as everyone else's and take it for granted that others will simply understand them without any special effort to adapt. Inability and unwillingness to modify one's thinking in response to others leads to egocentric speech—soliloquies and collective monologues—instead of real social exchange of ideas.

Challenges to Piaget

As Piaget's work spread to other European countries and the United States, several professionals challenged his ideas based on their own observations of children. In Britian, Isaacs (1930) collected detailed observations of young children's language while serving as director and teacher in a small experimental school. Isaac's naturalistic records suggest that genuine social communication involving coordination of viewpoints—exchange of opinion, argument, and mutual correction—abound in preschool children's speech, while egocentric speech is extremely rare. Although she observed instances of monologue, these seemed to be quite separate from children's relations with people. Monologues appeared to arise "from certain situations or moods." Younger children engaged in more monologues than older children, but they also used social language extensively. Very few instances of collective monologue were observed.

In the United States, McCarthy (1930) came to similar conclusions. A sample of 140 children between the ages of 18 and 54 months was observed while playing with toys. Mc-

Reprinted by permission from *Young Children*, July 1985, pp. 46-52. © 1985 by the National Association for the Education of Young Children, 1834 Connecticut Avenue, N.W., Washington, D.C. 20009.

Carthy acknowledged that her observations were based on speech when children were in the presence of an attentive adult, while Piaget's were made during free play. Nevertheless, she found that less than 5% of children's remarks were egocentric. She emphasized the importance of individual differences, as a few children showed extremely high rates of egocentric responses, but for most egocentric speech was so infrequent that no age or sex differences could be detected.

In his work on the relationship between children's language and thought in the Soviet Union, Vygotsky (1934/1962) proposed a major challenge to Piaget's view that young children's language is egocentric and nonsocial, and that egocentric speech plays no useful role in the child's development. Vygotsky found that children's egocentric speech varied according to the situation. When children engaged in tasks in which they encountered obstacles and difficulties, the incidence of egocentric speech nearly doubled. It appeared that children were trying to solve problems by talking to themselves: *"Where's the pencil? I need a blue pencil. Never mind, I'll draw with a red one and wet it with water; it will become dark and look like blue."* (p. 16).

Vygotsky viewed egocentric speech as the link in the transition from vocal speech to inner verbal thought:

• Early in development egocentric speech follows children's actions, occurring as an afterthought.

• Then it simultaneously accompanies children's activities.

• Later, it precedes children's actions and becomes externalized thought which helps children to control, order, and guide their actions.

• During the school years, it becomes internal thought.

Vygotsky viewed the development of egocentric speech as similar to the sequence in which children name their drawings. The youngest children draw first and then describe what they have drawn. Later, children name their drawings as they work. Finally, they decide beforehand what they will draw. At this last stage, speech precedes action and enables children to plan an idea. In this way, children come to use language to solve problems, to overcome impulsive action, to plan solutions ahead of time, and to master their behavior.

Finally, in sharp contrast to Piaget, Vygotsky believed that language even in the very youngest children, is inherently social and that egocentric speech has its origins in social communication. When young children are unable to solve a problem, they often ask for help from an adult. The adult may verbally describe a method that the chid may not have been able to implement alone. Vygotsky reasoned that eventually this social speech with another person is turned toward the self. Therefore he believed that egocentric speech was "communication with the self" for self-guidance and self-direction.

New directions

More recently, Kohlberg, Yaeger, and Hjertholm (1968) tried to resolve the differences between the theories of Piaget and Vygotsky. Their position mirrored neither Piaget nor Vygotsky completely. Still, it leaned strongly in the direction of the Russian view. They concluded that private speech arises out of children's early social communcation and is a conversational dialogue with the self. Its purpose is to help children guide themselves verbally and control their own actions.

In trying to understand how private speech develops, Kohlberg and his colleagues drew heavily from Mead (1934). Mead explained that children become more aware of the meaning of their actions when they attempt to communicate. As children engage in social interaction, they gradually call out in themselves the responses they get from others. This process leads to a "two-part self"—"a speaking self" and a "self talked to," which communicate with each other. Gradually this externalized conversation becomes internal thought, a developmental transition which Kohlberg labeled as movement from outer-directed to inner-directed private speech.

Kohlberg et al. (1968) expanded the types of private speech categories and predicted that they emerge in a five-level developmental sequence (see Table 1). This developmental trend, as well as several aspects of Vygotsky's theory, is documented in a series of empirical studies. In support of its social origins, private speech is positively related to social participation among preschool children and happens more frequently when children are in the company of peers than adults. The self-guiding function of private speech is also evident because it increases when children are faced with difficult tasks.

New challenges

Although the studies conducted by Kohlberg and his colleagues were more comprehensive than previous research, they were criticized for several reasons (Berk & Garvin, 1984; Fuson, 1979; Zivin, 1979). One major objection was that the private speech categories still seemed incomplete. Remarks involving **affect expression** (such as *"Ouch, my knee!"*); **solitary fantasy play; reading aloud or sounding out words;** and the Piagetian category of **collective monologue** were not included. These omissions raised questions about Kohlberg's conclusion that all private speech serves a single developmental purpose and indicated a need for further inquiry into Piaget's explanation.

Nevertheless, other studies in the 1970s continued to support Vygotsky's predictions. The most consistent finding was that private speech increases with task difficulty or when children make errors or are confused about how to proceed (Deutsch & Stein, 1972; Dickie, 1973; Goodman, 1975; Zivin, 1972). Klein (1964) and Beaudichon (1973) found that task-relevant private speech increases during the preschool years and seems to reflect the child's growing cognitive maturity.

Recently, Berk and Garvin (1984) observed the private speech of 5- to 10-year-old children using a revision of the Kohlberg categories which cover the full range of children's private speech types (see Table 1). Our findings once again support Vygotsky's view. Self-guiding forms of private speech occur most often. Private speech is correlated with social speech at the youngest ages and increases in the classroom when children are confronted with demanding academic tasks, if the adults remain at a distance and do not substitute for children's self-control. Also, the Appalachian children in our sample show a pattern of private speech development which resembles the Kohlberg sequence, although their rate of development is somewhat slower than Kohlberg's largely middle-class sample.

Finally, collective monologues rarely occur among these school-age children, or among 3- to 5-year-olds observed in a follow-up study (Berk, 1985). Thus, Piaget's theory does not

Table 1. **Representative private speech category systems.**

	Piaget (1926)	Vygotsky (1962)	Kohlberg, Yaeger, & Hjertholm (1968)	Berk & Garvin (1984)	Diaz (1984)
Generic label(s) used	Egocentric speech	Egocentric (self-guiding) speech	Private speech	Private speech	Self-regulatory private speech

Categories, subtypes	1. Repetition
	2. Monologue
	3. Collective monologue
		No separate subtypes	(Hierarchically organized developmental sequence)
			A. Presocial, self-stimulating language
			  1. Word play and repetition
			B. Outward-directed private speech
			  2. Remarks addressed to nonhuman objects
			  3. Describing own activity
			C. Inward-directed, self-guiding speech
			  4. Questions answered by the self
			  5. Self-guiding comments
			D. External manifestations of inner speech
			  6. Inaudible muttering
			E. Silent inner speech or thought
				(Revision and expansion of Kohlberg, et al. categories)
				1. Affect expression
				2. Egocentric communication
				3. Word play/repetition
				4. Fantasy play
				5. Addressing nonhuman objects
				6. Describing own activity/self-guidance
				7. Self-answered questions
				8. Reading aloud
				9. Inaudible muttering
					1. Labeling and describing
					2. Focusing attention
					3. Regulation of motor activity
					4. Facilitating transitions
					5. Ending uncertainty and perseveration
					6. Abstraction of distinctive features
					7. Praise and self-reinforcement
					8. Whispers
					9. Play and relaxation

seem to adequately explain most instances of young children's private speech.

Although our overall findings seem to confirm Vygotsky's theory, we also found instances of children's private speech which seem to serve functions other than self-guidance. Nor is private speech as unitary in purpose as Kohlberg and his colleagues suggest. Rather, children's private speech seems to be multifaceted and diverse. For example, the youngest children engaged in much word play and self-directed fantasy play especially in peer social contexts. Affect expression, though low in frequency, occurred to some degree in children of all ages. Besides self-guidance, children seemed to use private speech for verbal stimulation; play and relaxation; and to express feelings and emotionally integrate thoughts and experiences. These different types of private speech can best be viewed as independent language forms, each serving a distinct developmental purpose and each running its own unique developmental course.

This complex view of private speech is reflected in the work of the other recent investigators. As Table 1 indicates, the distinctions have evolved from relatively simple to elaborate classifications.

For example, Diaz (1984) examined the speech of nearly 100 preschool children while they performed tasks in a laboratory. He identified nine different self-regulatory functions of private speech (see Table 1), and called for more research on the way in which the use of these varied forms of self-guiding private speech facilitate children's task performance.

Diaz also studied the early developmental origins of self-guiding private speech by asking mothers to teach their young children to build a three-dimensional puzzle. The mothers' verbal teaching behaviors were strikingly similar to the preschoolers' private speech when they did the task by themselves. These findings confirm Vygotsky's belief that young children's private speech grows out of social experiences involving support and assistance in guiding the child's behavior by a significant adult.

An extensive study of the occurrence of private speech and children's task-facilitating and tension-reducing motor behavior in elementary school classrooms was conducted by Pechman (1978). In both open and traditional classrooms, Pechman found that self-guiding private speech and task-facilitating motor activity (pointing to follow a line or read a word; counting on fingers) increases when children work on assignments in peer contexts. In contrast, working under the supervision of adults or alone is associated with tension-reducing motor activity (self-manipulation; chewing on pencils). The research also found that those children who are just entering the Piagetian stage of concrete operations, use both more self-guiding private speech and task-facilitating motor activity when they work on individualized written assignments than when assignments are made to large groups.

The Pechman research suggests that school environments that provide peer contexts for learning and tailor experiences to individual needs and abilities will encourage task-related verbal and motor behaviors that may enhance learning and achievement.

Individual differences

Although only a few studies have investigated the extent to which individual differences affect private speech, recent research indicates that children's intelligence, sex, and personality characteristics are important.

Verbal ability. High verbal ability affects the quantity and maturity of private speech (Frauenglass & Diaz, 1985; Fuson, 1979), especially among preschool children for whom private speech is undergoing a rapid period of development. Kohlberg et al. (1968) and Kleiman (1974) found higher rates of private speech among preschoolers who score above the mean on measures of verbal intelligence.

Klein (1964) concluded that preschool children who are good problem-solvers use more private speech. Deutsch and Stein (1972) reported that preschool children who score above the sample median on the Peabody Picture Vocabulary Test show more mature forms of private speech according to the Kohlberg categories.

Sex differences. Several studies have found more mature forms of private speech among girls than boys. Pechman (1978) reported that elementary school boys use more externalized private speech, while girls use more thought without observable speech (instances in which either no vocalization, muttering, or lip movement accompanied tasks). Berk and Garvin (1984) reported similar findings. Boys show higher rates of word play and egocentric communication, while girls show a higher incidence of reading aloud and inaudible muttering. Dickie's (1973) observations of 2- to 8-year-olds also reveal more presocial self-stimulating speech by boys.

Since few studies have examined sex differences, more research is needed. We do not know for sure whether girls' greater maturity in private speech is explained by their earlier and more rapid growth in verbal abilities; family socialization practices stressing verbal skills for girls; or perhaps girls' greater willingness to conform to teacher demands that children not talk aloud in classrooms (Gump & Kounin, 1961).

Personality differences. Only one study (Rubin, 1982) systematically examined the private speech of children who vary in sociability. Observing isolate and highly sociable children at play with a "normal" peer in a laboratory setting, Rubin found that children who do not play often with their peers show higher rates of fantasy statements—comments to nonpresent or inanimate others. Rubin suggested that this form of private speech may serve a special coping function for children who do not interact often with peers. Isolate children may use such speech to provide themselves with a comfortable, nonthreatening playmate and to practice communication skills.

The largest quantity of research on children's personality characteristics and private speech concerns impulsive and hyperactive children. Studies of these children are important, because private speech is involved in the child's development of verbal self-control over behavior, and the fragmented, disorganized behavioral styles of impulsive children indicate a marked inability to control their own actions. One major therapeutic approach has been to teach impulsive children to use self-directed verbal commands to regulate and direct their own behavior (Camp, Blom, Herbert, & vanDoornick, 1977; Meichenbaum & Goodman, 1971; Palkes, Stewart, & Kahana, 1968; Varni & Henker, 1979).

Most observational studies of impulsive children's private speech indicate that they show higher rates of the more immature, self-stimulating forms (Dickie, 1973; Meichenbaum, 1971; Zivin, 1972). Zivin concluded that impulsive children's use of private speech is self-distracting and does not help them channel their behavior constructively. Only Goodman (1977) found that impulsive children use greater quantities of private speech of the more mature, self-regulatory varieties—describing one's own activity, questions posed to the self, and self-guiding speech.

Although more research is needed, it is possible that impulsive children who fare best in cognitive development and academic performance use a high degree of externalized private speech over a long developmental period because of special difficulties in channeling their attention, organizing their own activities, and suppressing task-irrelevant behavior. It is especially important that teachers of these children understand why children talk to themselves and the crucial role that such speech plays in cognitive development.

Implications for practice

Private speech is an important way in which children organize, understand, and gain control over their environment. Developmentally oriented experiences which encourage children's private speech also promote active, independent, goal-directed learning.

1. **For the preschool child, play in social contexts is especially important.** Vygotsky's theory indicates that in-

Faith Bowlus

When children play together at meaningful, goal-directed tasks, they solve unexpected problems and increase the complexity of their social interaction with the guidance and collaboration of more capable peers.

ternalized verbal thinking originates in the give-and-take of social interaction, and research shows that the earliest forms of private speech emerge in circumstances rich in peer social opportunities. In particular, activities that involve role play, where children apply their language and communication skills as they experiment with different roles and settings, lay

Children need learning environments which permit them to be verbally active while solving problems and completing tasks.

the foundation for effective self-communication in problem-solving situations.

Vygotsky's theory also implies that play experiences should bring together children of different ages who have diverse skills and abilities. When children play together at meaningful, goal-directed tasks (building with blocks, cooperatively painting a mural, role playing a family), they solve unexpected problems and increase the complexity of their social interaction with the guidance and collaboration of more capable peers.

2. **In problem-solving situations, children often must rely on guidance and direction from adults before they can function independently.** When adults provide the strategic assistance to carry out new cognitive tasks—for the preschooler tying shoes, working a puzzle, counting by relating numbers and objects in one-to-one correspondence; for the elementary school child learning to read and comprehend new words, solving a math problem, or applying the rules of a new game—they ensure the children's successful movement through the steps of the task and help them realize that meaningful goals are reached as the result of this assisted effort. Once this happens, children naturally begin to organize and structure their own efforts in the same way.

3. **The assistance the adult provides must be of a nature that can eventually be applied by children themselves.** Zetlin and Gallimore (1983) provide an example of a child who is struggling to read the word *bus.* In this situation, for the adult to prompt the child by asking, "What do you ride to school?" is of little use, since children could never ask themselves this question to help figure out another unknown word!

Instead, if the adult uses statements and questions to identify **properties** relevant to the task at hand, the child can later use the same strategies to solve similar problems alone. Ask the child to decipher the word by examining the story context (*"Why is the boy in this story waiting at the corner?"*) and applying phonetic knowledge (*"Think of the* b *and* s *sounds and see if that also helps you."*).

4. **Adults are most helpful when their guidance occurs in tasks at the edge of the child's ability and experience, and when their assistance is coordinated with the child's current level of development.** For the very youngest child,

who may have little idea of what a particular activity is about, the adult may initially give directions, such as (in working a puzzle), *"Put this piece here, and the next piece goes there,"* until the child realizes the relationship of these actions to the activity. Once this happens, the adult's regulation should quickly change so that it reveals successful strategies the child can later use independently. With the puzzle the adult can then say, *"What shape is this piece? What empty space looks like that shape?"* In prompting the child, the adult calls attention to important aspects of the situation and helps the child focus attention on the next step (Wertsch, 1978).

What the adult does to guide the child depends upon the child's momentary behaviors and the obstacles to successful completion of the task encountered by the child. For example, if the child has difficulty sustaining attention, the adult might remind the child of the goal of the tasks (*"When the puzzle is finished you can see the whole picture"*) or point out that there is another step after the one the child has just completed (*"Where does this next part go?"*). If the child appears uncertain about what to do next, the adult might assist by reminding the child of the important features of the task (*"Remember, the red ones go here and the blue ones over there,"* or *"When the problem says* 'How many altogether?' *then does it mean to add or subtract?"*). When children start to use the regulating phrases previously given by the adult, this is a sign that more of the control is being assumed by children themselves.

5. **Children need learning environments which permit them to be verbally active while solving problems and completing tasks.** When formal learning experiences begin in elementary school, children are expected to sustain attention for longer periods of time, and more of the school day is devoted to mastery of specific skills of gradually increasing difficulty. One way children cope successfully with this change is through greater use of self-guiding private speech. Pupils first learning to read can be heard carefully and cautiously sounding out words to themselves, and new math skills are initially learned by reciting the problems and counting aloud to oneself.

In the early years of elementary school, when a great many children still need to use externalized private speech, it does not seem to bother others nearby, and self-directed speech should not be interpreted as indicating insufficient self-control or misbehavior on the part of the child. As children mature, and through experience and practice their skills become more routine and automatic, they transform their task-related vocalizations into less audible whispers, until finally they are internalized as verbal thought.

However, under conditions of increased stress or task difficulty, children again need self-directed vocalizations to assist performance. Also, children who are less mature developmentally (for example, children with special learning problems) require more adult guidance and may need to use audible private speech over an especially long developmental period. For these children, teachers can arrange conditions in the classroom, such as special study corners, where task-related verbal activity can take place freely.

Conclusion

Children's private speech can be an especially effective

learning tool if early childhood learning experiences are
- rich in peer social opportunities involving a diverse mix of children, and
- tailored to individual needs and abilities.

The research cited here suggests that when tasks are new and unfamiliar, teachers need to provide verbal guidance sensitively tuned to the child's level of development. This guidance should at first be specific and directive enough to assure that the child understands the goal of the task and successfully completes it. Gradually it becomes more general and less directive, permitting children to take over by regulating their own behavior.

Research on private speech calls special attention to the verbal environments adults provide for children. It also provides renewed support for what good teachers have known all along: Educational programs that involve a balance of adult control and pupil self-initiative, where the goal of adult direction is to help children build their own self-control, are optimal for fostering children's learning and development.

References

Beaudichon, J. (1973). Nature and instrumental function of private speech in problem solving situations. *Merrill-Palmer Quarterly, 19,* 117–135.

Berk, L.E., & Garvin, R.A. (1984). Development of private speech among low-income Appalachian children. *Developmental Psychology, 20,* 271–286.

Berk, L.E. (1985). Development of private speech among preschool children. Unpublished manuscript.

Camp, B., Blom, C., Hebert, F., & vanDoornick, W. (1977). Think aloud: A program for developing self-control in young aggressive boys. *Journal of Abnormal Child Psychology, 5,* 157–169.

Deutsch, F., & Stein, A.H. (1972). The effects of personal responsibility and task interruption on the private speech of preschoolers. *Child Development, 15,* 310–324.

Diaz, R.M. (1984). *The union of thought and language in children's private speech: Recent empirical evidence for Vygotsky's theory.* Paper presented at the International Congress of Psychology, Acapulco, Mexico.

Dickie, J.R. (1973). Private speech: The effect of presence of others, task and intrapersonal variables. *Dissertation Abstracts International, 34,* 3-B. (University Microfilms No. 73-20, 329).

Frauenglass, M.H., & Diaz, R.M. (1985). Self-regulatory functions of children's private speech: A critical analysis of recent challenges to Vygotsky's theory. *Developmental Psychology, 21,* 357–364.

Fuson, K.C. (1979). The development of self-regulating aspects of speech: A review. In G. Zivin (Ed.), *The development of self-regulation through private speech* (pp. 135–217). New York: Wiley.

Goodman, S. (1975). *Children's private speech and their disposition to use cognitive mediational processes.* Unpublished master's thesis, University of Waterloo, Ontario.

Goodman, S. (1977). *A sequential functional analysis of preschool children's private speech.* Paper presented at the meeting for the Society for Research in Child Development, New Orleans.

Gump, P.V., & Kounin, J.S. (1961). Milieu influences in children's concepts of misconduct. *Child Development, 32,* 711–720.

Isaacs, S. (1930). *Intellectual growth in young children.* New York: Harcourt Brace.

Kleiman, A.S. (1974). *The use of private speech in young children and its relation to social speech.* Unpublished doctoral dissertation, University of Chicago.

Klein, W.L. (1964). An investigation of the spontaneous speech of children during problem solving. *Dissertation Abstracts International, 25,* 2031. (University Microfilms No. 69-09, 240).

Kohlberg, L., Yaeger, J., & Hjertholm, E. (1968). Private speech: Four studies and a review of theories. *Child Development, 39,* 691–736.

McCarthy, D. (1930). The language development of the preschool child. *Monographs of the Institute of Child Welfare.* (Serial No. 4).

Mead, G. (1934). *Mind, self, and society.* Chicago: University of Chicago Press.

Meichenbaum, D.H. (1971). *The natural modification of impulsive children.* Paper presented at the Biennial Meeting and the Society for Research in Child Development, Minneapolis.

Meichenbaum, D.H., & Goodman, J. (1971). Training impulsive children to talk to themselves: A means of developing self-control. *Journal of Abnormal Psychology, 77,* 115–126.

Palkes, H., Stewart, M., & Kahana, B. (1968). Porteus maze performance of hyperactive boys after training in self-directed verbal commands. *Child Development, 39,* 817–826.

Pechman, E. (1978). Spontaneous verbalization and motor accompaniment to children's task orientation in elementary classrooms. *Dissertation Abstracts International, 39,* 786A. (University Microfilms No. DDK 78-05964).

Piaget, J. (1926). *The language and thought of the child.* London: Kegan Paul, Trench, and Trubner.

Rubin, K.H. (1973). Egocentrism in childhood: A unitary construct? *Child Development, 44,* 102–110.

Rubin, K.H. (1982). *The private speech of preschoolers who vary with regard to sociability.* Paper presented at the meeting of the American Educational Research Association, New York.

Varni, J.W., & Henker, B. (1979). A self-regulation approach to the treatment of three hyperactive boys. *Child Behavior Therapy, 1,* 171–192.

Vygotsky, L. (1962). *Thought and language.* Cambridge, MA: MIT Press. (Original work published 1934)

Wertsch, J. (1978). Adult-child interaction and and the roots of meta cognition. *Quarterly Newsletter of the Institute for Comparative Human Development, 1,* 15–18.

Zetlin, A., & Gallimore, R. (1983). The development of comprehension strategies through the regulatory function of teacher questions. *Education and the Training of the Mentally Retarded, 18,* 176–184.

Zivin, G. (1972). Functions of private speech during problem solving in preschool children. *Dissertation Abstracts International, 33.* (University Microfilms No. 72-26, 224).

Zivin, G. (1979). Removing common confusions about egocentric speech, private speech, and self regulation. In G. Zivin (Ed.), *The development of self-regulation through private speech* (pp. 13–19). New York: Wiley.

Family, School, and Cultural Influences on Child Development

- Family Structure and Child Development (Articles 25-28)
- School, Culture, and Child Development (Articles 29-32)

Child rearing advice has changed over the ages. In early 1900 the view held by behaviorists such as John Watson was that children should be reared strictly in order to correctly shape their behavior. Post-World War II advice emphasized rearing children in a democratic and permissive atmosphere. Today, child rearing advice seems to have struck a middle road between the behavioral and permissive approaches. Thus, parents are encouraged to provide their children with ample love, to cuddle their infants, to use reason as the major disciplinary technique, and to encourage verbal interaction—all in an environment where rules are clearly spelled out and enforced. Suggestion, persuasion, and explanation have become the preferred techniques of rule enforcement, rather than spanking or withdrawal of love.

Perhaps modern day opinions on child rearing merely reflect the ebb and flow of advice that has been brought forth over the ages. On the other hand, contemporary child-rearing advice may reflect a growing awareness of effective child rearing based upon knowledge gained from the scientific study of human development, and a growing aversion to the excessive violence, aggression, and alienation in contemporary American society. Weighted against this breath of optimism are daily reports of sexual and physical abuse of children, extreme neglect, sexism, racism, and countless instances of family stress that contribute to teenage pregnancy, suicide, and delinquency.

Family Structure and Child Development. Parenting is not an easy task. At minimum parents must be flexible and willing to try different approaches, while constantly evaluating these approaches against "expert" opinion and against their own common sense. However, such factors as the marked increase in single parent families, the decline of intergenerational families, and the number of teenage mothers have stressed the available support services for such families. Even when knowledgeable about "expert" opinion, families that are stressed, either because of external or internal factors, may turn to expedient child-rearing techniques instead. One such expediency is the use of physical punishment in order to "discipline" children. Bettelheim points out, however, that physical punishment is an ineffective form of discipline which teaches lack of self-control and use of aggression in order to control someone else's behavior. Cross-cultural studies suggest that Japanese children differ from their American counterparts in self-control and self-discipline, a finding that he attributes to differences in parental discipline techniques.

Although it is easy to bemoan the decline in the nuclear family, the fact of its decline remains. Increasing numbers of American women have sole responsibility for child care and child support; that is, most single parent families are father-absent families to one degree or another. In the "Importance of Fathering," Alice Sterling Honig reviews evidence that suggests that both father absence and father presence have important consequences for children's sex-role development, social adjustment, and cognitive achievement. "Women at Odds" draws attention to one outcome of the increase in the number of women in the labor force—the clash between working and non-working mothers of school-age children. Working women have difficulty finding quality supplemental care for their children, while non-working women resent demands placed on them because "they do not work." This subsection concludes with an article that looks at the children of the 1960s as parents in the 1980s. The analysis may be surprising!

School, Culture, and Child Development. The family is the first major socialization force to which the child is exposed. However, as early as infancy children become involved with the second major force in socialization—the schools. Infants attend day care centers or supplementary family care homes, preschoolers attend nursery school, and older children attend public or private elementary school. Today, many states are preparing guidelines for the introduction of formal school experiences for the pre-kindergarten-age child. School often represents children's first extended separation from the home and first experiences with significant caregivers other than parents. Most children experience little difficulty adapting to school, but for others the adjustment is more problematic. School expands the child's social network beyong the neighborhood peer group and often presents new social adjustment problems which can interfere with learning.

B.F. Skinner takes the American educational system to task for its failure to clarify the goals of education and to use effective behavioral technologies to enhance student learning. One aspect of his proposed solution to the problem of American education, is to teach teachers how to teach more effectively. Stengel argues that schools should take a more active role in the teaching of moral education and values. Her solution is to teach children coping strategies based on their understanding of the stages of moral reasoning. In "Rumors of Inferiority," differences in black-white performance are linked to self-doubt, feelings of inferiority, and fear of intellectual abilities which

Unit 4

are by-products of cultural racism. Finally, Urie Bronfenbrenner draws attention to the increase in disorganized families and environments which contribute to alienation from family, friends, school, and work. His solution is to provide greater linkages among home, school, and community in a coordinated effort to combat alienation.

Looking Ahead: Challenge Questions

Skinner argues that behaviorism provides a basis for correcting the ills of American education. Underlying his approach seems to be an attitude that one can teach anything to anybody at anytime. Do you agree with the premise that behavioral technology can revitalize the educational system? What problems do you see with Skinner's proposals?

Do you agree with the premise that sex role attitudes about marriage, parenting, and family relationships are fragile and correlate poorly with actual behavior? If not, what kind of evidence would you require in order to be convinced?

Honig advocates development of programs to teach paternal behavior similar to those now available to teach maternal behavior. Why, do you suppose, is it necessary in the twentieth century to have programs to teach parenting behavior to either mothers or fathers? Isn't parenting behavior essentially instinctual? If not, how did parents of previous generations accomplish the task without formal instructional classes or the benefits of parent effectiveness training?

With which side of the "Women at Odds" battle do you agree? Is it fair for working mothers to expect non-working mothers to assume a greater share of volunteerism responsibilities for school functions? What effect does work have on the development of the attachement relationship between mother and infant, or on the effectiveness of discipline in school-age children? What responsibility does society have to provide supplementary care for children of working mothers? What is industry's responsibility?

It is relatively easy to blame teen pregnancy and teen substance abuse on disorganization of the American family. It is far more difficult to suggest effective solutions to the problem. If you were King of the World for a day or two, what changes in American society would you institute to resolve such problems as divorce, child abuse, teen pregnancy, racism, sexism, and substance abuse?

A child can be expected to behave well
only if his parents live by the values they teach

PUNISHMENT VERSUS DISCIPLINE

BRUNO BETTELHEIM

M ANY PARENTS WONDER WHAT IS THE BEST WAY TO teach their children discipline. But the majority of those who have asked my opinions on discipline have spoken of it as something that parents impose on children, rather than something that parents instill in them. What they really seem to have in mind is punishment—in particular, physical punishment.

Unfortunately, punishment teaches a child that those who have power can force others to do their will. And when the child is old enough and able, he will try to use such force himself—for instance, punishing his parents by acting in ways most distressing to them. Thus parents would be well advised to keep in mind Shakespeare's words: "They that have power to hurt and will do none. . . . They rightly do inherit heaven's graces." Among those graces is being loved and emulated by one's children.

Any punishment sets us against the person who inflicts it on us. We must remember that injured feelings can be much more lastingly hurtful than physical pain.

A once common example of both physical and emotional punishment is washing out a child's mouth with soap because the child has used bad language. While the procedure is only uncomfortable, rather than painful, the degradation the child experiences is great. Without consciously knowing it the child responds not only to the obvious message that he said something bad but also to the implicit message that the parent views his insides as dirty and bad—that the child himself is vile. In the end the parent's goal—to eliminate bad language from the child's vocabu-

lary—is rarely achieved. Instead, the punishment serves to convince the child that although the parent is very much concerned with overt behavior, he is completely uninterested in whatever annoyance compelled the child to use bad language. It convinces him that the parent is interested only in what he wants, and not in what the child wants. If this is so, the child in his inner being reasons, then why shouldn't he too be interested only in what he wants, and ignore the wishes of his parent?

I have known children who, upon having their mouths washed out with soap, stopped saying bad words out loud but continually repeated the words to themselves, responding to even the slightest frustration with streams of silent vituperation. Their anger made them unable to form any good relationships, which made them angrier still, which made them think up worse swear words.

Even if a child feels he has done wrong, he senses that there must be some better way to correct him than by inflicting physical or emotional pain. When we experience painful or degrading punishment, most of us learn to avoid situations that lead to it; in this respect punishment is effective. However, punishment teaches foremost the desirability of not getting caught, so the child who before punishment was open in his actions now learns to hide them and becomes devious. The more hurtful the punishment, the more devious the child will become.

Like the criminal who tries to get a more lenient sentence by asserting that he knows he has done wrong, our children learn to express remorse when we expect them to. Usually they are sorry only that they have been found out and may be punished. Thus we should not be fooled when they tell us that they know they did wrong, and we certainly should not extract such an admission from them, since it is essentially worthless—made to pacify us or to get the reckoning over with.

It is much better to tell a child that we are sure that if he had known he was doing wrong he would not have done so. This is nearly always the case. The child may have thought, "If my father finds out, he will be angry," but this is very different from believing that what one is doing is wrong. At any moment a child believes that whatever he is doing is fully justified. If he takes a forbidden cookie, to his mind the intensity of his desire justifies the act. Later, parental criticism or punishment may convince him that the price he has to pay for his act is too high. But this is a realization after the event.

When we tell a child that we disapprove of what he has done but are convinced that his intentions were good, our positive approach will make it relatively easy for him to listen to us and not close his mind in defense against what we have to say. And while he still might not like our objecting, he will covet our good opinion of him enough to want to retain it, even if that entails a sacrifice.

Although we may be annoyed when our children do wrong, we ought to remember Freud's observation that the voice of reason, though soft, is insistent. Shouting will not help us. It may shock a child into doing our will, but he knows and we know that it is not the voice of reason. Our task is to create situations in which reason can be heard. If we become emotional, as we are apt to do when we are upset about our child's undisciplined behavior and anxious about what it may foretell about his future, then we are not likely to speak with this soft voice of reason. And when the child is upset by fear of our displeasure, not to mention when he is anxious about what we may do to him, then he is in no position to listen well, if at all, to this soft voice.

Even the kindest and most well-intentioned parent will sometimes become exasperated. The difference between the good and the not-so-good parent in such situations is that the good parent will realize that his exasperation probably has more to do with himself than with what the child did, and that showing his exasperation will not be to anyone's advantage. The good parent makes an effort to let his passions cool. The not-so-good parent, in contrast, believes that his exasperation was caused only by his child and that therefore he has every right to act on it.

The fundamental issue is not punishment at all but the development of morality—that is, the creation of conditions that not only allow but strongly induce a child to wish to be a moral, disciplined person. If we succeed in attaining this goal, then there will be no occasion to think of punishment. But even setting aside the goal of inspiring ethical behavior, punishing one's child is, I believe, undesirable in every respect but one: it allows the discharge of parental anger and aggression.

There is little question that when a child has seriously misbehaved, a reasonable punishment may clear the air. By acting on his annoyance and anxiety, the parent finds relief; freed of these upsetting emotions, he may feel some-

what bad about having punished the child, maybe even a bit guilty about having done so, but much more positive about his child. The child, for his part, no longer feels guilty about what he has done. In the eyes of the parent he has paid the penalty; in his own eyes, usually, he has more than paid it.

In this manner parent and child, freed of emotions that bothered them and stood between them, can feel that peace has been restored to each of them and between them. But is this the best way to attain the long-range goal: to help the child become a person who acts ethically? Does the experience of having a parent who acts self-righteously or violently produce in the child the wish to act ethically on his own? Does that experience increase the child's respect for and trust in his parent? Would it not have been better, from the standpoint of deterrence and moral growth, if the child had had to struggle longer with his guilt? Isn't guilt—the pangs of conscience—a much better and more lasting deterrent than the fear of punishment? Acting in line with the urgings of one's conscience surely makes for a more responsible and sturdy personality than acting out of fear.

Punishment is a traumatic experience not only in itself but also because it disappoints the child's wish to believe in the benevolence of the parent, on which his sense of security rests. Therefore, as is true for many traumatic experiences, punishment can be subject to repression.

A good case can be made that adults who remember childhood punishments as positive experiences do so because the negative aspects were so severe that they had to be completely repressed or denied. When the punished child reaches adulthood, he remembers only the relief that came with the re-establishment of positive feelings—with the reconciliation that followed the punishment. But this does not mean that at the time the punishment was inflicted it was not detrimental. As far as I know, no child claims right after being punished that it did him a lot of good.

PROBABLY NONE OF THE COMMON TRANSGRESSIONS OF childhood upsets parents more than stealing. What disturbs them most is usually not the thefts themselves—bad as they are. It is the idea that their children may grow up to be thieves. But a child has no intention of becoming a criminal when he takes some small item, and he can be deeply hurt when his parents react as if he might become one. The child nearly always knows that he has done wrong, and if his parents are dissatisfied with him, he understands, but if they are anxious about him, his self-esteem is shaken. We ought not to view what the child has done as a crime. According to law, a child cannot commit a crime. So why should we be more severe with our child than the law would be?

I am not saying that parents should disregard what their child does. The reactions of a child's parents strongly influence the formation of that child's personality. Any transgression that parents consider serious requires an appropriate response, so that the child can learn. If a child's error remains unrecognized, or is made light of, he is likely to feel encouraged to repeat what he has done, maybe even on a larger scale. (This is why it is important for parents to be aware of what their child is doing—what he's been up to when, say, he acquires a new possession of unknown origin.) But though parents should take seriously what their child does, they should not make more of it than the child can comprehend as justified.

Clearly, a child must not be permitted to enjoy ill-gotten goods. He must immediately restore what he has taken to its rightful owner, with the appropriate apologies. If some damage has been done, the owner must be adequately compensated. Every child can understand the necessity of this, even though he might be afraid to approach the owner.

Having the child see the owner all by himself is usually not the best idea. When we supervise, we can be sure of the manner in which he returns what he has taken. More important, the child can observe directly how embarrassed we are by what he has done. One of the worst experiences a child can have is to realize that he has embarrassed his parents in front of a stranger. If we punish the child in addition to putting him through such an experience, we may considerably weaken the impression we have made. The child's sense of guilt usually centers more on the pain he has caused us than on the misdeed itself. For this reason punishment is a weak deterrent: it makes the offender so angry at those who inflict it that his sense of guilt is diminished.

It may make little difference to parents who fear for their child's future whether the child has stolen from them or from others. Parents tend to lump those two actions together in their minds. But for the child, taking things from a member of the family and taking things from a stranger are entirely different matters. We do him an injustice if we do not discriminate between these two situations. We also hamper our efforts to set things straight in the present and to prevent repetitions in the future.

Most children are occasionally tempted to take some small change from their parents. The reasons are manifold. The child wishes to buy something he longs for; he wishes to find out how observant his parents are, with respect both to their own possessions and to those the child acquires; he wishes to make his parents aware of how desperately he wants something. He may wish to keep up with his friends, or to buy their friendship. He may wish to punish the person from whom he takes something.

Parents ought to be careful not to be satisfied with the idea that their child took something, such as money, simply to indulge himself. In my experience, whenever a child—especially a middle-class child, whose needs are well taken care of—takes something from a relative, the attitude of the child toward that relative is always an important factor. For example, the child may take from a sibling because he thinks that this sibling receives more from his parents than he does. Or perhaps the child thinks that his parents have deprived him unnecessarily, or that they have shortchanged him in some way. In such cases the child thinks that he is merely correcting an unfair situation. Simply asking why the child took from one family member rather than from another may be instructive—revealing, for example, that he was angry at this person, or

that the person's negligence tempted him. But parents can elicit such important information only if they remain calm. A child is not likely to be able to discover or reveal his motives when pressed to do so by people who are very angry with him or who think they know what he is going to say.

The deepest concern of most parents is their child and his future development, not their loss, which in most cases is relatively small. But a child has a hard time realizing this unless his parents go out of their way to make it clear, by trying to understand what motivated his action. Only the child's conviction that his parents care a great deal for him—not for his future, but for him right now—will strongly motivate him to preserve their good opinion of him, by striving to do nothing that is wrong in their eyes.

Children tend not to view family possessions the way their parents do. So much that is around the house is free to be used by all family members that children may have a hard time drawing the line. Moreover, if parents play loose with their child's possessions, they ought to expect that the child might be at least tempted to do the same with their possessions.

Because of his dependence on the family, the child often has a keener sense of family—on an intuitive, subconscious level—than do his parents. Being a more primitive person, he experiences things in much more primitive and direct ways. It is *his* family; it must be *his* for him to feel secure and to be able to grow up well. If so, is not then everything that is the family's property also his? If he belongs to his parents, and they belong to him, then don't silly objects—silly by comparison with the importance of persons—such as money or other valuables that belong to his parents also belong to him? When all family possessions were really that—possessions of the family, not the private property of individual family members—perhaps the sense of family was stronger and gave each family member more security than people now experience.

We can instill in our children a much deeper feeling of family cohesiveness if we make it clear that—within reason—family property is for everyone's use. This includes relatively small amounts of money or minor valuables, the expenditure or loss of which cannot jeopardize the family's future.

THE ORIGINAL DEFINITION OF THE WORD *DISCIPLINE* refers to an instruction to be imparted to disciples. When one thinks about this definition, it becomes clear that one cannot impart anything, whether discipline or knowledge, that one does not possess oneself. Also it is obvious that acquiring discipline and being a disciple are intimately related.

Most of us when hearing or using the word *disciple* are likely to be reminded of the biblical Apostles. Their deepest wish was to emulate Christ. They made him their guide not just because they believed in his teachings but because of their love for him and his love for them. Without such mutual love the Master's teaching and example, convincing though they were, would never have persuaded the disciples to change their lives and beliefs as radically as they did.

The story of Christ's disciples suggests that love and admiration are powerful motives for adopting a person's values and ideas. By the same token, the combination of teaching, example, and mutual love is most potent in preventing one from going against what this admired individual stands for, even when one is tempted to do so. Thus the most reliable method of instilling desirable values and a discipline based on these values into the minds of our children should be obvious.

Probably the only way for an undisciplined person to acquire discipline is through admiring and emulating someone who *is* disciplined. This process is greatly helped if the disciple believes that even if he is not *the* favorite of the master, at least he is one of the favorites. Such a belief further motivates the disciple to form himself in the image of the master—to identify with him.

Fortunately, the younger the child, the more he wishes to admire his parents. In fact, he cannot do other than admire them, because he needs to believe in their perfection in order to be able to feel safe himself. And in whose image can the young child form himself but in that of one or both of the parents, or whoever functions in their place? Nobody else is as close and as important to him as they are; nobody else loves him as much or takes such good care of him. It is for these reasons, too, that the child wishes to believe that he is his parents' favorite. Sibling rivalry is caused by the fear that he might not be the favorite, and that one of his siblings is. How acutely a child suffers from sibling rivalry is a clear indication of how great his wish is to be the parents' favorite, and how consuming his fear is that he is not.

It is natural and probably unavoidable for parents sometimes to prefer one of their children to others. Parents sometimes fool themselves into believing that they love all their children equally, but this is rarely the case. At best, a parent will like each of his children very much—most parents do—but he will like each child in different ways, and for different reasons. Most parents love one child more at one time and another more at another time, which is only natural, because children behave differently at various moments in their lives and thus evoke different emotional reactions in their parents. But if a child has reason to feel that he is the favorite some of the time, he is likely to believe that he is the favorite most of the time. In this situation, as in so many others, the wish is father to the thought. All of this works, of course, only if the child is not too often and too severely disappointed by the attitudes of his parents.

As the child grows older, he will cease to admire his parents so single-mindedly. By comparison with the wider circle of people he gets to know as he grows up, his parents will seem deficient in some respects. However, while the child may admire his parents less and question aspects of their behavior, his need to admire them unconditionally is so deeply rooted that it will be powerfully present in his unconscious for a long time—at least until he reaches maturity, if not longer. Thus, fortunately, in most families there is a solid basis for the child's wish to be his parents' disciple—to be able to love and admire them, and to emulate them, if not in all then certainly in some very important respects, and if not in his conscious then certainly in his unconscious mind.

We all know families in which this is not the case—in which the parents do not like their child very much, are disappointed in him, or do not behave so that the child can love and admire them. When a child neither admires his parents nor wishes to emulate them, he will not become disciplined under their influence. How can he, when they are not suitable models?

Such a child often finds some other person to admire, whose favorite he wishes to be, and whom he therefore comes to emulate, acquiring discipline in order to find favor in this person's eyes. The trouble is that the child is likely to seek and find an undisciplined master. An example of this syndrome is the member of a delinquent gang who is so impressed by its delinquent or otherwise asocial leader that he admires and emulates him, with disastrous consequences for the youngster and for society. On some level the youngster may know that he has not chosen well, but his need to attach himself to someone whom he can admire, and who seems to offer acceptance and security in return, is so great that it drowns out the voice of reason. It is on their child's need for such an attachment that parents can and must build in order to promote not just disciplined behavior of the child around particular issues—this is not all that difficult to obtain—but a lasting inner commitment to be, or at least to become, a disciplined person.

It is by no means easy for a child to become disciplined. Often part of the reason is that his parents are not very well disciplined themselves and thus do not provide clear models for their child to follow. Another difficulty is that parents try to teach self-discipline to their child in ways that arouse his resistance rather than his interest. And still another difficulty is that a child responds to his parents most readily—both positively and negatively—when he sees that their emotional involvement is strong. When parents act with little self-discipline, they show their emotions. When they get their emotions under control, they are nearly always again able to act in line with their normal standards of discipline. Rare as it may be for a parent to lose control, those are the times that impress a child most. Disciplined behavior, while pleasing and reassuring to the child and likely to make life good for him in the long run, does not make such a strong impression on him.

For these and many other reasons teaching discipline requires great patience on the part of the teacher. The acquisition of true inner discipline, which will be an important characteristic of one's personality and behavior, requires many years of apprenticeship. The process is so slow that in retrospect it seems unremarkable—as if it were natural and easy. And yet if parents could only remember how undisciplined they themselves once were and how hard a time they had as children in disciplining themselves—if they could remember how put upon, if not abused, they felt when their parents forced them to behave well against their will—then they and their children would be much better off. One of the world's greatest teachers, Goethe, wrote an epigram that turns on this very point: "Tell me how bear you so comfortably/The arrogant conduct of maddening youth?/Had I too not once behaved unbearably,/They would be unbearable in truth." Goethe could write these lines and enjoy their humor because he had achieved great inner security, which made it possible for him to understand with amusement the otherwise "unbearable" behavior of the young. The same feeling of security allowed him to remember how difficult—unbearable, even—he himself had been in his younger years, which many of us are tempted to forget, if our self-love does not compel us to repress or deny it.

Despite all the obstacles that parents encounter in trying to impart discipline to their children, they are the logical persons to do so, because the learning has to start so early and continue for so long. But while most parents are ready to teach their children discipline and know that they are the ones to do so, they are less ready to accept the idea that they can teach only by example. Unfortunately, the maxim "Do as I say, not as I do" won't work with children. Whether they obey our orders or not, deep down they are influenced less by what we tell them than they are by who we are and what we do. Our children form themselves in reaction to us: the more they love us, the more they emulate us, and the more they respond positively to our consciously held values and to those of which we are not conscious but which also influence our actions. The less they like and admire us, the more negatively they respond to us in forming their personalities.

A study conducted in Sweden demonstrates how persuasive the example set by the parents can be to a child. Some years ago the Swedish government became concerned because undisciplined behavior among Swedish teenagers—as indicated by alcoholism, vandalism, delinquency, drug use, and criminal behavior—had become prevalent. To find out why some children become troublesome and others did not, researchers compared the homes of law-abiding teenagers with those of delinquents. They found that neither material assets nor social class exercised a statistically significant influence on the behavior of these young people. Instead, what was decisive was the emotional atmosphere of the home.

Teenagers who behaved well tended to have parents who were themselves responsible, upright, and self-disciplined—who lived in accord with the values they professed and encouraged their children to follow suit. When the good teenagers were exposed, as part of the investigation, to problem teenagers, their behavior was not permanently affected. They had far too securely internalized their parents' values. While some, out of curiosity, joined the activities of the delinquent or drug-using group, such experimentation was always tentative and short-lived. By the same token, when problem teenagers were forced to associate solely with "square" peers, they showed no significant improvement. Indeed, they did not even temporarily adopt non-delinquent ways of living.

The Swedish researchers found that undisciplined, asocial, problem teenagers did not necessarily come from what one would consider undisciplined or disorganized homes, nor did they have visibly asocial parents. But the parents of the asocial youngsters did tend to have conflicting values or to be inconsistent in putting their values into practice. And they tended to try to hold their children to values that they themselves did not live by. As a result the children had not been able to internalize those values. Expected by their parents to be more disciplined than the examples set, most of the children turned out to be much less so.

Further study of the family backgrounds of these youngsters revealed that it hardly mattered what specific values the parents embraced—whether the parents were conservative or progressive in the views they held, strict or permissive in the ways in which they brought up their children. What made the difference was how closely the parents lived by the values that they tried to teach their children.

A parent who respects himself will feel no need to demand or command respect from his child, since he feels no need for the child's respect to buttress his security as a parent or as a person. Secure in himself, he will not feel his authority threatened and will accept it when his child sometimes shows a lack of respect for him, as young children, in particular, are apt to do. The parent's self-respect tells him that such displays arise from immaturity of judgment, which time and experience will eventually correct.

Demanding or commanding respect reveals to the child an insecure parent who lacks the conviction that his way of life will, all by itself, over time, gain him the child's respect. Not trusting that respect will come naturally, this parent has to insist on it right now. Who would wish to form himself in the image of an insecure person, even if that person is his parent? Unfortunately, the child of insecure parents often becomes an insecure person himself, because insecure parents cannot inculcate security in their children or create an environment in which the children can develop a sense of security on their own.

To be disciplined requires self-control. To be controlled by others and to accept living by their rules or orders makes it superfluous to control oneself. When the more important aspects of a child's actions and behavior are controlled by, say, his parents or teachers, he will see no need to learn to control himself; others do it for him.

How parents in other cultures try to inculcate self-control in their children can be instructive. Consider, for example, a study designed to find out why young Japanese do much better academically than Americans. When the researchers studied maternal behavior they saw clear differences between the Japanese and the Americans. Typically, when young American children ran around in supermarkets, their mothers—often annoyed—told them, "Stop that!" or "I told you not to act this way!" Japanese mothers typically refrained entirely from telling their children what to do. Instead they asked them questions, such as "How do you think it makes the storekeeper feel when you run around like this in his store?" or "How do you think it makes me feel when my child runs around as you do?'' Similarly, the American mother, wanting her child to eat what he was supposed to eat, would order the child to do so or tell him that he ought to eat it because it was good for him. The Japanese mother would ask her child a question, such as "How do you think it makes the man who grew these vegetables for you to eat feel when you reject them?'' or "How do you think it makes these carrots that grew so that you could eat them feel when you do not eat them?'' Thus from a very early age the American child is told what to do, while the Japanese child is encouraged not only to consider other persons' feelings but to control himself on the basis of his own deliberations.

The reason for the higher academic achievement of Japanese youngsters may well be that the Japanese child in situations important to his mother is invited to think things out on his own, a habit that stands him in good stead when he has to master academic material. The American child, in contrast, is expected to conform his decisions and actions to what he is told to do. This expectation certainly does not encourage him to do his own thinking.

The Japanese mother does not just expect her child to be able to arrive at good decisions. She also makes an appeal to her child not to embarrass her. In the traditional Japanese culture losing face is among the worst things that can happen to a person. When a mother asks, "How do you think it makes me—or the storekeeper—feel when you act this way?" she implies that by mending his ways the child does her, or the storekeeper, a very great favor. To be asked to do one's own thinking and to act accordingly, as well as to be told that one is able to do someone a favor, enhances one's self-respect, while to be ordered to do the opposite of what one wants is destructive of it.

WHAT IS A PARENT TO DO IN THE SHORT RUN TO prevent a child from misbehaving, as children are apt to do from time to time? Ideally, letting a child know of our disappointment should be effective and should lead the child to abstain from repeating the wrongdoing in the future. Realistically, even if a child has great love and respect for us, his parents, simply telling him of our disappointment, or showing him how great it is, will not always suffice to remedy the situation.

When our words are not enough, when telling our child to mend his ways is ineffective, then the threat of the withdrawal of our love and affection is the only sound method to impress on him that he had better conform to our request. Subconsciously recognizing how powerful a threat this is, some parents, with the best of intentions, destroy its effectiveness by assuring their children that they love them no matter what. This might well be true, but it does not sound convincing to a child, who knows that he does not love his parents no matter what, such as when they are angry at him; so how can he believe them when he can tell that they are dissatisfied, and maybe even angry at him? Most of us do not really love un-

conditionally. Therefore any effort to make ourselves look better, to pretend to be more loving than we are, will have the opposite effect from the one we desire. True, our love for our child can be so deep, so firmly anchored in us, that it will withstand even very severe blows. But at the moment when we are seriously disappointed in the child, our love may be at a low point, and if we want the child to change his ways, he might as well know it.

The action to take is to banish the child from our presence. We may send him out of the room or we ourselves may withdraw. Whatever, the parent is clearly indicating, "I am so disappointed in you that I do not wish, or feel unable, to maintain physical closeness with you." Here physical distance stands for emotional distance, and it is a symbol that speaks to the child's conscious and unconscious at the same time. This is why the action is so effective.

Sending the child out of sight permits both parent and child to gain distance from what has happened, to cool off, to reconsider. And that does help. But it is the threat of desertion, as likely as not, that permanently impresses the child. Separation anxiety is probably the earliest and most basic anxiety of man. The infant experiences it when his prime caretaker absents herself from him, an absence that, should it become permanent and the caretaker not be replaced, would indeed lead to the infant's death. Anything that rekindles this anxiety is experienced as a terrible threat. Hence, as long as a child believes, however vaguely, that his very existence is in danger if his prime caretaker deserts him, he will respond to this real, implied, or imagined threat with deep feelings of anxiety. Even when he is old enough to know that his life is not in real danger, he will respond to separation from a parent with severe feelings of dejection, because to some degree he will feel as if he were endangered. The difference is that at an older age the fear is not of physical but of emotional starvation.

If we should have any doubt that physical separation can be an effective expression of our disgust with a child's behavior, we can look to our children themselves to set us straight. The worst that a child can think of when he is disgusted with his parents is that he will run away. He makes such a threat because he is convinced that it is so terrible that it will compel us to mend our ways. Clearly, a child understands very well that when we threaten to distance ourselves from him physically we are threatening to distance ourselves from him emotionally. That threat makes a very deep impression.

We must be honest about our strong emotional reactions to our children's behavior, showing our children how deeply we love them, on the one hand, and, on the other, letting them know when we are disappointed in them, provided we do not become critical or punitive. This is all just part of being ourselves. We need not make any claim to be perfect. But if we strive as best we can to live good lives ourselves, our children, impressed by the merits of living good lives, will one day wish to do the same.

The Importance of Fathering

Alice Sterling Honig

Alice Sterling Honig is Associate Professor, Department of Child and Family Studies, at Syracuse University, Syracuse, New York.

Margaret Mead once observed that fathers are a "biological necessity but a social accident." Little appreciation of the direct role and importance of fathering was expressed scarcely a quarter of a century ago by the pioneer advocate of infant attachment (to mother), John Bowlby:

> In the young child's eyes, father plays second fiddle and his value increases only as the child's vulnerability to deprivation decreases. Nevertheless, as the illegitimate child knows, fathers have their uses even in infancy. Not only do they provide for their wives to enable them to devote themselves to the care of the infant and toddler, but, by providing love and companionship, they support her emotionally and help her maintain that harmonious contented mood in the aura of which the infant thrives (1958, p. 363).

Hand-in-hand with this belief in the minimal direct influence of fathers went societal views that gave approval to those fathers who were at their desks by 7:30 a.m. and who often did not return home from business until well past the bedtime of small children. Thus, society's "theory" of what constitutes a good father focused one-dimensionally on the role of provider and excluded the role of nurturer and socializer of the young.

Research has indeed sometimes tended to confirm the peripheral nature of the fathering role. Knox & Gilman (1974) questioned 102 first-time fathers regarding their preparation for, and adaptation to, fatherhood. Most fathers had had little preparation for their role. They participated minimally in the day-to-day care of the new baby. In another study of middle-class families in Boston, 43 percent of the fathers had never changed a diaper. Rebelsky & Hanks (1971) reported that new fathers spent 37 seconds per day in one-to-one verbal interaction with baby. Finally, to complete this picture of father as "second-class" parent, 40 percent of American children surveyed preferred television to their fathers (Dodson, 1979)!

Father Absence

More than six million children live in fatherless families. Much research on fathering has concentrated on the effects of father absence. The main question seems to be "How, and how much, are children harmed by growing up in a fatherless home?" Reviews of research and bibliographic searches suggest that psychosexual and emotional maladaptive functioning occurs more often in conjunction with father absence than presence (Herzog & Sudia, 1973; Honig, 1977; Lynn, 1974). Father absence has been found to be negatively related to sex role development, moral development, cognitive competence and to social adjustment.

The impact of father absence has to be assessed in terms of total family functioning under conditions of that absence. Much of the research seems to imply that when a father is missing, only that variable alone is affecting the child's development. Yet, family climate and clusters of family attributes concomitant with father absence may be far more important for a child's development than the actual number of parents present in the home. For example, in a supportive neighborhood or in a three-generation household, grandfather or older relatives may be available as masculine and fathering role models for young children.

Another aspect that is critical in reviewing the findings on fatherless children has to do with the mother's expressed attitudes toward the child and toward the absent father. A mother who derogates the absent father and/or the child will very likely cause profound self-doubt in a son and distrust of males in a daughter.

Thus, out of the father-absence stud-

From *Dimensions*, October 1980, pp. 33-38, 63. Reprinted by permission of the author and Dimensions, the journal of the Southern Association of Children Under Six, Little Rock, Arkansas.

ies comes a fairly consistent picture. Fathering is important.

Research on Effects of Fathers Present

Recent societal changes have triggered an upsurge of interest in fatherhood. Twenty-five years ago, only 1.5 million mothers worked, compared to 14 million today. More than one-third of all mothers of children under six work outside the home. Almost 1,400,000 children live in single-parent families headed by fathers (Mendes, 1976). A re-examaination of father roles and the importance of fathering seems then to flow from a rapid increase in working wives and mothers; increasing flexibility in divorce custody arrangements; and increasing pressure on fathers to shoulder some of the child-rearing responsibilities traditionally "relegated" to mothers.

Interviews with college-educated young men and women turned up almost unanimous agreement among these young people that the decision to parent in their lives was closely tied to an egalitarian expectation of fathers' role in sharing the care of the infant and in assuming an important role in child-rearing (Honig, 1980).

Fathers and Infants

A wide variety of recent studies focuses on fathering in the infancy period — that period formerly left to maternal ministrations. Father-infant research of the past few years has consistently shown that fathers can be quite competent. They are sensitive to infant cues and responsive to the signals of newborns. Fathers are quite as likely, if given the chance through sympathetic and facilitative hospital birthing procedures, to bond lovingly with their babies.

When low-income fathers who had not participated in the birth experience were observed in the days after delivery, they proved just as nurturant and stimulating with their infants as the mothers — if the fathers were observed alone with their babies (Parke & Sawin, 1980). The moral may be that father needs time alone with baby to build a love relationship. Mothers have traditionally always had such time together. An intimate father-infant relationship may require the same "twosome" quality that love relationships usually need to grow and deepen.

Infants 12-21 months, brought up by both parents, did not register any preference for either parent, when separation protest, vocalizing and smiles were the measures used (Kotelchuck, 1972; Lamb, 1976). In the former study,

Babies with fathering get more variety in life.

those few infants who did not relate well to father (that is, spend at least 15 seconds near him on his arrival), came from families with the lowest amount of father caregiving. Thus, it looks as if baby attachment to fathers follows a general rule that you get what you give. Fathers who spend loving attentive time with infants will have infants who attach well to them and under ordinary conditions prefer paternal company equally. Data indicate, however, that when distressed, infants may still turn to mother as primary comforter.

Strong differences are reported in the ways in which fathers interact with infants compared to mothers. "When they have the chance, fathers are more visually attentive and playful (talking to the baby, imitating the baby) but they are less active in feeding and caretaking activities such as wiping the child's face or changing diapers" (Parke & Sawin, 1980, p. 204).

Lamb reports that "when both parents are present, fathers are more salient persons than mothers. They are more likely to engage in unusual and more enjoyable types of play, and hence, appear to maintain the infant's attention more than the mothers do" (1976, p. 324). In short, "fathers seem to be more fun for babies! Fathers play different kinds of games with infants, more vigorous games" (Honig, 1979a, p. 247).

Split-screen motion picture work by Daniel Stern and by T. Berry Brazelton at Harvard Medical School reveals distinctive patterns of interaction of father and mother with baby. Mothers and infants play more reciprocal vocalization games. Baby limb movements tend to be smooth and more rounded with mothers. Fathers use more bursts of a tapping kind of touch. Babies show more angular and abrupt body movements to father's touch. What turns out, delightfully, is that the contributions of maternal and paternal touch and vocal reciprocity patterns from earliest infancy differ.

Babies with fathering get more variety in life. There is the priceless redundancy that paternal loving, extra stimulation and sensitivity to signals afford for advancing development. There is also the variation on human interaction themes and variation in gender and style that help a child grow up learning more perceptively to deal with differing patterns of adult interaction, expectancies and styles.

When their fathers have also tended to their needs, babies cope better with strangers. Babies seem to learn more skills to put them at ease socially in strange situations. Father-infant interactions may add social resilience to an infant's social repertoire. The fact that fathers, albeit limited to after-work interaction time, are salient figures in interaction with infants is well illustrated in Friendlander, et al.'s analysis of infants' natural language environments in the home (1972). Systematic analysis of tape-recorded utterances of fathers to infants revealed that one infant, raised primarily by an English-speaking mother, but talked to regularly and prompted abundantly by father in Spanish daily, was able to demonstrate good understanding of many Spanish words at one year of age.

Quality of father's interaction time, as so well predicted by Eriksonian theory for mothers, seems to count strongly for the positive special influence that fathers can have not only as attachment figures but as boosters of infant language learning.

Fathers and the Development of Intellective Achievements

Ainsworth's pioneer work in attachment has delineated the ways in which baby's secure attachment to mother by one year is related to the organization of socio-emotional behavior up to at least five years and to early achievement of developmental milestones (1979). Sroufe and his colleagues (1979) have demonstrated that securely attached infants are better problem solvers and tool users when challenged with somewhat difficult tasks as toddlers and preschoolers.

Just how involved are fathers in teaching their children, and how important is fathering for the development of child intellectual competence?

Social class and sex of child have been found as significant confounding variables in studies that attempt to assess father impact on cognitive achievement. Deal & Montgomery (1971) observed the techniques used by fathers, from professional and non-professional families, to teach their five-year-old sons two sorting tasks. Professional fathers verbalized more, used more complete sentences and more verbal rewards than non-professional fathers.

An interview study with black fathers from three social classes revealed that middle-class black fathers' scores in the domains of "provision of developmental stimulation" and "qual-

. . . Paternal nurturance has been found to be positively related to high child achievement.

ity of the language environment" were significantly higher than scores for fathers in the upper-lower or lower-lower social class groups. And the amount of enriching home stimulation was significantly highest for middle-class fathers of daughters (Honig & Main, 1980).

Paternal verbal and cognitive interactions seem to differ as a function of child sex. The research results are often confusing. McAdoo (1979) found that middle- and working-class black fathers interacted more with sons than with daughters. There was a significant positive relationship between fathers' warmth and nurturance and amount of interaction with child. Epstein & Radin (1975) observed social class differences related to cognitive achievement among male children. Among middle-class boys, there was a positive relationship between paternal nurturance and child's Binet I.Q. Among lower-class boys, paternal restrictiveness had a negative impact on the child's cognitive explorations.

The researchers speculate that "it may be that in the working class, where sex role stereotypes are strongest, intellectual and academically-oriented activities are viewed as feminine and hence not appropriate for boys" (p. 838). They report that fathers seem to interfere with daughters' task motivation by restrictiveness and by offering mixed messages that they will both meet and ignore daughters' explicit needs.

Lamb & Frodi (1980) have speculated that, in effect, when a warm father encourages femininity in a daughter and yet believes, traditionally, that femininity and achievement are incompatible, a girl may have grave doubts or conflict about the appropriateness of achievement for women. Alternatively, fathers may not be as strong role models as mothers for girls' intellective development. Research provides too little and conflicting evidence to decide this point. Crandall and colleagues (1964), for example, found that daughters who demonstrated excellence in reading and arithmetic had mothers who often praised and rewarded their intellectual efforts and seldom criticized them. Yet, in Bing's (1963) study, father's strictness was related to verbal achievements of daughters even more than sons'.

More often than not, paternal nurtur-

ance has been found to be positively related to high child achievement. High-achieving college students reported that their fathers were more accepting and somewhat less controlling than fathers described by low-achievers (Cross & Allen, 1969). When parents are too anxiously intrusive in trying to foster intellectual achievement, their efforts may have unfortunate effects.

Teahan (1963) administered a questionnaire to low-achieving college freshmen (who had previously done well in school) and their parents. These sons had fathers who felt that children should make only minor decisions, that they should always believe their parents, that they should be under their parents' complete control and that it was wicked to disobey parents. The picture emerged of a clash between a domineering, punitive, overprotective father — and his underachieving son.

It is possible that the relationship among the variables of paternal nurturance, high academic expectations, sex-role attitudes, quality of cognitive facilitation, and marital congruence with spouse may interact in subtle and complex ways to foster a child's intellectual competence.

Rapidly changing beliefs about the roles of fathers may increase the confusion among research findings as to paternal influence on achievement. Talcott Parsons' theory (Parsons & Bales, 1955), for example, that fathers were "instrumental" and mothers "expressive" with children may be outdated by evolving new beliefs. Parsons hypothesized that father comes to be seen by the child as representative of the outside world. Father is the significant major parent to make strong demands that expand a child's horizons for achievement. Such a simplified role dichotomy is no longer acceptable in the value system of many families.

But there is some research support for this conceptualization. Cox, in a 1962 dissertation with gifted fourth to sixth graders, found that their fathers and mothers showed a predominant pattern of affection, setting firm limits and positive relationships. Fathers, especially, set high expectations for their gifted high-achieving children.

A summary of the main thrust of the research so far would counsel that a father who wishes to have positive

academic influence should have high academic expectations, be a helpful teacher, remain nurturant, but respond with flexibility, perceptivity and sensitivity over time to a child's changing needs for assuming autonomous responsibility for the child's own learning career and social life.

Fathers and Sex Role Development

Research indicates that fathers are more focused on sex role differences. They influence sex stereotyping more than mothers. Fathers have preference for male offspring. By one year, fathers prefer boy babies. Fathers talk more to sons than to daughters.

The father's character and the extent to which he has made a success of his personal life, plus his easy affection with his son and his loving relationship with the mother, have been suggested as the foundation for a son's ready acceptance of being male and, in turn, confident acceptance of the role of husband and father in adulthood (Green, 1976).

A father who is violent, contemptuous toward women and overbearing may impair his daughter's ability to grow up "feminine." She may grow up without a basic sexual understanding that men and women can be equally accepting of and tender with each other. Young girls who have grown up with a good relationship with an admiring, nurturant father have been found to relate easily and well in college relationships with young men (Johnson, 1963).

Two-thirds of the world's 800 million illiterates are females. Dominant males in many third-world societies devalue the potential for intellectual development of their daughters and wives. Yet, even in the United States, subtle, intellectual devaluating of daughters occurs at many levels. Research on conversational management techniques reveals that fathers interrupt their children more than mothers do, and daughters are interrupted more than sons. Fathers engage in simultaneous speech more with daughters than with sons. Boys and girls get different messages about their status and role in society. Girls are more interruptible than their brothers — which suggests in a not-so-subtle way that they are less important (Greif, 1980).

Such prejudices run deep. Perhaps

Fathers are important for adequate sex role development of daughters as well as sons.

only as this nation commits itself to teaching parenting skills and family life courses at all educational levels will prospective parents begin to become aware of the impelling and compelling influences they can have on the growth of emotionally healthy and self-actualized daughters and sons in the future.

Fathers are important for adequate sex role development of daughters as well as sons. Daughters' conceptualizations of the worth of fathers may be persistently distorted by being raised in households bereft of a father. Girls then may in some way expect that only women are strong and can take care of families and men are weak and cannot be expected to do so.

What does such sex role learning presage for tomorrow's families? Green has summarized the important role that fathers play in a daughter's sex role development: "A young girl learns how to be a female from her similarly shaped mother. But she will learn how to be a girl who likes men, or does not trust or feel affection for men by the way she responds to her father" (p. 165).

Recent studies on androgynous roles for both mothers and fathers are beginning to reveal some of the fathering patterns that may emerge as new beliefs about masculinity and the fathering role become more widespread. Fathers classified as androgynous by the Bem Sex Role Inventory have been found to be more involved in day-to-day care, activities and play with their children than those classified as masculine. Fathers classified as masculine, married to women classified as androgynous or masculine provided the next highest level of involvement with children. The least involved were masculine fathers married to feminine women (Russell, 1978).

Fathering and Prosocial Child Behaviors

School vandalism, rising delinquency at younger ages and crime in the streets have all helped to spur interest in the area of prosocial behaviors such as empathy, altruism, generosity and helpfulness. Most of the research has focused on female parents, models and teachers. In a study that inquired about fathering patterns, Rutherford & Mussen (1968) played a game with nursery school boys who then had an opportu-

nity to share some of their winnings with friends. The most generous boys, by action and by teacher rating, much more frequently described their fathers as nurturant and warm parents and as models of generosity, sympathy and compassion.

Hoffman (1975) used sociometric questionnaries to assess fifth-grade pupils' reputations for altruism and consideration of others. Children nominated the three same-sexed classmates who were most likely to "stick up for a kid that the other kids are making fun of or calling names" and "to care about how other kids feel and try not to hurt their feelings." Parents of the children were then asked to rank 18 life values. The fathers of those boys rated as most helpful and considerate (and the mothers of similar girls) ranked altruism high in their own hierarchy of values.

Yarrow & Scott (1972) found a child's consideration of others, as assessed by classmates' nominations, to be related to maternal and not paternal affection among middle-class children. However, lower-class boys' (but not girls') consideration for others was significantly related to father and mother affection.

Middle-class six- and eight-year-old boys were found to be more generous when high paternal affection and high maternal child-centeredness were present (as measured by parental Q Sorts). These relationships did not hold for girls (Feshbach, 1973). The level of altruism modeled by fathers appears to be a factor in the development of sons' prosocial behaviors. In a review of studies of the development of prosocial behaviors in children, Mussen & Eisenberg-Berg (1977) conclude that "nurturance is most effective in strengthening predispositions toward prosocial behavior when it is part of a pattern of child-rearing that prominently features the modeling of prosocial acts" (p. 92).

The converse has been found also. Where fathers are relatively unaffectionate and controlling, authoritarian and rejecting, and not likely to trust their sons, boys have been found high in aggression (Feshbach, 1973; Stevens & Mathews, 1978).

Fathering in Alternate Life Styles

Just beginning to receive the research attention they so critically de-

serve are stepfathers, divorced fathers with custody, divorced fathers with only visitation rights, and single unwed fathers*. In 1974, slightly over six million children were living with a stepparent. Rallings (1976) has summarized some of the sociological findings on the extent of stepparenting and the adjustment problems faced by stepfathers in particular, and Pannor, et al. (1971) have focused on studies of unwed fathers.

Hetherington, Cox & Cox (1976) have documented the extent of the disruption on children's lives where father is the non-custodial parent. Divorced parents are less consistent with children; they are less likely to use reasoning and explanation. There is a steady decline in nurturance expressed by divorced fathers toward their children. Two years after divorce, negative affect and distressful symptoms were diminished in girls. Yet boys from divorced families were still more hostile and less happy than boys from nuclear families.

When the father was emotionally mature, then frequency of father's contact with the child was associated with more positive mother-child interactions. When the father was poorly adjusted or there was disagreement and inconsistent attitudes toward the child, or there was ill will between the former spouses, then frequent visitation was associated with poor mother-child functioning and disruptions in the children's behavior.

Wallerstein and Kelly (1980) followed children of divorced families for five years. Boys particularly reported depressed and difficult feelings when there was not consistent attentive relationship maintained with the father. One-third of this middle-class sample of children were still considerably disturbed in functioning after five years.

The intimate relationship of fathers and children seems to be particularly crucial for positive adjustment of sons after divorce. Yet, parental conflict and immaturity can vitiate positive effects of frequent contact. Clearly more urgent is sensitivity to children's needs by both parents after divorce. The effect of divorce per se on children

*All of the articles in the Special Issue of the October 1979 *Family Coordinator* (Volume 28) are addressed to "Men's Roles in the Family."

When nourished by father love and intimate responsive care, babies become well attached to their fathers.

is not as critical as parental conflict and immaturity.

Gasser & Taylor (1976) inquired into the role adjustments faced by single-parent fathers. The middle-class fathers interviewed assumed major responsibility for all child-care activities, sought outside supports, and curtailed club meetings and educational attainments. These fathers felt that they were able to cope with the responsibilities of home management although they may have formerly assumed little responsibility while married. Single fathering no more seems to guarantee unhappiness than does nuclear family living (Katz, 1979). The ways in which stresses and burdens are handled seems to be more indicative of whether a family functions fairly happily than whether one or two parents are present.

Father Involvement: Intervention Programs

Although the overwhelming number of programmatic efforts to enhance parenting skills in the past decade has focused on mothers, some programs have been involved with fathers as part of a family focus of intervention. Middle-class mothers and fathers of babies under 12 months were trained to increase their social competence with infants (Dickie & Carnahan, 1979). Post-training home observations showed that training affected trained fathers the most. They were superior to trained mothers and to control mothers and fathers in anticipating infants' needs, responding more appropriately to the infants' cues and providing more frequent verbal and non-verbal contingent responses. Infants sought interaction least with untrained fathers and most with trained fathers. An extra benefit accrued to the marital partners: trained mothers and fathers thought their spouses were more competent than did untrained mothers and fathers.

An experimental program in Chicago with a small number of low-income families in an urban housing project found that fathers could be more actively involved in their children's educational experiences when male workers tailored the home visitation program specially for the fathers. Tuck (1969) has described this model for working with black fathers.

The Importance of Fathering

The importance of fathering has become more and more evident as research in this area proliferates. Some comments are appropriately representative of major findings to date and some as suggestions for needed research.

Men who traditionally have rejected expression of tender feelings or a range of emotional responsiveness as unmanly may need to rethink "what is masculinity" in light of the needs of infants for fathers and the delights of intimacy with infants for fathers. As Lamb elegantly expresses this: "It is important not to confuse conformity to traditional sex role prescriptions with the security of gender identity or with mental health." Provided an individual's gender identity is secure, a wide range of gender roles can be assumed (1979, p. 942).

The myth that only mothers can nurture an infant seems just that — a myth. Fathers are just as upset by squalling babies as are mothers. Fathers can be as attentive to infant cues as mothers. When nourished by father love and intimate responsive care, babies become well attached to their fathers.

Paternal nurturance may be related in complex ways to cognitive achievement in children. The family serves as a nurturing matrix that allows a child's natural curiosity and exploration to flourish into developmental learnings. A child filled with anxiety or despair that he or she is neither cared for nor cared about cannot focus well on learning tasks whether a father is absent or present. Future research should focus on the complex interweave of factors in family and community that facilitate intellectual engagement and achievement, rather than on putative effects of father absence or presence conceived of as a single variable of an heuristically critical nature.

Process rather than status variables have proved more relevant to child intellective attainments (Honig, 1979b). Indeed, in a father-present family where book-learning is considered sissyish, it is not difficult to predict that despite father presence, neither cognitive strivings nor academic excellence may be a goal of father or son.

Because the effects of fathering may

be related to marital harmony, economic stress and a host of social and cultural variables, future emphasis in fathering research needs to enquire into the covariation of factors that affect father influence on child development. For example, Park & Sawin's (1977) studies demonstrated that both mother and father show more interest when they are together with the newborn. They count toes, check ears, and smile at the infant more.

Clarke-Stewart (1979) found, in a small sample-size but provocative study, that as mothers rear a contented, interesting baby, fathers, after the first year, are lured by such an attractive infant into increased interactions. Triadic effects of fathers and mothers and infants need to be examined. Pederson's (1975) finding that father's warmth and affection helps support the mother and make her more effective with the baby is relevant. Research must be sensitive both to direct and to indirect effects of fathering.

The relation of fathers to the development of altruism, empathy and the gentler arts of positive relations with others deserves far more research effort. If a parent preaches "love thy neighbor" but father models proudly his "he-man" imperviousness and insensitivity to the feelings and rights of others, particularly wife and children, then present research suggests that children will practice what they live rather than what they are told.

As divorce statistics increase, more and more ways to help children weather parental storms and uncouplings must be found. Divorce findings reveal that the role of a well-adjusted father who can communicate without strong rancor with his ex-spouse can do much through intimate, consistent contact with children, post-divorce, to help heal the distress and anger that divorce entails for children. Otherwise, what "frees" the parents may engender possible long-lasting grief and academic difficulty, particularly for sons.

Single parents may have a harder job rearing children well because of the extra stresses that may ensue when there is lack of a supporting other person. Yet, single fathers may have a very high motivation to parent. Strong positive motivation has been known to overcome "handicaps" far more severe than those involved in single fathering.

Despite the many studies which sug-

... Fathering may still be a profound and deeply satisfying experience in human intimacy and engagement.

gest that diapering and child care in early infancy are not the occupation of choice for many fathers (and some mothers too), fathering may still be a profound and deeply satisfying experience in human intimacy and engagement. It may be well to remember the impressive findings from the long-term study of Terman's gifted children. When gifted boys were reinterviewed decades later at age 62, they agreed that the greatest source of satisfaction in their lives was their families.

Sears (1977) has commented that in spite of autonomy and great average success in their occupations, these men placed greater importance on having achieved satisfaction in their family life than in their work. Furthermore, these men believed that they had found such satisfaction. May it be so for fathers of the future. Such a deep conviction and satisfaction would augur well for the children of tomorrow.

Fathering education is not yet politically an "in" issue for society. Yet, a man needs to learn fathering the way he would learn to play ball or set up a business or cook a gourmet meal — early and with lots of practice, patience and encouragement. Communities must become alert to the ways schools and service organizations can provide opportunities for boys to learn about and to nurture younger children responsively and responsibly.

REFERENCES

Ainsworth, M.D.S. Attachment: Retrospect and prospect. Presidential address presented at the Biennial Meeting of the Society for Research in Child Development, San Francisco, March, 1979.

Bing, E. Effect of childrearing practices on development of differential cognitive abilities. *Child Development*, 1963, *34*, 631-648.

Bowlby, J. The nature of the child's tie to his mother. *International Journal of Psychoanalysis*, 1958, *39*, 350-373.

Clarke-Stewart, A. The father's impact on mother and child. Paper presented at the Biennial Meeting of the Society for Research in Child Development, New Orleans, March, 1979.

Crandall, V. J., Dewey, R., Katkovsky, W. & Preston, A. Parents' attitudes and behaviors and grade-school children's academic achievements. *Journal of Genetic Psychology*, 1964, *104*, 53-66.

Cross, H. J. & Allen, J. Relationship between memories of parental behavior and academic achievement motivation. Proceedings of the 77th Annual Convention. American Psychological Association, Washington, D.C., September, 1969, 285-286.

Deal, T. N. & Montgomery, L. L. Techniques fathers use in teaching their young sons. Paper presented at the Meeting of the Society for Research in Child Development, Minneapolis, April 1971.

Dickie, J. R. & Carnahan, S. Training in social competence: The effect on mothers, fathers and infants. Paper presented at the Biennial Meeting of the Society for Research in Child Development, San Francisco, March, 1979.

Dodson, F. How to make your man a great father. *Harper's Bazaar*, April, 1979, p. 155, 194.

Epstein, A. S. & Radin, N. Motivational components related to father behavior and cognitive functioning in preschoolers. *Child Development*, 975, *46*, (No. 4), 831-839.

Feshback, N. The relationship of child rearing factors to children's aggression, empathy, and related positive and negative social behaviors. Paper presented at the NATO Conference on the Determinants and Origins of Aggressive Behavior, Monte Carlo, Monaco, July, 1973.

Friedlander, B. Z., Jacobs, A. C., Davis, V. B. & Wetstone, H. S. Time-sampling analysis of infants' natural language environments in the home. *Child Development*, 1972, *43*, 730-740.

Gasser, R. D. & Taylor, C. M. Role adjustment of single parent fathers with dependent children. Family Coordinator, 1976, *25*, (No. 4), 397-402.

Green, M. *Fathering*. New York: McGraw-Hill, 1976.

Greif, E. Sex differences in parent-child conversations, ERIC, ED 174 337, 1980.

Herzog, E. & Sudia, C. Children in fatherless families. In B. M. Caldwell & H.N. Ricciuti, (Eds.) *Review of Child Development Research Vol. 3*, Chicago: University of Chicago Press, 1973.

Hetherington, E. M., Cox, M. & Cox, R. Divorced fathers. *Family Coordinator*, 1976, *25*, 417-428.

Hoffman, M. L. Altruistic behavior and the parent-child relationship. *Journal of Personality and Social Psychology*, 1975, *31*, 937-943.

Honig, A. S. *Fathering: A bibliography*. Urbana, Illinois: ERIC (Document Reproduction Service No. 142293: (Cat No. 164) 1977.

Honig, A. S. A review of recent infancy research. *The American Montessori Society Bulletin*, 1979, *17* (No. 3 & 4) (a)

Honig, A. S. *Parent involvement in early childhood education*. 2nd Edition. Washington, D.C.: National Association for the Education of Young Children, 1979. (b)

Honig, A. S. Choices: To parent or not to parent. Paper presented at the 6th Annual Symposium on Sex Education, Toulouse, France, July, 1980.

Honig, A. S. & Main, G. Black fathering in three social class groups. Manuscript submitted for publication, 1980.

Johnson, M. M. Sex role learning in the nuclear family. *Child Development*, 1963, *34*, 319-333.

Katz, A. J. Lone fathers: Perspectives and implications for family policy. *The Family Coordinator*, 1979, *28*, 521-528.

Knox, I. D. & Gilman, R. C. The first year of fatherhood. Paper presented at the National Council on Family Relations, Missouri, 1974.

Kotelchuck, M. *The nature of the child's tie to his father*. Unpublished doctoral dissertation, Harvard University, 1972.

Lamb, M. E. The role of the father: An overview. In M. E. Lamb (Ed.) *The role of the father in child development*, New York: Wiley, 1976.

Lamb, M. E. Paternal influences and the father's role: A personal perspective. *American Psychologist*, 1979, *34*, 938-943.

Lamb, M. E. & Frodi, A. M. The role of the father in child development. In R. R. Abidin (Ed.) *Parent education and intervention handbook*. Springfield, Illinois: Charles C. Thomas, 1980.

Lynn, D. B. *The father: His role in child development*. Belmont, California: Brooks Cole, 1974.

McAdoo, J. L. Father-child interaction patterns and self esteem in black preschool children. *Young Children*, 1979, *34*, 46-53.

Mendes, H. A. Single fatherhood. *Social Work*, 1976, *21*, (No. 4), 308-312.

Mussen, P. & Eisenberg-Berg, N. *Roots of caring, sharing, and helping*. San Francisco: W. H. Freeman, 1977.

Pannor, R., Evans, B. W. & Massarik, F. *The unmarried father*. New York: Springer Publishing Co., 1971.

Parke, R. D. & Sawin, D. B. Fathering: It's a major role. *Psychology Today*, 1977, *11*, 108-113.

Parke, R. D. & Sawin, D. B. Fathering: It's a major role. H. E. Fitzgerald (Ed.) *Human Development 80/81*. Guilford, Connecticut: Dushkin, 1980.

Parsons, T. & Bales, R. F. *Family, socialization and interaction process*. Glencoe, Illinois: Free Press, 1955.

Pederson, F. A. Mother, father, and infant as an interactive system. Paper presented at the Symposium Fathers and Infants at the meetings of the American Psychological Association, Chicago, August, 1975.

Rallings, E. M. The special role of stepfather. *Family Coordinator*, 1976, *25*, 445-450.

Rebelsky, F. & Hanks, C. Fathers verbal interactions with infants in the first three months of life. *Child Development*, 1971, *42*, 63-68.

Russell, G. The father role and its relation to masculinity, femininity, and androgyny. *Child Development*, 1978, *49*, 1174-1181.

Rutherford, E. & Mussen, P. Generosity in nursery school boys. *Child Development*, 1968, *39*, 755-765.

Sears, R. R. Sources of life satisfactions of the Terman gifted men. *American Psychologist*, 1977, *32*, 119-128.

Sroufe, L. A. The coherence of individual development: Daily care, attachment, and subsequent developmental issues. *American Psychologist*, 1979, *34*, 834-341.

Stevens, J. H., Jr. & Mathews, M. (Eds.) *Mother/child, father/child relationships*. Washington, D. C.: National Association for the Education of Young Children, 1978.

Teahan, J. E. Parental attitudes and college success. *Journal of Educational Psychology*, 1963, *54*, 104-109.

Tuck, S. A model for working with black fathers. Paper presented at the Annual Meeting of the American Orthopsychiatric Association, San Francisco, 1969.

Wallerstein, J. S. & Kelly, J. B. Divorce counseling: A community service for families in the midst of divorce. In R. R. Abidin (Ed.) *Parent education and intervention handbook*. Springfield, Illinois: Charles C. Thomas, 1980.

Yarrow, M. R. & Scott, R. M. Imitation of nurturant and non-nurturant models. Journal of Personality and Social Psychology, 1972, *23*, 259-270.

WOMEN AT ODDS

The battle between the women who work and the women who stay at home is heating up. The issue is the question of the decade: Who makes the better mother?

Barbara J. Berg

Barbara J. Berg is the author of The Crisis of the Working Mother, *published by Summit Books.*

"The world is soon to be divided into two enemy camps, and one day they may not be civil toward each other."

"I feel a tremendous anger toward them—they want to have it all and give up nothing."

These two quotes may sound like snatches of conversation between the major world powers, but the cold war is not the subject here. The subject is motherhood, and "they" and "them" are employed and at-home mothers describing each other—sometimes with envy, often with misunderstanding, and, increasingly, with rancor.

With more than half the mothers of children under six employed outside the home, an unexpected and dramatic rift has appeared among women who traditionally have felt a strong identification with one another. As recently as 15 years ago, women who were mothers shared a sense of concern and interest, no matter what their status as bread-winners. Individual differences existed, to be sure, but in general the textures of their lives—the routines, the problems, the hopes for their families—were strikingly similar, creating a feeling of camaraderie and

a community of purpose. "In the past, women with children felt an affinity for one another, often cutting across class lines," observes Nina Cobb, historian and consultant on women's issues in New York City.

The opening up of new employment opportunities for women and the resulting exodus from the home, however, have unwittingly promoted a division among women. That those in the labor force and those who are not should now have different interests and imperatives is both predictable and understandable, but that they should harbor such deep animosity toward one another is neither.

That anger centers on a question as old as the century: Who makes the better mother, the one who stays home or the one who works? Consider the experience of 35-year-old Susan S., who returned to her job soon after the birth of her son. She began to have a difficult time with women friends at her firm. "And I thought you were a good mother," one of her colleagues said to her. "Well, maybe I am. Why don't we wait and see?" replied Susan, who privately confided later: "I resent the assumption that mothers at home are doing a good job, and those who are not at home aren't."

In part, the rift has been deepened by the current political climate, by the ideological split between what may be called the old liberalism on

one side and the new conservatism on the other. Many of the economically secure women who are leaving the work force for full-time motherhood are caught up in the wave of conservatism sweeping the country. President Reagan's opposition to the idea of comparable worth and to most of the enforcement provisions of the Equal Employment Opportunity Commission regulations and his stand on abortion encourage a return to traditional gender roles. So does the fundamentalist movement, which continues to wage a relentless campaign for social retrenchment. The Reverend Tim La Haye, a fierce advocate of the religious right, recently moved to Washington, D.C., where, praised by Reagan, he operates the new American Coalition for Traditional Values. His wife, Beverly, is president of Concerned Women, the largest counterfeminist organization in the country. (With its reported 600,000 members, Concerned Women has a bigger following than the National Organization of Women, the National Women's Political Caucus, and the League of Women Voters combined.)

American Mothers, Inc., a group pledged to strengthen the moral and spiritual foundation of the family and home, has also benefitted from the nation's swing to the right. Phyllis B. Marriott, honorary president of the 50-year-old organization,

which exalts the values of homemaking, announced a recent surge in membership and interest.

And, of course, the press has entered the fray. Although the women leaving the boardroom for the nursery are only a minority of working mothers, they are attracting a disproportionate amount of media attention. Magazines and television programs that once waxed euphoric over the energy and capabilities of the "superwoman" are now running features on "Getting off the Fast Track" or "Packing It In." Even an article about children's school clothing in a recent edition of *The New York Times* ended with "it's still nice to have Mom around when school lets out"—a sentiment that might have come right out of the 1950s.

All of this—the political shift to the right, the vocal outbursts of fundamentalist groups, the pressure from the press—filters down to the local level, where both employed and at-home mothers feel the reverberations. Within their communities, working mothers often say they feel discriminated against. High on their list of complaints are insensitive school policies. Barbara K. Docs, an administrative assistant at C.M. Offray and Son, a New Jersey-based ribbon manufacturer, finds it impossible to attend all her child's art shows and concerts, which are almost always held during the day. And when professional women cannot come to these events, "the 'at-home' mothers twist it around and condemn them for not 'being there' for their children," says Docs.

Most working mothers say they try to be at the schools as often as they can, but arbitrarily assigned parent-teacher conferences and little advance notice for such schedule changes as early dismissals and alterations in bus routes wreak havoc with their routines. At-

tempts to communicate with the school personnel about the difficulties these policies may impose on the family do not necessarily bring about a change in programs or attitudes. When Susan S., a director of human resource planning at one of the Big Eight

accounting firms, called her son's teacher to discuss a lengthy homework assignment that required parental assistance during afternoon hours, the teacher said, "Oh, I didn't realize that *you* had a problem—that you worked." Susan tried to point out to the

Uneasy Choices

I recently read an article that contained the phrase, ". . . the almost limitless opportunities for women in the 80s." I'm not sure what women that refers to, but I know this much: It isn't those of us with two or three young children at home.

The truth is that these "limitless opportunities" for women are illusory. They are available only to two narrow groups: those who remain child-free, as most of the high-achieving women in history have; and those have-it-all mothers on the cover of *People* magazine. (You know, the ones with six-figure incomes, one lone infant, and a live-in au pair—in short, those women who can afford to hire a full-time wife to do the donkey work.) The rest of us are caught between the devil and the deep blue sea. We can do what we need, and neglect our children; or do what they need, and neglect ourselves. Some choice.

Years ago, when my son was several months old and my maternity leave was up, I returned to my job. The first day I woke him up in the dark and dropped him off at the sitter's, I wept all the way to work. Later, when my two daughters were born, I left my job because I thought it would be better for them, because we could (almost) afford it, because I believed that in the 80s women can have it all. Working and motherhood seemed interchangeable options. Then the doors began closing in my face; I discovered the inescapable fact that children don't need less parenting as they grow older, they need more.

In all the current news items about working mothers and their babies, one fact tends to be ignored: Toddlers grow older. "The baby" becomes a six-year-old who begins to stutter or wet the bed; an eight-year-old who snitches something from the corner store; a ten-year-old who's falling behind in math; a 12-year-old who needs to be driven to the orthodontist every few weeks; a 14-year-old who's cutting school. The mother who believes she's home free as soon as her child gets into first grade hasn't been there yet.

The current cliché that quality time is more important to a child than quantity time is a rationale that we have invented to justify the fact that nobody is taking care of the children. It presumes that the two concepts are mutually exclusive, that the less time a mother has to spend with her children, the better-quality time it will be. I fear the opposite is true. The idea that a busy working mother will spend her off-work time singing nursery rhymes to her children is laughable.

Far more typical, I think, is the situation described by a friend of mine when she said, "My two brothers and I were latchkey kids, and I don't think it hurt us, because we all more or less looked out for each other. But the main memory I have of my mother all those years is of her coming home from work and screaming because the house was a mess and our chores weren't done and dinner wasn't started."

Another acquaintance, the mother of two, told me: "The first two or three weeks after I went back to work my husband was just great about sharing the household load. Then things began to slide until I was doing it all again. So I just didn't do it. Nobody did." While there are some token moves by husbands "helping out," we know that the majority of working mothers still have 90% of the responsibility for child care and household work.

The real trap of motherhood is that children need to see their mothers doing something other than servitude. Girls need it as an example to follow and boys need it so they won't grow up to expect servitude from the women in their own lives. Yet—and I never thought I'd say it—children also need the servitude.

Parenting is, after all, a learned behavior. We know that a child, even an abused or neglected one, grows up to treat his or her own children in the same way he or she was treated. What are we teaching the present generation of children—stockpiled in day-care centers—about families, parenting, and a child's relative importance in the scheme of things?

It isn't the occasional trip to Disneyland or the bike at Christmas, but the most trivial threads of everyday living that weave together the fabric of a childhood. What are our children learning from us when we aren't there to respond to the broken bone, the unexpected A, the party they weren't invited to, the team they did or didn't earn a place on—the dozen details on a kid's mind when he bursts in the door after school and says, "Guess what!"

There is a myth in our society that motherhood is honored and children are valued. It strikes me as sheer hypocrisy. Except for a few weeks' maternity leave, there is no allowance made and no quarter given to a woman who takes child-rearing time. The so-called lost opportunity time that mothers sacrifice in their professions when they take a break to raise children, is no myth. They must begin again at square one with both hands tied behind their backs.

The mothers in my two-year-old daughter's play group, all housewives, included a pharmacist, a speech therapist, a stockbroker, a nursing administrator, a software programmer, an actress with an MFA. They are women who would be making valuable professional contributions to their community, if that community provided them with some option other than the either-or choice they now face. Imagine expecting a man in those careers to make such a choice: either be a father, or practice your profession. It is unthinkable.

I have no perfect answers, but I know that we need more tracks for women with children. We need to begin to value children in fact as well as in the Coke commercials; to stop expecting mothers to pretend that their children don't exist as a precondition of employment; to offer child-care days and paternity leaves to both parents; to loosen up job-sharing opportunities for mothers.

We need large companies to fund in-house child-care facilities, and to make professional day care something more than the underpaid, haphazard provision it is now. We need to begin seeing childlessness as an acceptable option, and we need for those rare and wonderful women who are really gifted at mothering to be compensated in some tangible way so the word "housewife" isn't synonymous with "vulnerable."

I am fortunate beyond measure to have three happy, healthy children and a husband who plays fair. But however nice they are, they aren't enough. I am torn between my love for my children and my need to do something in addition to raising them. And I am angry about a society that forces half its members into what author Anne Crompton has called "this cheat, this half-lie": either work, or children.

—Saralee Kucera

teacher that a career was not a problem.

Professional women also complain about "innuendoes" and "subtle criticisms" from homemakers, designed, they claim, to make them feel guilty about being away from their children. Before 29-year-old Nancy Spice of Albuquerque, New Mexico, returned to her position as an investment assistant at a large trust company, she tried to get her three-month-old son on a good eating schedule, only to have a neighbor tell her that "he really should be fed on demand." Teresa Hernandez, a 32-year-old office manager for Design Gifts International, a West Coast company, told of friends who always ask her, "Don't your children miss you?" or "Isn't it too bad that you didn't get to see their first steps?" And Nicole Young Klein, an assistant product manager with a discount brokerage house in San Francisco, complained that one of her neighbors likes to tell her about all the enriching things she can do with *her* child because she's at home.

Common among many working mothers is the feeling that they are being "pushed from their communities." When 33-year-old Marjorie Richter, a systems analyst from Plainview, New York, asked to join two of her neighbors who were taking a walk with their children, she was told, "There's no room next to us for you." When Marjorie mentioned that she was returning to work, they asked her why she bothered to have a child at all. Other women say they are made to feel unwelcome at school or community events. They tell of icy receptions and snide comments ("I never expected to see *you* here"). "Those remarks are very hurtful," says a project director at Metropolitan Life Insurance. "I feel I'm being put in a separate category."

It was precisely those feelings of separation and exclusion that outraged the working mothers in a popular preschool program in New York City. When the director of the program suggested that they send their children to a special class just for kids brought by housekeepers, the mothers were indignant. So indignant, in fact, that the class

offered on Mondays at 10:45 A.M., usually the first slot filled, received no enrollment. What annoyed them, the director reported, was the implication that the care their housekeepers were giving wasn't comparable to their own, and that their children were being isolated, and perhaps even stigmatized, because of it. Other mothers, like those in the preschool program, voiced concern that their children are being excluded from play dates because at-home mothers do not want to make plans with their housekeepers or baby-sitters.

Nonworking mothers counter by saying that *they* often end up being baby-sitters for their working friends' children. "Every time one of their kids has a half day at school, they call me up," says 34-year-old Beth Finkel of Manhattan, a mother of three. "It's not fair. I'm not a baby-sitting service."

Though most are glad to help out occasionally, many at-home mothers accuse their working counterparts of constantly "dumping their kids" on them. They do not want to be the community suppliers of after-school milk and cookies. They do not want their homes used as hangouts for children who prefer them to the baby-sitter, and they do not like being called by the school to pick up the sick child of a mother who is "too busy" at the office to come.

"I'm always feeling like I'm the one left holding the bag," protests Beth L., a former sales manager for a lingerie company. Whether it's running a school book fair, organizing a church bazaar, or helping out with a class trip, homemakers are inevitably the ones who shoulder the burden of school and community events.

Many also say they're frustrated by the loss of intellectual outlets. Traditionally, voluntary associations have been important sources of personal gratification for full-time mothers. But now there's a trend toward increased responsibility and decreased recognition within these organizations because working mothers get the "glamour" jobs. Thirteen years ago, when 41-year-old Shirley Nook of Shaker Heights, Ohio, stopped working and began to

raise her son, the Junior League provided her with much of the stimulation and collegial support she had enjoyed as a college English teacher. "It was mostly full-time moms then, all college graduates, very bright, very articulate. We had so much in common," Shirley recalls. "Then the Junior League began to reach out to the working professional, and suddenly we mommies were told we couldn't meet at ten in the morning. We had to meet at noon downtown, or at night." Those times were not convenient for the homemakers, however. As Shirley explains: "Part of the reason I don't work is so I can be there when my eight-year-old comes home for lunch and be available in the evenings to help my thirteen-year-old with his homework." But even more annoying to Shirley and the other full-time mothers were the tasks they were asked to perform—such as vacuuming and cleaning the toilets in a house that the Junior League was in the process of renovating—while the working mothers "simply breezed in and out of meetings."

As the war between the homemakers and the working mothers escalates, many community organizations find themselves drawn into the fray. At the 92nd Street Y in Manhattan, a popular preschool program—Park Bench—became a focus of dispute. "Park Bench used to be attended mostly by children and their mothers," said Fretta Reitzes, director of the Parenting Center at the Y. Over the last two years, however, the numbers of paid "caregivers"—housekeepers, babysitters, and others hired by working women to look after their children—who brought kids to the program increased to the point where they were sometimes the majority in a class of 12. The full-time mothers were irate. They joined the class to meet other mothers, not caregivers, they said, and called and demanded their money back.

To try to appease everyone, the Parenting Center started a Park Bench class for working mothers in the evening, one just for caregivers in the morning, and decided to limit to four the

number of caregivers enrolled in other classes. Although the overt complaints against the presence of caregivers usually focused on the way they treated the children in their charge, Reitzes believes there is a hidden grievance—the full-time mothers feel devalued when they have to consort with the hired help.

"I know what people pay their housekeepers," says Beth Finkel, the New York mother of three. When she attends a class where most of the adults are caregivers, she says, "it's as if the working mothers are saying I'm only worth $150 a week. It galls me."

The suspicion that working mothers devalue the homemaker—her intellect, her interests, her involvements—is what most irritates the at-home mothers. Luuk Oleson from Clifton, Virginia, a community outside Washington, D.C., told of a legislative assistant, a new mother, who came up to her at a cocktail party and asked her what she did. When Oleson, a former financial analyst with Pan American World Airways, replied that she was raising her three children, the "woman did a 180-degree turn in midsentence and went off to find someone she obviously found more interesting," Oleson recalls, adding "I resent being looked down upon. People seem to think I have nothing to discuss."

That attitude is not limited strictly to toddlers' groups and cocktail parties. Like working mothers, homemakers have to put up with their share of humiliation from the school systems. Ellen Cohen, an at-home mother in New York City, for example, recalls standing outside her daughter's classroom as the teacher dismissed the children in the following way: "All those whose mommies are lawyers can get in line. All those whose mommies are in business can go..." The teacher continued down the roster of professions until only one child was left. "All those whose mommies walk around all day and shop can now leave," she said, and Ellen felt herself turn purple with rage as her daughter stood up. "This is a very sore topic with me," Ellen says. "I feel that people don't respect me, but I really do believe in my heart that I belong with my children."

For all their anger and bitterness, working and at-home mothers want the same things for their children: a warm, nurturing environment and quality care. They disagree, however, on how those goals are attained. Although studies find the children of both groups to be strikingly similar in their overall emotional adjustment (see box, "Work or Home: The Consequences"), full-time mothers contend that those who work neglect their children in selfish pursuit of a career. Working mothers counter by saying that homemakers overindulge their offspring and selfishly burden their husbands with the sole responsibility for supporting the family. Professional women further charge that homemakers make it more difficult for them on the job. "When the wife of my husband's boss didn't resume her career after having a baby, I knew he would never understand the complexities of my life," says Susan S. To which at-home mothers, like Luuk Oleson, respond that they are often snubbed at social and office functions by their husbands' associates who are working mothers.

Because of the animosity and hurt feelings on both sides,

Work or Home: The Consequences

What really happens to the children of mothers who work—and of mothers who stay at home full-time? The psychiatric community has been trying to answer that question definitively for a long time, but statistics are hard to come by in an area so clouded by subjectivity and emotion—and so influenced by the popular opinion of the times. For instance, studies conducted earlier in the century seemed to show that children suffered irreparable damage from maternal separation, but new interpretations of the same data now indicate that those conclusions may have been misleading.

Most of the studies being done today are yielding fairly good news for the working mother: Children in adequate day-care programs are scoring high on sociability and adjustment and have broader ideas of sex roles. On the other hand, they also tend to be more aggressive and independent than home-raised children, and sometimes more rebellious against adult authority.

These results are a far cry from those obtained in an early series of studies conducted by French psychoanalyst Renée Spitz in 1946, which showed that infants separated from their parents and raised in sterile institutionalized settings failed to thrive, showed arrested development, and sometimes died. English child developmentalist John Bowlby, working at London's Tavistock Clinic in 1960, found a strong link between maternal care and mental health, postulating that interference with the mother–infant bonding process that begins at birth and is reinforced during early childhood development can result in juvenile delinquency and later emotional disturbance. Other studies done by West German ethnologist Konrad Lorenz in 1952 and a series of investigations conducted by psychologist Harry Harlow in the 60s added to the negative evidence. In a series of studies with monkeys, Harlow demonstrated that behaviors similar to neurosis in humans occurred when maternal bonding was interrupted.

But there's a different way of looking at all of these studies: The emotional disturbances noted may well have been due to the lack of stimulation and the total family separation, not to the simple absence of the mother for periods of time each day, as is the case with working mothers. Researchers Jay Belsky at Pennsylvania State and Laurence Steinberg at the University of California at Irvine are convinced, after reviewing recent studies of the effects of day care—where the child receives stimulation and experiences social interaction, not total isolation—that daily absence does not interfere with the mother–child attachment process or hinder the child's emotional tie to the mother.

That's not to suggest that there is no difference at all between day-care children and home-care children. But some of the differences may bode well for the future. Two Canadian investigators, Delores Gold and David Andres, studied three age groups—preschool children, school-age children, and adolescents—and found that the children of working mothers had broader conceptions of sex roles than children of nonemployed mothers. Girls of employed mothers felt that activities like household chores, child care, discipline, and decision making were equally appropriate for both sexes, while children of at-home mothers divided those activities into male and female camps.

Children who spend some of their early developmental years in day care have also been found to be more sociable, to relate more readily with other children, and to be better socially adjusted in school than children of nonemployed mothers, according to Belsky and Steinberg. The flip side is that children in day care have less tolerance for frustration, higher levels of impulsivity, and are somewhat less cooperative with adults.

A 1975 study done in Denmark by psychologist Terrance Moore fleshes out Belsky and Steinberg's findings, and helps draw up character profiles of the difference between day-care and home-care kids. In comparing boys raised with mothers at home and those who had alternative care like nursery school, day care, or a baby-sitter, Moore found that boys who had home mothering until age five were more sensitive, conforming, self-controlled, timid, school-oriented, and had a rather strict conscience. Moore projected that in the future these boys could be expected to accept self-blame more easily and, under severe conditions of stress, to experience bodily symptoms like migraine headaches or ulcers. The other group of boys were generally more aggressive, nonconforming to parental requirements, and more influenced by their peer group, suggesting that when they're grown up they may show their feelings more directly, be more open with others in interpersonal situations, and, under extreme pressure of stress or frustration, act out feelings aggressively.

Another factor in all of this, however, is the mother's feeling about the choices she has made. Psychologist Lois Wladis Hoffman, at the University of Michigan, cites research that shows that the mother's satisfaction with her role can make a big difference in how good she is at parenting. One study, which ranked mothers on an adequacy-of-mothering scale, suggested that women who were unemployed and satisfied with their roles scored highest, followed by working mothers satisfied with their roles. Of women who were *unhappy* with their roles, at-home mothers were rated at the bottom of the scale, while working ones came in a little higher.

Hoffman cited other studies as well, one of which found that employed mothers tend to feel better about themselves and generally have higher morale. Hoffman also found that working mothers encouraged more independence in their children. Our own studies at the Rochester Mental Health Center have shown that employed mothers were significantly more enthusiastic, disciplined, exacting, and less tied to traditional ideas—being more experimental and free-thinking—than nonworking mothers.

So how does it all add up? The evidence is varied enough to suggest that the best answer lies in moderation: Good mothers come in all shapes and sizes. The happier you are with the overall shape of your life, the better a parent you will be, whether you're home from nine to five or not.

—Nelson W. Freeling and Stanley Kissel

many women tell of losing good friends over the issue of who makes the best mother. The intensity of emotion is such that some women speak of feeling personally betrayed by those who make a decision different from theirs. Says one executive: "I don't think of it as two women choose this, two women choose that. I think of it as that's two who say I'm right and two who say I'm wrong."

The reason the issue elicits such powerful and judgmental responses seems to stem from insecurity. "Women are constantly looking for validation and confirmation of their choices, because they are insecure in those choices," explains clinical psychologist Rosemary Jennings of New York City. Many homemakers voice concern that they have sacrificed self-development and economic independence. Those who left jobs early in their careers wonder if they would have been successful had they remained. They admit to feeling envious of the glamorous lives they imagine professionals leading and feel a diminished sense of confidence as they watch fashionably suited women rush off to work.

Some also worry about the role model they are offering their children. One woman said she burst into tears as she listened to her five-year-old telling her doll that when she grew up she would buy her clothing with the money she got from her husband.

This is not to say that working mothers do not also have feelings of anxiety and uncertainty. They are concerned about having so little time to develop strong community ties or close

friendships. But above all they worry about not enough time with their children. "Every day I struggle with whether I work or I mother," says an assistant manager. "When I see women who are home all day, it makes me feel guilty. I wonder how I'm going to give my son all those benefits like computer lessons when he's in day care ten hours a day. And his fantasy mother is one who stays home."

"Some of the uneasiness on both sides is natural to parenting," explains Ellen Galinsky, project director of the Work and Family Life Studies at the Bank Street College of Education in New York. All parents begin with expectations of perfection for their children. When reality doesn't meet these expectations, they pin the blame on the specifics of their lives: "My child has problems because I work" or "My child is spoiled because I'm home all day." And then the mother begins to question her choice.

As Galinsky points out, that uncertainty is more profound in the present generation. Living through a time of cultural transition, with no antecedents for guidance, women do not have the comfort of knowing the long-term implications of their choices. If they choose one way, there's always the discomforting possibility of having chosen the other. That can cause anger and resentment toward those women who have taken a different path, because it leads to heightened ambivalence about one's own choices. To resolve the conflict they feel and to justify the correctness of their choice, many women try to find some irrefutable reason for their decision.

That the self-righteousness of each group springs from a need somehow to validate each choice is borne out by the fact that many women admit to changing sides in the argument as their roles change. As one professional quipped, "Now that I'm on maternity leave, I hate working mothers."

TIME AND AGAIN in the course of researching this article, women made references to the 1950s, either as an ideal time when women knew that their place was in the home—or as a stifling period when women had little chance to exercise their skills and talents. "People keep talking about the fifties," said one woman in exasperation. "But that was only one moment in history. Women have always worked." And, indeed, when female labor was required to keep the economy moving, as it was in the Colonial period and during major wars, women— the majority of them mothers—were encouraged to work.

During World War II, for example, the U.S. Department of Labor reported, "it can hardly be said that any occupation is absolutely unsuitable for the employment of women." Rosie was not only a riveter, but a truck driver, a tractor operator, and a stevedore. When the war began, 95% of the women with jobs said they planned to quit when the men came home, but by 1945, 80% wanted to continue working. Although their hopes were eventually dashed, these wartime workers wanted employment for the same reasons as women today: for financial need, for fulfillment, and for a better life for their children.

Says one woman in senior

management, "I'm doing this as much for my kids as for me. I'm teaching them always to participate for the good of the family, that no one person should ever be in the position of having everything done for him or her. I feel that my children are getting a feeling of choice and control over their lives from me."

As for the split between women, she adds, "You can't hide how you feel about this issue. You can't say, 'I stay home, but I think it's wonderful that you work' or vice versa, because if it were so great, why aren't you doing it? You can fake orgasms, but you can't fake your priorities on this issue."

Yet whatever their priorities, women do not have to remain so intensely at odds. "Women will respect one another more when they realize that their lives are actually similar in many ways," remarks Bank Street College's Ellen Galinsky. Staying at home or going to work each has its drawbacks. Neither mother has as much time to accomplish all that she thinks important, and one is as likely as the other to be doing laundry in the middle of the night.

Finally, women must cope with the insecurity of living in a society that accepts, but does not completely support, either choice. The homemaker knows how easily she can be "displaced" and how difficult it might be for her to find a job if she is. The professional must deal with inadequate maternity leaves and the problems of finding child care. If, however, women accept one another as allies instead of as adversaries, we can try to bring about the changes that will enhance all our lives.

The Children of the '60s as Parents

*BEING A 'NATURAL' PARENT IS HARDER
THAN IT SEEMS; IT TAKES WORK TO BREAK
OUT OF CULTURAL HABITS AND ASSUMPTIONS.*

Thomas S. Weisner
and Bernice T. Eiduson

Bernice T. Eiduson, who died in July 1985, was professor of medical psychology in the department of psychiatry at the University of California, Los Angeles (UCLA), where she started and codirected the long-term family study project. Thomas S. Weisner, codirector of the project, is professor of anthropology in the UCLA departments of psychiatry and anthropology.

Not everyone who has children follows the conventional middle-class pattern: Mom and Dad, married and living with their kids in their own home. Some differ by default—because poverty, divorce or other adversity makes such goals unattainable. Others differ by design—because they believe that the prevailing values and practices of conventional middle-class American life are flawed. Deliberately unconventional families have chosen to work hard at creating an innovative life-style that they believe will produce a better life for them and their children.

In 1974 we began a long-term study of more than 200 families (about 50 conventional and 150 unconventional) to see how their values differed, how they put these values to work in their actual child-rearing practices and whether these differences would affect the development of their children (see "The Families" box for details on the four groups: conventional, married two-parent families and three groups of unconventional families—single mothers, unmarried "social-contract" couples and people living in communal groups).

It is too soon to tell about the children's long-term development, but we already know a great deal about how the kids were doing when they reached school age. We also can describe quite precisely how their parents put many of their values and child-rearing expectations into action during those early years.

When our project began, we had some concerns about the developmental well-being of the children reared in the alternate life-styles. Although many of the parents thought they were innovating in their children's best interests, we were less sure. However, judging by our developmental data through the first year of school, our concerns were not borne out. By school age, on average, the children in all four groups were doing well, and there were no major group differences in their physical or mental development, as indicated by overall health, school measures, cognitive and IQ scores and social and emotional development scores on standardized tests. We do expect the values and life choices of children raised conventionally to differ from those raised unconventionally as they grow older, however, and we intend to track them into young adulthood to see whether such differences emerge.

Although we find little evidence that children in any family group are particularly prone to severe mental or social problems, we have identified some families, about 17 percent, who have continuing difficulties that include drinking, drugs or personal problems and a somewhat chaotic, changeable quality in their lives. Such troubled families are more often found among the single-mother group, but every group has some, and their problems usually preceded the birth of their children.

How did parents whose values and family arrangements were often highly unconventional raise children who developed in normal patterns and entered public schools with no unusual difficulties? We think this happened because they were only selectively unconventional, because concerns for their children's culturally normal development became stronger as the children grew and because they played out many conventional American cultural "scripts" for parenting and child development along with their more innovative practices.

When we first interviewed the parents, we found that many value differences separated the conventional and unconventional families, such as beliefs about equality between the sexes, materialism and conventional achievement striving. We will concentrate on yet another value difference: the set of beliefs we call "pronaturalism."

Many of the unconventional parents (but relatively few of the conventional ones) were committed to what they saw as a more natural, emotionally expressive style of family life than is typical in middle-class America. Their plans for putting these pronatural beliefs into practice had three facets:

• **Natural-organic:**
Parents wanted to deemphasize materialism and to use "nonplastic" products, including making their own foods (especially baby foods), using natural herbs and medicines and not buying toys, particularly plastic ones or those with commercial logos.

• **Warm, emotionally expressive:**
Parents emphasized the importance of teaching their children to show their feelings, to be honest and open in their emotional expression. They emphasized the importance of warmth, intimacy and expressiveness, and preferred soft, chest- or back-carry ("Snugli"-type) devices for better mobility and closer physical contact with their child. Nudity was not to be discouraged or dealt with negatively. Parents desired a long breastfeeding period and virtually no bottle feeding.

• **"Laid-back," low conflict:**
Parents stated that they did not want to "lay a trip" on their child, preferring a loose, relaxed family style emphasizing low conflict and no physical punishment or aggression. They saw this style as natural to them and to their children.

Some parents wanted to follow child-care practices that were (and are) widespread throughout the world if not within our culture at that time, such as extensive breastfeeding and

skin-to-skin contact between caretakers and baby. They saw practices such as breastfeeding, late weaning and feeding only additive-free foods as ways to ensure the child's health and safety and to protect the child from the dangers of industrial society.

Most pronatural families were aware that the values and practices they espoused were not natural in the sense of being widespread and taken for granted in our culture. They felt like a minority vanguard, fighting off the tendencies of dominant, commercial, unhealthful cultural beliefs about how families should raise young children.

Some of the same values were also found among the conventional families but much less frequently. We found that these pronatural values were highest among social-contract families and lowest among conventional families. Among the other family life-style groups, pronatural values were more important to living-group dwellers than to single mothers.

Did pronatural parents actually put their goals into action in their childrearing? Did they use "natural" practices more often and longer than other parents? Indeed they did. For a sense

---

*O*NE MOTHER WHO LET HER INFANT CRY FOR 20 MINUTES SAID, 'IT WAS GOOD FOR HIM TO WORK THROUGH HIS FEELINGS.'

---

of the extremes in pronatural versus conventional child-rearing, consider these two families in our study, described when their children were quite small:

Andrea, her "old man," Jake, and their daughter, Sunrise, lived together by social contract in what we have termed the "earthmother" life-style. Sunrise was "birthed" at home, in the presence of her father, close friends and a midwife. The family lived in a small wooden cabin in a heavily forested area of Mendocino, California. An-

drea and Jake, both from upper-middle-class homes in large metropolitan areas, had left their urban, fast-paced existence in search of a simpler, more natural life-style. Their purposely primitive cabin had a potbellied stove; water had to be drawn from a well. Much of their food came from their large garden and their chickens and goats.

Sunrise slept in her parents' bed at night, and during the day slept in the main room in a cradle made by Jake. She was nursed on demand until she was 18 months old. Her first solid foods, begun at 6 months, were homemade and often homegrown. Andrea strongly believed in quickly gratifying Sunrise's every physical and emotional need. She and Jake brought her with them whenever they went on outings, taking turns carrying her in a front sling or backpack.

In contrast, Cynthia, her husband, John, and their daughter, Joanne, lived in a large suburban house where Joanne had a crib in her own room. For privacy at night, Joanne's parents closed the door to her room. John watched Joanne's birth in a hospital, where she was delivered by a doctor, but he went home that night while Cynthia and Joanne remained in the hospital. Joanne was breastfed for a few weeks, but early on was given supplemental bottles, and by 2 months of age was introduced to commercial baby foods. She was fed according to a pediatrically prescribed schedule with a concern that she not be spoiled. When her parents went out, Joanne was either left at home with a sitter or taken along in a car bed, an infant seat or a stroller.

Most pronatural families we studied did not live in rural settings and did not make life choices as unusual as those of Andrea and Jake. More typical were George and Vicky, who lived together as an unmarried social-contract couple in a small house in Los Angeles, which they shared intermittently with one or two other families. They took their baby boy, Shawn, with them almost everywhere, using a sling or backpack, and mixed frequent parental attention and involvement with some periods of "independence"—"to get him to explore and be on his own." Vicky breastfed Shawn frequently until he was more than 12 months old, and they used natural, homemade toys and foods. Both parents cut down on all drugs during Vicky's pregnancy and afterwards, because they were seen as health hazards.

Their home was an active, lively place, with people, music and schedules constantly changing. Since neither parent worked at a conventional career or job, both were at home and available much of the time. While George helped Vicky take care of Shawn, Vicky was his primary caretaker. Like many parents with pronatural values, George and Vicky tried to put these into practice in their daily life but were somewhat constrained by the many changes in their lives—moves, new people living with them, personal problems—and by other goals, such as independence for both parents and children.

The pronatural parents' child-rearing behavior and their own "folk interpretations" of its naturalness varied widely. For example, one mother who let her young infant cry for 20 minutes said, "It was good for him to get in touch with and work through his feelings"; another mother responded immediately to her child's first whimper, believing this to be more natural and a way to "help him to feel loved and secure." Yet both mothers placed a premium on naturalism and emotional involvement.

Unconventional families in general, especially those who particularly valued being natural, were more prone than conventional families to use slings and chest-carrying packs and to follow several other practices: breastfeeding, late weaning, late introduction of solid foods, allowing children to sleep in the parents' bed and permissiveness toward nudity. Some conventional families also followed these same practices but less frequently, for less time or with less conviction and effort.

Late weaning provides a good example. Many pronatural mothers deliberately tried to wean their children late and saw this as an important, difficult, bold and controversial innovation in rearing their child. They hoped that children who were breastfed often, on demand and for a long period would be healthier and sustain a more secure, warm, empathic emotional bond with them and with others.

When the babies were 18 months old, about 30 percent of the unconventional families were still doing at least some breastfeeding, compared with

THE FAMILIES

In 1974 and 1975, we chose for our "unconventional" sample three types of families with mothers in their last three months of pregnancy: approximately 50 unmarried "social-contract" couples; 50 single mothers; and 50 couples who lived in communes or in domestic living-group arrangements.

Some of the social-contract couples were highly committed to their "nonmarried marriage" and felt that it would lead to even stronger bonds than legal marriage, since they had to continually negotiate their relationship. Others were uncertain about marriage, in the process of divorcing a prior spouse or unable to divorce for religious reasons.

The single mothers included women who hoped to marry, women who planned to continue on their own and were adjusting to their single status and women who elected not to live with men.

The families living in communes (either married, social-contract couples or single parents) included 23 living in various "creedal-based" communities such as fundamentalist Christian, Hari Krishna or Zen Buddhist groups. Another 28 families were living in "domestic residential communities," which consisted of large houses or pieces of land with a diverse group of loosely connected families and individuals. Unlike the creedal-based communities, these groups had diverse philosophies and reasons for living together, economic and social ties that extended outside of the residential group and flexible boundaries between members and nonmembers.

The conventional group consisted of 50 two-parent, married couples, 80 percent selected from a random sample of pregnant patients of California obstetricians and 20 percent staff contacts.

All the families we have studied were young, white and from varied educational backgrounds (high school through Ph.D.). They came from both working-class and middle-class but not poverty-level backgrounds. All the parents were initially between 18 and 35. These general similarities allowed us to concentrate on life-style choices unconfounded by differences in race, and to control for socioeconomic status, age and educational level. We have tracked them since 1974 and 1975 using a combination of interviews, questionnaires, direct observation in their homes and formal testing. Without their enduring patience and cooperation our study could never have happened.

fewer than 10 percent of the conventional families. Further, all the unconventional family groups had higher, although very small, percentages of mothers who breastfed until 2 years and 9 months. As shown in the chart, "Breastfeeding Practices Compared," the social-contract parents were the most likely to wean late.

But it's important to put these differences in perspective. Group differences of 10 to 30 percent, such as we found for breastfeeding duration, are fairly small. Further, most societies in the world do not begin weaning much before 18 months. Thus, while some pronatural families weaned later than did conventional American families,

compared with most cultures around the world they weaned early.

The same pattern holds for many other practices. For instance, 48 percent of all unconventional groups used a sling-type carrying device, compared with 18 percent of conventional families. But few families of any type carried their infants and young toddlers more than two hours a day. In other cultures, infants and young children are routinely carried by their mothers, older siblings or cousins for six to eight hours a day or longer.

A study by psychologist Herbert Barry and then-research assistant Leonora Paxson of the University of Pittsburgh found that typical parents in more than 70 percent of 186 societies sleep with their infant children in the same bed or same room. In contrast, only 20 percent of our communal-living parents slept with their babies, and fewer than 10 percent did so in other groups.

We wondered whether pronatural parents were as emotionally open and expressive with their children as their values suggested they would be. Did they touch, hold, cuddle and exchange smiles with their children more frequently than did conventional families? Did they show a more laid-back attitude toward their children's discipline or respond differently to their children's crying?

Based on ratings by trained observers, who visited families at home when the babies were 6 and 18 months old and measured interactions between caretakers and the infants, the answer seems to be a qualified no. Conventional and pronatural parents differed significantly on only 3 of the 15 measures we used: Pronatural mothers more often talked to and interacted with their children in ways that seemed friendly and positive, and their children smiled more in response to their presence. But even on these measures, the differences, although significant, were small. For instance, on average, babies in pronatural families smiled 3.8 times in a 25-minute sampling period versus 3.3 times in the conventional families.

We found virtually no differences between the two in their child health and safety measures. The children had their shots, were given well-baby checkups and were taken to the pediatrician for health problems about the same number of times. Parents' knowledge of medical danger signals and their health monitoring were similar as well. Pronatural parents often used unusual diets and home remedies, but they backed this up with conventional health care for their children.

Overall, although pronatural parents tried hard to put their new ideas into practice, their behavior was tempered by our culture and their backgrounds. Despite their different values, both conventional and unconventional families adjusted to the cultural and environmental constraints of our society. What might have become extensive, even extreme, innovations, judging from the parents' goals and plans at the time their child

was born, were limited in actual practice. A few families, especially some in communes, did use highly unusual (for our culture) innovations in caretaking and daily scheduling, but most changes were more modest and selective.

Two broad factors influence the way families actually practice what they preach: the "environmental niche"—made up of goals, people, organizations, work and daily routines and schedules—and the many "scripts" people acquire from their culture about parenthood and children. The niche makes some kinds of change—such as balancing time-bound work schedules, nuclear families and childcare—structurally difficult. The cultural scripts shape the possibilities people imagine, the choices they make and the countless small, repeated episodes of interaction in families.

The impact of such scripts is seen in the fact that only a few communal families in our study ever routinely used their own older children or those of others as primary caretakers of their child, although this kind of childcare is among the most common forms in societies around the world and also occurs in the United States among some black and other subcultures. Further, only a few families in our sample ever "loaned out" their baby to another household for even a limited

time, and only a few asked another household to lend them an older child or young adult to live with them and help care for their child, as is done in many other cultures. All these practices could have helped the families to share care (a pronatural value), but cultural scripts and the environmental niche apparently prevented parents from seriously considering or using them.

Similarly, unconventional parents were strongly influenced by the dominant American model for parenting, which emphasizes one-on-one helping and direction. They made direct, active efforts to be warmer, more intimate, looser and more laid back than conventional parents in relating to their children and others. But, as a group, they did not reduce their direct efforts to control and shape their child's future; indeed, they often were more controlling than other parents. They seldom allowed others to take primary responsibility for their child.

Unconventional parents also seemed to accept the widespread middle-class idea that responsibility for changing the family and the child rests on the parent, a belief based on the American cultural theme of individualism and personal responsibility. Many innovative parents relied on social networks of friends, kin and like-minded souls for aid. However, they saw such ex-

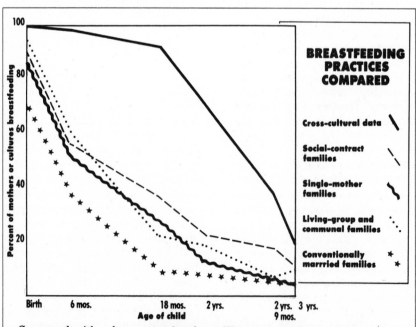

Compared with a large sample of non-Western societies, American families wean early. But among American families, those with "unconventional" life-styles wean later than do their "conventional" counterparts.

panded networks as extending their control and decision-making authority, not reducing it.

Conventional and unconventional parents acted as if they shared the assumption that children have some choice and autonomy in family activities, and that their points of view, feelings and needs must be taken into account. Both types of parents were likely to ask children what food they would like to eat, when and where they would like to play and even which playsuit they would like to wear. This characteristic American script for a child's autonomy and power in sharing decision-making started in early childhood and continued throughout later periods. Further, nearly all parents shared beliefs that early experience powerfully influences later life, and that talking with very young children is the critical way to stimulate their development.

Pronatural parents often did try to encourage a wide circle of social support, and many fought against the encapsulation of the nuclear family and the burden of exclusive responsibility for child-care. Thus, they tried to form "communities" and to reach out for new networks and nonexclusive marital, economic, familial and social attachments. But except for certain communal and a few other families, parents still made final decisions and remained in charge of subsistence, discipline and major decisions about safety, schooling and health. They would have thought it "unnatural" not to.

Despite the constraining features of the environmental niche and cultural scripts, pronatural parents did make changes that affected both their lives and those of their children. Many continue to emphasize equality between the sexes as both a political and a relationship goal. Many practice their politics and religious beliefs with fervor. The pronatural parents have a conserving and appreciative view of the environment and have tried to transmit these ideals to their children.

If some of these families' goals have been implemented only modestly and selectively, this does not suggest that the families are not different, for they certainly are. Indeed, the transmission of their new cultural ideals may prove to be among the most successful steps such parents have taken in changing their children's futures. They may have provided their children with the skill to adapt to differing circumstances with more than one model and script for conduct.

The new kinds of family life-styles and child-care forged by the parents in our study and others like them seem to have influenced the spread of new scripts for child-rearing to many other families in our society. Given the added impetus of sanctioning by medical and mental-health professionals, many of the then-new ideals of pronatural families—such as breastfeeding, natural childbirth and concern for nutrition—have already been widely adopted by large sectors of the population and are becoming part of the accepted patterns of American family life.

This process—of family innovation and experimentation followed by partially successful implementation, diffusion through imitation, transformation into more widely shared, stable scripts for family life-styles and child-rearing patterns—reveals the cultural evolution of the family in action.

THE SHAME OF AMERICAN EDUCATION

B.F. Skinner
Harvard University

ABSTRACT: *Recent analyses of American schools and proposals for school reform have missed an essential point: Most current problems could be solved if students learned twice as much in the same time and with the same effort. It has been shown that they can do so (a) when the goals of education are clarified, (b) when each student is permitted to advance at his or her own pace, and (c) when the problem of motivation is solved with programmed instructional materials, so designed that students are very often right and learn at once that they are. The theories of human behavior most often taught in schools of education stand in the way of this solution to the problem of American education, but the proposal that schools of education simply be disbanded is a step in the wrong direction. Teachers need to be taught how to teach, and a technology is now available that will permit them to teach much more effectively.*

On a morning in October 1957, Americans were awakened by the beeping of a satellite. It was a Russian satellite, Sputnik. Why was it not American? Was something wrong with American education? Evidently so, and money was quickly voted to improve American schools. Now we are being awakened by the beepings of Japanese cars, Japanese radios, phonographs, and television sets, and Japanese wristwatch alarms, and again questions are being asked about American education, especially in science and mathematics.

Something does seem to be wrong. According to a recent report of the National Commission on Excellence in Education (1983), for example, the average achievement of our high-school students on standardized tests is now lower than it was a quarter of a century ago, and students in American schools compare poorly with those in other nations in many fields. As the commission put it, America is threatened by a "rising tide of mediocrity."

The first wave of reform is usually rhetorical. To improve education we are said to need "imaginative innovations," a "broad national effort" leading to a "deep and lasting change," and a "commitment to excellence." More specific suggestions have been made, however. To get better teachers we should pay them more, possibly according to merit. They should be certified to teach the subjects they teach. To get better students, scholarship standards should be raised. The school day should be extended from 6 to 7 hours, more time should be spent on homework, and the school year should be lengthened from 180 to 200, or even 220, days. We should change what we are teaching. Social studies are all very well, but they should not take time away from basics, especially mathematics.

As many of us have learned to expect, there is a curious omission in that list: It contains no suggestion that teaching be improved. There is a conspiracy of silence about teaching as a skill. The *New York Times* publishes a quarterly survey of education. Three recent issues (Fisk, 1982, 1983a, 1983b) contained 18 articles about the kinds of things being taught in schools; 11 articles about the financial problems of students and schools; 10 articles about the needs of special students, from the gifted to the disadvantaged; and smaller numbers of articles about the selection of students, professional problems of teachers, and sports and other extracurricular activities. Of about 70 articles, only 2 had anything to do with how students are

From *American Psychologist*, September 1984, pp. 947-954. Reprinted by permission of the author.

taught or how they could be taught better. Pedagogy is a dirty word.

In January 1981, Frederick Mosteller, president of the American Association for the Advancement of Science, gave an address called "Innovation and Evaluation" (Mosteller, 1981). He began with an example of the time which can pass between a scientific discovery and its practical use. The fact that lemon juice cures scurvy was discovered in 1601, but more than 190 years passed before the British navy began to use citrus juice on a regular basis and another 70 before scurvy was wiped out in the mercantile marine—a lag of 264 years. Lags have grown shorter but, as Mosteller pointed out, are often still too long. Perhaps unwittingly he gave another example. He called for initiatives in science and engineering education and said that a major theme of the 1982 meeting of the association would be a "national commitment to educational excellence in science and engineering for all Americans" (p. 886).

When Mosteller's address was published in *Science*, I wrote a letter to the editor (Skinner, 1981) calling attention to an experiment in teaching algebra in a school in Roanoke, Virginia (Rushton, 1965). In this experiment an eighth-grade class using simple teaching machines and hastily composed instructional programs went through *all* of ninth-grade algebra in *half* a year. Their grades met ninth-grade norms, and when tested a year later the students remembered rather more than usual. Had American educators decided that that was the way to teach algebra? They had not. The experiment was done in 1960, but education had not yet made any use of it. The lag was already 21 years long.

A month or so later I ran into Mosteller. "Did you see my letter in *Science* about teaching machines?" I asked. "Teaching machines?" he said, puzzled. "Oh, you mean *computers*—teaching machines to *you*." And, of course, he was right. Computers is the current word. But is it the right one? Computers are now badly misnamed. They were designed to compute, but they are not computing when they are processing words, or displaying Pac-Man, or aiding instruction (unless the instruction is in computing). "Computer" has all the respectability of the white-collar executive, whereas "machine" is definitely blue-collar, but let us call things by their right names. Instruction may be "computer aided," and all good instruction must be "interactive," but machines that teach are teaching machines.

I liked the Roanoke experiment because it confirmed something I had said a few years earlier to the effect that with teaching machines and programmed instruction one could teach what is now taught in American schools in half the time with half the effort. I shall not review other evidence that that is true. Instead I shall demonstrate my faith in a technology of teaching by going out on a limb. I claim that the school system of any large American city could be so redesigned, at little or no additional cost, that students would come to school and apply themselves to their work with a minimum of punitive coercion and, with very rare exceptions, learn to read with reasonable ease, express themselves well in speech and writing, and solve a fair range of mathematical problems. I want to talk about why this has not been done.

The teaching machines of 25 years ago were crude, of course, but that is scarcely an explanation. The calculat-

ing machines were crude, too, yet they were used until they could be replaced by something better. The hardware problem has now been solved, but resistance to a technology of teaching survives. The rank commercialism which quickly engulfed the field of teaching machines is another possible explanation. Too many people rushed in to write bad programs and make promises that could not be kept. But that should not have concealed the value of programmed instruction for so many years. There is more than that to be said for the marketplace in the selection of a better mousetrap.

PSYCHOLOGICAL ROADBLOCKS

I shall argue that educators have not seized this chance to solve their problems because the solution conflicts with deeply entrenched views of human behavior, and that these views are too strongly supported by current psychology. Humanistic psychologists, for example, tend to feel threatened by any kind of scientific analysis of human behavior, particularly if it leads to a "technology" that can be used to intervene in people's lives. A technology of teaching is especially threatening. Carl Rogers has said that teaching is vastly overrated, and Ivan Illich has called for the de-schooling of society. I dealt with the problem in *Beyond Freedom and Dignity* (Skinner, 1971). To give a single example, we do not like to be told something we already know, for we can then no longer claim credit for having known it.

To solve that problem, Plato tried to show that students already possess knowledge and have only to be shown that they possess it. But the famous scene in Plato's *Meno* in which Socrates shows that the slaveboy already knows Pythagoras's theorem for doubling the square is one of the great intellectual hoaxes of all time. The slaveboy agrees with everything Socrates says, but there is no evidence whatsoever that he could then go through the proof by himself. Indeed, Socrates says that the boy would need to be taken through it many times before he could do so.

Cognitive psychology is causing much more trouble, but in a different way. It is hard to be precise because the field is usually presented in what we may call a cognitive style. For example, a pamphlet of the National Institute of Education (1980) quotes with approval the contention that "at the present time, modern cognitive psychology is the dominant theoretical force in psychological science as opposed to the first half of the century when behavioristic, anti-mentalistic stimulus-response theories of learning were in the ascendance" (p. 391). (The writer means "ascendant.") The pamphlet tells us that cognitive science studies learning, but not in quite those words. Instead, cognitive science is said to be "characterized by a concern with understanding the mechanisms by which human beings carry out complex intellectual activities including learning" (p. 391). The pamphlet also says that cognitive science can help construct tests that will tell us more about what a student has learned and hence how to teach better, but here is the way it says this: "Attention will be placed on two specific topics: Applications of cognitive models of the knowledge structure of various subject matters and of learning and problem solving to construction of tests that identify processes underlying test answers, analyze errors and provide information about what students know and don't know, and strategies for integrating testing information with instructional decisions" (p. 393). Notice especially the cognitive style in the last phrase—the question is not "whether test results can suggest better ways of teaching" but "whether there are strategies for integrating testing information with instructional decisions."

The Commission on Behavioral and Social Sciences and Education of the National Research Council (1984) provides a more recent example in its announcement of a biennial program plan covering the period 1 May 1983 to 30 April 1985. The commission will take advantage of "significant advances . . . in the cognitive sciences" (p. 41). Will it study learning? Well, not exactly. The members will "direct their attention to studies of fundamental processes underlying the nature and development of learning" (p. 41). Why do cognitive psychologists not tell us frankly what they are up to? Is it possible that they themselves do not really know?

Cognitive psychology is certainly in the ascendant. The word *cognitive* is sprinkled through the psychological literature like salt—and, like salt, not so much for any flavor of its own but to bring out the flavor of other things, things which a quarter of a century ago would have been called by other names. The heading of an article in a recent issue of the APA *Monitor* (Turkington, 1983) tells us that "cognitive deficits" are important in understanding alcoholism. In the text we learn simply that alcoholics show losses in perception and motor skills. Perception and motor skills used to be fields of psychology; now they are fields of cognitive science. Nothing has been changed except the name, and the change has been made for suspicious reasons. There is a sense of profundity about "cognitive deficits," but it does not take us any deeper into the subject.

Much of the vogue of cognitive science is due to advances in computer technology. The computer offers an appealing simplification of some old psychological problems. Sensation and perception are reduced to input; learning and memory to the processing, storage, and retrieval of information; and action to output. It is very much like the old stimulus-response formula patched up with intervening variables. To say that students process information is to use a doubtful metaphor, and how they process information is still the old question of how they learn.

Cognitive psychology also gains prestige from its alignment with brain research. Interesting things are certainly being discovered about the biochemistry and circuitry of the brain, but we are still a long way from knowing what is happening in the brain as behavior is shaped and maintained by contingencies of reinforcement, and that means that we are a long way from help in designing useful instructional practices.

Cognitive science is also said to be supported by modern linguistics, a topic to which I am particularly sensitive. Programmed instruction emerged from my analysis of verbal behavior (Skinner, 1957), which linguists, particularly generative grammarians, have, of course, attacked. So far as I know they have offered no equally effective practices. One might expect them to have improved the teaching of languages, but almost all language laboratories still work

in particularly outmoded ways, and language instruction is one of the principal failures of precollege education.

Psycholinguistics moves in essentially the same direction in its hopeless commitment to development. Behavior is said to change in ways determined by its structure. The change may be a function of age, but age is not a variable that one can manipulate. The extent to which developmentalism has encouraged a neglect of more useful ways of changing behavior is shown by a recent report (Siegler, 1983) in which the number of studies concerned with the development of behavior in children was found to have skyrocketed, whereas the number concerned with how children learn has dropped to a point at which the researcher could scarcely find any examples at all.

There are many fine cognitive psychologists who are doing fine research, but they are not the cognitive psychologists who for 25 years have been promising great advances in education. A short paper published in *Science* last April (Resnick, 1983) asserts that "recent findings in cognitive science suggest new approaches to teaching in science and mathematics" (p. 477), but the examples given, when expressed in noncognitive style, are simply these: (a) Students learn about the world in "naive" ways before they study science; (b) naive theories interfere with learning scientific theories; (c) we should therefore teach science as early as possible; (d) many problems are not solved exclusively with mathematics; qualitative experience is important; (e) students learn more than isolated facts; they learn how facts are related to each other; and (f) students relate what they are learning to what they already know. If these are *recent* findings, where has cognitive science been?

Cognitive psychology is frequently presented as a revolt against behaviorism, but it is not a revolt; it is a retreat. Everyday English is full of terms derived from ancient explanations of human behavior. We spoke that language when we were young. When we went out into the world and became psychologists, we learned to speak in other ways but made mistakes for which we were punished. But now we can relax. Cognitive psychology is Old Home Week. We are back among friends speaking the language we spoke when we were growing up. We can talk about love and will and ideas and memories and feelings and states of mind, and no one will ask us what we mean; no one will raise an eyebrow.

SCHOOLS OF EDUCATION

Psychological theories come into the hands of teachers through schools of education and teachers' colleges, and it is there, I think, that we must lay the major blame for what is happening in American education. In a recent article in the *New York Times* (Botstein, 1983), President Leon Botstein of Bard College proposed that schools of education, teachers' colleges, and departments of education simply be disbanded. But he gave a different reason. He said that schools of that sort "placed too great an emphasis on pedagogical techniques and psychological studies" (p. 64), when they should be teaching the subjects the teachers will eventually teach. But disbanding such schools is certainly a move in the wrong direction. It has long been said that college teaching is the only profession for which there is no professional training. Would-be doctors go to medical schools, would-be engineers go to institutes of technology, but would-be college teachers just start teaching. Fortunately it is recognized that grade- and high-school teachers need to learn to teach. The trouble is, they are not being taught in effective ways. The commitment to humanistic and cognitive psychology is only part of the problem.

Equally damaging is the assumption that teaching can be adequately discussed in everyday English. The appeal to laymanship is attractive. At the "Convocation on Science and Mathematics in the Schools" called by the National Academies of Sciences and Engineering, one member said that "what we need are bright, energetic, dedicated young people, trained in mathematics . . . science . . . or technology, mixing it up with 6- to 13-year-old kids in the classroom" (Raizen, 1983, p. 19). The problem is too grave to be solved in any such way. The first page of the report notes with approval that "if there is one American enterprise that is local in its design and control it is education" (p. 1). That is held to be a virtue. But certainly the commission would not approve similar statements about medicine, law, or science and technology. Why should the community decide how children are to be taught? The commission is actually pointing to one explanation of why education is failing.

We must beware of the fallacy of the good teacher and the good student. There are many good teachers who have not needed to learn to teach. They would be good at almost anything they tried. There are many good students who scarcely need to be taught. Put a good teacher and a good student together and you have what seems to be an ideal instructional setting. But it is disastrous to take it as a model to be followed in our schools, where hundreds of thousands of teachers must teach millions of students. Teachers must learn how to teach, and they must be taught by schools of education. They need only to be taught more effective ways of teaching.

A SOLUTION

We could solve our major problems in education if students learned more during each day in school. That does not mean a longer day or year or more homework. It simply means using time more efficiently. Such a solution is not considered in any of the reports I have mentioned — whether from the National Institute of Education, the American Association for the Advancement of Science, the National Research Council, or the National Academies of Sciences and Engineering. Nevertheless, it is within easy reach. Here is all that needs to be done.

1. Be clear about what is to be taught. When I once explained to a group of grade-school teachers how I would teach children to spell words, one of them said, "Yes, but can you teach spelling?" For him, students spelled words correctly not because they had learned to do so but because they had acquired a special ability. When I told a physicist colleague about the Roanoke experiment in teaching algebra, he said, "Yes, but did they learn algebra?"

For him, algebra was more than solving certain kinds of problems; it was a mental faculty. No doubt the more words you learn to spell the easier it is to spell new words, and the more problems you solve in algebra the easier it is to solve new problems. What eventually emerges is often called *intuition*. We do not know what it is, but we can certainly say that no teacher has ever taught it directly, nor has any student ever displayed it without first learning to do the kinds of things it supposedly replaces.

2. Teach first things first. It is tempting to move too quickly to final products. I once asked a leader of the "new math" what he wanted students to be able to do. He was rather puzzled and then said, "I suppose I just want them to be able to follow a logical line of reasoning." That does not tell a teacher where to start or, indeed, how to proceed at any point. I once asked a colleague what he wanted his students to do as a result of having taken his introductory course in physics. "Well," he said, "I guess I've never thought about it that way." I'm afraid he spoke for most of the profession.

Among the ultimate but useless goals of education is "excellence." A candidate for president recently said that he would let local communities decide what that meant. "I am not going to try to define excellence for them," he said, and wisely so. Another useless ultimate goal is "creativity." It is said that students should do more than what they have been taught to do. They should be creative. But does it help to say that they must acquire creativity? More than 300 years ago, Molière wrote a famous line: "I am asked by the learned doctors for the cause and reason why opium puts one to sleep, to which I reply that there is in it a soporific virtue, the nature of which is to lull the sense." Two or three years ago an article in *Science* pointed out that 90% of scientific innovations were accomplished by fewer than 10% of scientists. The explanation, it was said, was that only a few scientists possess creativity. Molière's audiences laughed. Eventually some students behave in creative ways, but they must have something to be creative with and that must be taught first. Then they can be taught to multiply the variations which give rise to new and interesting forms of behavior. (Creativity, incidentally, is often said to be beyond a science of behavior, and it would be if that science were a matter of stimulus and response. By emphasizing the selective action of consequences, however, the experimental analysis of behavior deals with the creation of behavior precisely as Darwin dealt with the creation of species.)

3. Stop making all students advance at essentially the same rate. The phalanx was a great military invention, but it has long been out of date, and it should be out of date in American schools. Students are still expected to move from kindergarten through high school in 12 years, and we all know what is wrong: Those who could move faster are held back, and those who need more time fall farther and farther behind. We could double the efficiency of education with one change alone — by letting each student move at his or her own pace. (I wish I could blame this costly mistake on developmental psychology, because it is such a beautiful example of its major principle, but the timing is out of joint.)

No teacher can teach a class of 30 or 40 students and allow each to progress at an optimal speed. Tracking is too feeble a remedy. We must turn to individual instruments for part of the school curriculum. The report of the convocation held by the National Academies of Sciences and Engineering refers to "new technologies" which "can be used to extend the educational process, to supplement the teacher's role in new and imaginative ways" (Raizen, 1983, p. 15), but no great enthusiasm is shown. Thirty years ago educational television was promising, but the promise has not been kept. The report alludes to "computer-aided instruction" but calls it the latest "rage of education" and insists that "the primary use of the computer is for drill" (p. 15). (Properly programmed instruction is *never* drill if that means going over material again and again until it is learned.) The report also contains a timid allusion to "low-cost teaching stations that can be controlled by the learner" (p. 15), but evidently these stations are merely to give the student access to video material rather than to programs.

4. Program the subject matter. The heart of the teaching machine, call it what you will, is the programming of instruction—an advance not mentioned in any of the reports I have cited. Standard texts are designed to be read by the student, who will then discuss what they say with a teacher or take a test to see how much has been learned. Material prepared for individual study is different. It first induces students to say or do the things they are to learn to say or do. Their behavior is thus "primed" in the sense of being brought out for the first time. Until the behavior has acquired more strength, it may need to be prompted. Primes and prompts must then be carefully "vanished" until the behavior occurs without help. At that point the reinforcing consequences of being right are most effective in building and sustaining an enduring repertoire.

Working through a program is really a process of discovery, but not in the sense in which that word is currently used in education. We discover many things in the world around us, and that is usually better than being told about them, but as individuals we can discover only a very small part of the world. Mathematics has been discovered very slowly and painfully over thousands of years. Students discover it as they go through a program, but not in the sense of doing something for the first time in history. Trying to teach mathematics or science as if the students themselves were discovering things for the first time is not an efficient way of teaching the very skills with which, in the long run, a student may, with luck, actually make a genuine discovery.

When students move through well-constructed programs at their own pace, the so-called problem of motivation is automatically solved. For thousands of years students have studied to avoid the consequences of not studying. Punitive sanctions still survive, disguised in various ways, but the world is changing, and they are no longer easily imposed. The great mistake of progressive education was to try to replace them with natural curiosity. Teachers were to bring the real world into the classroom to arouse the students' interest. The inevitable result was a neglect of subjects in which children were seldom naturally interested—in particular, the so-called basics. One solution is to make some of the natural reinforcers—goods or privileges—artificially contingent upon basic behavior, as in a token economy. Such contingencies can be justified if they correct a lethargic or disordered classroom, but there should

be no lethargy or disorder. It is characteristic of the human species that successful action is automatically reinforced. The fascination of video games is adequate proof. What would industrialists not give to see their workers as absorbed in their work as young people in a video arcade? What would teachers not give to see their students applying themselves with the same eagerness? (For that matter, what would any of us not give to see ourselves as much in love with our work?) But there is no mystery; it is all a matter of the scheduling of reinforcements.

A good program of instruction guarantees a great deal of successful action. Students do not need to have a natural interest in what they are doing, and subject matters do not need to be dressed up to attract attention. No one really cares whether Pac-Man gobbles up all those little spots on the screen. Indeed, as soon as the screen is cleared, the player covers it again with little spots to be gobbled up. What is reinforcing is successful play, and in a well-designed instructional program students gobble up their assignments. I saw them doing that when I visited the project in Roanoke with its director, Allen Calvin. We entered a room in which 30 or 40 eighth-grade students were at their desks working on rather crude teaching machines. When I said I was surprised that they paid no attention to us, Calvin proposed a better demonstration. He asked me to keep my eye on the students and then went up on the teacher's platform. He jumped in the air and came down with a loud bang. Not a single student looked up. Students do not have to be made to study. Abundant reinforcement is enough, and good programming provides it.

THE TEACHER

Individually programmed instruction has much to offer teachers. It makes very few demands upon them. Paraprofessionals may take over some of their chores. That is not a reflection on teachers or a threat to their profession. There is much that only teachers can do, and they can do it as soon as they have been freed of unnecessary tasks.

Some things they can do are to talk to and listen to students and read what students write. A recent study (Goodlad, 1983) found that teachers are responding to things that students say during only 5% of the school day. If that is so, it is not surprising that one of the strongest complaints against our schools is that students do not learn to express themselves.

If given a chance, teachers can also be interesting and sympathetic companions. It is a difficult assignment in a classroom in which order is maintained by punitive sanctions. The word *discipline* has come a long way from its association with *disciple* as one who understands.

Success and progress are the very stuff on which programmed instruction feeds. They should also be the stuff that makes teaching worthwhile as a profession. Just as students must not only learn but know that they are learning, so teachers must not only teach but know that they are teaching. Burnout is usually regarded as the result of abusive treatment by students, but it can be as much the result of looking back upon a day in the classroom and wondering what one has accomplished. Along with a sense of satisfaction goes a place in the community. One proposed remedy for American education is to give teachers greater respect, but that is putting it the wrong way around. Let them teach twice as much in the same time and with the same effort, and they will be held in greater respect.

THE ESTABLISHMENT

The effect on the educational establishment may be much more disturbing. Almost 60 years ago Sidney Pressey invented a simple teaching machine and predicted the coming "industrial revolution" in education. In 1960 he wrote to me, "Before long the questions will need to be faced as to what the student is to do with the time which automation will save him. More education in the same place or earlier completion of full-time education?" (Sidney Pressey, personal communication, 1960). Earlier completion is a problem. If what is now taught in the first and second grades can be taught in the first (and I am sure that it can,) what will the second-grade teacher do? What is now done by the third- or fourth-grade teacher? At what age will the average student reach high school, and at what age will he or she graduate? Certainly a better solution is to teach what is now taught more effectively and to teach many other things as well. Even so, students will probably reach college younger in years, but they will be far more mature. That change will more than pay for the inconvenience of making sweeping administrative changes.

The report of the National Commission on Excellence in Education (1983) repeatedly mistakes causes for effects. It says that "the educational foundations of our society are being eroded by a rising tide of mediocrity," but is the mediocrity causing the erosion? Should we say that the foundations of our automobile industry are being eroded by a rising tide of mediocre cars? Mediocrity is an effect, not a cause. Our educational foundations are being eroded by a commitment to laymanship and to theories of human behavior which simply do not lead to effective teaching. The report of the Convocation on Science and Mathematics in the Schools quotes President Reagan as saying that "this country was built on American respect for education. . . . Our challenge now is to create a resurgence of that thirst for education that typifies our nation's history" (Raizen, 1983, p. 1). But is education in trouble because it is no longer held in respect, or is it not held in respect because it is in trouble? Is it in trouble because people do not thirst for education, or do they not thirst for what is being offered?

Everyone is unhappy about education, but what is wrong? Let us look at a series of questions and answers rather like the series of propositions that logicians call a *sorites:*

1. Are students at fault when they do not learn? No, they have not been well taught.

2. Are teachers then at fault? No, they have not been properly taught to teach.

3. Are schools of education and teachers' colleges then at fault? No, they have not been given a theory of behavior that leads to effective teaching.

4. Are behavioral scientists then at fault? No, a culture too strongly committed to the view that a technology of behavior is a threat to freedom and dignity is not supporting the right behavioral science.

5. Is our culture then at fault? But what is the next step?

Let us review the sorites again and ask what can be done. Shall we:

1. Punish students who do not learn by flunking them?

2. Punish teachers who do not teach well by discharging them?

3. Punish schools of education which do not teach teaching well by disbanding them?

4. Punish behavioral science by refusing to support it?

5. Punish the culture that refuses to support behavioral science?

But you cannot punish a culture. A culture is punished by its failure or by other cultures which take its place in a continually evolving process. There could scarcely be a better example of the point of my book *Beyond Freedom and Dignity.* A culture that is not willing to accept scientific advances in the understanding of human behavior, together with the technology which emerges from these advances, will eventually be replaced by a culture that is.

When the National Commission on Excellence in Education (1983) said that "the essential raw materials needed to reform our educational system are waiting to be mobilized," it spoke more truly than it knew, but to mobilize them the commission called for "leadership." That is as vague a word as excellence. Who, indeed, will make the changes that must be made if education is to play its proper role in American life? It is reasonable to turn to those who suffer most from the present situation.

1. Those who pay for education—primarily taxpayers and the parents of children in private schools—can simply demand their money's worth.

2. Those who use the products of grade- and high-school education—colleges and universities on the one hand and business and industry on the other—cannot refuse to buy, but they can be more discriminating.

3. Those who teach may simply withdraw from the profession, and too many are already exercising their right to do so. The organized withdrawal of a strike is usually a demand for higher wages, but it could also be a demand for better instructional facilities and administrative changes that would improve classroom practices.

But why must we always speak of higher standards for students, merit pay for teachers, and other versions of punitive sanctions? These are the things one thinks of first, and they will no doubt make teachers and students work harder, but they will not necessarily have a better effect. They are more likely to lead to further defection. There is a better way: Give students and teachers better reasons for learning and teaching. That is where the behavioral sciences can make a contribution. They can develop instructional practices so effective and so attractive in other ways that no one—student, teacher, or administrator—will need to be coerced into using them.

Young people are by far the most important natural resource of a nation, and the development of that resource is assigned to education. Each of us is born needing to learn what others have learned before us, and much of it needs to be taught. We would all be better off if education played a far more important part in transmitting our culture. Not only would that make a stronger America (remember Sputnik), but we might also look forward to the day when the same issues could be discussed about the world as a whole—when, for example, all peoples produce the goods they consume and behave well toward each other, not because they are forced to do so but because they have been taught something of the ultimate advantages of a rich and peaceful world.

REFERENCES

Botstein, L. (1983, June 5). Nine proposals to improve our schools. *New York Times Magazine*, p. 59.

Fisk, E.B. (Ed.). (1982, November, 14). Fall survey of education [Supplement]. *New York Times*.

Fisk, E.B. (Ed.). (1983a, January 9). Winter survey of education [Supplement]. *New York Times*.

Fisk, E.B. (Ed.). (1983b, April 24). Spring survey of education [Supplement]. *New York Times*.

Goodlad, J.L. (1983). *A place called school*. New York: McGraw-Hill.

Mosteller, F. (1981). Innovation and evaluation. *Science, 211*, 881-886.

National Commission on Excellence in Education. (1983, April). *A nation at risk: The imperative for educational reform.* Washington, DC: U.S. Department of Education.

National Institute of Education. (1980) Science and technology and education. In *The five-year outlook: Problems, opportunities and constraints in science and technology* (Vol. 2, 391-399). Washington, DC: National Science Foundation.

National Research Council, Commission on Behavioral and Social Sciences and Education. (1984). Biennial program plan, May 1, 1983-April 30, 1985. Washington, DC: National Academy Press.

Raizen, S. (1983). *Science and mathematics in the schools: Report of a convocation.* Washington, DC: National Academy Press.

Resnick, L.B. (1983). Mathematics and science learning: A new conception. *Science, 220,* 477-478.

Rushton, E.W. (1965). *The Roanoke experiment.* Chicago: Encyclopedia Britannica Press.

Siegler, R.S. (1983). Five generalizations about cognitive development. *American Psychologist, 38,* 263-277.

Skinner, B.F. (1957). *Verbal behavior.* New York: Appleton-Century-Crofts.

Skinner B.F. (1971). *Beyond freedom and dignity.* New York: Alfred A. Knopf.

Skinner, B.F. (1981). Innovation in science teaching. *Science, 212,* 283.

Turkington, C. (1983, June). Cognitive deficits hold promise for prediction of alcoholism. *APA Monitor*, p. 16.

Moral Education for Young Children

Susan R. Stengel

Susan R. Stengel, Ed.D., is a kindergarten teacher at the Colorado Academy in Denver.

The form that children's moral education should take and the values to be taught are controversial among parents and educators. Some feel that a few simple behaviors such as how to share and how to say thank you are sufficient. Others focus on the development of self-esteem. Still others believe in providing a maximum amount of independence for children to work out their own social arrangements and disputes.

Decisions about how to discipline and teach young children to cope in the world should be based on an understanding of the development of morality. Parents and teachers of young children can benefit from knowing about the stages of moral development through adulthood so that they can provide early experiences upon which later development can flourish.

What Is Morality?

Morality has been defined (Berkowitz 1964) as action that conforms to socially determined standards of right behavior. Moral education based on this framework involves identifying those right behaviors and then training children to act accordingly, either by rewarding, punishing, modeling, lecturing, or some combination of these.

Some philosophers assert that morality involves more than prescribed action (Wilson 1967, p. 60). The right action must be freely chosen, must be based on reasons that take into account the interests of others, and must be accompanied by appropriate feelings or attitudes. Moral education derived from this viewpoint suggests exposing children to a variety of moral codes from which they select one. Teaching the right reasons and how to apply those reasons to situations involving moral choice would be included.

The cognitive-developmental view of morality (Kohlberg 1969) maintains that morality is a developmental process. One's reasons for acting morally, and the kinds of motivation required to act morally, change with time and experience in a predictable way. For example, while rewarding appropriate action may be sufficient to elicit moral behavior in young children, good reasons are usually necessary to convince teenagers or adults that a given action is moral. For the developmentalist, moral education means facilitating children's progress through the stages of moral reasoning, leading to the adoption of freely chosen moral values.

The Development of Moral Reasoning

Inspired by Piaget, Kohlberg (1969) has spent 20 years investigating how people think about moral problems by presenting moral dilemmas and noting what kinds of reasons people use to back up their resolutions to the dilemmas.

Kohlberg found that moral reasoning develops in stages. The stages as summarized by Fenton (1976, pp. 189-190) are:

Stage 1. The punishment and obedience orientation: The physical consequences of doing something determine whether it is good or bad without regard for its human meaning or value. People at Stage 1 think about avoiding punishment or earning rewards, and they defer to authority figures with power over them.

Stage 2. The instrumental relativist orientation: At Stage 2 right reasoning leads to action which satisfies one's own needs and sometimes meet the needs of others. Stage 2 often involves elements of fairness, but always for pragmatic reasons rather than from a sense of justice or loyalty. Reciprocity, a key element of Stage 2 thought, is a matter of "you scratch my back and I'll scratch yours."

Stage 3. The interpersonal sharing orientation: At this stage, people equate good behavior with whatever pleases or helps others and of whatever others approve. Stage 3 thinkers often conform to stereotypical ideas of how the majority of people in their group behave. They often judge behaving by intentions, and they earn approval by being nice.

Stage 4. The societal maintenance orientation: Stage 4 thought orients toward authority, fixed rules, and the maintenance of the social order. Right behavior consists of doing one's duty, showing respect for authority, or maintaining the given social order for its own sake.

Stage 5. The social contract, human rights, and welfare orientation: Stage 5 thinkers tend to define right action in terms of general individual rights and standards which have been examined critically and agreed upon by the entire society. The legal point of view is stressed, but emphasis is placed on the possibility of changing laws after rational consideration of the welfare of the society. Free agreement and contract bind people together where no laws apply.

Kohlberg's theory describes a sixth stage, the universal ethical principle orientation, which was dropped from the measurement because he did not find examples of it in his experimental populations. However, the sixth stage which few people seem to attain, still is part of his moral development theory—a theory he is expanding to include a seventh stage (Kohlberg 1981).

The primary educational method suggested by Kohlberg is to produce cognitive conflict in the mind of an individual that will force her or him to use higher stage reasoning to solve a problem. If individuals are exposed to reasoning one stage higher than their own, the growth process will be enhanced. One of the best ways to promote children's thinking about moral problems is through peer interaction in the form of dilemma discussions or class meetings in which students can exchange and justify their points of view.

Since Kohlberg published the original formulation of his stages, many researchers and educators have studied, elaborated upon, and used stage theory in their own work. Damon (1977) investigated children's ideas of justice, friendship, rules, manners, sex-role conventions, and authority. His work supports stage theory and describes in greater detail the thought processes of young children.

Damon considers the development of the concept of positive justice to be the central aspect of morality. He interviewed 144 children four through eight years old and presented a problem to them in which four children are given ten candy bars in exchange for making bracelets. One child makes the most bracelets; one child is the biggest; and one is younger than the others. The subject is asked how to distribute the candy bars. The results show distinct stages that correlate with age.

Level O-A: Justice is equated with the child's wishes. "I should get it because I want it."

Level O-B: The child is still self-centered, but feels the need to justify her or his choices. Illogical external factors such as size, sex, or other physical characteristics are used as justification. "I should get four because I'm four."

Level 1-A: Strict equality. Everyone gets the same. No special consideration is given to anyone.

Level 1-B: Reciprocity and merit are important factors. Those who work harder deserve more.

Level 2-A: Different but equally valid claims to justice are weighed. Special consideration is given for special needs. "She shouldn't have to work as hard because she's little."

Level 2-B: The claims of the various people (need, merit, etc.) involved and the demands of the specific situation are considered. Everyone should be given a fair share.

"Justice level 0-A was found predominantly at age four; 0-B, at ages four and five; 1-A, at age five (with some subjects scattered at higher ages); 1-B, at ages six and seven; 2-A, at age eight; and 2-B, at age eight" (Damon 1977, p. 91). Damon's results indicate that young children are capable of making moral judgments and that they should be called upon to do so in a variety

of circumstances. However, when he tried to administer Kohlberg's interviews to the same subjects, the younger children did not comprehend the dilemmas, and the older children exhibited nothing above Stage 1 reasoning.

It appears, then, that children need to be given meaningful opportunities to take the viewpoints of others (role taking). They also need to learn that reasons are important in stating their own viewpoint. Before suggesting ways to accomplish these goals in an early childhood program, we first need to look at conditioning—the commonly used method of controlling behavior (if not teaching morality).

How to Facilitate Moral Development

Why Conditioning Is Inadequate

Why not simply teach children good reasons for moral actions and then reward them for behaving morally in accordance with those reasons? Certainly in the early years children are not able to take into account all the facts of a situation, so they must learn some behaviors that have general applicability. However, conditioning children to behave in response to a reward is not acceptable as a long-term solution for several reasons.

Parents and teachers do not control rewards for very long. As soon as children begin to interact with other children, they begin responding to rewards outside of the home and family. How will children make decisions on their own and decide among the many rewards available to them outside the home? If adult reward is the only method used to elicit behavior, children may change their behavior as soon as the reward is no longer present or if a more enticing reward is offered.

Another problem with conditioning is that children cannot be conditioned to respond appropriately in all situations. For example, if a child has been conditioned not to take food from other people, this behavior may be totally inappropriate if the child sees a baby eating poisonous berries. Clearly, our aim should be to help children understand principles and reasons rather than to teach specific actions, because the appropriate action will depend upon the situation. Creating a moral atmosphere in the home or classroom is thus the first step to moral education.

Create a Moral Atmosphere

How do teachers or parents refrain from indoctrinating and conditioning children in an environment that offers optimal learning experiences? Teachers must exercise some control in classrooms in order to ensure justice for individuals, creating an atmosphere in which moral education (and other learning) can take place. Children need examples of fair treatment and organized social arrangements (e.g., ensuring that everyone who wishes gets a turn at easel painting, or allowing only four to play at a time in the block corner) if they are to understand justice and choose it for themselves when they get older. Such structure gives children the idea of a moral framework—a set of rules organizing human behavior. This structure reduces anxiety for both children and teachers because it ensures that each child will have the maximum opportunity to learn without preventing other children from having the same opportunity. "To try to impose values is immoral, but to fail to create frameworks within which people can choose their own values is just as bad" (Wilson 1967, p. 143).

The way in which the framework is established, however, can affect the attainment of the long-range goals of moral education. Therefore, the long-range goals should be kept in mind when formulating and enforcing basic rules (see Stone 1978).

Teachers can begin by asking what kinds of ground rules will facilitate a safe atmosphere conducive to moral education and other kinds of education for all of the children. Early childhood teachers may want to begin with four or five simple rules stated positively, such as "In this classroom, we walk," or "We put away our materials at this school." Too many ground rules may confuse the children and make them afraid to do anything for fear of breaking a rule. Later, rules for specific activities can be introduced, such as "We wear smocks to paint." Ground rules must be based on good reasons that are aimed at optimum learning and development for all children and that take into account the interest of teachers and parents. Ground rules should promote justice in the classroom.

Once the ground rules are chosen, the rules and the reason for each rule should be explained clearly to the children. For example, "Here, we put our materials away so they will be ready for the next person to use." Try to acquaint the children with the thought processes you went through to determine the rule. Rules will be more meaningful if adults mention the viewpoints of others and how the rules will serve to meet the needs of everyone. A demonstration of the rule in action and opportunities to implement the rule may be helpful.

After the ground rules have been introduced, the adults must follow them and enforce them consistently. Unless the adults model appropriate behavior, children will test the limits constantly. For example, one of the ground rules may be "When we are indoors, we speak in low voices." If the teacher calls loudly across the room to a child, everyone will probably begin speaking louder, and it will become difficult to change the pattern of speaking louder in order to be heard.

However, even with modeling, there will be infractions of the rules. These are handled best through patient and consistent guidance, and, if necessary, the imposition of logical consequences (Dreikurs and Grey 1970). For example, suppose Kaylene has not put away her materials after she has finished working with them. The teacher would first remind Kaylene of her responsibility. Often a child simply will forget a rule if she is engrossed in activity, until she has repeated the action a number of times and it becomes a habit. If she does not respond to a reminder, the teacher may take her hand and guide her to the material, perhaps explaining again the reason for the rule as they walk. However, a child

may refuse outright to comply, in which case a consequence logically related to the misbehavior will have to be imposed. For example, the teacher may require that a child not work with anything else until the materials are put away. Patient and consistent enforcement of the rules initially will result in a program in which children can work alone or in groups and can have the maximum opportunity for learning and cooperating with one another.

The logical consequences approach to discipline is based on the idea that children should experience the result of their own actions, whether pleasant or unpleasant. Depending on the situation, the parent or teacher may have to select a result that is distasteful or annoying to the child, but one that is not harmful. Logical consequences are not used as threats but are imposed in a matter-of-fact, even friendly, manner.

Enforcing the ground rules gives children some perspective on the social group and their relationship to the group. Teachers will probably want to explain to the child how her or his behavior affects others and will perhaps ask the child what would happen if everyone disobeyed that particular rule. Reasoning with children may help convince them to comply with the rules. But even if they do not change their behavior, giving reasons teaches children that reasons are important, and that some reasons are better than others, and it shows respect for the child as a person. Discipline by logical consequences appeals to children's sense of fairness and maintains their integrity.

Using reasoning and consistent and logical discipline prepares children for the times when they do not have coaches to tell them how to act. They soon establish appropriate behavior patterns, make some decisions for themselves, and begin to use reasoning to convince others of their viewpoint. As young adults, these children will probably begin to examine their own values and the values of others and will look for criteria to use in choosing their own values. Children who use reason as the ultimate authority become less egocentric more quickly than those who rely on adult authority.

Learning to take the point of view of another (role taking) can be facilitated also in the ways we communicate with children and by helping them communicate with each other effectively.

Use Communication Techniques

Based on some ideas from counseling psychology (Rogers 1951; Truax and Carkhuff 1967), Gordon's (1970; 1974) Parent Effectiveness Training (P.E.T) assists parents and teachers in becoming skilled communicators. The first skill is active listening, which requires not only listening, but also communicating with the speaker that you have heard her or him and that you understand what was said. The listener tries to repeat the essence of what the speaker has said in an attempt to check out the listener's perception. If the listener has interpreted the speaker correctly, the speaker will be encouraged to continue talking or to come up with a solution. Parents or teachers can help children think through and resolve a difficulty without telling them what to do. Note the differences between this typical exchange between student and teacher and the active listening alternative.

Typical exchange:

Situation: First day of school. Eddie is gazing out of the window of his second-grade classroom.

Teacher: What are you doing, Eddie?

Eddie: Nothing. *(Pause)* Do we have to go to school all day?

Teacher: Certainly we go all day. You went all day last year. Whatever made you think we didn't have to go all day? (Gordon 1974, pp. 102-103)

Active listening:

Teacher: What are you doing, Eddie?

Eddie: Nothing. *(Pause)* Do we have to go to school all day?

Teacher: You really wish you were out there instead of sitting here in the classroom.

Eddie: Uh-huh. There's nothing to do in here. You just have to sit in a seat all day and do papers and read.

Teacher: You miss being able to play outside like you did all summer.

Eddie: Yes. Playing games and swimming and climbing trees.

Teacher: That was fun, so it's hard to give all that up and come back to school again.

Eddie: Yes. I hope summer comes back soon.

Teacher: You're really looking forward to next summer.

Eddie: Yes, when next summer comes I can do what I want to do. (Gordon 1974, pp. 103-104)

Active listening respects children's integrity and their ability to solve their own problems. It demonstrates empathy and caring by showing children that their point of view is understood or, at least, that it is worth trying to understand. Furthermore, it provides a communication model for children that will assist them to be sensitive to others' points of view as they begin to use active listening themselves.

Active listening can be used with infants and toddlers as well as older children.

Child: *(Crying)* Truck, truck—no truck.

Parent: You want your truck, but you can't find it. (Active listening.)

Child: *(Looks under sofa, but doesn't find truck)*

Parent: The truck's not there. (Feeding back non-verbal message.)

Child: *(Thinks; moves to back door)*

Parent: The truck's not there. (Feeding back non-verbal message.)

Parent: Maybe the truck's in the back yard. (Feeding back non-verbal message.)

Child: *(Runs out, finds truck in sandbox, looks proud)* Truck!

Parent: You found your truck yourself. (Active listening.) (Gordon 1970, pp. 101-102)

Note that the parent does not praise the child when he finds his truck. The child already feels happy because he knows he has done a good job. He has good feelings because the parent has not intervened but has allowed him to take responsibility and assisted him with the use of active listening to fulfill his responsibility by himself. In this case praise is not necessary and might be interpreted as patronizing. Active listening affirms children's sense of competence and helps them see that they feel good because they did it themselves. This child will begin to associate good feelings with finding something himself rather than with having a truck.

The second communication skill Gordon advocates is called sending an I-message. I-messages are used when there is a conflict between the teacher and child. An I message serves the purpose of describing as clearly as possible the speaker's point of view without threatening or blaming anyone for causing the problem. I-messages carry with them the assumption that the listener is willing to help the speaker solve the problem if possible. They include a feeling, a nonblameful description of the behavior or problem that is causing the feeling, and the tangible effect of the behavior on the parent or teacher. For example:

> Child comes to the table with very dirty hands and face. (Teacher:) "I can't enjoy my [lunch] when I see all that dirt. It makes me feel kind of sick and I lose my appetite (Gordon 1970, pp. 132-133)

I-messages can be sent to young children nonverbally as in these examples.

> Rob is squirming while Mother is putting his clothes on. Mother gently but firmly restrains him and continues to dress him. (Message: "I can't dress you when you are squirming.")
>
> While Dad is carrying Tim in the supermarket, he starts to kick Dad in the stomach. Dad immediately puts him down. (Message: "I don't like to carry you when you kick me.") (Gordon 1970, p. 134)

Both active listening and I-messages put the teacher and the child on the same side of the problem rather than on opposing sides. They promote trust and empathy and they help children do exactly what they need to do to develop moral reasoning, i.e., take the viewpoints of others. Active listening helps children clarify their own point of view; I-messages help them perceive another's point of view. Using these communication skills promotes dialogue and thought and an attitude of solving problems by reasoning. Because this attitude usually results in harmonious human interaction, the children feel good about themselves and others in the process, thus promoting cooperation in the classroom. In such an atmosphere children are given optimal opportunity for increasing their understanding of social relationships and moral issues.

Lead Group Discussions

Group meetings with children at least four years of age also can be used to promote moral growth in several ways. The basic principle is to stimulate children to think. A question asking *what* (What did you do over the weekend?) is not as thought provoking as a question asking *why* (Why did your father want you to help him with the dishes?)

The teacher will have to take the lead in most discussions, but the content and interaction will come from the children. Children will first need to learn how to have an effective group discussion, speaking in turn and listening to each other. One way of taking turns is to have a talk ticket. The child holding the ticket may talk, and when she or he is finished, gives the ticket to another child who wishes to speak. To promote listening and understanding, children can occasionally summarize the remarks of each speaker before the next child speaks.

Once the ground rules for talking and listening have been established, the teacher keeps the discussion going by enforcing the rules and keeping the discussion focused on one topic. New rules may have to be formulated from time to time. As the children find that some people make lengthy speeches, for example, they may see the need for a time limit rule. If they do not think of it, the teacher may want to suggest the idea.

Small groups—five or six children—are more effective at first. Teachers should speak as little as possible and encourage the children to explain *why* they hold a certain position, e.g., "Why do you think we should share the crayons?" Teachers may share their own point of view in an objective manner ("When we share, then everyone gets a turn.") as long as they do not monopolize the discussion. Asking a child "Do you agree with the reason Sara gave?" is a good way of helping children to think and to focus on reasons. In this way the children will begin to see that their ideas count, that listening to each other is productive, and that they can solve problems.

To encourage children to express different points of view on a subject, teachers might want to ask questions such as "Does anyone have a different suggestion to make?" or "What do you think your mother (brother, friend) would say about this?" or "What would you think if you were the person who had to clean up everything?" Because children think more like each other than like adults, they will be able to understand their classmates' ideas, and perhaps their own thinking and ways of looking at the world will be challenged.

Simple social dilemmas also can be presented (*First Things: Social Reasoning*). Those that arise spontaneously in the classroom are usually more effective than hypothetical dilemmas because the children are more likely to understand the situation. Problems relating to sharing, taking turns, cleaning up, treating each other kindly, etc., can be discussed productively with young children.

Conclusion

The development of moral reasoning is an important goal for the education of young children. While teaching a specific set of values often is not acceptable to parents and is not particularly effective with children, ignoring moral education in early childhood programs is irresponsible and impossible.

Although young children tend to operate egocentrically, they can begin to apply their emerging intellectual abilities to moral and social issues, thus laying the groundwork for their future moral development. The pedagogical techniques suggested here will facilitate emergence from egocentricity into a world of satisfying human relationships.

Moral education is a part of every early childhood education program; we teach moral values with every action, every rule, and every activity. It is therefore necessary to plan consciously for moral education just as we plan in other areas of the curriculum.

References

Berkowitz, L. *Development of Motives and Values in a Child.* New York: Basic Books, 1964.

Damon, W. *The Social World of the Child.* San Francisco: Jossey-Bass, 1977.

Dreikurs, R., and Grey, L. *A Parents' Guide to Child Discipline.* New York: Hawthorn, 1970.

Duska, R., and Whelan, M. *Moral Development: A Guide to Piaget and Fenton, E. "Moral Education: The Research Findings." Social Education* 40 (April 1976): 189-193.

Gordon, T. *P.E.T.: Parent Effectiveness Training.* New York: New American Library, 1970.

Gordon, T. *T.E.T.: Teacher Effectiveness Training.* New York: David McKay, 1974.

Kohlberg. New York: Paulist Press, 1975.

First Things: Social Reasoning. Filmstrip series. Pleasantville, N.Y.: Guidance Associates.

Kohlberg, L. "Stage and Sequence: The Cognitive-Developmental Approach to Socialization." In *Handbook of Socialization Theory and Research,* ed. D.A. Goslin. Chicago: Rand McNally, 1969.

Kohlberg, L. *Philosophy of Moral Development.* New York: Harper & Row, 1981.

Rogers, C.R. *Client-Centered Therapy.* Boston: Houghton Mifflin, 1951.

Stone, J.G. *A Guide to Discipline.* Rev. ed. Washington, D.C.: National Association for the Education of Young Children, 1978.

Truax, C.B., and Carkhuff, R.R. *Toward Effective Counseling and Psychotherapy.* Chicago: Aldine Publishing Co., 1967.

Wilson, J. In *Introduction to Moral Education,* ed. J. Wilson, N. Williams, and B. Sugarman. Baltimore: Penguin, 1967.

My thanks to Tom Lickona for providing me with a summary of Damon's stages that I have adapted.

The hidden obstacles to black success.

RUMORS OF INFERIORITY

JEFF HOWARD AND RAY HAMMOND

Jeff Howard is a social psychologist; Ray Hammond is a physician and ordained minister.

TODAY'S black Americans are the beneficiaries of great historical achievements. Our ancestors managed to survive the brutality of slavery and the long history of oppression that followed emancipation. Early in this century they began dismantling the legal structure of segregation that had kept us out of the institutions of American society. In the 1960s they launched the civil rights movement, one of the most effective mass movements for social justice in history. Not all of the battles have been won, but there is no denying the magnitude of our predecessors' achievement.

Nevertheless, black Americans today face deteriorating conditions in sharp contrast to other American groups. The black poverty rate is triple that of whites, and the unemployment rate is double. Black infant mortality not only is double that of whites, but may be rising for the first time in a decade. We have reached the point where more than half of the black children born in this country are born out of wedlock—most to teenage parents. Blacks account for more than 40 percent of the inmates in federal and state prisons, and in 1982 the probability of being murdered was six times greater for blacks than for whites. The officially acknowledged high school dropout rate in many metropolitan areas is more than 30 percent. Some knowledgeable observers say it is over 50 percent in several major cities. These problems

not only reflect the current depressed state of black America, but also impose obstacles to future advancement.

The racism, discrimination, and oppression that black people have suffered and continue to suffer are clearly at the root of many of today's problems. Nevertheless, our analysis takes off from a forward-looking, and we believe optimistic, note: we are convinced that black people today, because of the gains in education, economic status, and political leverage that we have won as a result of the civil rights movement, are in a position to substantially improve the conditions of our communities using the resources already at our disposal. Our thesis is simple: the progress of any group is affected not only by public policy and by the racial attitudes of society as a whole, but by that group's capacity to exploit its own strengths. Our concern is about factors that prevent black Americans from using those strengths.

It's important to distinguish between the specific circumstances a group faces and its capacity to marshal its own resources to change those circumstances. Solving the problems of black communities requires a focus on the factors that hinder black people from more effectively managing their own circumstances. What are some of these factors?

Intellectual Development. Intellectual development is the primary focus of this article because it is the key to success in American society. Black people traditionally have understood this. Previous generations decided that segregation had to go because it relegated blacks to the backwater of American society, effectively denying us the opportunities, exposure, and competition that form the basis of intellectual development. Black intellectual development was one of the major benefits expected from newly won

From *The New Republic*, September 9, 1985, pp. 17-21. Reprinted by permission of *The New Republic*, © 1985 The New Republic, Inc.

access to American institutions. That development, in turn, was expected to be a foundation for future advancement.

YET NOW, three decades after *Brown v. Board of Education*, there is pervasive evidence of real problems in the intellectual performance of many black people. From astronomical high school dropout rates among the poor to substandard academic and professional performance among those most privileged, there is a disturbing consistency in reports of lagging development. While some black people perform at the highest levels in every field of endeavor, the percentages who do so are small. Deficiencies in the process of intellectual development are one effect of the long-term suppression of a people; they are also, we believe, one of the chief causes of continued social and economic underdevelopment. Intellectual underdevelopment is one of the most pernicious effects of racism, because it limits the people's ability to solve problems over which they are capable of exercising substantial control.

Black Americans are understandably sensitive about discussions of the data on our performance, since this kind of information has been used too often to justify attacks on affirmative action and other government efforts to improve the position of blacks and other minorities. Nevertheless, the importance of this issue demands that black people and all others interested in social justice overcome our sensitivities, analyze the problem, and search for solutions.

The Performance Gap. Measuring intellectual performance requires making a comparison. The comparison may be with the performance of others in the same situation, or with some established standard of excellence, or both. It is typically measured by grades, job performance ratings, and scores on standardized and professional tests. In recent years a flood of articles, scholarly papers, and books have documented an intellectual performance gap between blacks and the population as a whole.

• In 1982 the College Board, for the first time in its history, published data on the performance of various groups on the Scholastic Aptitude Test (SAT). The difference between the combined median scores of blacks and whites on the verbal and math portions of the SAT was slightly more than 200 points. Differences in family income don't explain the gap. Even at incomes over $50,000, there remained a 120-point difference. These differences persisted in the next two years.

• In 1983 the NCAA proposed a requirement that all college athletic recruits have a high school grade-point average of at least 2.0 (out of a maximum of 4.0) and a minimum combined SAT score of 700. This rule, intended to prevent the exploitation of young athletes, was strongly opposed by black college presidents and civil rights leaders. They were painfully aware that in recent years less than half of all black students have achieved a combined score of 700 on the SAT.

• Asian-Americans consistently produce a median SAT score 140 to 150 points higher than blacks with the same family income.

• The pass rate for black police officers on New York City's sergeant's exam is 1.6 percent. For Hispanics, it's 4.4 percent. For whites, it's 10.6 percent. These are the results *after* $500,000 was spent, by court order, to produce a test that was job-related and nondiscriminatory. No one, even those alleging discrimination, could explain how the revised test was biased.

• Florida gives a test to all candidates for teaching positions. The pass rate for whites is more than 80 percent. For blacks, it's 35 percent to 40 percent.

This is just a sampling. All these reports demonstrate a real difference between the performance of blacks and other groups. Many of the results cannot be easily explained by socioeconomic differences or minority status per se.

WHAT IS the explanation? Clear thinking about this is inhibited by the tendency to equate performance with ability. Acknowledging the performance gap is, in many minds, tantamount to inferring that blacks are intellectually inferior. But inferior performance and inferior ability are not the same thing. Rather, the performance gap is largely a behavioral problem. It is the result of a remediable tendency to avoid intellectual engagement and competition. Avoidance is rooted in the fears and self-doubt engendered by a major legacy of American racism: the strong negative stereotypes about black intellectual capabilities. Avoidance of intellectual competition is manifested most obviously in the attitudes of many black youths toward academic work, but it is not limited to children. It affects the intellectual performance of black people of all ages and feeds public doubts about black intellectual ability.

I. INTELLECTUAL DEVELOPMENT

The performance gap damages the self-confidence of many black people. Black students and professional people cannot help but be bothered by poor showings in competitive academic and professional situations. Black leaders too often have tried to explain away these problems by blaming racism or cultural bias in the tests themselves. These factors haven't disappeared. But for many middle-class black Americans who have had access to educational and economic opportunities for nearly 20 years, the traditional protestations of cultural deprivation and educational disadvantage ring hollow. Given the cultural and educational advantages that many black people now enjoy, the claim that all blacks should be exempt from the performance standards applied to others is interpreted as a tacit admission of inferiority. This admission adds further weight to the questions, in our own minds and in the minds of others, about black intelligence.

The traditional explanations—laziness or inferiority on the one hand; racism, discrimination, and biased tests on the other—are inaccurate and unhelpful. What is required is an explanation that accounts for the subtle influences people exert over the behavior and self-confidence of other people.

Developing an explanation that might serve as a basis for corrective action is important. The record of the last 20 years suggests that waiting for grand initiatives from the outside to save the black community is futile. Blacks will have to rely on our own ingenuity and resources. We need local and national political leaders. We need skilled administrators and creative business executives. We need a broad base of well-educated volunteers and successful people in all fields as role models for black youths. In short, we need a large number of sophisticated, intellectually developed people who are confident of their ability to operate on an equal level with anyone. Chronic mediocre intellectual performance is deeply troubling because it suggests that we are not developing enough such people.

The Competitive Process. Intellectual development is not a fixed asset that you either have or don't have. Nor is it based on magic. It is a process of expanding mental strength and reach. The development process is demanding. It requires time, discipline, and intense effort. It almost always involves competition as well. Successful groups place high value on intellectual performance. They encourage the drive to excel and use competition to sharpen skills and stimulate development in each succeeding generation. The developed people that result from this competitive process become the pool from which leadership of all kinds is drawn. Competition, in other words, is an essential spur to development.

Competition is clearly not the whole story. Cooperation and solitary study are valuable, too. But of the various keys to intellectual development, competition seems to fare worst in the estimation of many blacks. Black young people, in particular, seem to place a strong negative value on intellectual competition.

Black people have proved to be very competitive at some activities, particularly sports and entertainment. It is our sense, however, that many blacks consider intellectual competition to be inappropriate. It appears to inspire little interest or respect among many youthful peer groups. Often, in fact, it is labeled "grade grubbing," and gives way to sports and social activity as a basis for peer acceptance. The intellectual performance gap is one result of this retreat from competition.

II. THE PSYCHOLOGY OF PERFORMANCE

Rumors of Inferiority. The need to avoid intellectual competition is a psychological reaction to an image of black intellectual inferiority that has been projected by the larger society, and to a less than conscious process of internalization of that image by black people over the generations.

The rumor of black intellectual inferiority has been around for a long time. It has been based on grounds as diverse as twisted biblical citations, dubious philosophical arguments, and unscientific measurements of skull capacity. The latest emergence of this old theme has been in the controversy over race and IQ. For 15 years newsmagazines and television talk shows have enthusiastically taken up the topic of black intellectual endowment. We have watched authors and critics debate the proposition that blacks are genetically inferior to whites in intellectual capability.

Genetic explanations have a chilling finality. The ignorant can be educated, the lazy can be motivated, but what can be done for the individual thought to have been born without the basic equipment necessary to compete or develop? Of course the allegation of genetic inferiority has been hotly disputed. But the debate has touched the consciousness of most Americans. We are convinced that this spectacle has negatively affected the way both blacks and whites think about the intellectual capabilities of black people. It also has affected the way blacks behave in intellectually competitive situations. The general expectation of black intellectual inferiority, and the fear this expectation generates, cause many black people to avoid intellectual competition.

OUR HYPOTHESIS, in short, is this. (1) Black performance problems are caused in large part by a tendency to avoid intellectual competition. (2) This tendency is a psychological phenomenon that arises when the larger society projects an image of black intellectual inferiority and when that image is internalized by black people. (3) Imputing intellectual inferiority to genetic causes, especially in the face of data confirming poorer performance, intensifies the fears and doubts that surround this issue.

Clearly the image of inferiority continues to be projected. The internalization of this image by black people is harder to prove empirically. But there is abundant evidence in the expressed attitudes of many black youths toward intellectual competition; in the inability of most black communities to inspire the same commitment to intellectual excellence that is routinely accorded athletics and entertainment; and in the fact of the performance gap itself—especially when that gap persists among the children of economically and educationally privileged households.

Expectancies and Performance. The problem of black intellectual performance is rooted in human sensitivity to a particular kind of social interaction known as "expectancy communications." These are expressions of belief—verbal or nonverbal—from one person to another about the kind of performance to be expected. "Mary, you're one of the best workers we have, so I know that you won't have any trouble with this assignment." Or, "Joe, since everyone else is busy with other work, do as much as you can on this. When you run into trouble, call Mary." The first is a positive expectancy; the second, a negative expectancy.

Years of research have clearly demonstrated the powerful impact of expectancies on performance. The expectations of teachers for their students have a large effect on academic achievement. Psychological studies under a variety of circumstances demonstrate that communicated expectations induce people to believe that they will do well or poorly at a task, and that such beliefs very often trigger responses that result in performance consistent with the expectation. There is also evidence that "reference group expectancies"—directed at an entire category of people rather than a particular individual—have a similar impact on the performance of members of the group.

EXPECTANCIES do not always work. If they come from a questionable source or if they predict an outcome that is too inconsistent with previous experience, they won't have much effect. Only credible expectancies—those that come from a source considered reliable and that address a belief or doubt the performer is sensitive to—will have a self-fulfilling impact.

The widespread expectation of black intellectual inferiority—communicated constantly through the projection of stereotyped images, verbal and nonverbal exchanges in daily interaction, and the incessant debate about genetics and intelligence—represents a credible reference-group expectancy. The message of the race/IQ controversy is: "We have scientific evidence that blacks, because of genetic inadequacies, can't be expected to do well at tasks that require great intelligence." As an explanation for past black intellectual performance, the notion of genetic inferiority is absolutely incorrect. As an expectancy communication exerting control over our present intellectual strivings, it has been powerfully effective. These expectancies raise fear and self-doubt in the minds of many blacks, especially when they are young and vulnerable. This has resulted in avoidance of intellectual activity and chronic underperformance by many of our most talented people. Let us explore this process in more detail.

The Expectancy/Performance Model. The powerful effect of expectancies on performance has been proved, but the way the process works is less well understood. Expectancies affect behavior, we think, in two ways. They affect performance behavior: the capacity to marshal the sharpness and intensity required for competitive success. And they influence cognition: the mental processes by which people make sense of everyday life.

Behavior. As anyone who has experienced an "off day" knows, effort is variable; it is subject to biological cycles, emotional states, motivation. Most important for our discussion, it depends on levels of confidence going into a task. Credible expectancies influence performance behavior. They affect the intensity of effort, the level of concentration or distractibility, and the willingness to take reasonable risks—a key factor in the development of self-confidence and new skills.

Cognition. Expectations also influence the way people think about or explain their performance outcomes. These explanations are called "attributions." Research in social psychology has demonstrated that the causes to which people attribute their successes and failures have an important impact on subsequent performance.

All of us encounter failure. But a failure we have been led to expect affects us differently from an unexpected failure. When people who are confident of doing well at a task are confronted with unexpected failure, they tend to attribute the failure to inadequate effort. The likely response to another encounter with the same or a similar task is to work harder. People who come into a task expecting to fail, on the other hand, attribute their failure to lack of ability. Once you admit to yourself, in effect, that "I don't have what it takes," you are not likely to approach that task again with great vigor.

Indeed, those who attribute their failures to inadequate effort are likely to conclude that more effort will produce a better outcome. This triggers an adaptive response to failure. In contrast, those who have been led to expect failure will attribute their failures to lack of ability, and will find it difficult to rationalize the investment of greater effort. They will often hesitate to continue "banging my head against the wall." They often, in fact, feel depressed when they attempt to work, since each attempt represents a confrontation with their own feared inadequacy.

THIS COMBINED EFFECT on behavior and cognition is what makes expectancy so powerful. The negative expectancy first tends to generate failure through its impact on behavior, and then induces the individual to blame the failure on lack of ability, rather than the actual (and correctable) problem of inadequate effort. This misattribution in turn becomes the basis for a new negative expectancy. By this process the individual, in effect, internalizes the low estimation originally held by others. This internalized negative expectancy powerfully affects future competitive behavior and future results.

The process we describe is not limited to black people. It goes on all the time, with individuals from all groups. It helps to explain the superiority of some groups at some areas of endeavor, and the mediocrity of those same groups in other areas. What makes black people unique is that they are singled out for the stigma of genetic intellectual inferiority.

The expectation of intellectual inferiority accompanies a black person into each new intellectual situation. Since each of us enters these tests under the cloud of predicted failure, and since each failure reinforces doubts about our capabilities, all intellectual competition raises the specter of having to admit a lack of intellectual capacity. But this particular expectancy goes beyond simply predicting and inducing failure. The expectancy message explicitly ascribes the expected failure to genes, and amounts to an open suggestion to black people to understand any failure in intellectual activity as confirmation of genetic inferiority. Each engagement in intellectual competition carries the weight of a test of one's own genetic endowment and that of black people as a whole. Facing such a terrible prospect, many black people recoil from any situation where the rumor of inferiority might be proved true.

For many black students this avoidance manifests itself in a concentration on athletics and socializing, at the expense of more challenging (and anxiety-provoking) academic work. For black professionals, it may involve a tendency to shy away from competitive situations or projects, or an inability to muster the intensity—or commit the time—necessary to excel. This sort of thinking and behavior certainly does not characterize all black people in competitive settings. But it is characteristic of enough to be a serious problem. When it happens, it should be understood as a less than conscious reaction to the psychological burden of the terrible rumor.

The Intellectual Inferiority Game. There always have been constraints on the intellectual exposure and development of black people in the United States, from laws prohibiting the education of blacks during slavery to the Jim Crow laws and "separate but equal" educational arrangements that persisted until very recently. In dismantling these legal barriers to development, the civil rights movement fundamentally transformed the possibilities for black people. Now, to realize those possibilities, we must address the mental barriers to competition and performance.

The doctrine of intellectual inferiority acts on many black Americans the way that a "con" or a "hustle" like three-card monte acts on its victim. It is a subtle psychological input that interacts with characteristics of the human cognitive apparatus—in this case, the extreme sensitivity to expectancies—to generate self-defeating behavior and thought processes. It has reduced the intellectual performance of millions of black people.

Intellectual inferiority, like segregation, is a destructive idea whose time has passed. Like segregation, it must be removed as an influence in our lives. Among its other negative effects, fear of the terrible rumor has restricted discussion by all parties, and has limited our capacity to understand and improve our situation. But the intellectual inferiority game withers in the light of discussion and analysis. We must begin now to talk about intellectual performance, work through our expectations and fears of intellectual inferiority, consciously define more adaptive attitudes toward intellectual development, and build our confidence in the capabilities of all people.

THE expectancy/performance process works both ways. Credible positive expectancies can generate self-confidence and result in success. An important part of the solution to black performance problems is converting the negative expectancies that work against black development into positive expectancies that nurture it. We must overcome our fears, encourage competition, and support the kind of performance that will dispel the notion of black intellectual inferiority.

III. THE COMMITMENT TO DEVELOPMENT

In our work with black high school and college students and with black professionals, we have shown that education in the psychology of performance can produce strong performance improvement very quickly. Black America needs a nationwide effort, now, to ensure that all black people—but especially black youths—are free to express their intellectual gifts. That effort should be built on three basic elements:

• Deliberate control of expectancy communications. We must begin with the way we talk to one another: the messages we give and the expectations we set. This includes the verbal and nonverbal messages we communicate in day-to-day social intercourse, as well as the expectancies communicated through the educational process and media images.

• Definition of an "intellectual work ethic." Black communities must develop strong positive attitudes toward intellectual competition. We must teach our people, young and mature, the efficacy of intense, committed effort in the arena of intellectual activity and the techniques to develop discipline in study and work habits.

• Influencing thought processes. Teachers, parents, and other authority figures must encourage young blacks to attribute their intellectual successes to ability (thereby boosting confidence) and their failures to lack of effort. Failures must no longer destroy black children's confidence in their intelligence or in the efficacy of hard work. Failures should be seen instead as feedback indicating the need for more intense effort or for a different approach to the task.

The task that confronts us is no less challenging than the task that faced those Americans who dismantled segregation. To realize the possibilities presented by their achievement, we must silence, once and for all, the rumors of inferiority.

Who's Responsible? Expectations of black inferiority are communicated, consciously or unconsciously, by many whites, including teachers, managers, and those responsible for the often demeaning representations of blacks in the media. These expectations have sad consequences for many blacks, and those whose actions lead to such consequences may be held accountable for them. If the people who shape policy in the United States, from the White House to the local elementary school, do not address the problems of performance and development of blacks and other minorities, all Americans will face the consequences: instability, disharmony, and a national loss of the potential productivity of more than a quarter of the population.

However, when economic necessity and the demands of social justice compel us toward social change, those who have the most to gain from change—or the most to lose from its absence—should be responsible for pointing the way.

It is time that blacks recognize our own responsibility. When we react to the rumor of inferiority by avoiding intellectual engagement, and when we allow our children to do so, black people forfeit the opportunity for intellectual development that could extinguish the debate about our capacities, and set the stage for group progress. Blacks must hold ourselves accountable for the resulting waste of talent—and valuable time. Black people have everything to gain—in stature, self-esteem, and problem-solving capability—from a more aggressive and confident approach to intellectual competition. We must assume responsibility for our own performance and development.

ALIENATION

AND THE FOUR WORLDS OF CHILDHOOD

The forces that produce youthful alienation are growing in strength and scope, says Mr. Bronfenbrenner. And the best way to counteract alienation is through the creation of connections or links throughout our culture. The schools can build such links.

Urie Bronfenbrenner

Urie Bronfenbrenner is Jacob Gould Shur-man Professor of Human Development and Family Studies and of Psychology at Cornell University, Ithaca, N.Y.

To be alienated is to lack a sense of belonging, to feel cut off from family, friends, school, or work—the four worlds of childhood.

At some point in the process of growing up, many of us have probably felt cut off from one or another of these worlds, but usually not for long and not from more than one world at a time. If things weren't going well in school, we usually still had family, friends, or some activity to turn to. But if, over an extended period, a young person feels unwanted or insecure in several of these worlds simultaneously or if the worlds are at war with one another, trouble may lie ahead.

What makes a young person feel that he or she doesn't belong? Individual differences in personality can certainly be one cause, but, especially in recent years, scientists who study human behavior and development have identified an equal (if not even more powerful) factor: the circumstances in which a young person lives.

Many readers may feel that they recognize the families depicted in the vignettes that are to follow. This is so because they reflect the way we tend to look at families today: namely, that we see parents as being good or not-so-good without fully tak-ing into account the circumstances in their lives.

Take Charles and Philip, for example. Both are seventh-graders who live in a middle-class suburb of a large U.S. city. In many ways their surroundings seem similar; yet, in terms of the risk of alienation, they live in rather different worlds. See if you can spot the important differences.

CHARLES

The oldest of three children, Charles is amiable, outgoing, and responsible. Both of his parents have full-time jobs outside the home. They've been able to arrange their working hours, however, so that at least one of them is at home when the children return from school. If for some reason they can't be home, they have an arrangement with a neighbor, an elderly woman who lives alone. They can phone her and ask her to look after the children until they arrive. The children have grown so fond of this woman that she is like another grandparent—a nice situation for them, since their real grandparents live far away.

Homework time is one of the most important parts of the day for Charles and his younger brother and sister. Charles's parents help the children with their homework if they need it, but most of the time they just make sure that the children have a period of peace and quiet—without TV—in which to do their work. The children are allowed to watch television one hour each night—but only after they have completed their homework. Since Charles is doing well in school, homework isn't much of an issue, however.

Sometimes Charles helps his mother or father prepare dinner, a job that everyone in the family shares and enjoys. Those family members who don't cook on a given evening are responsible for cleaning up.

Charles also shares his butterfly collection with his family. He started the collection when he first began learning about butterflies during a fourth-grade science project. The whole family enjoys picnicking and hunting butterflies together, and Charles occasionally asks his father to help him mount and catalogue his trophies.

Charles is a bit of a loner. He's not a very good athlete, and this makes him somewhat self-conscious. But he does have one very close friend, a boy in his class who lives just down the block. The two boys have been good friends for years.

Charles is a good-looking, warm, happy young man. Now that he's beginning to be interested in girls, he's gratified to find that the interest is returned.

PHILIP

Philip is 12 and lives with his mother, father, and 6-year-old brother. Both of his parents work in the city, commuting more than an hour each way. Pandemonium strikes every weekday morning as

From *Phi Delta Kappan*, February 1986, pp. 430-436. Reprinted by permission of the author and Phi Delta Kappan.

the entire family prepares to leave for school and work.

Philip is on his own from the time school is dismissed until just before dinner, when his parents return after stopping to pick up his little brother at a nearby day-care home. At one time, Philip took care of his little brother after school, but he resented having to do so. That arrangement ended one day when Philip took his brother out to play and the little boy wandered off and got lost. Philip didn't even notice for several hours that his brother was missing. He felt guilty at first about not having done a better job. But not having to mind his brother freed him to hang out with his friends or to watch television, his two major after-school activities.

The pace of their life is so demanding that Philip's parents spend their weekends just trying to relax. Their favorite weekend schedule calls for watching a ball game on television and then having a cookout in the back yard. Philip's mother resigned herself long ago to a messy house; pizza, TV dinners, or fast foods are all she can manage in the way of meals on most nights. Philip's father has made it clear that she can do whatever she wants in managing the house, as long as she doesn't try to involve him in the effort. After a hard day's work, he's too tired to be interested in housekeeping.

Philip knows that getting a good education is important; his parents have stressed that. But he just can't seem to concentrate in school. He'd much rather fool around with his friends. The thing that he and his friends like to do best is to ride the bus downtown and go to a movie, where they can show off, make noise, and make one another laugh.

Sometimes they smoke a little marijuana during the movie. One young man in Philip's social group was arrested once for having marijuana in his jacket pocket. He was trying to sell it on the street so that he could buy food. Philip thinks his friend was stupid to get caught. If you're smart, he believes, you don't let that happen. He's glad that his parents never found out about the incident.

Once, he brought two of his friends home during the weekend. His parents told him later that they didn't like the kind of people he was hanging around with. Now Philip goes out of his way to keep his friends and his parents apart.

THE FAMILY UNDER PRESSURE

In many ways the worlds of both

Institutions that play important roles in human development are rapidly being eroded, mainly through benign neglect.

---

teenagers are similar, even typical. Both live in families that have been significantly affected by one of the most important developments in American family life in the postwar years: the employment of both parents outside the home. Their mothers share this status with 64% of all married women in the U.S. who have school-age children. Fifty percent of mothers of preschool children and 46% of mothers with infants under the age of 3 work outside the home. For single-parent families, the rates are even higher: 53% of all mothers in single-parent households who have infants under age 3 work outside the home, as do 69% of all single mothers who have school-age children.[1]

These statistics have profound implications for families — sometimes for better, sometimes for worse. The determining factor is how well a given family can cope with the "havoc in the home" that two jobs can create. For, unlike most other industrialized nations, the U.S. has yet to introduce the kinds of policies and practices that make work life and family life compatible.

It is all too easy for family life in the U.S. to become hectic and stressful, as both parents try to coordinate the disparate demands of family and jobs in a world in which everyone has to be transported at least twice a day in a variety of directions. Under these circumstances, meal preparation, child care, shopping, and cleaning — the most basic tasks in a family — become major challenges. Dealing with these challenges may sometimes take precedence over the family's equally important child-rearing, educational, and nurturing roles.

But that is not the main danger. What

threatens the well-being of children and young people the most is that the external havoc can become internal, first for parents and then for their children. And that is exactly the sequence in which the psychological havoc of families under stress usually moves.

Recent studies indicate that conditions at work constitute one of the major sources of stress for American families.[2] Stress at work carries over to the home, where it affects first the relationship of parents to each other. Marital conflict then disturbs the parent/child relationship. Indeed, as long as tensions at work do not impair the relationship between the parents, the children are not likely to be affected. In other words, the influence of parental employment on children is indirect, operating through its effect on the parents.

That this influence is indirect does not make it any less potent, however. Once the parent/child relationship is seriously disturbed, children begin to feel insecure — and a door to the world of alienation has been opened. That door can open to children at any age, from preschool to high school and beyond.

My reference to the world of school is not accidental, for it is in that world that the next step toward alienation is likely to be taken. Children who feel rootless or caught in conflict at home find it difficult to pay attention in school. Once they begin to miss out on learning, they feel lost in the classroom, and they begin to seek acceptance elsewhere. Like Philip, they often find acceptance in a group of peers with similar histories who, having no welcoming place to go and nothing challenging to do, look for excitement on the streets.

OTHER INFLUENCES

In contemporary American society the growth of two-wage-earner families is not the only — or even the most serious — social change requiring accommodation through public policy and practice in order to avoid the risks of alienation. Other social changes include lengthy trips to and from work; the loss of the extended family, the close neighborhood, and other support systems previously available to families; and the omnipresent threat of television and other media to the family's traditional role as the primary transmitter of culture and values. Along with most families today, the families of Charles and Philip are experiencing the unraveling and disintegration of social institutions that in the past were central to the health and well-being of children and their parents.

Notice that both Charles and Philip come from two-parent, middle-class families. This is still the norm in the U.S. Thus neither family has to contend with two changes now taking place in U.S. society that have profound implications for the future of American families and the well-being of the next generation. The first of these changes is the increasing number of single-parent families. Although the divorce rate in the U.S. has been leveling off of late, this decrease has been more than compensated for by a rise in the number of unwed mothers, especially teenagers. Studies of the children brought up in single-parent families indicate that they are at greater risk of alienation than their counterparts from two-parent families. However, their vulnerability appears to have its roots not in the single-parent family structure as such, but in the treatment of single parents by U.S. society.[3]

In this nation, single parenthood is almost synonymous with poverty. And the growing gap between poor families and the rest of us is today the most powerful and destructive force producing alienation in the lives of millions of young people in America. In recent years, we have witnessed what the U.S. Census Bureau calls "the largest decline in family income in the post-World War II period." According to the latest Census, 25% of all children under age 6 now live in families whose incomes place them below the poverty line.

COUNTERING THE RISKS

Despite the similar stresses on their families, the risks of alienation for Charles and Philip are not the same. Clearly, Charles's parents have made a deliberate effort to create a variety of arrangements and practices that work against alienation. They have probably not done so as part of a deliberate program of "alienation prevention" — parents don't usually think in those terms. They're just being good parents. They spend time with their children and take an active interest in what their children are thinking, doing, and learning. They control their television set instead of letting it control them. They've found support systems to back them up when they're not available.

Without being aware of it, Charles's parents are employing a principle that the great Russian educator Makarenko employed in his extraordinarily success-

ful programs for the reform of wayward adolescents in the 1920s: "The maximum of support with the maximum of challenge."[4] Families that produce effective, competent children often follow this principle, whether they're aware of it or not. They neither maintain strict control nor allow their children total freedom. They're always opening doors — and then giving their children a gentle but firm shove to encourage them to move on and grow. This combination of support and challenge is essential, if children are to avoid alienation and develop into capable young adults.

From a longitudinal study of youthful alienation and delinquency that is now considered a classic, Finnish psychologist Lea Pulkkinen arrived at a conclusion strikingly similar to Makarenko's. She found "guidance" — a combination of love and direction — to be a critical predictor of healthy development in youngsters.[5]

No such pattern is apparent in Philip's family. Unlike Charles's parents, Philip's parents neither recognize nor respond to the challenges they face. They have dispensed with the simple amenities of family self-discipline in favor of whatever is easiest. They may not be indifferent to their children, but the demands of their jobs leave them with little energy to be actively involved in their children's lives. (Note that Charles's parents have work schedules that are flexible enough to allow one of them to be at home most afternoons. In this regard, Philip's family is much more the norm, however. One of the most constructive steps that employers could take to strengthen families would be to enact clear policies making such flexibility possible.)

But perhaps the clearest danger signal in Philip's life is his dependence on his peer group. Pulkkinen found heavy reliance on peers to be one of the strongest predictors of problem behavior in adolescence and young adulthood. From a developmental viewpoint, adolescence is a time of challenge — a period in which young people seek activities that will serve as outlets for their energy, imagination, and longings. If healthy and constructive challenges are not available to them, they will find their challenges in such peer-group-related behaviors as poor school performance, aggressiveness or social withdrawal (sometimes both), school absenteeism or dropping out, smoking, drinking, early and promiscuous sexual activity, teenage parenthood, drugs, and juvenile delinquency.

This pattern has now been identified in a number of modern industrial societies, including the U.S., England, West Germany, Finland, and Australia. The pattern is both predictable from the circumstances of a child's early family life and predictive of life experiences still to come, e.g., difficulties in establishing relationships with the opposite sex, marital discord, divorce, economic failure, criminality.

If the roots of alienation are to be found in disorganized families living in disorganized environments, its bitter fruits are to be seen in these patterns of disrupted development. This is not a harvest that our nation can easily afford. Is it a price that other modern societies are paying, as well?

A CROSS-NATIONAL PERSPECTIVE

The available answers to that question will not make Americans feel better about what is occurring in the U.S. In our society, the forces that produce youthful alienation are growing in strength and scope. Families, schools, and other institutions that play important roles in human development are rapidly being eroded, *mainly through benign neglect*. Unlike the citizens of other modern nations, we Americans have simply not been willing to make the necessary effort to forestall the alienation of our young people.

As part of a new experiment in higher education at Cornell University, I have been teaching a multidisciplinary course for the past few years titled "Human Development in Post-Industrial Societies." One of the things we have done in that course is to gather comparative data from several nations, including France, Canada, Japan, Australia, Germany, England, and the U.S. One student summarized our findings succinctly: "With respect to families, schools, children, and youth, such countries as France, Japan, Canada, and Australia have more in common with each other than the United States has with any of them." For example:

• The U.S. has by far the highest rate of teenage pregnancy of any industrialized nation — twice the rate of its nearest competitor, England.
• The U.S. divorce rate is the highest in the world — nearly double that of its nearest competitor, Sweden.
• The U.S. is the only industrialized society in which nearly one-fourth of all infants and preschool children live in families whose incomes fall below the

poverty line. These children lack such basics as adequate health care.

• The U.S. has fewer support systems for individuals in all age groups, including adolescence. The U.S. also has the highest incidence of alcohol and drug abuse among adolescents of any country in the world.[6]

All these problems are part of the unraveling of the social fabric that has been going on since World War II. These problems are not unique to the U.S., but in many cases they are more pronounced here than elsewhere.

WHAT COMMUNITIES CAN DO

The more we learn about alienation and its effects in contemporary post-industrial societies, the stronger are the imperatives to counteract it. If the essence of alienation is disconnectedness, then the best way to counteract alienation is through the creation of connections or links.

For the well-being of children and adolescents, the most important links must be those between the home, the peer group, and the school. A recent study in West Germany effectively demonstrated how important this basic triangle can be. The study examined student achievement and social behavior in 20 schools. For all the schools, the researchers developed measures of the links between the home, the peer group, and the school. Controlling for social class and other variables, the researchers found that they were able to predict children's behavior from the number of such links they found. Students who had no links were alienated. They were not doing well in school, and they exhibited a variety of behavioral problems. By contrast, students who had such links were doing well and were growing up to be responsible citizens.[7]

In addition to creating links within the basic triangle of home, peer group, and school, we need to consider two other structures in today's society that affect the lives of young people: the world of work (for both parents and children) and the community, which provides an overarching context for all the other worlds of childhood.

Philip's family is one example of how the world of work can contribute to alienation. The U.S. lags far behind other industrialized nations in providing child-care services and other benefits designed to promote the well-being of children and their families. Among the most needed benefits are maternity and paternity leaves, flex-time, job-sharing

*C*aring is surely an essential aspect of education in a free society; yet we have almost completely neglected it.

arrangements, and personal leaves for parents when their children are ill. These benefits are a matter of course in many of the nations with which the U.S. is generally compared.

In contemporary American society, however, the parents' world of work is not the only world that both policy and practice ought to be accommodating. There is also the children's world of work. According to the most recent figures available, 50% of all high school students now work part-time — sometimes as much as 40 to 50 hours per week. This fact poses a major problem for the schools. Under such circumstances, how can teachers assign homework with any expectation that it will be completed?

The problem is further complicated by the kind of work that most young people are doing. For many years, a number of social scientists — myself included — advocated more work opportunities for adolescents. We argued that such experiences would provide valuable contact with adult models and thereby further the development of responsibility and general maturity. However, from their studies of U.S. high school students who are employed, Ellen Greenberger and Lawrence Steinberg conclude that most of the jobs held by these youngsters are highly routinized and afford little opportunity for contact with adults. The largest employers of teenagers in the U.S. are fast-food restaurants. Greenberger and Steinberg argue that, instead of providing maturing experiences, such settings give adolescents even greater exposure to the values and lifestyles of their peer group. And the adolescent peer group tends to emphasize immediate gratification and consumerism.[8]

Finally, in order to counteract the

mounting forces of alienation in U.S. society, we must establish a working alliance between the private sector and the public one (at both the local level and the national level) to forge links between the major institutions in U.S. society and to re-create a sense of community. Examples from other countries abound:

• Switzerland has a law that no institution for the care of the elderly can be established unless it is adjacent to and shares facilities with a day-care center, a school, or some other kind of institution serving children.

• In many public places throughout Australia, the Department of Social Security has displayed a poster that states, in 16 languages: "If you need an interpreter, call this number." The department maintains a network of interpreters who are available 16 hours a day, seven days a week. They can help callers get in touch with a doctor, an ambulance, a fire brigade, or the police; they can also help callers with practical or personal problems.

• In the USSR, factories, offices, and places of business customarily "adopt" groups of children, e.g., a day-care center, a class of schoolchildren, or a children's ward in a hospital. The employees visit the children, take them on outings, and invite them to visit their place of work.

We Americans can offer a few good examples of alliances between the public and private sectors, as well. For example, in Flint, Michigan, some years ago, Mildred Smith developed a community program to improve school performance among low-income minority pupils. About a thousand children were involved. The program required no change in the regular school curriculum; its principal focus was on building links between home and school. This was accomplished in a variety of ways.

• A core group of low-income parents went from door to door, telling their neighbors that the school needed their help.

• Parents were asked to keep younger children out of the way so that the older children could complete their homework.

• Schoolchildren were given tags to wear at home that said, "May I read to you?"

• Students in the high school business program typed and duplicated teaching materials, thus freeing teachers to work directly with the children.

• Working parents visited school classrooms to talk about their jobs and about how their own schooling now helped them in their work.

WHAT SCHOOLS CAN DO

As the program in Flint demonstrates, the school is in the best position of all U.S. institutions to initiate and strengthen links that support children and adolescents. This is so for several reasons. First, one of the major — but often unrecognized — responsibilities of the school is to enable young people to move from the secluded and supportive environment of the home into responsible and productive citizenship. Yet, as the studies we conducted at Cornell revealed, most other modern nations are ahead of the U.S. in this area.

In these other nations, schools are not merely — or even primarily — places where the basics are taught. Both in purpose and in practice, they function instead as settings in which young people learn "citizenship": what it means to be a member of the society, how to behave toward others, what one's responsibilities are to the community and to the nation.

I do not mean to imply that such learnings do not occur in American schools. But when they occur, it is mostly by accident and not because of thoughtful planning and careful effort. What form might such an effort take? I will present here some ideas that are too new to have stood the test of time but that may be worth trying.

Creating an American classroom. This is a simple idea. Teachers could encourage their students to learn about schools (and, especially, about individual classrooms) in such modern industrialized societies as France, Japan, Canada, West Germany, the Soviet Union, and Australia. The children could acquire such information in a variety of ways: from reading, from films, from the firsthand reports of children and adults who have attended school abroad, from exchanging letters and materials with students and their teachers in other countries. Through such exposure, American students would become aware of how attending school in other countries is both similar to and different from attending school in the U.S.

But the main learning experience would come from asking students to consider what kinds of things *should* be happening — or not happening — in American classrooms, given our na-

tion's values and ideals. For example, how should children relate to one another and to their teachers, if they are doing things in an *American* way? If a student's idea seems to make sense, the American tradition of pragmatism makes the next step obvious: try the idea to see if it works.

The curriculum for caring. This effort also has roots in our values as a nation. Its goal is to make caring an essential part of the school curriculum. However, students would not simply learn about caring; they would actually engage in it. Children would be asked to spend time with and to care for younger children, the elderly, the sick, and the lonely. Caring institutions, such as day-care centers, could be located adjacent to or even within the schools. But it would be important for young caregivers to learn about the environment in which their charges live and the other people with whom their charges interact each day. For example, older children who took responsibility for younger ones would become acquainted with the younger children's parents and living arrangements by escorting them home from school.

Just as many schools now train superb drum corps, they could also train "caring corps" — groups of young men and women who would be on call to handle a variety of emergencies. If a parent fell suddenly ill, these students could come into the home to care for the children, prepare meals, run errands, and serve as an effective source of support for their fellow human beings. Caring is surely an essential aspect of education in a free society; yet we have almost completely neglected it.

Mentors for the young. A mentor is someone with a skill that he or she wishes to teach to a younger person. To be a true mentor, the older person must be willing to take the time and to make the commitment that such teaching requires.

We don't make much use of mentors in U.S. society, and we don't give much recognition or encouragement to individuals who play this important role. As a result, many U.S. children have few significant and committed adults in their lives. Most often, their mentors are their own parents, perhaps a teacher or two, a coach, or — more rarely — a relative, a neighbor, or an older classmate. However, in a diverse society such as ours, with its strong tradition of volunteerism, potential mentors

abound. The schools need to seek them out and match them with young people who will respond positively to their particular knowledge and skills.

The school is the institution best suited to take the initiative in this task, because the school is the only place in which all children gather every day. It is also the only institution that has the right (and the responsibility) to turn to the community for help in an activity that represents the noblest kind of education: the building of character in the young.

There is yet another reason why schools should take a leading role in rebuilding links among the four worlds of childhood: schools have the most to gain. In the recent reports bemoaning the state of American education, a recurring theme has been the anomie and chaos that pervade many U.S. schools, to the detriment of effective teaching and learning. Clearly, we are in danger of allowing our schools to become academies of alienation.

In taking the initiative to rebuild links among the four worlds of childhood, U.S. schools will be taking necessary action to combat the destructive forces of alienation — first, within their own walls, and thereafter, in the life experience and future development of new generations of Americans.

1. Urie Bronfenbrenner, "New Worlds for Families," paper presented at the Boston Children's Museum, 4 May 1984.
2. Urie Bronfenbrenner, "The Ecology of the Family as a Context for Human Development," *Developmental Psychology*, in press.
3. Mavis Heatherington, "Children of Divorce," in R. Henderson, ed., *Parent-Child Interaction* (New York: Academic Press, 1981).
4. A.S. Makarenko, *The Collective Family: A Handbook for Russian Parents* (New York: Doubleday, 1967).
5. Lea Pulkkinen, "Self-Control and Continuity from Childhood to Adolescence," in Paul Baltes and Orville G. Brim, eds., *Life-Span Development and Behavior*, Vol. 4 (New York: Academic Press, 1982), pp. 64-102.
6. S.B. Kamerman, *Parenting in an Unresponsive Society* (New York: Free Press, 1980); S.B. Kamerman and A.J. Kahn, *Social Services in International Perspective* (Washington, D.C.: U.S. Department of Health, Education, and Welfare, n.d.); and Lloyd Johnston, Jerald Bachman, and Patrick O'Malley, *Use of Licit and Illicit Drugs by America's High School Students — 1975-84* (Washington, D.C.: U.S. Government Printing Office, 1985).
7. Kurt Aurin, personal communication, 1985.
8. Ellen Greenberger and Lawrence Steinberg, *The Work of Growing Up* (New York: Basic Books, forthcoming).

Development During Adolescence and Early Adulthood

The onset of adolescence is demarcated by the emergence of secondary sex characteristics and the achievement of reproductive maturity. The onset of adulthood is more difficult to distinguish, particularly in modern industrial societies. In some cultures, a ritualistic ceremony marks the transition of adulthood—a transition that occurs quickly, smoothly, and relatively problem free. In American culture the transition is vague. Does one choose the age at which the adolescent achieves the right to vote, the ability to legally order an alcoholic drink in a bar, or the right to volunteer for the armed forces? The first article discusses the question of "rites of passage" from adolescence to adulthood in American society, noting several contradictions between what adolescents say and what they mean. Some researchers argue that much of the storm and stress attributed to adolescence is a myth, created from an overemphasis on adolescent fads and rebelliousness and an underemphasis on obedience, conformity, and cooperation. Focusing on the negative aspects of adolescent behavior may create a set of expectations which the adolescent strives to achieve. On the other hand, for some adolescents the transition to adulthood can be fraught with despair, loneliness, and interpersonal conflict. The pressures of peer group, school, and family may produce conformity or may lead to rebellion or withdrawal directed against friends, parents, or society at large. These pressures may peak as the adolescent prepares to separate from the family and assume the independence and responsibilities of adulthood.

"The Sibling Bond: A Lifelong Love/Hate Dialectic" notes that adolescents who have experienced poor family stability may develop strong emotional relationships with their siblings that last long into adulthood. Separation from the family may be an especially difficult time. No more severe example of family instability and separation can be provided than that which occurred among the many millions of people sent to concentration camps during World War II. "Holocaust Twins: Their Special Bond" reports interview data from fraternal and identical twins who, as chil-

dren, struggled valiantly and successfully to survive the horrors of Auschwitz. The close bonds that were formed then still remain, and twins separated during childhood continue to search for their siblings.

Although much attention has been given to the problems of adolescence and the transition to adulthood, developmentalists have shown far less interest in the early years of adulthood. During early adulthood, however, many individuals experience significant changes in their lives. Marriage, parenthood, divorce, single parenting, employment, and the effects of sexism may have powerful influences on ego development, self-concept, and personality.

For increasing numbers of young adults in Western Culture, stress is related to issues of gender. Challenges to traditional sex role stereotypes are countered by arguments emphasizing biological differences between the sexes. Although there are obvious differences between the sexes in hormonal influences on behavior, there are less clear-cut differences in brain organization, brain function, and behavior. Where such differences exist, it is not at all clear whether they matter. Some individuals seem to grow stronger when confronted by the stresses of daily living, whereas others have great difficulty coping. Alarming increases in the suicide rates suggest that increasing numbers of adolescents and young adults are losing the battle against feelings of hopelessness, despair, alienation, and lack of self-control. The challenge for developmentalists is to discover the factors that contribute to the individual's ability to cope with stress and to deal with the natural crises of life with minimum disruption to the integrity of the individual's personality.

Looking Ahead: Challenge Questions

Describe the adolescent's ability to make use of moral and political principles in organizing his or her thinking about social issues. What can be done within the educational system to promote a broader and more complete sense of the adolescent's social environment?

What parenting practices will you use to ward off development of sibling relationships that result in life-long hatreds? Do you think that family collapse is as responsible for the nature of sibling relationships as is suggested in the Adams article, or do siblings themselves play the major role in structuring their emotional relationship?

Many individuals who separate themselves from the home seem to find only temporary relief in communal-cult groups. What kinds of child-rearing practices might have given such teens sufficient self-esteem and coping skills to combat their self-doubts and loneliness? How might schools effectively combat psychic surrender when they have been criticized as contributing to the boredom, values abdication, permissiveness, and drug use that promotes surrender of personal responsibility? On the other hand, why do some teens seek alternative family groupings while others turn to suicide in response to their sense of hopelessness, despair, and alienation from others and society?

What can be learned about development from such bizarre events as described in "Holocaust Twins"? What aspects of the twins' early childhood do you believe were absolutely essential to their survival? Do you think survivers of Auschwitz will ever be capable of truly resolving the basic conflict of trust vs. mistrust?

If scientists are obliged to use caution in interpreting findings in an emergent area of inquiry, shouldn't the same standards hold true for those who seek to apply knowledge obtained through basic research? If so, why do you suppose the field of education seems especially vulnerable to premature application of the kind of information that is emerging from the study of lateral specialization of the cerebral hemispheres? For example, even if preliminary findings on sex differences in brain organization are confirmed in future studies, isn't it a conceptual leap to translate evidence of these differences into different educational curricula for boys and girls? How might you react if your child's teacher said, "Your daughter does not have algebra this year because we know that girls' brains are not able to process this level of mathematical reasoning at this stage of their development"?

RITES OF PASSAGE

JOSEPH ADELSON

Joseph Adelson is professor of psychology at the University of Michigan at Ann Arbor. This article is adapted from a speech given in March by Adelson at the Urban Development Forum, which was sponsored by Research for Better Schools in Philadelphia.

HOW DO youngsters in the vital transitional period of pre- and early adolescence deal with the ideas of the social sciences and the humanities? How do they cope with the concepts they must absorb in learning about history or civics or political science or literary studies? Does psychology have anything useful to tell us about how to teach those subjects during that difficult age? Do we know something that would help us accelerate learning or deepen it or strengthen the child's grasp on what he has been taught?

The work I will report here is based on two major investigations, one cross-national, comparing over three hundred youngsters in our country, England, and Germany, ranging in age from ten or eleven to eighteen, from the fifth grade to the twelfth. The second study, in which we interviewed about 450 adolescents, covered the ages from eleven and twelve to eighteen. This study was directed and analyzed by my colleague, Judith Gallatin. The second study concentrated upon youngsters in an urban area, largely blue collar in origin, with an equal number of blacks and whites.

Our research instrument was the open-ended interview. After a great deal of trial and error, we hit upon an interview format that began with the following premise: a thousand people leave their country and move to a Pacific island to start a new society. We hoped that the use of an imaginary society would help free some of the children, the young ones particularly, from their preoccupation with getting "the right answer." Given this framework, we then offered our youngsters a great many questions on a wide variety of political, social, and moral issues: the scope and proper limits of political authority; the reciprocal obligations of the individual and the community; the nature of crime and justice; the collision between personal freedom and the common good; the prospects for

utopia; and so on. Put this way, it all sounds rather formidable, but the questions themselves were straightforward and generally quite concrete. In the second of the studies, we also introduced a number of questions having to do with urban tensions: the sources and outcomes of poverty, the relations between citizens and the police, and the proper channels for citizen protest. The interviews took, on the average, an hour to complete—the older the child, the longer the interview. We tape-recorded and then transcribed faithfully, including silences, uhs, "you knows," and grammatical incoherence, since we felt that the process of achieving a response might in some cases be as interesting as the response itself.

Since there are far too many findings to report even in summary form, I have identified five topics that I think are of central importance, since they influence so many other areas of social thought: the conceptions of community and of law, the growth of principles, and the grasp of human psychology and of social reality. In each of these topics we see some significant and at times startling changes in children's understanding during the adolescent years. I will concentrate here particularly upon those taking place in the earlier part of that period.

The Community

The first piece of advice to give any teacher preparing to work with ten, eleven, and twelve-year-olds is that one ought not to assume the child is talking about the same things you are. With respect to such concepts as "government" or "society" or "the state," the youngster may talk in a seemingly appropriate fashion; yet, when you extend the conversation or query him a bit, you may likely find something close to a conceptual void. At the threshold of adolescence, children find it difficult to imagine impalpable social collectivities; they do not yet enjoy the sense of community.

We can illustrate this graphically by looking at the answers eleven and twelve-year-olds give to the question "What is the purpose of government?" To begin with, many of them cannot answer the question at all. Either they fall mute entirely or provide obviously confused or irrelevant responses. In our cross-national study, we found that 15 percent of eleven-year-olds could give no answer at all to that question. More revealing yet is the number who are unable to give adequate answers—that is, answers of sufficient coherence and complexity to allow their being coded. The category "Simplistic, Missing the Point, Confused, Vague" accounts for 43 percent of responses among twelve-year-olds. A certain confusion about politics, government, law, and society is endemic among pre-adolescent youngsters. But the failure to understand the idea of government—and similar concepts of the collectivity—is especially significant because these are the regnant ideas in thinking about social, moral, and historical issues, and confusion, murkiness, error, and failure to grasp these concepts makes itself felt throughout a much larger domain of cognition.

BUT TO say that these youngsters are mistaken or confused does not take us very far, since it does not tell us about the specific nature of the cognitive flaw. To understand that, it may be best to turn to some specific responses, chosen at random, from eleven-year-olds of average intelligence, to the question on the purpose of government:

> To handle the state or whatever it is so it won't get out of hand, because if it gets out of hand you might have to...people might get mad or something.
> Well...buildings, they have to look over buildings that would be...um, that wouldn't be any use of the land if they had crops on it or something like that. And when they have highways the government would have to inspect them, certain details. I guess that's about all.
> So everything won't go wrong in the country. They want to have a government because they respect him and they think he's a good man.

What strikes us first about these statements is that, in each case, the speaker seems unable to rise securely above the particular. The child feels most comfortable in remaining concrete, by turning to specific and tangible persons, events, and objects—hence "government" becomes a "him," or the child talks about crops and buildings and highways. Of course an effort is made to transcend particularity, to discover a general principle or idea, but the reach exceeds the grasp, as we can see vividly in the first of these excerpts in which the speaker, seeking a general principle ("to handle the state"), gives up and subsides into concreteness ("people might get mad or something").

The shift from concrete to abstract modes of expression during the course of adolescence is a dramatic one. In our cross-national study, no eleven-year-old child was able to attain high-level abstractness in discussing the purpose of government; and no eighteen-year-old gave an answer as entirely concrete. Most eleven-year-olds (57 percent) can give only concrete responses. At thirteen and fifteen, a low level of abstractness is the dominant mode of conceptualizing government. And at eighteen, a strong majority of youngsters achieve a high level of abstractness.

The findings immediately above are based on our cross-national survey. In other studies we have tried different ways of categorizing responses, but the pattern remains essentially the same.

Unable to imagine "the community"—that is, the invisible network of rules and obligations binding citizens together—the child at the threshold of adolescence does not quite understand the mutuality joining the individual and the larger society. He does understand power, authority, coercion; indeed, he understands those all too well, in that his spontaneous discourse on "government" and the like relies heavily—at times exclusively—on the idea of force, authority being seen as the entitlement to coerce. Yet even that is imagined only concretely: it is the *policeman* who pursues and arrests the criminal, the *judge* who sentences him, and the *jailer* who keeps him. The less punitive purposes of the state are less readily discussed in large part, we believe, because the child, lacking a differentiated, textured view of collectivities, cannot quite grasp how they function or what their larger goals might be. The child at this stage may know that the government does things—fixes the streets, let us say—and that it does so in order to benefit the citizenry as a whole. But beyond such tangible activities leading to such tangible benefits, the need and purposes of the community remain a mystery, impenetrable.

Perhaps the most consistent finding we have is that the adolescent years witness a shift from a personalized, egocentric to a sociocentric mode of understanding social, political, historical, and moral issues. The sociocentric outlook is essentially absent at the beginning of adolescence—that is, when the child is ten, eleven, or twelve; yet, it is more or less universal by the time the child is seventeen or eighteen, with most of the movement taking place in the period we are talking about, somewhere between thirteen and fifteen years of age. The shift is dramatic in that it involves a fairly complete reorganization of how these issues are perceived and interpreted. We have here an expanding *capacity* to think in terms of the community. It does not mean that the youngster, having achieved that capacity, is held captive by it. It does not mean that discourse about society, from that point on, ignores individual needs and perspectives. It does mean, however, that the youngster, having achieved sociocentrism, is able to weigh the competing claims for ego and other, of the individual and the state, or the larger community. Until that point is achieved, social perceptions tend to be truncated, and social judgments and ratiocenation are vulnerable to the distortion of a narrow individualism.

The Law

Perhaps the most unnerving discovery we made upon first reading the interview transcripts was that a substantial minority of our youngest respondents were capable, on occasion, of the moral purview of Attila the Hun. On questions of crime and punishment, they were able—without seeming to bat an eyelash—to propose the most sanguinary means of achieving peace and harmony across the land. Here are three examples, all from the discourse of nice, clean-cut middle American thirteen-year-old boys, telling us their views on the control of crime:

> [On the best reason for sending people to jail]: Well, these people who are in jail for about five years must still own the same grudge, then I would put them in for triple or double the time. I think they would learn their lesson then.

[On how to teach people not to commit crimes in the future]: Jail is usually the best thing, but there are others...in the nineteenth century they used to torture people for doing things. Now I think the best place to teach people is in solitary confinement.

[On methods of eliminating or reducing crime]: I think that I would...well, like if you murder somebody you would punish them with death or something like this. But I don't think that would help because they wouldn't learn their lesson. I think I would give them some kind of scare or something.

These excerpts are *not* randomly chosen, since we have selected cases marked by colorful language and thought. Yet neither are they altogether atypical, in this sense—they represent only the more extreme expressions of a far more general social and moral outlook: the tendency to see law, government, indeed most other social institutions, as committed *primarily* to the suppression of wayward behavior. In this view, human behavior tends toward pillage and carnage, and the social order is characteristically on the brink of anarchy. That may overstate it a bit, but not by much. Gradually but steadily, however, an entirely different view of the purpose of law emerges in later adolescence. Toward the end of the period we are dealing with, and certainly by the time children are fifteen and sixteen, the dominant stress upon violence and injury has begun to diminish markedly, and it will more or less vanish by the time the child reaches the age of eighteen.

Two other motifs similarly signal the end of the pre- and early adolescent period. One of these is the tendency to see laws as *benevolent* as against restrictive, as designed to help people. A characteristic statement: "The purpose of laws is to protect people and help them out." Another motif, somewhat related, we suspect, is one that links law to the larger notion of community, that sees law as providing a means for interpersonal harmony, either among competing social groups or in the nation or the state as a whole ("...so that the country will be a better place to live"). These changes, from a purely restrictive to a benevolent or normative view of law, are as fundamental and quantitatively decisive as a shift from the concrete thinking to the abstract.

Principles

We have so far observed two major developments in political thought from the onset of adolescence to its end: the achievement of a sociocentric perspective, the ability to think about social and moral and philosophical issues while keeping the total community in mind; and the gradual abandonment of an authoritarian, punitive view of morality and the law. We now add a third theme: the youngster's capacity to make use of moral and political principles—ideas and ideals—in organizing his thinking about social issues. Once available, that capacity alters—decisively and irrevocably—the youngster's definition of social issues, and at the same time it alters the child's sense of himself as a social and political actor. Most current theories of political attitudes and thinking stress the central significance of more or less stable, more or less complex systems of belief, the presence of which allows the person to organize his understanding of social and political reality. It is in the period we now

have under consideration that we first see the emergence of those systems, as the child begins to use principles in coming to legal, moral, political and social judgments. To judge by our interviews, however, it is a rather late development in adolescence. We seem to see the first signs of it when the child is between fourteen and sixteen, and the use of principles does not make itself felt fully until the end of the adolescent period.

Perhaps we best begin by showing just how the older adolescent makes use of principles in making judgments on social issues. Here is an eighteen-year-old who has just been asked what the government ought to do about a religious group opposed to compulsory vaccination:

Well, anyone's religious beliefs have to be tolerated if not respected, unless it comes down to where they have the basic freedoms. Well, anyone is free until he starts interfering with someone else's freedom. Now, they don't have to get their children vaccinated, but they shouldn't have anything to say what the other islanders do, if they want their children vaccinated. If they're not vaccinated, they have the chance they may infect some of the other children. But then that's isolated, that's them, so if they don't get vaccinated, they don't have anyone else to blame. (Do you think that the government should insist these people go along with what the majority has to say, since they're such a small minority?) No, I don't think that the government should insist, but I think that the government should do its best to make sure that these people are well informed. A well-informed person will generally act in his own interest. I never heard of religion that was against vaccination. (There are religions that are against blood transfusions.) If they want to keep their bodies pure...well, like I said, I think that a well-informed citizen will act in his own best interest. If he doesn't, at least he should know what the possibilities are, you know, the consequences. So I think the government's job is to inform the people. In that case, at least, to inform them and not force them.

Younger children, when faced by a question of this type, find it difficult to reason on the issue. They come down hard on one side or the other or cannot make up their minds and therefore hedge; in support of their position, they may put forward a principle-like phrase, such as "freedom of religion," but they cannot do much with the idea except to assert it. What we see in the excerpt we have given—which we choose not because it is "brilliant" but because it is characteristic in late adolescence—is the capacity to advance a general and generalizing principle, which then allows the youngster to talk about specific issues with some flexibility. These formulae need not be absolute in nature, nor rigidly applied; indeed, in many cases the youngster brings forward circumstances that call for a suspension or modification of the principle.

H OW DOES the youngster come into possession of these principles? As far as we can tell, they are not constructed *de novo* but are acquired by the most mundane processes of learning, in the classroom or through the media, in the church or at home. At moments one can almost see the civics or history textbook before the child's inner eye as he struggles with the question. Here is a youngster trying to answer a question as to which law should be made permanent and unchangeable:

Well, freedom of speech is one, as you said. And then one law, well, I don't think you should be in prison for a longer time than twenty-four hours without them telling the charge against you. Or freedom of the press or freedom of the religion, that should never be changed, because anybody can pick any religion they want. There's no certain religion that everybody has to go by. (Can you think of any other kind of law that should not be changed or is that about it?) There are some more laws, but I know what they are, but I can't really put it into words because...you know, I really know what they are, like the laws, the Bill of Rights, you know, the first ten amendments of the Constitution, uh, them laws, you know, that I haven't mentioned. They should be put in there, in the United States Constitution. I can't remember what they were exactly, but if I had a history book, I'd look them up, you know.

Obviously, he has absorbed some of the principles of constitutional democracy, albeit a bit imperfectly. Nevertheless, it is almost certain that the mode of discourse we see here is not exclusively a function of learning; it depends also upon the growth of cognitive capacity. If we take a look at the interviews of average children in the early and middle-adolescent period, we get some sense of the limits of learning before the child is intellectually ready. Ninth and tenth graders have also been exposed to the fundamental ideas of constitutional government, at least in the students we worked with; yet, it seemed to us that the learning does not quite "take," not completely, not sufficiently to allow the child to make use of it in ordinary conversation. The principles do not "come to mind," even when the child is primed by how the question is phrased. In writing the interview item on permanent laws, we were aware that younger children would not spontaneously think of laws or constitutional provisions guaranteeing fundamental freedoms, and so we decided to prime the pump, so to speak, by mentioning "freedom of speech" as an example. Nevertheless, very few of our younger subjects took the hint. Instead, they concentrated on those issues—crime and punishment, violence and injury—that most concerned them and generally in the straightforwardly authoritarian manner we mentioned earlier:

> They should have a law, like people should stop stealing, and if they do steal, they would have to stay in jail for about a year until they settled down and stopped doing that. And they should stop killing each other because that's not right.

And even when the child is not entirely obsessed with fantasies of danger, the response to this question usually betrays an inability to make general statements:

> Don't litter. Don't steal. Keep off the grass. Don't break windows. Don't run up the stairs. Don't play with matches. Keep matches out of reach of little children.

We do not want to make either too much or too little of the child's acquisition of principle. It does not usher in a golden era of humanistic wisdom. The ordinary youngster acquires the conventional ideas and ideals of the world about him, and unless he is intensely interested in social or philosophical or literary topics, he is unlikely to have ideas that are discernibly unique or penetrating. Yet on the other hand, it is a development of some importance. One obvious reason is that until the child acquires a capacity for general ideas, he does not understand most of the language of social and moral discourse that envelopes him. He is in that sense like the tourist in a foreign land, unable to speak or read the indigenous language, and not quite

sure what the customs signify. If he is facile enough, he may be able to mimic some of the argot and conduct of the natives around him, yet studied inquiry would soon reveal the lacunae and confusions. Time and again in our interviews with pre- and early adolescents—those, let us say, between eleven and fourteen—we come upon such instances wherein the child's mimetic talent allowed him to talk as though he knew the language when, in fact, he did not. The majority rules, the child says. Ah, we say in turn, so tell us about the majority. Then the child replies, oh, that's when everybody agrees.

Achieving a grasp of principle also means that the child can resist the appeal of the immediate, hence is less vulnerable to mere sentiment. The government wants to build a highway and needs some farm land. The farmer resists; the authorities insist. Who is right? Without some general idea to aid him—either the virtues of property or the common good or eminent domain or some such—the youngster is not far from helpless in telling us what ought to be done, and why. Either he sides with the farmer, sentimentalized as the underdog, or with the government, sentimentalized as the guardian of the public weal. Without the guidance of principle, he is, we feel, so subject to the tug of emotion, and thus of demagoguery, that he cannot make reasoned—and hence reliable—decisions. He is much too responsive to the *evident* good.

One more comment before we leave this topic. It may be worth repeating that the term "principles" refers to both ideas and ideals. The increasing conceptual grasp of the adolescent allows him to come to an understanding of the conventions of social and moral reality as understood by the community at large. At the same time he becomes capable of cognizing the "irreal" as well, and hence of being in touch with the values, hopes, and utopian beliefs of the culture as a whole. Hence the grasp of principles means that the child can become both more "realistic" and "idealistic." It has been our unfortunate habit to concentrate upon "adolescent idealism" as though that were a dominant moral outlook of the adolescent period. In fact, the child's realism, the child's becoming socialized to the conventions of the culture, is a far more conspicuous feature of this era. But what is perhaps most important is that we see a dialectic between these attitudes, between being realistic and being idealistic.

Understanding Human Behavior

Near the beginning of the interview schedule we introduced a series of questions about law and laws, some of which we have already mentioned. What is the function of law? What would happen in a world without law? How and why do people get into trouble with the law? In developing the topic, we want to get some sense of how youngsters understand the psychology of malfeasance. One of our questions put forth the following proposition: some percentage of people need laws to keep them from getting into trouble, while others "follow their consciences naturally and do not need laws." We then asked what accounted for the difference between these two types of people.

What interests us here are not the particular theories proposed—these are fairly commonplace—but rather the somewhat abrupt shift in the child's capacity to talk about human psychology, a shift that in its rudiments seems to take place fairly early in adolescence—most of the time it is visible between the ages of eleven and thirteen. Here are some typical eleven-year-olds trying to distinguish between those who are naturally law abiding and those who need laws to guide them:

> Well...most people, some people they don't like, like speeding, they don't like to do this, but some people like...maybe...grownup people some people like to speed a lot.

> Well about the person I think he had been pushed around and people don't like him and stuff. The people that do not like the laws—well they probably had friends and he didn't get into much trouble so they just got used to it.

> Well...(pause, question repeated) well, it could be that the person who thinks that they were law abiding, I mean the criminals, they see things wrong. (How do you mean?) Well I mean they see...I can't explain it.

One is struck immediately by the sheer confusion of these comments: ideas—even phrases—do not quite connect to each other. There are gaps in discourse. Our experience has been that this sort of confusion suggests not so much ignorance, or fool's knowledge, as it does the child's earnest attempts to reach something just out of his grasp. He does not quite have the conceptual means to achieve a dimly sensed end. We sense that our third respondent is trying to say something about the social outlook of the delinquent ("they see things wrong"), while the second is speaking psycho-historically, that is, trying to link miscreancy to past experience ("he had been pushed around...and stuff"). In these instances we feel that the child's essential problem is a difficulty linking part to whole, particular to general, and vice versa. We may imagine that given the category "law abiding," the child's mind hits upon "speeding" as an instance of that larger category but cannot go beyond that, that is, cannot yet link speeding to other forms of social malfeasance, nor can he develop a differentiated view of the category "law abiding" that will allow him to classify different instances within it.

Even when the eleven-year-old's response is not quite so confused, it generally reveals some distinct limitations in the appraisal of human behavior. Here is a more typical response from a child at this age—it is neither the least nor the most advanced:

> Oh, well, someone—their mom and dad might separate or something and neither one wanted them or something like that, didn't like them very much and oh, if they happened to turn bad, I mean just, and they had trouble—pretty soon if they keep doing that and pretty bad conditions they'll probably get in a lot of trouble.

ONCE WE get into this long, meandering sentence, we discover that it contains not one but two theories of miscreancy and its sources—the first of these having to do with parental rejection, the second suggesting that trivial sins that go uncorrected lead implacably to larger ones. But here we see even more

clearly the problem in being unable to find a suitable language. Our youngster speaks only about specific acts or feelings—as though he were the most naive type of behaviorist, one who had vowed to avoid all speculation about internal states of mind. In a year or two this very youngster, proposing the same theory, will almost certainly be able to tell us that kids who come from broken families feel bad about themselves and become trouble makers; but at this moment, although the child seems to have that general idea in mind, even the concept "broken family" may be a bit too abstract (or too unfamiliar) to state. Similarly, even such familiar denominatives as "trouble maker" or "delinquent" may be difficult either to understand or to express comfortably. At any rate, we note at this age level—although not universal even here—a common reliance on action language, the child being unable to talk about "traits" or "character" or other structures or tendencies of the personality. Instead he talks about specific acts of malfeasance.

Children at this age have no stable idea of the personality nor an understanding of motives beyond the most simple (getting mad, getting even, teaching a lesson). The youngster cannot think in terms of *gradations* of motives nor of *variations* in personality. Nor can he formulate the impact of the situation upon the personality. Nor can he propose a theory of incentives beyond simple coercion, nor can he recognize the symbolic or indirect effects of rewards and punishment.

What we have, in short, is a markedly impoverished conception of the personality. Motives are few and starkly simple—fear, anger, revenge, envy, the wish to be liked. Motives tend to be either/or in character—the child cannot easily think in terms of conflict of motives, of compromises among them, or of other dialectical processes that would ultimately determine behavior.

We also see a sharp limitation in time perspective. The child at this age seems unable to grasp fully the effect of the past upon the present, in that he does not seem to consider the effect of personal history upon current conduct. That statement needs some qualification. The child may mention the immediate precipitants of a course of current conduct but finds it difficult to link the present to more remote events in the person's past. Equally striking is the difficulty the youngster shows in tracing out spontaneously the potential effect of current conduct upon later events. Again, we do not want to overstate this: if the question clearly asks for future consequences (what would happen if there were no laws?), the child will imagine those consequences. But in ordinary discourse, the "time window" seems quite narrow. Beyond that, the youngster is rarely able to imagine dialectical processes taking place in the future as the result of decisions taken today—that, for example, an unpopular law may ultimately generate law breaking or other forms of underground opposition.

It may seem to be loading the dice somewhat to take our examples so exclusively from the realm of crime and punishment, given the child's obsessive involvement with these issues. Yet we see these difficulties elsewhere, even when the child is discussing virtue or merit, and for some of the same reasons—an uncertain sense of major and minor, relevant and irrelevant.

Appraising Social Reality

There are some surprising similarities between the preadolescent patterns in learning to understand human psychology and the gradual, at times faltering, steps he takes in developing a sense of social reality. In both instances we come across problems in classification: what belongs to what; how to construct a hierarchy of types and functions; how to specify boundaries and limits. In both instances we perceive a shortness of time perspective, the youngster being unable initially to imagine the effect of the past upon the present, or more than the immediate effect of current social events upon the more or less remote future. And in both instances we note what can only be called a thinness of texture; the child does not seem to grasp ambiguity, complexity, or interaction.

We want to begin by looking at a specific social institution in order to describe the changes that take place in the youngster's grasp of a structure and function and of its relation to larger social processes. We chose the idea of "political party" for several reasons; to begin with, almost all children raised in democratic countries are exposed to information about political parties, and in the fullness of time, achieve an adequate understanding of them; secondly, as an institution, it is neither so diffuse nor so various that different youngsters may have had entirely different experiences of it.

It comes as a surprise to most people how little children at the onset of adolescence actually understand about the nature and purpose of the political party. Since the knowledge of parties seems to be so ubiquitous and since the child is exposed to that knowledge regularly in the mass media, at home, or in school, we are likely to assume that that exposure has resulted in some learning, especially so if the child is the kind who is alert to current events. Nevertheless, a distinct majority of children at the age of eleven, twelve, and thirteen cannot give satisfactory answers to straightforward questions on the purpose and functioning of political parties—and by "satisfactory" we mean no exalted standards of comprehension. Either they cannot answer the questions at all (about 15 percent at age eleven) or they give answers that are either too diffuse to be coded or plainly in error. What is of particular interest is the kind of mistake the child is liable to make when he does venture an opinion. The most common of these is the tendency to confuse the functions of the political party with those of government as a whole. The party is seen as making laws or carrying out either the general or specific tasks of the state. But here are some characteristic expressions of that misunderstanding from some twelve-year-old boys chosen randomly:

Ah, what, like the United States? I think they have these parties because they want to help the United States be a better state, I mean a better country and things like that. And then that's why they have one every one or two years.

I guess because if they wanted a law a certain way then they could have it that way. (probe) I guess if they had a law that people couldn't kill, I guess they didn't like it that way. (Didn't like what?) Some people don't like laws and some people do.

To keep people in order. (What else?) That's all I have to say. (Further probe) To keep people in order like I just said.

In these examples we sense that the child cannot yet classify, that is, cannot yet establish boundaries between the separate functions and structures of the political process. Since he has heard that parties are involved in elections, he may see them as carrying out elections; since he grasps vaguely that they are connected to government, he imputes to them some of the functions of government.

We might mention here, somewhat parenthetically, that these confusions and errors are by no means limited to the topic of the political party. We find much the same pattern in the early years of adolescence, when the child is addressing more general questions about governing. They can find it difficult to distinguish among the legislative, executive, and judicial apparatuses of the state; for that matter, they can find it difficult to distinguish between the government, the state, and the nation, all of which seem to blend into each other. That confusion of element, part standing for wholes and vice versa, characterizes the child's early apprehension of social and governmental institutions.

The next stage is marked by an accurate, although rudimentary, grasp of institutional function. It is a distinct advance over the confusion and error we have seen in the examples just given, and yet compared with what the child will later be capable of, it is marked by what we will call *thin* texture. The child will fasten upon a single, at most two, aspect of structure or function. With respect to political parties, we will be told that the party puts forward candidates or stands for certain ideas or supports candidates. From the interview:

To help the candidates running to have a better chance of getting the office.

Well, so that the people can express their views.

It's to help the people find their candidates and to back the people when they are candidates.

The change from thin to thick texture is difficult to describe succinctly, since it may involve somewhat different processes. In the most simple form, we find a capacity to describe multiple aspects or functions of the institution being discussed. Thus, in relation to political parties, the youngster may tell us that parties both represent positions *and* support candidates, or that they both finance *and* organize for issues *and* their nominees. A step beyond that level is the ability to synthesize several ideas in a single statement. Here is an eighteen-year-old speaking on the advantages of political parties:

A well, if you have a whole bunch of people with different ideas but have a government that's to be run, you are not going to get much accomplished, but if you put them together in a group, and then they pool their assets and ideas, then they have enough power to do something about what they want, than everybody just talking about what they want.

Now this is by no means an extraordinary statement;

the ordinary citizen would make it. And yet its very ordinariness may conceal from us that an important conceptual advance has taken place. She is telling us that parties are both efficient and potent in that they are able to unify otherwise disparate political voices: ideas in unison can be powerful, as they are not when voiced separately.

For reasons that are still obscure, at least to me, the degree of achievement of hypothetico-deductive reasoning that Piaget and other cognitive theorists have demonstrated to be involved in advance modes of reasoning in relation to scientific problems seems to be far less widespread in the social and philosophical reasoning of adolescents. When this degree of achievement occurs, it seems to take place much later in the child's development. The kind of cognitive operations that many children can perform at the ages of thirteen to fifteen when confronted with the mathematical and scientific problems seem to elude the grasp of all but the most exceptional youngsters when they confront problems of equivalent difficulty in the realm of social and humanistic ideas, and even among that exceptional group the level is not achieved until the age of eighteen.

Some Conclusions for Teaching

To return to the question we began from: Can the teacher of adolescents learn something from these findings? Can they improve the way we teach social and humanistic subjects?

In the course of preparing this essay, I read a good deal of the technical literature on learning, on concept formation, on whatever seemed germane, giving especially close scrutiny to those writings—few in number, alas—that make some effort to apply what we have learned in the laboratory to the actualities of teaching the young. It is not an edifying experience. The will is there, the earnestness, even a certain bumptiousness. Yet almost invariably something seems to be lost in translation, and with the best will in the world, we seem generally unable to use empirical findings, even reliable ones, to provide useful counsel to the educator. I think it can be done, but it will not be done easily, and it will certainly not be done by those who, like myself, are not directly engaged in teaching primary and secondary school youngsters. For that reason, what follows is offered modestly, indeed timidly.

When I first began doing the studies reported in this paper, my next-door neighbor was a man who taught social studies at our local junior high school. I soon found myself trying out my findings on him, and although I don't know whether my observations improved his teaching, his observations on my findings certainly sharpened my research. One day I consulted him about the following problem. The interview schedule contained several questions on taxes through which we had hoped to explore the child's understanding of the larger social functions of taxation, for example, to provide incentives or deterrents for certain economic or social activities, or to redistribute income. The power to tax is the power to destroy, as we all have been told; when does the

youngster grasp this and equivalent ideas about the indirect functions of the taxing authority?

As soon as we began doing the interviews we became aware that we had overshot the mark, in that the child's understanding of taxes was far less developed than we had expected it to be. Some of the younger children among the ten and eleven-year-olds understood next to nothing, only that the tax was something collected at the store when you bought something or something that one's parents had to pay to someone. More commonly, children did understand that the function of taxes was to raise revenue for government, but few of them could tell us more than that, and only a handful understood much about the use of taxes as a means of channeling economic and other behavior. One day I mentioned to my neighbor the general nature of these findings and how surprised our research group had been to discover how little children understood about this topic. He thought for a moment, then said that he himself was not surprised. Taxation was a required subject matter in the ninth grade civics course he taught, and he had found that children had trouble with it, indeed so much so that he tended to give the topic short shrift, moving on to more engaging issues as soon as he had covered the fundamentals. But why do the children have trouble, I asked. He wasn't sure, but he suspected it was because they did not find taxes to be of any direct importance to them. It was seen as an "adult" concern, and as a consequence they were bored. Being bored, they would not learn the information. That was, I should say, a characteristic formulation by my neighbor; he tended strongly to a motivational theory of learning, holding that if the child's interest could be captured, learning would follow as the night the day. As for myself, I was then in the first flush of a newly acquired Piagetism and urged that perspective on him, suggesting that the youngsters were not cognitively ready for those materials and that their boredom and inability to learn reflected an underlying confusion due to conceptual immaturity.

I am now not at all sure that I was right and my neighbor wrong, or vice versa. I suspect that we were both partly right, in that we had touched upon the right dimensions: interest or motivation, cognitive capacity, and information (or knowledge). In this essay I have stressed cognitive growth almost to the exclusion of other determinants of learning. I think that stress legitimate given the general neglect of that outlook until recently. Yet it must be understood to represent only one element of a more complex process wherein capacity, knowledge, and motivation interact continuously. If the child is not ready cognitively to grasp a particular concept, he will be unsteady in his grasp of related information, and he will also fail to show much interest in the general topic; at the same time, a high level of interest may stimulate the acquisition of knowledge and enhance cognitive capacity. Within limits, the mind stretches to fulfill its intellectual needs. In that sense the approach represented here—cognitive developmental—does not represent anything new so far as education is concerned. To the contrary, if one reads Piaget's writings on education,

The way that an adolescent or pre-adolescent processes information in a classroom setting can only be described as literal. To break away from this either/or perspective of events, structures, people, and learning experiences is an important phase of development.

for example, one is immediately struck by its closeness in spirit to the work of John Dewey.

WHAT, THEN, can this approach do for us? With respect to practical teaching it can alert us to the sources of specific difficulties the child is likely to experience in learning new information and ideas. Conversely, it may alert us to otherwise unrecognized intellectual opportunities the child is ready for and may teach us how to teach the child to grasp those opportunities. Let me offer an example. We found that at the outset of adolescence the youngster cannot adopt an as-if or conditional attitude to social or psychological phenomena. What is, is, now and forever. Bad people are bad and good people are good. If a law is passed, the child assumes it will stay in place

eternally, and he has a hard time understanding that it can be overturned; he has an even harder time grasping that it might be amended, that one part of a law might be retained and another part rejected: it is all or nothing. One of the unrecognized achievements of the adolescent period is the acquisition of the concept of amendment, which is itself part of a larger movement of the mind away from static, either/or conceptions of events, structures, and persons. The more inclusive concept of *mutability*—for example, of persons changing or institutions in flux, is not easily grasped until middle to late adolescence.

Now it seems to be vitally important that a teacher charged with the instruction of young adolescents would do well to keep that knowledge in mind, particularly since he is charged with teaching dynamic processes—that is, processes involving change—

relating to persons and societies. If he is teaching about "laws" he ought, at the least, remain aware that although he may have in mind modifiable statutes passed by a legislative body, the eleven-year-olds he is talking to have in mind something like the Ten Commandments. One might, in general, want to avoid certain topics as being too difficult conceptually; or one might try to develop methods of finessing those limitations, doing an end run around them; or one might want to develop methods of overcoming them. That choice is up to the teacher, and to the deviser of curricula.

Probably the most common problem the child experiences in dealing with social and humanistic materials is achieving the proper degree of abstractness; and the most common error the teacher makes comes from a failure to recognize the child's problem or to take account of it. As I suggested earlier in this essay, the child has a remarkable mimetic capacity, an ability to use the language of abstractness without genuine understanding. He may use a word like "majority" confidently, yet once we begin to query him we find he has only the vaguest idea of its meaning. Another such word is "government." Another is "election." By the former term, the ten- or eleven- or twelve-year-old child may very well have in mind the governor or the mayor or some other figure cloaked in the robes of authority. The child at the same age may not really know what it means to be "elected." He does not necessarily connect it with an electoral process but confuses it with being appointed, or perhaps even being anointed, that is, with having somehow assumed the cloak of authority.

Looking back, it is painfully clear that many of our first interviewees did not understand the meaning of these and other terms; nevertheless, it took us a long time to realize it. A youngster would half recognize a term and answer with some appropriate cliche or stark response, one sufficiently plausible to allow the conversation to continue. After we had examined several of these half-on, half-off responses, it would dawn on us that something was not quite right, and we would then discern that there was a concept present somewhat beyond the ken of the youngsters in question.

Why did we not see this immediately? Because the language of social and humanistic disciplines so largely overlaps common parlance, and its principles so largely overlap both common sense and common experience. That is not likely to happen in more technical disciplines. If I quiz a youngster on the properties of the isosceles triangle, his ignorance and confusion will be evident immediately; but if I quiz him about law and government, he may well be able to improvise sufficiently to conceal these states of mind. It is not that the youngster aims to deceive his interlocutor; rather, he may only be aiming to please, to give the answers that are wanted. It is the examiner who does the rest, filling in the gaps and elisions, imputing to the child a level of understanding that is largely in the mind of the beholder.

I might say here, a bit parenthetically, that there seems to be a general tendency among adults to inflate the understanding of the child in these areas. I have no firm idea why this is so, but I've seen this tendency in myself—it took me a long time to accept what the transcripts were clearly saying about the cognitive capacities of the children. I have since seen other adults, with few exceptions, make the same error, generally saying something along these lines: the findings may be true for this particular sample of children but would not be true for the children they knew, referring tacitly to their own children. But if they were to give the interview to their own children, as I did to mine, they would discover, as I did, that the intellectual gestalt that the child offers, via an overall aura of brightness, simply conceals the actual (lower) level of cognitive capacity. I suspect that classroom teachers, who deal with a variety of youngsters through the day, are less likely to misappraise cognitive level quite so often or to the usual degree; yet, I also suspect that the direction of error is similar, that they perceive in the child a more advanced grasp than is truly the case.

That may not be a bad thing, so far as education is concerned, to teach up rather than down in terms of cognitive level. It seems to me it may be helpful to introduce concepts just beyond the easy reach of the youngster. The cautions here are obvious: the concepts should not be too advanced nor should there be so many of them to cope with that the child feels overwhelmed. But keeping these cautions in mind, the teacher ought not to refrain from the use of, let us say, abstract ideas, notions of historical influence, or any of the other concepts or perspectives we have found to be difficult for children at the threshold of adolescence. In some cases, these are helpful in providing a framework—albeit a loose or hazy one—to help the child organize the more concrete ideas he is more comfortable with.

Take as an example the concept of democracy. If a youngster between the ages of ten and twelve is asked to give a definition of that word he will almost certainly be unable to do so satisfactorily. He may address the question in strictly emotive terms, pronouncing on its merits, or he may mix up specific aspects of democratic systems—elections of the legislature or the presidency—with the system itself. Yet, if you extend the conversation with the child, you may find that he has in his grasp most of the specific elements that make up democratic modes of government. It seems to me that the teacher would at this point do well to help the child connect what he can grasp—the more or less concrete aspects of government—to the more general concepts, such as democracy. Often the problem is less in the child than it is in the adult, because adults—almost reflexively—think abstractly when thinking about abstract matters, and when faced with incomprehension, tend to explain things by piling abstraction upon abstraction.

THERE IS another reason why we may want to teach concepts the child is not quite prepared to grasp fully—when they embody values we deem vital. Many American youngsters at this age will, when prompted, use such phrases as "freedom of speech" or "freedom of religion" or—in a few cases—"Bill of

Rights." Further discussion reveals that their understanding is incomplete or incorrect in important ways. They are certainly unable to grasp these ideas as abstractions. Yet these concepts are by no means empty of meaning to them. The child may well have an idea of First Amendment rights that is overblown or absurd; he may, for example, think that it means an utterly untrammeled tolerance for freedom of expression; but what is more important is that he has grasped, in however inchoate a fashion, the kernel of the idea of rights, and in time that idea will be placed in context, given resonance, qualified, and so on. What is more important is that some of the American reverence for "rights" has been communicated to the child.

Much the same can be said for the democratic rituals that the child is exposed to as part of his schooling. In trying to discuss the electoral process, some of our children adverted to the elections for student council or class president or most popular boy or girl that they had experienced. It was clear enough that the younger ones had only the dimmest notion of the connections, if any, between those processes and the electoral politics they learned about in the mass media. It is tempting to dismiss those exercises, precisely because they seem to be so hollow, so absent of genuine understanding. But talking to so many dozens of adolescent children myself and reading so many hundreds of their interviews has persuaded me that these presumably empty rituals do have an important socializing effect in habituating the child to the practices of democratic politics.

THE SIBLING BOND
A Lifelong Love/Hate Dialectic

VIRGINIA ADAMS

The link between brothers and sisters is in some ways the most unusual of family relationships. It is the longest lasting, often continuing for 70 or 80 years or more, and the most egalitarian. It is also the least studied. Researchers have been more interested in the relationship between parents and offspring than in sibling interaction. Even when psychologists have focused on the children in a family, they have paid attention chiefly to the effects of birth order on personality and intelligence, or to the role young siblings play in early development when they act as caretakers for still younger brothers and sisters. From the nature of most studies, one might almost deduce that sibling relationships end with childhood.

Over the past few years, however, a growing number of researchers have begun to ask questions about brothers and sisters in adulthood. Do siblings drift apart when they leave home, or do they stay in touch? If they maintain a real relationship, what is it like, and how long is it likely to last?

Some answers have already appeared in print or been reported at professional meetings. Others will be published over the next few months in three book-length studies now in preparation. Most of the new work falls under one of three headings: *fervent sibling loyalty* (never before systematically studied) that arises only under certain family conditions and is so extreme as to produce unexpectedly negative effects in some cases; *sibling rivalry* that can persist into adulthood and even into old age; and *sibling solidarity*, a sense of closeness that leads some siblings to turn to each other for understanding.

Overall, the latest research makes it clear that siblings can exert an important mutual influence, for good or ill, throughout the life span; this is sometimes so even when they are geographically separated. Some psychologists expect sibling relationships to become even more significant as the divorce rate increases, the number of one-parent families grows, and family size declines.

Research confirms what common sense suggests: that the degree and nature of sibling interaction varies greatly, not only from one set of siblings to another but within the same pair at different times in their lives. Some brothers and sisters become and remain best friends; others heartily detest each other.

The most interesting fact, although it is not often noted by researchers or acknowledged by siblings, is that love and hate may exist side by side, in very uneasy equilibrium. One writer speaks of "the delicate balance of competition and camaraderie among all sisters." Her words would seem to apply equally well to brothers. Whether the balance can be maintained, or whether it tips dramatically to one side or the other, depends partly on parental behavior and attitudes and partly on certain critical experiences in the lives of siblings themselves.

Fervent Sibling Loyalty

Among the most innovative of the psychologists now studying adult brothers and sisters are Stephen Bank, adjunct associate professor at Wesleyan University, and Michael D. Kahn, associate professor at the University of Hartford. In a seven-year period they have audiotaped more than 100 interviews with siblings in pairs or in larger groups, and in the course of their work as psychotherapists they have studied many other siblings. Bank and Kahn will describe some of their findings in *Sibling Relationships: Their Nature and Significance Across the Lifespan*, a book edited by Michael E. Lamb and Brian Sutton-Smith that will be published next summer. They will give a more detailed account of their work in their own book, *The Sibling Bond*, due out next February.

Bank and Kahn are particularly interested in "Hansels and Gretels," siblings as intensely loyal as the fairy-tale children whose loving concern for each other saved them from the wicked witch when their father and stepmother turned them out into the forest to starve. The two psychologists have observed the Hansel and Gretel phenomenon in numerous sibling groups, and they are at pains to distinguish it from sibling solidarity, the kind of cohesiveness that leads adult siblings to offer each other a modest degree of support in time of trouble. In an ordinary relationship, one sister might invite another in the throes of divorce to come and visit for a weekend. An extremely loyal sister, Bank said, might take a sibling into her home indefinitely, acting as "a kind of parent to someone who has been wounded."

Extreme loyalty, Bank says, "involves an irrational and somewhat blind process of putting one's sibling first and foremost," with a willingness to make enormous sacrifices. Pointing out that loyalty comes from the French word *loi* ("law"), he notes that intense loyalty is an unwritten contract providing that "we will stick together."

One of two brothers in their 20s told Bank, "If I ever got in any trouble, I wouldn't go to my wife—I wouldn't go to a friend—I wouldn't go to my boss; I'd go to my brother." One of four brothers aged 36 to 45 said of the other three, "If I knew they needed it, I'd give any one of these guys my last buck, despite my obligations to my wife and children."

The adult Hansels and Gretels that Bank and Kahn studied tried to be together as often as they could and were unhappy when necessity kept them apart. The two young brothers, students at the same university, shared a fantasy that they would not let even marriage part them. They talked about buying a joint homestead, "where their wives and children would blend with them into a big, happy household."

Some of the extremely loyal siblings were bound together by a private code—a special way of exchanging glances, for example, or a word or phrase that meant nothing to outsiders. "The four brothers," Bank said, "repeatedly broke into raucous laughter after one had made what seemed to the interviewer a perfectly neutral comment. They sarcastically 'apologized' for the 'silly' behavior of their brother as if to say, 'This is our sense of humor; *you'll* never understand it, since you didn't grow up with us.'"

How do such intense attachments develop? The critical factor is a kind of family collapse as the children are growing up, with the parents becoming actually or psychologically unavailable. In each of the families studied, the parents were hostile, weak, or absent, and the external circumstances of the children's lives were threatening. "Confronted with a hostile environment," Bank said, "they clung together as the only steady and constant people in one another's lives." The parents of the two college-age brothers died within two years of each other, when the boys were nine and 11 and then 11 and 13, and both boys were seriously abused by a psychotic foster mother. The mother of the four middle-aged brothers died when they were teenagers, and their father was emotionally exhausted, sometimes abusive, and unable to form a close relationship with his

children. The family was in a slum in which physical danger was a constant.

Parents who want their children to become and remain fervently loyal to each other do not understand the paradox that Bank underscores: "The *deep* bond between siblings will not develop if parents are real good parents; with good parents, you'll get caring, and solidarity, but not intense loyalty, because there's not much need for it."

Bank's assertion that intense sibling loyalty presupposes a kind of parental abandonment draws indirect support from a number of studies. One of these was done in 1965 by Albert I. Rabin, a Michigan State University psychologist, who compared kibbutz children in Israel with children raised in traditional families there. He found much less sibling jealousy in the kibbutz children, whose parents were with them only two hours a day, than in youngsters whose parents were available to them around the clock.

Even more persuasive evidence comes from the classic study by Anna Freud of six unrelated children orphaned by the Nazis before they were a year old. Kept together in the Terezin concentration camp and, after the war, in a succession of hostels, they had no chance to form consistent relationships with caring adults. When they were about three, they were flown to England and looked after in a special nursery for months to help them make the transition to a more normal life. For a long time, they appeared almost incapable of jealousy.

"It was evident," Anna Freud wrote, "that they cared greatly for each other and not at all for anybody or anything else. They had no other wish than to be together and became upset when they were separated. . . . There was no occasion to urge the children to 'take turns'; they did it spontaneously. . . . They were extremely considerate of each other's feelings. . . . When one of them received a present from a shopkeeper, they demanded the same for each other child, even in their absence. On walks they were concerned for each other's safety in traffic, looked after children who lagged behind, helped each other over ditches, turned aside branches for each other to clear the passage in the woods, and carried each other's coats. In the nursery they picked up each other's toys. . . . At mealtimes handing food

to the neighbor was of greater importance than eating oneself."

Among many examples of the children's mutual concern, this one is typical: "John, daydreaming while walking, nearly bumps into a passing child [who does not belong to the group]. Paul shouts at the passerby: '*Blöder Ochs; meine John; blöder Ochs Du*' ('Stupid fool; that's my John; you stupid fool')."

There can, of course, be too much of even such a good thing as loyalty. If parents thought more about it, they might worry a bit less about signs of rivalry and a bit more about excessively close ties. They might look up the story of Dorothy Wordsworth, sister of the poet, who gave her life entirely to her brother William, renouncing any independent existence of her own. They might also remember the reclusive Collyer brothers, who gave their lives to each other—and cannot have had much joy from the sacrifice. In the less extreme cases that Bank and Kahn studied, intense loyalty often imposed "the burden of an obligatory responsibility to one another." Sometimes it stifled friendships with outsiders whom siblings did not happen to like. Often it conflicted with loyalty to spouse and children.

Some very loyal brothers had difficulty admitting wives to their group at all and would submit their choice of spouse to their brothers for approval as if to a review board. At times the system worked well. The first wife admitted to the four-brother group mentioned earlier was a physical-education major who proved eminently suited to life among these macho brothers. "She had grown up with brothers," Bank said. "She knew how to handle brothers. She adapted by becoming 'one of the boys,' accepting their male humor and participating in sports with them. They, in return, adopted her almost as a sister. It was the perfect fit." In other cases, though, wives were made to feel "as if they were on the outside of a very exclusive club."

There were times when extreme loyalty made difficulties for the siblings themselves. The younger of the two college-age brothers felt he was a nobody who existed only as an extension of his brother. Despite the warmth the two felt for each other, Bank said, he decided to get away by himself for a few years so he could discover his own identity. When loyalty was one-way rather than reciprocal, with one sibling doing most of the giving, recipients often felt resentful—despite their gratitude—at being subjected to a kind of domination, while perpetual givers often found themselves shunting aside their own legitimate interests. "If you keep helping and helping, that is in a sense neurotic," said Bank.

But he emphasized that he "would not want to be quoted as calling these people only neurotic." Many siblings reported that their loyalty brought them advantages: never feeling entirely alone, learning skills from each other, having a chance to practice parenthood.

Loyal siblings can be deeply sensitive to each other's needs. According to Bank, "They show some of the qualities of a good psychotherapist; they know when to shut up and when to push." And, he went on, "I don't want to put down the absolute altruism we've seen in these very loyal siblings, because there is such a thing." Citing Jane Goodall's observation that when parent chimpanzees die, a "sibling" takes over as caretaker, Bank suggested that somewhat analogous behavior, having important survival value, may occur in human beings.

Rivalry in Adulthood

Bank and Kahn believe that the degree of rivalry between siblings has been vastly exaggerated and that what looks like rivalry often masks other feelings, for instance, dependency, or simply a need for some kind of intense relationship with a brother or sister. Most psychologists, however, argue that some degree of sibling rivalry is almost inevitable in childhood, and there is plenty of research evidence to support that view. Many people assume that this early rivalry dissipates in adulthood, but on that point the evidence is far from conclusive. Last September two psychologists, both associate professors of education at the University of Cincinnati, made a strong case for the persistence of rivalry into adulthood in a paper presented to the American Psychological Association. Helgola G. Ross and Joel I. Milgram (the brother of psychologist

Stanley Milgram) recruited 65 subjects aged 25 to 93 from a Midwestern university community, two senior citizen centers in town, and a Methodist retirement home in the suburbs.

Most of the data came from interviews that were tape-recorded, transcribed, and content-analyzed so that recurrent topics could be noted. In group interviews, all the subjects met in age groups (20s, 30s, 40s, 50s, or 60s) of four to six people. Individual follow-up interviews were conducted with 10 of the subjects so that rivalry could be explored in depth.

Seventy-one percent of the subjects reported that they had sometimes felt rivalrous toward brothers or sisters. Of those, 36 percent claimed that they had been able to overcome the feelings in adolescence or adulthood, but 45 percent admitted that the feelings were still alive.

Of subjects who admitted to experiencing rivalry at some time, 40 percent said it had begun in childhood, 33 percent believed it had not surfaced until adolescence, and 22 percent said they could not remember feeling it at all until adulthood.

About half of the subjects said that adults had initiated their rivalrous feelings when they were children. Parents were most often cited, although grandparents came in for their share of blame, and so, at times, did teachers.

The key stimulus for rivalry was favoritism. Some parents openly compared siblings, asking a child to match standards of behavior, skill, or personal characteristics achieved by a favored brother or sister. In other cases the comparison was covert; the child observed the parents' preferential treatment of one or more favored brothers or sisters.

Roughly 50 percent of all rivalry was said by the subjects to have been initiated by siblings, in half of these cases by brothers, in a third by sisters, in a tenth by the subjects themselves.

When competition began in childhood, Ross and Milgram said in their paper, it appeared to be "a vying for the parents' attention, recognition, and love, but also a more general juggling for power and position among the siblings." Sometimes the precipitating factor was the experience of being cared for by an older sibling. (According to Brenda K. Bryant, a psychologist at the University of

California at Davis who has studied siblings as caretakers, sibling babysitters are perceived by their charges as rougher disciplinarians than parents. Young babysitters, Bryant suggests, seem to pattern themselves after the comic strip character Lucy, "whose approach to caretaking with peers such as Charlie Brown is to resort to the sweet reason of the mailed fist.")

Ross and Milgram found that a pattern of rivalry sometimes took shape when a young sibling accepted an older one as mentor and then discovered that the mentor was less disposed to praise the younger person's accomplishments than to depreciate them.

The two experimenters emphasized that inequalities in levels of accomplishment were not by themselves sufficient to generate rivalry. Siblings did not necessarily believe that more is better unless parents thought so.

Some subjects indicated that rivalry could actually be fun. Others said it served to motivate them. "If siblings have the ability to live up to high standards, comparative expectations are not necessarily debilitating," said Ross and Milgram. Many subjects told them that when they managed to find their own areas of expertise, competitiveness gave way to pride in the former rival's successes.

More often than not, however, rivalry was destructive. Some subjects felt deeply hurt. "Many siblings felt excluded from valued sibling or family interactions and the sense of wholeness they can provide," Ross and Milgram said. "Some dissociated themselves psychologically and geographically from particular siblings or the family on a semi-permanent basis; two broke relations completely."

If competition was hurtful, what kept it going? Most often, parental favoritism continuing into adulthood. Also mentioned frequently was provocatively competitive behavior by the siblings themselves.

At times, the difficulty was family gossip that made too much of differences between brothers and sisters, assigning constricting roles or labels. Yet another problem was that siblings rarely talked about their rivalry. Successful siblings often did not even know that their achievements were causing envy, and siblings who felt inferior did not want to say so. "Admitting sibling rivalry may be experienced as equivalent to admitting maladjustment," Ross and Milgram believe. Besides, "To reveal feelings of rivalry to a brother or sister who is perceived as having the upper hand increases one's vulnerability in an already unsafe situation."

Psychologists often observe that even intense feelings of rivalry can coexist with feelings of affection and solidarity. One of the most telling examples is the case of William James, the philosopher-psychologist, and his brother Henry, the novelist. Henry's perceptive biographer, Leon Edel, gives a fascinating account of what he calls their "long-buried struggle for power," which began from the moment of Henry's birth and did not end until William's death.

Whatever praise William voiced over the years for Henry's writing was overbalanced by barbed criticism, especially of Henry's complex literary style. The older brother called Henry "a curiosity of literature." In a letter to the younger man he once wrote, "for gleams and innuendos and felicitous verbal insinuations you are unapproachable" and exhorted him to "say it *out*, for God's sake, and have done with it." William was not quite at ease with his own sharp words, though, and he urged Henry not to answer "these absurd remarks."

On that occasion, Henry complied, but to another critical letter he wrote, "I'm always sorry when I hear of your reading anything of mine, and always hope you won't—you seem to me so constitutionally unable to 'enjoy' it." Yet he did not appear resentful. Inscribing a gift copy of his novel *The Golden Bowl* to his brother, he called himself William's "incoherent, admiring, affectionate Brother."

Henry never knew of William's unkindest remarks about him. They were written in 1905, when William's election to the Academy of Arts and Letters followed Henry's by three months. Petulantly, William declined the offer. He was led to that course, he explained in a letter to the academy secretary, "by the fact that my younger and shallower and vainer brother is already in the Academy and that if I were there too, the other families represented might think the James influence too rank and strong."

As Edel points out, William was being both untruthful with himself

and inconsistent: "He had not considered that there was a redundancy of Jameses when he and Henry had been elected to the Institute [of Arts and Letters, the Academy's parent body] in 1898. He was having this afterthought only now, when Henry had been elected before him to the new body." Indeed, William's conscience was again not quite clear about his own behavior; his letter acknowledged that he was being "sour."

The relationship of the brothers was more complex than these quotations suggest. When William was dying, Henry wrote to a close friend, "At the prospect of losing my wonderful beloved brother out of the world in which, from as far back as in dimmest childhood, I have so yearningly always counted on him, I feel nothing but. . .weakness. . .and even terror."

To someone else, Henry confided that "William's extinction changes the face of life for me." Yet the change was perhaps not wholly unwelcome. Henry "had always found himself strong in William's absence," Edel writes. "Now he had full familial authority; his nephews deferred to him; his brother's wife now became [in effect] wife to him, ministering to his wants, caring for him as she had cared for the ailing husband and brother. Henry had ascended to what had seemed, for 60 years, an inaccessible throne."

Sibling Solidarity

Even rivalrous brother-sister relationships may show a good deal of the quality known as sibling solidarity, or cohesiveness. One of the most important investigators of this quality is Victor G. Cicirelli, a professor of developmental and aging psychology at Purdue University. The burden of Cicirelli's 12 years of sibling research is that brothers and sisters remain important to each other into old age, often becoming more important in time.

In a study published last year in the *Journal of Marriage and the Family*, Cicirelli compared the feelings of 100 college women toward their parents and toward their brothers and sisters and found that his subjects felt significantly closer to siblings than to fathers. For the most part, they also felt closer to siblings than to mothers;

that difference was not significant.

For a study of siblings in midlife, Cicirelli visited 140 Midwesterners, most of them aged 30 to 60, and interviewed them about their relationships with their 336 siblings. His results, as yet unpublished, show little conflict between brothers and sisters.

More than two-thirds of the subjects (68 percent, to be exact) described their relationships with siblings as close or very close; only 5 percent said they did not feel at all close. Sisters felt closer than brothers or cross-sex pairs. In general, cohesiveness increased slightly with age.

Asked how well they got along with brothers and sisters, 78 percent said well or very well; 4 percent said not very well or poorly. As to the degree of satisfaction the relationship brought, 68 percent called it considerable or very great, while 12 percent reported little or no satisfaction.

Other questions brought fewer positive responses. Asked how much interest they thought their siblings had in their—the subjects'—activities, 59 percent said they perceived the interest as very great or moderate, but 21 percent saw little or none. Well under half—41 percent—said they felt free to discuss personal or intimate matters with a sibling; 36 percent said they rarely or never did so. Only 8 percent usually or frequently talked over important decisions with a sibling; 73 percent rarely or never did so.

Cicirelli found little overt conflict: 93 percent maintained that they never or only rarely felt competitive; 89 percent asserted that their siblings were never or only rarely bossy; and 88 percent said they never argued with siblings, or did so only once in a while. Speculating on the far higher percentage of people admitting to rivalry in the Ross-Milgram study, Cicirelli suggested that the small-group interview method those psychologists used "might be more likely to stimulate self-disclosure about rivalry than traditional interview methods."

Cicirelli has administered questionnaires to 300 men and women over 60 in order to learn about sibling relationships in later life. Some 17 percent of his elderly subjects saw the sibling with whom they were in closest touch at least once a week, while 33 percent saw that sibling a minimum of once a month. There was some drop in the frequency of visits with increasing age, but the diminished contact did not seem to lessen closeness.

Overall, 53 percent used the words "extremely close" to describe their relationship, and another 30 percent characterized it as "close." The 83 percent for the two responses combined is striking, compared with the 68 percent reported above for the middle-aged. Cicirelli believes that as older people witness the aging and death of parents and see the effects of the years on themselves and their siblings, their sense of belonging is threatened. Strengthening ties with those who remain, he theorizes, may be partly an attempt "to preserve the attachment to the family system of childhood."

Even so, when Cicirelli asked his elderly subjects which family member they turned to for emotional support and practical assistance, he found that most relied mainly on their children. Only about 7 percent said a sibling was a major source of psychological support.

In a very different kind of study, Cicirelli administered a projective measure, the Gerontological Apperception Test (GAT), to 64 men and women aged 65 to 88 to elicit deep attitudes and feelings of a kind that are not likely to emerge in interviews. The GAT is made up of 14 pictures depicting ambiguous situations such as an older woman on a park bench observing a young couple. Subjects are asked to make up a story about each picture, telling "what the people are thinking and feeling and how it will come out."

From his analysis of responses, Cicirelli learned that sisters had a greater impact on the feelings and concerns of siblings than did brothers. He also discovered that sisters had different effects on sisters than they did on brothers. More often than not, the stories told by female subjects who had one or more sisters showed concern about keeping up their social skills and relationships with people outside the family, helping others in the community, and being able to handle criticism from younger people. Cicirelli interpreted that to mean that "sisters appear to stimulate and challenge the elderly woman to maintain her social activities, skills, and roles."

The stories told by male subjects with sisters tended to be happier in tone than the stories of men with brothers, and they revealed less worry about money, jobs, family relationships, and criticism from the young. "The more sisters the elderly man has, the happier he is," Cicirelli concluded. "Sisters seem to provide the elderly male with a basic feeling of emotional security."

Solidarity may be of major significance to some siblings, but to others it counts for nothing; it may not even develop. Michael Kahn says that if siblings have "low emotional access" to each other—if there are no important interactions early in life because the siblings are many years apart in age, or for other reasons—they cannot become truly important to each other.

Nor are they likely to be close as adults when sibling incest has occurred in the early years. "Feelings of guilt, shame, and anxiety linger on," Kahn says, and sisters, more deeply affected by incest than brothers, experience profound feelings of betrayal, sometimes even if they were willing participants in the experience. "They don't trust their brothers, ever. Sometimes they don't want to be in the same room with them."

Bank also observes that siblings may avoid each other as adults if there are unresolved conflicts left over from childhood because the parents were "conflictophobes," people who forbade sibling quarrels. "If you've been taught that it's dangerous to fight with your sibling, you may freeze around him when you're adults and sit in the same room hating his guts."

Both Bank and Kahn stress the lasting harm of "frozen misunderstandings" developed in childhood. These are distorted perceptions of a sibling that brothers and sisters may carry uncorrected into their adult relationship. A sister whose brother often hit her when they were children may infer, and believe for the rest of her life, that he did it because he was hateful. But it may be that hitting was the only way he could make the physical contact for which he was emotionally starved. And even if a child's image of a sibling was correct in youth, it may need later revision. "We change, but we may transact our relationship as if there had been no change," Bank says.

Adult Turning Points: Satisfactions and Stresses

Circumstances create a great diversity of adult sibling relationships, from hatred to detachment to love. "I have always experienced her as a vicious individual," a 28-year-old woman said of her sister in the course of a study by Elizabeth Fishel, author of *Sisters*. Helene S. Arnstein, author of *Brothers and Sisters, Sisters and Brothers*, quotes a sister who said of her brother: "If we weren't siblings I'd never see David again, because I don't care for the kind of person he has become."

History is filled with examples of more satisfying sibling relationships than that. Freud and his brother, Alexander, enjoyed each other's company all their lives. In his 1957 book on his father, *Glory Reflected*, Freud's son Martin wrote, "My father and Alexander could not have been more different in their outlook on life, but they were always good friends." Alexander often went along on holiday trips with Freud and his wife and children, and was Freud's frequent companion on climbing expeditions. The brothers had a particularly good time swimming together during an Adriatic holiday in 1895. "Uncle Alexander and father were seldom out of the water," wrote Martin. "When, as sometimes happened, they refused to come ashore even for lunch, a waiter would wade or swim out to them balancing a tray with refreshments and even cigars and matches."

In her autobiography *Blackberry Winter*, Margaret Mead tells of her lifelong warm relationship with her sister Elizabeth. "Her perceptions . . . have nourished me through the years. Her understanding of what has gone on in schools has provided depth and life to my own observations on American education. And her paintings have made every place I have lived my home."

Thinking of the women in her mother's family, Mead was struck by the way in which, generation after generation, pairs of sisters became good friends. "In this," she said, "they exemplify one of the basic characteristics of American kinship relations. Sisters, while they are growing up, tend to be very rivalrous, and as young

The relationship between siblings is a unique and long lasting one with both love and hate existing simultaneously. This bond has been found to be much more important and resilient than any other relationship people may develop during their lives.

mothers they are given to continual rivalrous comparisons of their several children. But once the children grow older, sisters draw closer together and often, in old age, they become each other's chosen and most happy companions."

If sibling relationships sometimes grow closer as siblings age, they may also founder at certain of life's turning points. Bank and Kahn say that siblings are best studied when under stress, because that is when the frequently submerged dynamics of their relationship are most likely to come to the surface, exposing the true nature, the real strength or fragility, of the bond. It is also at such moments that the "delicate balance" of a sibling relationship may tip decisively.

One turning point that can disrupt a previously close tie comes with a sibling's discovery that a brother or sister is getting a divorce, drinking too much, or going into psychotherapy. Under such circumstances, presumably untroubled siblings are often afraid that they will develop the same difficulty. Their reaction may be to make themselves feel better by offering the troubled sibling hostile advice on "how he could improve himself," Bank says. The mechanism, he explains, is projective identification: "The reason I don't like you is that I see in you what I know all too well I have inside me."

Of course, when a sibling goes into therapy, it does not inevitably damage a relationship. The outcome may be quite the opposite. "Under carefully arranged conditions, siblings can learn to cooperate with each other to resolve important conflicts in family relationships," Kahn and Bank write in a case report to be published in the journal *Family Process.* Called "In Pursuit of Sisterhood," the paper tells of Maureen, a 29-year-old nurse who entered treatment out of a kind of general unhappiness. Among other things, she hated the fact that her family had never taken her seriously. Eventually the therapist called a series of "sibling rallies," meetings of the four sisters. The three older women told Maureen that she had never let them know how she felt. They all agreed that none of them was as close to the others as they wanted to think they were. Over a period of weeks, the four managed to thaw some of the

"frozen misunderstandings" from their common past and to become a cohesive group; for the first time, Maureen considered herself the equal of her sisters. After a while, Kahn and Bank said, the group confronted their mother and father and "successfully parried their parents' attempts to avoid discussing feelings." The ultimate result: "Faced with a unified sibling group, the parents were forced to redress old grievances by being helpful and accepting."

Physical illness is another situation that sheds light on sibling relationships. One illness that has been studied by many psychiatrists and medical sociologists is end-stage renal disease, in which doctors may be considering a kidney transplant.

Genetically, the best donors are brothers and sisters, because they provide the closest tissue match, reducing the danger that the patient's body will reject the transplanted organ. Yet siblings volunteer to donate a kidney less often than do parents. In a study of more than 300 relatives of transplant patients, Roberta G. Simmons, a sociologist at the University of Minnesota, found that 52 percent of the patients' sisters and 54 percent of their brothers did not volunteer. In contrast, only 14 percent of the patients' parents and only 33 percent of their children failed to volunteer.

Norman B. Levy and Jorge Steinberg, New York psychiatrists who have interviewed many pairs of kidney donors and recipients, have described a patient who got a kidney from her mother and told a researcher, "My sister was better matched, but she got pregnant." The two psychiatrists believe that the pregnancy very likely came about "accidentally on purpose" because the sister did not really want to donate.

Another case Levy and Steinberg described illustrates the apparent power of the emotions siblings feel for each other. The family wanted the needed kidney to be given by the patient's sister, who was not on good terms with the family, because they imagined that the gift would bring family harmony. The patient let herself be persuaded to accept her sister as donor, but she told the psychiatrists that she was afraid her sister would somehow find a way to make her pay for the gift. The prospective donor was no more

enthusiastic. "After I give her the damned kidney, I never want to see her again," she said. In short, Levy said, "They hated each other," and he and Steinberg predicted—correctly—that the patient's body would soon reject the hated sister's kidney.

The dependency, illness, and death of parents provide other possible turning points in the relationship of brothers and sisters. Cicirelli finds that siblings' negative feelings toward each other often emerge as an elderly parent becomes increasingly dependent. "Why should I look after my father?" a sibling thinks. "Let my brother do it."

On the other hand, there are siblings who compete to do the most for a dependent parent. "A common observation for those of us who deal with families of aged people," writes Martin A. Berezin, a psychoanalyst, "is the irrational, hostile attitude that siblings express to each other as they quarrel about what should be the proper care for an aged parent. Each accuses the other of negligence, lack of sympathy, or avoidance of responsibility. They challenge each other with questions of who telephoned or visited how many times." In such cases, according to Barbara Silverstone, a social worker, each sibling claims to have the parent's welfare uppermost, "but the underlying motive is winning out in a family contest."

Many researchers say that sibling bonds often weaken or disintegrate entirely when the last parent dies. "The wounds that are given and received during a parent's illness and dependency," Silverstone says, are often "only the final blows ending a relationship which has been distant or seething for years." A brother or sister who sacrificed too much for a parent may resent those who did too little. Or, Silverstone suggests, "Siblings may resent the martyred caretaker who stood between them and their dying mother, making it impossible for them to share her final days."

In some instances, Bank and Kahn say, a parent's death "can set off fratricidal feelings and struggles as siblings jockey for positions of leadership within the adult sibling group." Yet there are times when a parent's illness brings siblings together. One sister said to Silverstone of her brother, "I never realized what a great person he'd turned into until I spent all that time at home when Mother was sick. We're real friends now." Generally speaking, Bank believes, previously bad sibling relationships get worse after the parents' death; good ones improve.

An Inescapable Bond

Bank and Kahn call the sibling relationship "a lifelong process, highly influential throughout the life cycle." Even when brothers and sisters drift apart as they enter careers and begin to raise families, they are likely to renew their relationship eventually. "It is extremely rare for siblings to lose touch with one another," Cicirelli says. As the British sociologist Graham Allan puts it, it is "permissible to forget about a neighbor or other associate, but less appropriate to 'drop' a sister or brother."

Throughout life, siblings are likely to serve each other as models, spurs to achievement, and yardsticks by which to measure accomplishments. In the best of circumstances, they become sources of practical help, non-judgmental advice, and true intimacy. Indeed, Cicirelli remarks, "The nature of the sibling relationship is such that intimacy between siblings is immediately restored even after long absences."

"One of the main theses of our book," Kahn says, "is that siblings are becoming more and more dependent upon one another in contemporary families because of the attrition in family size. If you have only one sibling, that one becomes enormously important." The decreased availability of parents in one-parent homes and in households where both mother and father work also makes siblings count for more with each other. The rising divorce rate, too, intensifies adult sibling relationships by making them more significant than a husband or wife in an ephemeral marriage.

Bank and Kahn also predict that the rising divorce rate will lead to increased sibling conflict, "with adolescent and adult siblings taking sides with each parent in a kind of proxy war." But with or without divorce, conflict occurs between many—some would say most—brothers and sisters.

"One can conceive of rivalry as a feeling which is always latent, appearing strongly in certain circumstances while closeness is elicited in others," Cicirelli said recently. "Possibly there is a balance between the functional and dysfunctional aspects of the sibling relation which is manifested in solidarity versus rivalry. There may be a love-hate dialectical process throughout the life span that leads to new levels of maturity or immaturity in sibling relationships."

Some siblings are well aware of their ambivalence. "My mixed feelings toward my sister, my irradicable love-hate mix, have been almost the longest-standing puzzle of my life," a woman of 30 told Elizabeth Fishel.

Even siblings who prefer not to confront their ambivalence probably do not wholly escape its effects. Whatever its nature in particular cases— warm or cool—the sibling relationship remains alive in some sense. Mental images of each other, vivid or half forgotten, can influence brothers and sisters even when they are oceans apart and not consciously preoccupied with each other. "It's a relationship for life," Bank says. "It's forever."

Holocaust Twins: Their Special Bond

THE SURVIVORS OF MENGELE'S EXPERIMENTS
AT AUSCHWITZ REUNITE 40 YEARS LATER
TO REAFFIRM THEIR TIES.

NANCY L. SEGAL

Nancy L. Segal, a psychologist, is associated with the Minnesota Study of Twins Reared Apart project at the University of Minnesota.

On January 27, 1945, Soviet armies liberated an excruciatingly emaciated group of survivors from the infamous Nazi concentration camp Auschwitz/Birkenau, in Poland. Among the survivors trudging through the snow toward freedom were 157 twin children, all that remained of the estimated 3,000 twins who had entered the camp since 1943. The twins had been specially selected to serve as subjects in a series of brutal, dehumanizing experiments engineered by the Auschwitz "Angel of Death," physician Josef Mengele.

Many of these twins have kept silent for 40 years, unable to confront the painful memories of their past. Some, suspecting that they were the only twin survivors, felt there was no one who could share their thoughts and feelings. But on January 27, 1985, the 40th anniversary of the death-camp liberation, more than half of the twins and their families, from six countries, joined together for eight days of hearings, discussions and dedications.

Eight came to Auschwitz and later joined 70 to 80 at Yad Vashem, the Holocaust memorial and research institute in Jerusalem, Israel.

They came to share memories and life histories with each other and with the world. At Yad Vashem, they were joined by psychologists, medical professionals and legal experts who came to counsel, to learn and to gather evidence on the nature and outcome of the experiments.

The twins call themselves CANDLES (Children of Auschwitz Nazi's Deadly Laboratory Experiments Survivors), an organization founded by Eva Kor and Marc Berkowitz, two unrelated twin survivors. CANDLES is dedicated to reuniting Auschwitz/Birkenau survivors, researching the experiments and their effects on the twins and bringing Mengele to trial. Its members, "children without a childhood," regard their unique prison-camp experience as a prominent, enduring component of their identities. In this, and in their fierce desire to see Mengele brought to justice, they are inextricably bound.

That the reunion took place at all, four decades after the event, attests to the strength of the social bonds between and among these twins. "Every time we meet another twin, it is like finding another piece of ourselves," one said. Their reactions to the reunion reflect the unique experience of being twins and concentration-camp

survivors. But they also speak to issues of human resilience and social ties that we can all share.

I attended the reunions at Auschwitz and Yad Vashem as a member of the Minnesota Study of Twins Reared Apart, a research project at the University of Minnesota. At Yad Vashem, I interviewed 12 identical and 12 fraternal twins, as well as several nontwin survivors. I was especially interested in exploring how the twin relationship affected the nature and outcome of the prison-camp experience, and whether identical and fraternal twins saw the significance of twinship at Auschwitz differently.

Twin researchers have shown clearly that identical twins are more behaviorally alike and emotionally closer than fraternal twins. In my own studies of social interaction between young twins, I found identical pairs to be more cooperative and altruistic. Indeed, an impressive body of experimental, clinical and biographical data portrays the identical-twin bond as the closest and most enduring of human relationships. I wondered whether these same tendencies also differentiate the twin children of the Holocaust, for whom daily survival was painful and frightening, and existence uncertain.

At Auschwitz, twins and people with genetic abnormalities such as dwarfism had been treated differently from other prisoners because Mengele

(Top) Twin portraits, taken as part of Auschwitz head-measurement studies; Auschwitz prisoners' barracks (bottom)

NANCY L. SEGAL (2)

CANDLES cofounder Eva Kor (left) and twin Miriam Czaigher point to themselves being liberated.

wanted to study them. Unlike other children, most of the twins were spared the gas chambers. Also, in contrast to the usual separation of family members, most twin pairs were placed together in special twin barracks, segregated by sex. Young members of male-female pairs were sometimes kept together in the girls' barracks, but those older than 10 were separated.

Aside from this special treatment, the twins endured the full horrors of life in a death camp: They had been wrenched from their parents and subjected to hunger, cold and the terror and humiliation of such experiments as twin-to-twin blood transfusions, chemical injections and even sterilization. Further, the older children realized that they were valued as twin pairs; if one died, the other probably would be killed. Mengele reportedly became quite frantic when any of the twins died. But this did not prevent him from killing the surviving twin, often ordering dissections so he could compare body organs.

All the twins I interviewed considered the twin relationship as the key factor in their psychological and physical survival at Auschwitz. This was true regardless of their sex or twin-type (identical or fraternal). Contrary to my expectations, being either an identical or fraternal twin was not important to the twins in their camp experience. What was most important was having a twin, a familiar, constant companion with whom to share support and assistance.

Their personal stories are subtle variations on the common theme of

mutual caring and concern: "I was not alone." "We always felt rather like a unit; therefore, parting would have been a greater disaster." "The fact that my twin was with me gave me confidence and feelings of family unity." "Having a member of my family with me gave me the strength to go on. I became a father to my sister."

Kor, as a 9-year-old, managed to obtain potatoes for her identical-twin sister, Miriam Czaigher, who was completely uninterested in living and had suffered extreme weight loss. Rachel Smadar admitted that without the in-

*H*OLOCAUST TWIN BONDS: A VIEW OF AUSCHWITZ AS A MAJOR, ENDURING PART OF THEIR IDENTITIES AND A FIERCE DESIRE TO SEE MENGELE BROUGHT TO JUSTICE.

CANDLES cofounder Marc Berkowitz before his prison-camp image (far right, front)

sistent prodding and encouragement of her fraternal-twin sister, Zehava Friedman, she could not have walked through the snow and cold after they were liberated at the age of 12. She still marvels at her twin's strength and her own ability to follow. Today, these twin sisters, who live in Israel,

E VERY TIME WE MEET ANOTHER TWIN IT IS LIKE FINDING ANOTHER PIECE OF OURSELVES.

describe their relationship as very close. "We sometimes ask, 'What would we do without one another?'"

Some twins explained that assuming responsibility for their "weaker" twin gave them a mission and a will to live. Some also admitted that knowing that their own survival depended on that of their twin gave them further incentive to live and give. Many of the twins were also sustained by the dream that when the war ended, they would re-

turn to their homes and families, which prolonged their hope.

In describing themselves as children, many twins used such terms as "determined," "rebellious" and "never willing to give in." One had vowed "never to yield to this inhumanity"; another kept the words "hope, freedom and home" constantly in mind. She bit the SS officer who tattooed her identification number on her arm; the crookedness of the numbers, attesting to her resistance, still pleases her.

Such inner resolve may explain, in part, why these twins survived while the majority perished. It may also underlie their many clever survival strategies (see "The Survival Game" box).

Many of the twins who came to Yad Vashem were interviewed by psychiatrist Shamai Davidson of Bar-Ilan University in Ramat Gan, Israel. Davidson, who heads the university's research on psychological and social trauma produced by the Holocaust, concluded, as I did, that the closeness and caring between the twins served to "protect and buffer" them from the surrounding atrocities.

He further points out that the twin relationship helped to "preserve their humanness." Despite parental separation, cold, hunger, painful medical procedures and the real possibility of death, the presence of a twin partner reminded the children that someone cared and, in turn, required caring.

This may have spelled the difference between identity and anonymity.

Of all the twins, separated male-female pairs may have fared the worst. Berkowitz recalled that the boys eagerly anticipated catching sight of their sisters and the other twin girls across a barrier during certain hours of the day. But this distant contact ended when new transports of prisoners arrived, since trains blocked the view of the girls' barracks. Another male twin, only 6 years old when he entered Auschwitz, remembers very little other than a glimpse of his twin sister through a barbed-wire fence. Peter Greenfeld continues to search for his twin sister, from whom he was separated after the war. Mendel Guttman, a triplet, is still searching for his brother and sister. Greenfeld and Guttman hoped that the twins' convention might reunite them with the siblings they have never forgotten, but they were disappointed.

Psychologists have commented on the "inner emptiness" and restlessness felt by many adoptees. My colleagues at the Minnesota Study of Twins Reared Apart project are continually impressed by the determination of separated twins to locate their twin brother or sister; all those we have studied have agreed that finding the twin was well worth the effort.

This desire helps to explain why some twin survivors overcame considerable anxiety and trepidation to come to the reunion, often at great financial and personal sacrifice. But what drove other twins who were not searching for siblings? Frank Klein, for example, had lost a kidney—probably due to experimental procedures at Auschwitz. He was on dialysis but arranged to attend the meeting.

One common thread was the twins' need to recover bits and pieces of their past so they could recreate the children they had been. As one told me, "After the war, there was no childhood for the children of Auschwitz. We appeared to play like children, but we were thinking like adults." Berkowitz, now 52, explains that he returned to Auschwitz "to see whether I am still the little boy who cared, who could be a friend and give a smile."

In the presence of the other twins they could remember and, most importantly, be remembered. Unlike other children, most did not have parents

WRENCHED FROM THEIR PARENTS, THE TWINS ENDURED HUNGER, COLD AND THE TERROR AND HUMILIATION OF TRANSFUSIONS, CHEMICAL INJECTIONS, EVEN STERILIZATION.

to recall and relate events about their growing-up years. Thus, recognition by other twins was vital to restoring their shattered identities, affirming their sense of self and comprehending their present attitudes and feelings.

The twins spent hours at the Auschwitz archives examining files and photographs in search of familiar names and faces. Finding one's name on a list of Auschwitz twins or locating one's face in a photograph was extremely important. But far more significant was being personally acknowledged by even one other twin.

At the reunion, the men appeared to recall one another's names and faces more readily than did the women; only small groups of women recognized one another. These contrasting reactions may stem, in part, from differences in how the boys and girls were organized at Auschwitz. The boys were assigned a leader—the extraordinarily devoted Zvi Spiegel—while the girls had none. Spiegel, who had a twin sister at the camp, was 29 when he arrived at Auschwitz from Hungary. Given the job of caring for 40 to 50 twin boys, he found his duties included accompany-

ing his charges to the washroom and seeing to it that he and they were available for experimentation.

"Uncle Spiegel," as he was called by his charges, said that he attempted to calm the boys and taught them geography and mathematics at odd moments. He reportedly saved the lives of two nontwin brothers by insisting to SS officers that they were twins. It is also reported that, at another time, he saved all the younger twin boys in his barracks by daring to inform Mengele that another physician had ordered them to the gas chambers. Mengele, appalled at the thought of losing his valuable twins, overrode these orders.

Spiegel saw to it that his boys knew each other's names and countries of origin. After liberation, he escorted 36 twin children from the camps to their hometowns throughout Poland and Czechoslovakia, along with numerous other homeless children he gathered along the way.

Lacking any such leader, the girls' group was probably less unified. In this context, the reunion experience of Irene Hizme is particularly poignant. Until the meeting's final day, no one remembered her. She found speaking with the other twins quite painful, especially since "Spiegel's Boys" readily recognized her twin brother, René Slotkin. She and Ruth Elias, a nontwin Auschwitz survivor, finally recognized one another. Elias reminded her of events such as the transport from Theresienstadt, another concentration camp, to Auschwitz. This helped Hizme to remember, a major reason that she had come.

Some twins felt a special need to honor their parents, from whom they had been separated—often at age 8 or younger, when children tend to idealize their parents and are highly dependent upon them. Few had suspected at the time that they would never again see their parents. Graves were never marked and funeral services never held.

To the adult twins of the Holocaust, a parent's "grave" might be a railroad platform, an entry in a list or a certificate of death. When we arrived at Auschwitz, Kor located the spot near the tracks where she recalls being dragged away from her mother: "I'm coming back here because this was the beginning of the end. . . . It was tough

---

THE SURVIVAL GAME

Many twins imprisoned at Auschwitz survived through creative teamwork. Their "schemes" were not unlike the pranks twins often devise for fun with family and friends, but the stakes were much higher: life or death.

After 9-year-old Eva Kor was injected with an unknown substance, she developed a high fever and was taken to the "hospital," a place from which, the children knew, one "never returned." Knowing that for the good of herself and her twin sister, Miriam Czaigher, she had to return to the barracks, she devised a clever strategy: She made her fever appear to drop gradually over many days by surreptitiously shaking down the thermometer. When she got a normal reading, she convinced the hospital staff that she was well enough to be released.

Nine-year-old Menashe Lorenczy and another twin boy, assigned to bring back soup for the children

from the camp kitchen, used the opportunity to "organize"—the inmates' term for obtaining goods and supplies, especially food, from the camp's storage. Menashe arranged to carry his twin, Leah, to the kitchen by hiding her in a large, empty soup container, allowing her access to extra bits of food. On the return trip, with the container full of soup, she held the lid, providing an excuse for her presence if an SS officer were to question them.

On another occasion, Menashe helped Leah escape the crematorium after she developed an infection in the hospital following an experimental operation. Knowing the hospital's reputation, Menashe faked a toothache so he could join his twin there. His healthy tooth was extracted, but the twins escaped the hospital by attaching themselves to a group of children in the hospital who were allowed to return to the barracks.

(Top) Railroad tracks used to bring prisoners to Auschwitz/ Birkenau.

growing up without you, mother, but we finally came back to say goodbye to you. Now we can get on with our lives." It was as if the unfinished business of 40 years had finally been completed and a huge burden lifted.

Twins Hizme and Slotkin, separated from each other and their mother at age 6, found their mother's death certificate in an Israeli museum honoring inmates of Theresienstadt. This was the nearest Slotkin had ever been to her grave; he found it difficult to leave that piece of paper. Hizme stroked it tearfully.

Berkowitz had watched his mother being led toward the gas chambers. He had promised her then that he would work for the good of humanity and, in coming back to Auschwitz and in cofounding the CANDLES group, could tell her at last that he was fulfilling this promise. For him and many others, coming in contact with the railroad platform or a piece of paper constituted a legitimate and meaningful reunion with their long-lost parents.

A few of the twins had been reunited with their parents after liberation. The wife of a fraternal twin observed that both brothers continue to "idolize" their father, who is still alive today. She explained, "They never experienced the normal adolescent rebellion against their parents."

The concentration-camp experience left deep emotional scars on many sur-

vivors. Some of the effects of imprisonment at Auschwitz are unique to the twins, while others may be more characteristic of Holocaust survivors in general.

The majority of those I interviewed indicated that they had been so humiliated by the experience ("We were made to feel like animals, garbage, pieces of meat") that they had tried to put this terrible part of their lives behind them. Many had never spoken of their past to anyone.

Their silence was often reinforced by their adoptive parents, many of whom attempted to dismiss Auschwitz as though the experience had never existed. One woman, as a little girl, had been dressed in long-sleeved blouses to conceal the identification number indelibly etched on her arm, causing her deep shame.

As these children grew up, they were anxious to establish themselves professionally and to raise families, trying to put the past to rest. In these efforts they have been enormously successful. While many still hurt deeply inside, they continue to display the optimism and determination that have been central to their survival, and they are surprisingly lacking in bitterness or hate.

Having children was a very critical event for both the men and women—a way of reestablishing severed biological links. Children also represented re-

venge against the Nazis, Mengele in particular, and their attempts to eradicate the Jews. Some women, however, found pregnancy and birth quite stressful; they feared that because of the injections they had received, they might pass on something very harmful to their children. (Despite their fears, however, most had normal births.) For those who had been experimentally sterilized, the results were overwhelming grief and hurt.

Only during the past four or five years have most of these twins started dealing overtly with their past. Having established families and careers, they can turn to issues long-buried but never forgotten. One twin observed, "You don't get up in the morning and think about it, but there are reminders." Stale bread tossed away or smoke rising from a distant chimney can vividly revive traumatic events and scenes at Auschwitz.

I found a number of twins quite willing to talk openly with me about their emotional reactions and problems. There was little expression of guilt about being saved because they were twins while family and friends were murdered. To many, the twins' situation was simply a matter of fate; they were not responsible.

One identical twin recalled being initially directed toward the gas chambers, since she was physically weak and considered unfit. But her elder sister, unaware of the twin experiments, pleaded with Mengele to keep the two twins together since they had never been separated. Upon learning that she was a twin, Mengele ordered the little girl back from the gas chambers.

It is hardly surprising that nearly

F OR ADULT TWINS, A PARENT'S GRAVE MIGHT BE A TRAIN PLATFORM, A NAME ON A LIST, A DEATH CERTIFICATE.

*D*ESPITE DEEP HURT, MANY STILL SHOW THE OPTIMISM AND DETERMINATION THAT HELPED THEM SURVIVE AUSCHWITZ, BUT SURPRISINGLY LITTLE BITTERNESS OR HATE.

half of the 24 twins I interviewed had received psychotherapy or felt a need for it. They were separated from their parents at an early age, given no preparation for the loss and little chance to mourn, denied an adequate support system and forced to live in extremely unstable, unfamiliar and hostile surroundings.

Some twins mentioned nightmares and suicidal tendencies. Others and their children mentioned marital difficulties, especially with spouses who had not been through the prison-camp experience, although spouses were usually described as being very understanding and supportive. One Israeli twin described an intermittent but overwhelming desire to get away—anywhere. She literally packs her bags and leaves, traveling to the home of her mother or fraternal-twin sister. Her twin has the same urges, but lacks the physical or mental strength to actually leave. Still other twins admitted to excessive worry over separation from their children.

Israeli researcher Davidson points out that many of these twins suffer even more from knowing that the ex-

periments in which they participated had absolutely no scientific value. Mengele apparently did not adequately identify actual twins (we know of three cases of "pseudo-twins") or apply any theory that could guide his experiments or their interpretation. Further, there is no evidence that he distinguished between identical and fraternal twins in his experiments. Some twins were apparently "sent for" more frequently than others, but the reasons are unknown.

After participating with the twins in a series of intimate discussion groups, I think that this may be the best way to help them come to terms with the past. The stressful events of Auschwitz can be most effectively confronted in an atmosphere of warmth, familiarity and understanding. Many of the twins agreed that they made their greatest personal progress by talking with other twins.

Self-help groups of all kinds are increasing in number and specificity; that specificity seems to be particularly important. Having worked many years with twins and their families, I am keenly aware of the useful pur-

pose served by the National Mothers of Twins Clubs. Its members insist that only mothers with twins can understand the unique situations that arise during their rearing.

It is, apparently, no different for the twin Holocaust survivors. CANDLES was formed when the twins finally felt ready and willing to talk about their past. Five of those I interviewed had participated in other Holocaust-survivor groups, but all were dissatisfied. They found CANDLES, which offers the special understanding only fellow-twins can provide, more responsive to their unique needs. As one member said, "I have hands to hold."

Through the efforts of CANDLES, approximately 110 of the 157 twin children released in 1945 have been located, and the search continues for the rest. Not all are willing or able to forge links with their past. But for those who did this year, one theme was dominant: Twinship was special 40 years ago, and it is special now.

FOR MORE INFORMATION

To contact CANDLES or to provide information on the twins or experiments write:

Eva Kor
24 West Lawrin Boulevard
Terre Haute, Indiana 47803

Nancy L. Segal
University of Minnesota
Department of Psychology
Minneapolis, Minnesota 55455

Men and Women: How Different Are They?

Cullen Murphy

The most perfectly organized societies in nature are sexless ones, or those where sex differences have been minimized or somehow suppressed. In America, during the turbulent late 1960s and '70s, feminists began to suggest, in effect, that our own complicated society ought to move in that direction. The role of housewife and mother was disparaged as "unfulfilling"; women entered the labor force by the millions; discriminatory laws were struck down; divorce rates soared. Yet, as scholars note, boys and girls still behave differently as youngsters. The call to motherhood remains strong even to ambitious career women. Males and females continue to look at the world through different eyes. In an odd way, the feminist drive for sexual equality has spurred rather than eroded scholarly efforts to examine "masculinity" and "femininity." Here, Cullen Murphy surveys the growing mass of research on sex differences in behavior.

God fashioned Eve from Adam's rib, the Bible says, but scholars these days would turn the metaphor on its head. As psychologist June Machover Reinisch has put it, nature "imposes masculinity against the basic feminine trend of the body." That may be part of the reason that there is a 500-percent-greater incidence of dyslexia in boys than in girls and that girls have more stamina. Then again, it may not. The scholars keep at it.

Stacked on a library table, the literature on sex differences in behavior and physiology, published in scholarly journals during the 1970s by chemists, sociologists, physicians and other researchers, would stand about 6 feet high. That does not include a dozen or so reputable books, such as Eleanor Maccoby and Carol Nagy Jacklin's *Psychology of Sex Differences* and John Money's *Love and Love Sickness*.

As they peel the onion of sex, scholars have scrutinized males and females in the workplace, in the army, in the schools and in the uterus. They have contemplated "deviance" as a clue to "normality" and drawn lessons from the experience of wallabies and coral-reef fish. Where the specialists have been less successful is in imposing theoretical order on our expanding body of knowledge. That men and women *do* differ, biologically, cognitively and behaviorally, no one disputes—although such differences, it must be stressed, are not usually absolute but apparent only as averages when groups of men and women are compared. Yet, as psychologist Jeanette

McGlone writes, "Questions such as 'Why?' and 'Does it matter?' remain unanswered."

Those two questions, of course, are what the fuss is all about. The staunchest believer in equal opportunities for both sexes will, if he or she is honest, concede that the real world is not Plato's cave. Rightly or wrongly, men and women have long assumed—and still assume—that differences in expectations and behavior exist between the sexes; over time, through countless adjustments and accommodations, they have learned to live with what they thought those differences were, constructed their societies accordingly, come to depend on one another in different ways, to behave in one way when with one's own sex and another in mixed company.

During the past decade, scientists have probably quadrupled what is "known" about *biological* differences between the sexes. Men's and women's brains seem to be dissimilar in certain respects, but the human brain remains a mystery, and drawing inferences is like writing on sand. In some ways, social scientists are more helpful, at least in explaining the broader implications of the way men and women behave. One fact that does emerge clearly—and here the research merely ratifies common sense—is that, regardless of their origin, gender-linked traits appear, and acquire significance, at varying ages for men and women.

Males, the Vulnerable Sex

It begins at fertilization. Men and women will never again be so much alike as they are during the first seven or eight weeks after conception. Until then, although the male possesses a "Y" chromosome in addition to an "X" (the female has a pair of X's), male and female embryos appear identical. Scientists debated for years whether it was the distinctive Y or the extra X that prompted sexual divergence. It turns out to be primarily the Y.

The mechanics of this process are still not entirely clear. In essence, though, midway through the first trimester, the male embryo secretes a hormone that incites his previously undifferentiated gonads to develop into testes. These produce another hormone, testosterone, which in turn programs further development of male sex organs. If an XY embryo cannot produce testosterone, or cannot metabolize it, it is in for trouble and will develop, however quirkily, along a preprogrammed female line. In a sense, then, all human beings are female until something acts to make some of them male.

The likelihood of error in male development is extraordinary. About 140 boys are conceived for every 100 girls, but various defects cause most of those extra boys to succumb before birth. A differential remains even then (about 106 boys are born for every 100 girls), but males are more susceptible to childhood diseases. Boys are also more likely to stutter and to be colorblind. Males may be, as Henry Higgins put it, a "marvelous sex," but they are also exceedingly vulnerable.

The first connection between hormones and behavior was made long ago, in 1849, by the German scientist Arnold Berthold. Berthold discovered that castrated roosters stopped fighting and lost their interest in hens. Research into hormones and their effects intensified during the 1970s. In some nonmammals, researchers discovered, the injection of male hormones (androgens) before birth can change a fe-

male into a male. Certain mature fish can change their sex when confronted with new environmental conditions. Nothing so extreme has been demonstrated in mammals, but female offspring of rhesus monkeys that have been heavily dosed with androgens do exhibit "male" behavior—"rough and tumble" play, for example, and the mounting of other females.

For ethical reasons, scientists do not conduct experiments on humans. Here, they have had to glean information from "experiments of nature"—e.g., children with brain damage, hermaphrodites—or by pondering the unexpected side effects of hormones administered to avoid toxemia in pregnancy. John Money of Johns Hopkins and Anke Ehrhardt at Columbia have studied girls with adrenal hyperplasia, an enzyme defect resulting in production of massive amounts of androgens. These girls, they found, became extreme "tomboys," were very athletic and rarely played with dolls. Most studies confirm that boys, on the average, are more aggressive than girls, and most studies indicate that testosterone probably has something to do with it. Hormones may not make certain types of behavior inevitable but merely, as John Money puts it, "lower the threshold so that it takes less of a push to switch you on to some behavior."

The male and female timetables continue to vary after birth. As neurologist Richard Restak has noted, girls at the age of four months are far more attentive than boys to "social contexts": faces, speech patterns and tones of voices. Girls begin to talk sooner. Boys, on average, are the first on their feet; they have better total body coordination throughout their lives but somewhat less stamina. They are more curious, more active and more mechanically inclined.

No one knows how much (if any) of this to attribute to chemistry and how much to child rearing. Parents treat boys and girls differently, and that difference rubs off. For example, if girls learn to talk earlier, it may be primarily because most mothers spend more time chatting to their infant girls than to their baby boys. Hormones do leave an imprint on men's and women's liv-

ers, kidneys and the nerve endings in their brains. They differentiate the hypothalamus into a male and female type. What scientists cannot establish is whether hormones account for the many observed differences in the way male and female brains work.

The most striking difference is in brain "lateralization." In right-handed people, the left hemisphere of the brain is primarily responsible for verbal skills, the right hemisphere for spatial-perceptual skills. But this lateralization is less pronounced in girls than boys—so much so that in girls, one side of the brain seems to be able to make up for deficiencies in the other. Thus, girls have a lower incidence of dyslexia, aphasia and infantile autism. Thanks to her neural "insurance," an adult woman will recover faster, and more completely, from a cardiovascular accident.

Different Creatures?

Women are far more sensitive than men to odors, tastes and touch, as well as to extremes of light and sound. For example, they can detect Exaltolide (a musklike odorant) when it is dispersed in quantities as low as one part per billion; the male threshold is 1,000 times higher. "It may be," conclude June Reinisch, Ronald Gandelman and Frances Spiegel, "that males and females are essentially quite different creatures, whose perceptions of the world differ markedly even when confronted with similar physical environments."

It is not necessary to understand the origins of these differences in order to glimpse some of their down-to-earth implications, particularly for boys and girls starting elementary school. As some scientists and educators are beginning to point out, throwing both sexes together in a classroom and teaching them in the same way may be doing each sex an injustice.

Because of boys' greater spatial-perceptual skills and girls' superior verbal ability, it may be better to use the "look-say" method of teaching reading with the former and the "phonics" method with the latter. Schoolboys tend to be far more "hyperactive" than girls (95 percent of all clinically hyperactive children are male). One reason could be that the classroom environment is oriented more aurally and verbally than visually. Opportunities for rambunctious young males to work off steam are few. In the early grades, at least, school is geared to skills that come naturally to girls. Ninety percent of the time, the teacher is also a woman. In later grades, when certain subjects with a heavy spatial-perceptual content are introduced—math and the sciences, for example—girls tend to lose their advantage. In these courses, too, the teacher is most often a man.

A radical overhaul of the educational system would cause more problems than it would solve. But some tinkering may be in order. "The nerves that feed the brain," Virginia Woolf speculated in 1928, "would seem to differ in men and women, and if you are going to make them work their best and hardest, you must find out what treatment suits them."

The onset of puberty generally coincides with the three years of junior high school, but again the male-female timetable differs. In most girls, estrogen begins to build up in the body between the ages of 10 and 12; boys get their hormonal burst on average two years later. In both sexes, one result is a period of rapid physical growth, lasting for two to four years in girls and for six years or longer in boys—on into the late teens.

Puberty is the second time in male and female lives that hormones exert a sudden, decisive and unquestionable impact. In women, they control the onset of menstruation and regulate it thereafter until menopause. They determine the shape of the female pelvis and the level of body fat. (About 25 percent of the body weight of mature women is fat, compared to 14 percent for men.) Hormones spur sexual maturity in men and promote the growth of body and facial hair. The males' bones grow longer, their shoulders broader; they acquire 10 percent more heart and lung capacity than do females.

Mirroring Society

During adolescence, the difference in verbal skills between men and women begins to narrow, but the gap in spatial-perceptual skills does not. Boys start getting better grades than girls do. Certain patterns in behavior and expectations continue to firm up. A window on these years is provided by the U.S. Department of Education's comprehensive *High School and Beyond* (1980), a survey of 58,000 secondary-school students.

Not surprisingly, boys and girls in high school mirror the larger society. Already, the males have taken

---

U.S. Men and Women: Some Comparisons

Health: Women have a marked advantage in longevity over men—77.1 versus 69.3 years in the United States. In any given year, twice as many men as women die of heart disease, 50 percent more die of cancer. However, the average American woman pays two more visits to the doctor than a man does every year, and as a group, females undergo 5 million more operations annually than do males. Throughout the industrial world, women evidence a far higher recorded incidence of depressive psychoses and psychoneuroses. But more alcoholics are men, and males have a 290 percent higher suicide rate than females.

Education: There are currently more women than men in college (5.9 versus 5.7 million), but somewhat more men than women in graduate or professional school (862,000 versus 709,000). While women stay numerically abreast of men through the master's degree level, males earn about 70 percent of all Ph.D.'s. Less than 13 percent of doctoral degrees awarded in 1980 in the fields of mathematics or the physical sciences were granted to women.

Crime: For all races, ages and income levels, men are far more likely to commit a criminal act than are women (except for prostitution); only one out of five serious crimes—murder, robbery, arson—is committed by women. In 1979, some 8 million arrests were made for various offenses; women accounted for 1.3 million of them. But women's arrest rates are growing in virtually all nonviolent categories and, overall, are rising faster than men's. Some of women's gains reflect increased employment opportunities—for instance, the 24 percent increase in embezzlement by females in 1979.

Employment: Of 98.8 million working Americans, 38.9 million are women. Men and women are represented in *every* occupational category, but the percentages vary. Only 1 percent of the nation's 48,000 kindergarten teachers are men; only 0.01 percent of the 554,000 auto mechanics are women. Contrary to popular belief, the earnings gap between men and women is greatest in the traditionally male jobs such as law and medicine, and smallest in the traditionally female jobs such as teaching and nursing.

Politics: Men were more likely to go to the polls than women until the 1980 election, when women cast slightly more than their share of the 86.5 million votes for President. On balance, women lean more toward the Democratic Party than do men and are more likely to consider themselves liberals. The margin, however, is slight. Whether a political candidate is a woman does not seem to affect the way men or women cast their ballots. This was not always so. Through the 1950s and '60s, women tended, disproportionately, to shun candidates of their own sex, for reasons that remain unclear.

---

after-school jobs and entered the labor force in greater numbers than have the females; they are working longer at their part-time jobs (22.5 hours a week versus 18.6) and making more money ($3.38 per hour versus $2.99). By a margin of 64 to 41 percent, the boys are more likely to participate in school athletics; they have far more disciplinary problems. Girls are the mainstay of extracurricular activities other than sports. They spend more time reading (unless the reading matter is a

213

newspaper) and talking on the phone. (According to Ma Bell, the girls will, as adults, initiate 60 percent of all nonbusiness telephone conversations.)

What about the future? Both sexes see themselves taking "traditional" jobs—the girls lean toward teaching and clerical work, for example; the boys indicate a taste for managerial and blue-collar jobs. High-school girls are more concerned than boys about "finding the right person to marry," high-school boys are more apt to envision "having lots of money." More boys than girls look forward to having no children at all; more girls than boys hope to have "four or more."

The Coeducation Paradox

Adults are a more diverse lot, their lives more complex. We have plenty of general statistics about men's and women's jobs and education. But in-depth research necessarily focuses on smaller, more cohesive groups of individuals. Here, the availability of funding and the "relevance" of the subject tend to favor some groups over others: men and women at "elite" universities rather than those at community colleges; women executives "climbing the corporate ladder" rather than women on the assembly line (and most people on assembly lines are women). Especially since the rise of the women's movement, researchers have been more interested in females than in males—a propensity that is less pronounced when boys and girls are the object of study.

That said, the existing studies do raise some intriguing questions.

One example involves higher education. By 1970, the historic education gap between men and women had virtually been eliminated. On average, both sexes finished high school and about half a year of college. At the same time, however, many of the nation's elite schools—ranging from small colleges such as Haverford to universities such as Princeton—remained "male bastions." Angry voices were raised, and, during the 1970s, despite alumni grumbling, nearly all of the elite all-male institutions that had not already done so opened their doors to women.

A decade later, scholars have begun to assess the impact. So far, at any rate, it appears that equality of opportunity is not necessarily the surest path to similarity of outcome.

The most comprehensive study of the effects of coeducation was sponsored by Brown University and published in 1980. It was based on a survey of 3,300 men and women at Barnard, Brown, Dartmouth, Princeton, the State University of New York at Stony Brook and Wellesley. One major finding was that women at coed schools tended, in effect, to lose much of their worldly ambition. They majored in fields where women had always done well—the humanities, the arts, the social sciences. While men and women aspired to graduate school in equal numbers, in practice the women aspirants experienced significant attrition. They seemed, in sum, "to be adjusting their plans downward" to a greater extent than men students.

Shortly after the release of the Brown study, the Women's College Coalition, a Washington-based association, reported that America's 118 women's colleges had recovered from a brief slump and recorded a net enrollment increase of 15 percent since 1970. Up to 30 percent of the women at many of these schools were majoring in math and science. The report's message, though never bluntly stated, was that women's colleges were still uniquely equipped to motivate women to excel in the courtroom, the operating room and the board room.

Dropping Out

The entry of large numbers of women into the labor force beginning in the late 1960s—whether in search of a "career" or just a "job"—is among the most significant phenomena of the postwar era. As Peter Drucker has written: "We are busily unmaking one of the proudest social achievements of the 19th century, which was to take married women *out* of the work force so they could devote themselves to family and children."

About 39 million adult women, including 55 percent of all mothers, now hold full- or part-time jobs. While half of them are still employed in "traditionally female"

jobs—those like stenography or teaching elementary school, where more than 80 percent of all workers are female—women have made extraordinary gains in virtually every occupation. One-third of all accountants today are female (versus one-sixth in 1960); one-half of all tailors and bus drivers are women, as are 33.5 percent of law-school students (compared to 3.6 percent in 1963). While women physicians (10 percent of all M.D.'s) still tend to shun careers in aerospace medicine or orthopedics, they are coming to dominate other medical specialties, such as obstetrics and gynecology.

The impact of all this on American society has been immense. One reason that the unemployment rate is so high—10.2 percent in April 1983—is not because women are taking jobs that would otherwise go to men, but because nearly 4 million women are out "looking for work," which is the U.S. Labor Department's threshold for inclusion in the labor force. For a full-time working mother, raising a family can become a severe challenge. No survey shows that menfolk do their full share of the housework. Of course, there may be compensations. Few intact families where both the husband and wife work are below the poverty line. (Some 51 percent of all married couples are "dual-earner" families.) But 21 percent of all working mothers are without husbands, and 44 percent of these are living below the poverty line.

The income of women who work full time is only 59 percent that of men—relatively less than it was in 1955. But it is by no means clear how much sex discrimination or, more important, the concentration of most women in low-paying occupations (e.g., nursing) can account for the earnings gap. France, West Germany and Sweden are all experimenting with programs that would diversify women's employment and thereby eliminate the "parallel labor market." But such experiments fail to address a central problem: Female labor-force participation slumps deeply between the ages of 25 and 35 as women bear and rear their children. As economist Lester C. Thurow observes, "If there is any one decade when it pays to work hard and be consistently in the labor force it is the decade between 25 and 35." This is when lawyers become partners, academics get tenure, blue-collar workers become supervisors or acquire new skills and businessmen move onto the "fast track." "For those who succeed," Thurow says, "earnings will rise rapidly. For those who fail, earnings will remain flat for the rest of their lives."

All of women's gains during the past decade have not erased this basic fact. Nor has the advent of effective contraception dampened the urge to bear children. Increasing numbers of women, who entered the labor force five or ten years ago telling pollsters and reporters that the most important thing to them was proving themselves on the job, can now be found proudly showing off their new babies.

Over time, at least two choices that working women must make have far more ramifications than the same choices when faced by men: whether to get married; whether to have children. It is probably no coincidence that a 1976 Harvard University survey of its junior faculty revealed that 61 percent of the institution's married women professors had no children, compared to only 32 percent of their male peers. It is perhaps no coincidence, either, that virtually every male chief executive officer of a major American company is currently married, while 54 percent of the female CEOs are divorced or never married.

There are, perhaps, other kinds of trade-offs. A recent study of 123 women who graduated from business schools in 1977 and 1978 found that they were "paying a price" for success. They demonstrated significantly more stress than their male colleagues, much of it due to worry about how things were going at home. (Other studies, however, suggest that holding a job may improve a woman's mental health.) Although it is impossible to say whether more employment has anything to do with it, women's overall *physical* health has deteriorated relative to men's during the past 30 years. They are suffering from more ulcers and respiratory ailments than ever before. They have not been as quick as men have been to quit smoking. "Adult women," writes the University of Michigan's Lois M. Verbrugge, "are adopting life styles which bode ill for their longevity." They are, in short, behaving more like men.

We do not live in an ideal world and rarely agree on what an ideal world would be. Even when we do agree on some incremental "improvement," it is generally difficult to bring about. For example, every bit of poll data indicates much rethinking by employers, employees and ordinary citizens about the relative capabilities of men and women. The old notion that "a woman's place is in the home" finds a dwindling number of adherents. If the Gallup Poll's measure of

people's ideals were an accurate reflection of their behavior, the National Organization for Women might have disbanded long ago for lack of new fields to conquer. In fact, as everyone knows, human beings take a more personal, less abstract approach to their own lives. "Give me chastity," St. Augustine prayed, "but not yet."

At a time when many popular attitudes are slowly, unevenly changing, when legal and social barriers to women's autonomy and advancement are falling and when American society is patiently absorbing the resultant aftershocks, it is sometimes easy to overlook the things that never change. Men and women still manage to fall in love, still seem to draw some special comfort from one another that they don't get from their own sex. They still get married and have children, and enjoy their little boys and girls in different ways. Having both a mother and a father at home is still the best way for a child to grow up; single-parent households are, statistically, candidates for trouble and, collectively, a troublesome burden on the larger society. Biology aside, despite the misunderstandings and injustices they have imposed, differences between the sexes contribute something vital to our lives and essential to our civilization. For most people, in the end, being male or female is not a circumstance to be overcome but one to be savored, and the odds are good that this useful sentiment will long survive.

Male Brain, Female Brain: The Hidden Difference

*GENDER DOES AFFECT HOW OUR BRAINS WORK—
BUT IN SURPRISING WAYS.*

Doreen Kimura

Doreen Kimura is a professor of psychology at the University of Western Ontario in London, Ontario, Canada.

The idea that male and female brains are organized differently has been around for a long time. After all, since men and women are dissimilar in size, appearance and sexual role, why shouldn't their brain organization differ too? Research has documented in the past 25 years that there are intellectual differences in the way the two sexes solve problems: On average, women do better in certain verbal skills and men in spatial and mathematical skills.

The notion that men's and women's brains are differently organized began to take hold in earnest as a result of experiments spanning the 1960s and 1970s. It began with the work of psychologist Herbert Lansdell, who studied neurosurgical patients. Others before him had found that, in general, removing the brain's left temporal lobe interfered with verbal skill, while removing the right impaired nonverbal skills. Lansdell found that although such injuries caused a similar overall pattern of impairment in women and men, women were less severely affected than men.

These preliminary findings were confirmed and given support by people working in my laboratory and in other laboratories. Studies of both brain-damaged and normal people revealed that while men and women tend to use one hemisphere more than the other for certain verbal tasks, such as recognizing spoken or seen words, women seem to rely less strongly on a single hemisphere than men do. Particularly striking was a finding by psychologist Jeanette McGlone, then a graduate student in my laboratory, that damage to the left hemisphere (the one usually dominant for language) caused less aphasia (language disorder) in women than in men.

These findings led to what seemed to be an obvious conclusion: Certain thinking skills are more lateralized—more dependent on one hemisphere—in the male brain than in the female. Or, putting the comparison the other way, women's brains are more diffusely organized than men's.

A number of explanations were of-fered for this apparent sex difference: Women were more verbal, meaning that both their hemispheres were given up to speech; women developed more quickly and lateralization required slower development; women were just as lateralized as men but used verbal strategies more often; connections between the hemispheres were stronger in women and, therefore, the asymmetrical organization of their brain was less detectable. And so on.

Anatomical studies began to suggest that there were structural differences between men's and women's brains, but results were inconsistent regarding the question of men's greater lateralization. After neurologist Norman Geschwind and coworker Walter Levitsky at Harvard Medical School had found that a verbal part of the brain was generally larger on the left than the right side (see "Of Hemispheres, Handedness and More," this issue), neuroscientist Juhn Wada reported that these anatomical differ-

From *Psychology Today*, November 1985, pp. 50-52, 54, 56-58. Copyright © 1985 by the American Psychological Association.

MAJOR AREAS OF THE LEFT HEMISPHERE

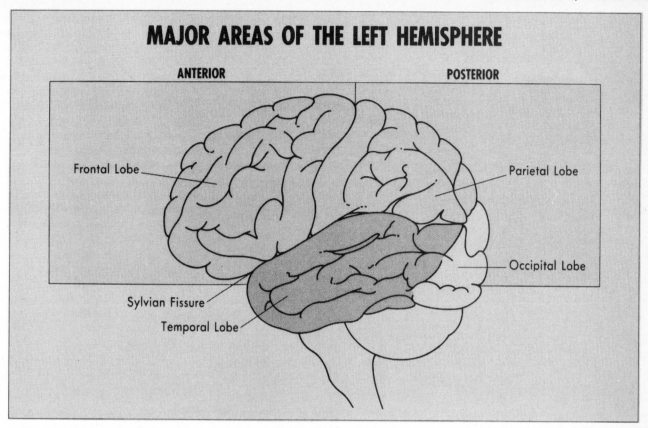

ANTERIOR   POSTERIOR

Frontal Lobe

Parietal Lobe

Occipital Lobe

Sylvian Fissure

Temporal Lobe

ences were smaller in women. In addition, biologist Christine de Lacoste-Utamsing, working with anthropologist Ralph Holloway at Columbia University, claimed that part of the corpus callosum, the major connection between the hemispheres, was slightly larger in women.

The problem with viewing men's brains as more lateralized than women's was that it left a lot of questions unanswered. Why, for example, are women more often right-handed than men? Even among right-handers, more women are purely right-handed, suggesting that, if anything, women rely more on one hemisphere than do men. And why was there no evidence that speech disorders occur more often in women following right-hemisphere brain damage, as one would expect if they, unlike men, depended on both hemispheres for speech?

In the course of looking at how damage to specific regions of one hemisphere affects speech and related functions, I came across some unexpected findings that, I believe, help to clarify some of these issues. One of the difficulties in doing research on neurological patients is finding enough people with various kinds of

THE OLDER VIEW: WOMEN'S BRAINS ARE MORE DIFFUSELY ORGANIZED THAN MEN'S.

brain damage to study—enough women, for example, with damage limited to particular regions of one hemisphere. It took me 10 years to gather enough data on brain-damaged patients to make meaningful comparisons. But an important and surprising sex difference emerged.

I was looking at people whose brain damage was restricted to either the front (anterior) or back (posterior) sections of the brain. I found that left-hemisphere damage could cause aphasia in both men and women, but different sites within that hemisphere

were involved in the two sexes. Men were equally likely to have a severe speech disorder with anterior or posterior damage, in keeping with the classical picture of "speech areas" in the left hemisphere. Women, however, were much less likely than men to become aphasic after restricted posterior damage, and so far, in all women with damage to the posterior area, the left temporal lobe was affected. No woman has lost her capacity for speech because of damage to the left parietal lobe, but several men have.

This seemed to suggest that the brain area involved in women's speech is, if anything, more localized than in men, at least in the left hemisphere. This idea was so radical that it took me some time to accept it. After all, according to the prevailing wisdom, speech was supposed to be more broadly represented in the female brain than in the male. But there was support for this idea from another type of research evidence: a study of how speech is affected when the cortex is electrically stimulated in awake patients during brain surgery.

Since the brain itself cannot feel pain, patients do not need a general anesthetic during brain surgery; local anesthesia to the scalp and skull will suffice. Catherine Mateer, a neuropsychologist who worked with neurosurgeon George Ojemann at the University of Seattle, found that electrical stimulation in the brain area near the Sylvian fissure interfered with a picture-naming task, but the particular brain areas responsible for such interference differed in men and women. In men, stimulation almost anywhere in the vicinity of the Sylvian fissure resulted in naming difficulties, while in women the pattern was more restricted. In particular, posterior parietal stimulation in women did not result in any naming problems, while it did in men. Thus, both my own data on brain damage and Mateer's on stimulation suggested that speech was not, as believed, more diffusely organized in women's left hemisphere than in men's.

What's more, the right hemisphere does not seem to contribute to speech any more in women than in men. Reviewing our own series of right-handed cases with damage restricted to the right hemisphere, we found that aphasic disorders after such damage are very rare (1 to 2 percent of cases), and there is absolutely no difference between men and women in this respect. So we are left with the very strong probability that speech is organized differently within women's left hemisphere compared with men's. It looks as though in women's brains—but not men's—speech favors anterior systems and avoids the parietal region.

Why, then, is aphasia less common after left-hemisphere damage in women? Presumably, it's just a matter of odds: If speech is localized in a more restricted area of women's left hemisphere, and we look at a random series of patients with left-hemisphere damage, the speech systems are simply less likely to be hurt in women and aphasia is less likely to occur.

The different representations of speech functions within the left hemisphere might also partially explain why normal women are slightly less dependent on a single hemisphere than men in dichotic listening tests (see "Listening in to the Hemispheres" box). (In the 1970s psychologists Richard Harshman and Philip Bryden found independently that women have a less pronounced advantage in the right ear, that is, in the left hemi-

THE VIEW NOW: WOMEN'S BRAINS MAY BE ORGANIZED LIKE MEN'S FOR SOME TASKS, MORE OR LESS DIFFUSELY FOR OTHERS.

sphere, on these tests.) In women, the left auditory cortex may be less directly connected to the speech centers than it is in men because the speech areas are differently located. It's also possible that by having a greater number of fibers in the corpus callosum, women have more effective transmission from the less-favored left ear to the left hemisphere.

Whatever the explanation, our findings that basic speech functions are quite focally organized in women mean that we have to give up the idea that women's brains are generally more diffusely organized than men's. But this could still be true, if not for speech, at least for other functions.

We have, in fact, found some verbal functions more related to abstract verbal ability than to speech production, such as defining words and using them appropriately, that do seem to be more bilaterally organized in women than in men, as several people had earlier suggested.

In particular, Harshman and I found, on analyzing data from people with damage to only one hemisphere, that regardless of which hemisphere was injured, women's vocabulary—the ability to define words—was impaired. I then found this was true whether I looked at anterior or posterior damage in either hemisphere, suggesting that defining words is a function of the

WHAT ARE LITTLE BOYS AND GIRLS MADE OF?

One of the most fascinating facts of biology to emerge in the past few years is that sex is not determined by the genetic makeup of a person in any simple, direct way. An individual can be born with an XY genetic make-up (the male chromosomal pattern), yet grow up to have female genitals and look and behave like a woman. Another individual may have XX chromosomes (the female pattern) and become a man.

What determines whether a group of cells carrying XY or XX chromosomes will turn into a male or female human being is the presence of critical sex hormones early in fetal life. The Y chromosome appears to be necessary for converting the gonads into testes. The testes, in turn, are responsible for secreting androgens, male hormones, which result in the development of a penis rather than a vagina. If no Y chromosome is present, the gonads become ovaries.

But curiously, no hormones are needed to develop the female reproductive tract. If there are no hormones, or if female hormones are present, a vagina rather than a penis develops. This means that we have a biological bias toward being female. It also means that through variations in fetal and perhaps even pubertal hormones, it is possible to have somewhat wider biological variations in "maleness" and "femaleness" than we previously suspected.

Psychologist June Reinisch of the Kinsey Institute in Indiana has shown that girls who have had a higher-than-usual exposure to androgens before birth tend to be tomboyish. And researcher Günther Dorner's work in East Germany suggests that even some instances of homosexuality may reflect variation in fetal hormones. These examples may mean that, although the two sexes differ sharply in genital appearance, each has a range of potential behavior broad enough to defy characterizing behavioral patterns as exclusively limited to one sex or the other.

Kimura's Model of Dichotic Listening

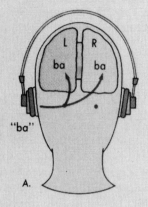

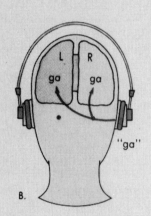

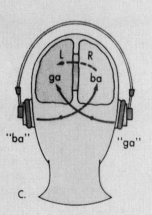

A. Syllable ("ba") sent to left ear goes to right and left hemispheres by different pathways. Subjects report syllable accurately. B. Syllable ("ga") sent to right ear also goes to both hemispheres by different pathways and is reported accurately. C. "Ba" sent to left ear and "ga" to right ear simultaneously. "Ga" goes only to the left (speech) hemisphere and "ba" to the right. So "ga" is usually reported more often and more accurately than "ba."

whole brain in women. Men had problems in defining words only after left-hemisphere damage. So for this kind of thinking at least, women's brains do indeed seem to be more diffusely organized.

I found different patterns of brain organization using other verbal tests, ones in which people were asked either to generate words beginning with a certain letter or to describe what they should do in various social situations. Other people have found that damage to the left anterior part of the brain causes the most difficulty in performing such tasks. I found this to be true for both men and women. So for this task, men's and women's brains were quite similarly organized.

In short, we are finding that, depending on the particular intellectual function we're studying, women's brains may be more, less or equally diffusely organized compared with men's. No single rule holds for all aspects of thinking. When it comes to speaking and making hand movements that contribute to motor skill, the brain seems to be very focally organized in women compared with men. This may relate to the fact that girls generally speak earlier, articulate better and also have better fine motor control of the hands. Also, a larger proportion of women than men are right-handed, and unequivocally so. But when it comes to certain, more-abstract tasks, such as defining words, women's brains are more diffusely organized than men's, although men and women don't differ in overall vocabulary ability.

I have been describing the average state of affairs. But there is reason to believe that there is a lot of variation in brain organization from person to person. We know, for example, that the brains of left-handers and right-handers are organized somewhat differently, yet on average they function quite similarly.

In addition to individual variations, there are some interesting combined effects. Harshman and his colleagues at the University of Western Ontario found, for example, when they looked separately at people with above-average reasoning ability that sex and hand preference interacted. Left-handed men with above-average reasoning ability showed poorer scores on certain spatial tests, as well as other tests, than did right-handed men; but

left-handed women were better at these tests than were right-handed women. When Harshman and coworkers looked at people with below-average reasoning ability, just the opposite happened: Now the left-handed men performed better than right-handed men on spatial tasks, but left-handed women did worse than the right-handed women.

What does this confusion suggest? It must mean that brain organization for such problem-solving abilities is related not only to sex and hand preference but also to overall intelligence level. And more to the point, it indicates that we have probably not one or two types of brain organization but several.

How are these different patterns of brain organization determined? There have been several suggestions in recent years that they may be related to the organism's rate of development both before and after birth. Biopsychologist Jerre Levy of the University of Chicago suggested some time ago that the two halves of the body, including the brain hemispheres, might grow at different rates in boys and girls, even before birth. The left hemisphere may develop more quickly in girls, and the right hemisphere in boys, thus favoring verbal skills in girls and spatial skills in boys. This idea has persisted in modified form in much of the literature on sex differences.

A recent report by biologist Ernest Nordeen and psychobiologist Pauline Yahr of the University of California at Irvine on the effect of injecting hormones into the brain of newborn rats suggested that even the hypothalamus, a very basic regulating system, is asymmetrically organized for sexual behavior; injections on the left or right side affected sex-typical behavior differently. So although it may seem a bit farfetched at first, there do appear to be basic asymmetries in the developing organism and these asymmetries may well have far-reaching repercussions for later differences between the sexes.

Functions such as speech and spatial ability traditionally have been thought to depend primarily on the cerebral cortex. Although we should not dismiss the idea that deeper brain structures contribute something to these abilities as well, it would be particularly interesting if there were sex-related differences in the structure of the cortex.

Neuropsychologist Marian Diamond of the University of California at Berkeley, comparing cortical thickness in male and female rats, did find that the right cortex is thicker in males at most ages, while the left cortex is thicker in females but only at some ages (see "A Love Affair with the Brain," *Psychology Today*, November 1984). Also very suggestive is her finding that, when ovaries are removed at birth, the female rat develops a pattern of hemispheric dominance more like that of the male.

These studies on anatomical asymmetries in the brain are in a very early stage, of course, but they indicate quite strongly that the biological sex differences in brain organization are probably dynamic, rather than a crystallized pattern that is laid down entirely by the genes. At various periods in life, different brain structures may be undergoing more- or less-rapid growth, and patterns of brain organization will vary from time to time as a result. This may very well go on

We CAN PREDICT VERY LITTLE ABOUT AN INDIVIDUAL'S MENTAL ABILITIES BASED ON HIS OR HER SEX.

throughout a person's life, in fact, since hormonal environments are in lifelong flux.

The role of sex hormones in prenatal development is quite dramatic and profound (see "What Are Little Boys and Girls Made of?" box). It may also be appreciable in adult life, even affecting cognitive abilities in men and in women. While hormonal changes occur in both sexes over a variety of

short and long cycles, the changes in women during stages of the menstrual cycle have been most thoroughly studied. For example, there is some evidence that spatial ability in women may vary monthly as natural levels of sex hormones in the bloodstream change; it may be best during the phase when the level of the female sex hormone estrogen is lowest.

In contrast to these findings, Elizabeth Hampson, one of my graduate students, has found that women perform best on tests of motor skill when their female sex hormones are at their highest level. So, as in brain organization, the pattern we see may very well depend on the particular function that we study.

What do all these findings tell us about the inherent capabilities of the two sexes? And what can we, as a result, deduce about the abilities of an individual man or woman? The fact seems inescapable that men and women do differ genetically, physiologically and in many important ways psychologically. This should not be surprising to us, since as a species we

BRAIN ORGANIZATION: MEN AND WOMEN COMPARED

FUNCTION	BRAIN LOCATION		SUMMARY	
	Men	Women	Men and women same	Men and women different
Producing speech	Left hemisphere, front and back	Left hemisphere, mostly front		X Women more focal
Hand movements for motor skill	Left hemisphere, front and back	Left hemisphere, mostly front		X Women more focal
Vocabulary-defining words	Left hemisphere, front and back	Both hemispheres, front and back		X Women more diffuse
Other verbal tests (Naming words beginning with certain letters; describing appropriate social behavior)	Left hemisphere, front	Left hemisphere, front	X	

LISTENING IN TO THE HEMISPHERES

THE FACT SEEMS INESCAPABLE THAT MEN AND WOMEN DO DIFFER GENETICALLY, PHYSIOLOGICALLY AND PSYCHOLOGICALLY.

Before the early 1960s, people interested in the differing roles of the left and right hemispheres depended almost entirely on evidence drawn from animal research, from studies of neurological patients with one-sided brain damage or from patients who had had their corpus callosum, the conduit connecting the two hemispheres, surgically severed. But I found that it was possible to detect which brain hemisphere was most involved in speech and other functions in normal people by having them listen to two different words coming to the two ears at the same time. This became known as the "dichotic listening" procedure. When several word pairs are given in a row, people are unable to report them all, and most right-handers prefer to report—and report more accurately—words given to their right ear. This seems to be related to the fact that signals from the right ear, although sent to both hemispheres, are preferentially sent to the left hemisphere, which controls speech. People who have speech represented in the right hemisphere, a very unusual occurrence even in left-handed people, more accurately report what their left ears hear.

In contrast to the right-ear (left-hemisphere) advantage for speech, there is generally a left-ear (right-hemisphere) advantage for another type of auditory signal: music. When right-handed people listen to melodic patterns, which neuropsychologist Brenda Milner at the Montreal Neurological Institute has shown depend more on the right hemisphere, they report them better from the left ear.

have a long biological history of having two sexual forms and have had a sexual division of labor dating back perhaps several million years. Men and women probably have been evolving different advantages for a wide range of activities for at least hundreds of thousands of years. In short, given two genetically different sexes, we can expect differing behavioral capabilities extending even beyond directly sexual roles.

But having said all that, I also have a number of important caveats. First, biological sex itself has turned out to be much more variable and dynamic than we ever imagined. And brain-organization patterns are even more variable from person to person, and probably even within the same person at different times. Further, on most tests of cognitive ability there is enormous overlap of men and women. We strain to look for differences and, of course, tend to emphasize the few we find.

Given these facts, it follows that while genital sex is related to our mental capabilities, it is going to be a very poor screening device for intellectual assessment. Numerous environmental events interact with our genetic heritage from prenatal development onward, and the human brain is extraordinarily malleable and variable. Thus, we can predict very little about an individual's mental capabilities based on his or her sex. A number of men and women can and do excel in activities that, on average, favor the other sex. There may be no inherent characteristics unique to the brains of either sex that necessarily limit the intellectual achievements of individual men or women.

221

For thousands of mental-health researchers and clinicians, "suicidology" is now a full-time specialty. ✓ *At some three hundred prevention centers, volunteers man hotlines.* ✓ *Yet experts still know too little about suicide, and its incidence in the U.S. increases each year.*

George Howe Colt

George Howe Colt is a contributing editor of this magazine. He graduated from Harvard College in 1976.

The rate remains highest among older people, but adolescent suicide has trebled over the past 25 years—a period that has also seen expanding prevention efforts, an outpouring of books and papers, and the creation of "suicidology" as a field of specialization. These developments are forcing a new awareness of the problem of self-destruction in American life.

At a recent high-school graduation in Weymouth, Massachusetts, a seventeen-year-old senior stepped up to the podium, said "This is the American way," and shot himself, though not fatally. His words provided a haunting corollary to Durkheim's theory that suicides tend to reflect not only personal failures but failures of society.

This year more than 30,000 Americans, an average of eighty per day, will take their own lives. The murder rate in this country is high, yet more people kill themselves than kill others. Suicide ranks ninth among major causes of death, and might rank as high as fifth if reporting procedures were stricter. Driven up by the alarming increase among adolescents, the U.S. suicide rate rose from 10.2 per hundred thousand in 1955 to 12.5 in 1980. Suicide now accounts for at least one percent of deaths in this country.

What are we doing about it? As a nation, almost nothing. Apparently satisfied with our rank near the middle of world statistics, the government no longer funds research on suicide. Funding for research on depression, such as it is, is biomedical, not socioeconomic.

As individuals and institutions, we are trying—but not concertedly. For the state of the art of suicide prevention is fragmented, controversial, and (worst of all) largely ignored. For more than two years I have been talking with men and women who have attempted suicide, with families of suicides, and with psychiatrists, social workers, sociologists, psychologists, and clergy. In these conversations I frequently felt we were talking about a dozen subjects at once. But often the talk was revealing, opening new perspectives on life as well as death.

EMMIE

Emmie* stares straight ahead as if at a drive-in movie. The windshield frames the kind of property advertised in the back pages of the New York Times Sunday Magazine. The six-bedroom house has enough angles and ivy to give it character. The yard grows into a meadow along the edge of a lake. Two blonde children in down vests chase each other in and out of view. The car windows are rolled up against the January chill and the ears of her children. I'm afraid to ask whether this car, the car she'll

*Asterisked names in this article are pseudonyms.

soon use to pick up a third blonde daughter from a piano lesson, is the car that two years ago, at age 47, she drove into a field at four a.m.; hooking up a vacuum-cleaner hose to the exhaust pipe, she zipped herself into a sleeping bag, stuck the hose in her mouth, and turned the key in the ignition.

When Emmie talks about it now—and she rarely does—she wonders how she could have reached that point, but the sadness in her voice suggests the forces aren't yet foreign to her. "It's easy be glib about it now," she interrupts herself, "but it wasn't that way at the time. I didn't realize what was happening to me. I was like a frog that can heat up so slowly, just adapting, adapting, and adapting. About the time he boils to death, he begins to wake up." She closes her eyes. "That's what happened to me, except unconsciously, I knew it."

Emmie had one of those "everything to live for" lives: good schools, an elite women's college, and a publishing job that sent her around the world from her Manhattan apartment. "But I had a few problems. I kept trying more things—flying an airplane, scuba diving, ballooning—to prove to myself I *was* somebody. Deep down, I didn't think I was very worthwhile."

She didn't marry till she was 36. Stan* was a paper manufacturer, divorced with two kids, a nice house in the suburbs, and a prefabricated life for her. "The whole situation changed," Emmie says. "I didn't have my job anymore, I didn't have my friends, I didn't have my name." She and Stan had three children, but as years passed, the marriage faltered. "I had trouble with Stan. He was not always taking my feelings into account." Emmie's voice rises in pitch but not volume. "I wasn't very assertive. I just let everything get eroded away, including my self-esteem."

When Emmie's father died, she was too busy with details of the funeral to confront his death. "I started to have what I now know to be anxiety attacks," she says. "It's as if you walked into the kitchen and your mother was on the floor in a pool of blood, you'd panic. I felt like that when the phone rang or I heard a car in the driveway." She saw several doctors, hoping for a diagnosis of physical illness. A psychiatrist told her she was having a normal reaction to her father's death—but that didn't stop the feelings. "I went through all the classic symptoms of depression," she says. "I couldn't sleep and would slip outside for walks at three in the morning. In the day I was doing the housework, but I wasn't able to concentrate. Our marriage had a lot to do with it—in many ways I was simply being ignored. I'd ask a question and the answers weren't forthcoming and it was as if I didn't exist. So I began to believe I didn't exist."

Emmie convinced Stan they should see a marriage counselor, but work kept him busy and appointments were irregular. "I had these wild dreams—'I'll get an M.B.A., then I'll be able to talk to him about business and he'll respect me.'" She took courses in the human-potential movement. "I kept thinking if I could just change, things would be better."

Things got worse. "We were on vacation when I first had a feeling I was going to kill myself. I was very vocal about it. I told Stan and I told my mother. They said, 'How terrible.' I saw a psychiatrist and I remember saying, 'I understand people who talk about suicide never do it.' That was my great out. I must be okay because I was talking about it. But he said, 'No, people who talk about it usually do it.' That sort of cranked something into my subconscious." Her doctor suggested the hospital. "I had wanted to go six months before," says Emmie, "but the marriage counselor pooh-poohed it. She thought I was too intelligent." The doctor also sug-

> *E*MMIE: "I wanted something that was final but wouldn't be messy. I didn't want to blow my head off . . . I kept thinking about what would be easiest for everybody else."

gested a meeting with her husband and the counselor. "The more I thought about it, the more I thought, 'They're going to gang up on me and I don't want any part of that.' That's when I really decided I was going to have the last word.

"You have mixed feelings when you come to that point because you're living half in this world and half in the next. I lived for a month in that double state, knowing I was going to do it but wondering when, and unable to organize my life because I didn't know whether I'd be dead or alive. It's curious—if someone calls you up for lunch, you think, 'No, I can't make it Wednesday, I'm going to be dead.'" She purses her lips. "I didn't have any control over it—that's the interesting part. It all seemed inevitable, as if I had no choice but to carry out the impulse officially.

"The difference between having thoughts of doing it and actually doing it is tremendous. It takes an amazing amount of energy to figure out how to end your life, if you have certain criteria. I wanted something that was final but wouldn't be messy. I didn't want to blow my head off . . . I kept thinking about what would be easiest for everybody else." She smiles. "Of course, the easiest thing for everybody else would be if I lived. But that didn't occur to me.

"I was very matter-of-fact about it. I tried it one night in the garage but chickened out. I got the whole deal set up and then said, 'Well, this is just a dry run. Now I know how to do it.'" Two nights later, her husband woke at four to find Emmie gone. The police discovered her at eight the next morning, curled in her sleeping bag, unconscious. She revived at the hospital. No one could figure out how she survived. Three weeks later, a friend realized the car had a catalytic converter, which screened out enough carbon monoxide to keep Emmie alive. She'd taken the wrong car.

THE X-FACTOR

In the last 25 years, more time, energy, and money have been devoted to the study of suicide than in the previous two centuries. A computer check on the literature yields a five pound, seven ounce pile of printout. "Every psychiatrist has to write a paper on suicide,'' jokes a psychiatrist who's written fourteen. The Encyclopedia Britannica gives the subject 25 times more space than it did 25 years ago. In the interim, an entire professional movement has sprung up: thousands of ''suicidologists'' now devote themselves to the study of suicide and its prevention, while thousands of trained volunteers tend hotlines at three hundred prevention centers across the country. Why, then, has the suicide rate continued to rise?

No one has a convincing answer. No one even has convincing statistics for the actual number of suicides in this country, because reporting procedures aren't uniform.* The accepted estimate for 1982 is 30,000, but some think 100,000 a year is more probable. Even the government-certified figures are only the tip of the ''suicide iceberg'': for every 30,000 suicides, runs the algorithm, there are 300,000 attempts, 3 million ''ideators'' (who threaten suicide), 30 million engaged in self-destructive behavior (from alcoholism, anorexia, and smoking to skydiving and compulsive gambling), and 120 million family members or ''survivors.''

Beyond the confetti of numbers, we know little about suicide, and less about preventing it. If we fail to understand why 30,000 people a year take their lives, it may be because we have such a limited understanding about one person's decision to do so. ''The problem of suicide cuts across all diagnoses,'' says John Mack, a psychiatrist at Cambridge Hospital. ''Some are mentally ill, most are not. Some are psychotic, most are not. Some are impulsive, most are not.''

Clinicians once tended to link suicide almost exclusively to depression, as if there were a threshold one could not sink below. This did not explain why many people who aren't depressed kill themselves, nor why the majority of depressed people do not take their own lives. For Aaron Beck, father of cognitive therapy, ''hopelessness'' is the key. Calvin Frederick, former chief of NIMH's Mental Health Disaster Assistance Program, goes two *h*'s further: ''helplessness, hopelessness, haplessness.'' Psychiatrist Joseph Richman, of the Albert Einstein College of Medicine, declares that ''suicide is almost always the expression of some malfunction in the family system.'' Says Harvard Medical School psychiatrist Leon Eisenberg, ''A deficiency of social connections.'' Biologically oriented clinicians talk of chemical imbalance: recent research has linked suicide to low serotonin levels in the brain.

Nonclinicians are also divided. Some in the clergy cite erosion of faith, while philosophers speak of alienation and existential angst. To sociologists it is ''lack of connection,'' the breakdown of the family, changing sex roles, unemployment. Yet ''if life, or life in our culture provides the stress, the vulnerability to a response by suicide usually has a lifelong history,'' writes psychiatrist Herbert Hendin. Some clinicians see a predisposition to suicide that may have its roots in a childhood loss, past

---

* Some coroners classify a death as suicide only when a note is found (about fifteen percent of all suicides); others report only hangings and overdoses, overlooking shooting, jumping, drowning, and a dozen other forms of exit that can be viewed as accidents. Single-car crashes are rarely included, though some claim that as many as five percent are deliberate.

suicide in the family, unexpected separations from parents, or even the quality of mothering during the first three months of life.

While it is often said that suicide may be committed by twelve different people for twelve different reasons, it may be just as true that suicide may be committed by one individual for twelve different reasons. No single theory can account for all forms and motivations of suicidal behavior, and there can be no single preventative. ''There is no pill for suicide,'' says psychologist Edwin Shneidman, whose definition of suicide may be the best we've got: ''Suicide is a biological, sociocultural, interpersonal, dyadic existential malaise.''

How do we prevent *that*?

THE PREVENTION MOVEMENT

When Emile Durkheim published *Suicide* in 1897, demonstrating that suicide rates varied with social structure and with social change, he did not hope to save suicidal people; suicide was merely a mannikin for sociological theory. When Freud suggested that suicide is anger toward a lost love object turned back on the self, his interest was largely abstract. Suicide was not seen as something that could or should be prevented until Baptist minister Harry Warren formed the National Save-a-Life League in New York City in 1906. About the same time, the Salvation Army founded the London Antisuicide Bureau. Both used lay volunteers as counselors. Physicians refused to treat suicidal people, who were believed to be insane, or doomed by heredity; suicide was a crime and a sin, and the medical profession did not wish to contaminate itself with such cases. Today, almost all medical authorities regard suicide as a medical problem.

''Call me Ishmael,'' says Ed Shneidman, explaining the nest of neckties, each sporting thirteen spouting whales, on the desk of his Manhattan hotel room. Coat and tie slung over his chair, shirt unbuttoned, Shneidman looks out at the Staten Island ferry as he waits for a call from his wife in Los Angeles. He's in town for the annual meeting of the American Association of Suicidology. On the bureau, a pair of dark glasses, a copy of the New York Times, and a box of Vanilla Wafers complete what could be any traveling salesman's still life, only Ed Shneidman is never still. He's antic as a roly-poly third base coach giving all his signs at once. ''Okay.'' He claps his hands. ''Now.'' Another clap. ''The *first* paragraph is entirely about suicide. It says 'Some time ago, never mind how long ago exactly,' and blah-blah-blah, 'when I feel as though I'm going to commit homicide—knock somebody's hat off—I consider it high time to take to the ship.' '' Shneidman squirms to the edge of his seat and back. ''This is his substitute for pistol and ball, he says, for putting a gun to his head. 'Cato, with a philosophic flourish, throws himself on his sword.' *He* kills himself, he says, *I* take to the ship.'' He furls and unfurls his shirtsleeves. ''The whole book is about suicide, direct and indirect. Ahab *makes* the whale kill him. Ahab *nudges* the whale: the whale wants to leave Ahab alone. Starbuck sees that, he says 'Look, fellow, you're making him kill you.' But Ahab *has* to pursue him. On the other hand, Ahab could not come back from that voyage. He had to be killed. A substitute for suicide—a subintentioned death.''

''Subintentioned death'' is one of dozens of words invented by Shneidman for the study of suicide. ''Suicidology'' is another. The growth of the field he named is difficult to separate from Shneidman's own career as a suicidologist—which began in 1949, when Shneidman, a psychologist working at the Los Angeles Veterans' Hospital, was asked to draft letters of condolence to the widows of two suicides. As preparation he visited the county coroner's office, where he found himself alone in an underground vault with two thousand suicide notes. He borrowed 721 of them and with psychologist Norman Farberow set out to see what last words might tell about people who kill themselves. Published articles led to grants, then to books, and then to the Los Angeles Suicide Prevention Center, a consortium of psychologists, social workers, psychiatrists, and trained volunteers, combining research and treatment. Opened in 1958, its guiding precept was that a suicidal person has a side that wants to live as well as a side that wants to die, and that a suicide attempt is a cry for help expressing that ambivalence.

The prevention movement gained federal sanction in 1966, when Shneidman was named first director of the Center for Studies of Suicide Prevention, a branch of NIMH: its goal was ''to effect a reduction of the suicide rate in this country.'' CSSP spurred the growth of prevention centers. There were three when Shneidman arrived in Washington; three years later there were 200 and counting. The centers offered caring voices and ears (by telephone or in person) and perhaps ''crisis intervention''—reality-oriented ''short cut'' therapy delivered by trained paraprofessionals.

Suicidology had been born. In 1967, CSSP sponsored a three-year postgraduate fellowship program in suicidology at Johns Hopkins. Social workers, sociologists, and psychologists took courses in crisis intervention, psychology of suicide, law and suicide, and did field work in prevention centers. In 1968, Shneidman promoted a ''supergalactic symposium on suicide'' in Chicago, with psychiatrists Karl Menninger and Erwin Stengel, and philosopher Jacques Choron (''the grand old men of suicidology''). The conference evolved into the American Association of Suicidology, an alliance of mental-health professionals, sociologists, clergy, and prevention-center volunteers. Shneidman was founding president.

Those were halcyon days. Regional centers were organized, symposia held, and ambitious ten-year plans outlined. At a 1970 CSSP summit conference in Phoenix, a task force proposed that government funding be apportioned among three groups: ''Discoverers'' (researchers), ''disseminators'' (playwrights, novelists, filmmakers, and journalists), and ''gatekeepers'' (clergy, police, bartenders, teachers, skid-row hotel managers, and anyone else who might usher a suicidal individual through a ''gate'' to prevention). The message was clear: suicide prevention is everybody's business.

But the growth of knowledge about suicide wasn't keeping pace with the zeal to prevent it. True, it was often pointed out that the rate of publications had more than tripled over ten years; in 1970, close to 500 articles

on suicide appeared, prompting that same CSSP task force to declare that ''we are far from a state of ignorance in suicidology.'' But a review of these publications reveals little of practical use. AAS journal articles included ''Lunar Associations with Suicide'' (reviewing four years of Cuyahoga County suicides, the authors found a slight increase during the full moon, which they could not explain) and ''Suicide on the Subway,'' which found ''important differences'' between those who lay in the train's path (''traumatic death'') and those who touched the third rail (''nontraumatic death''). ''One of the problems in the study of those who kill themselves is that the object of the study is deceased and hence not available for study,'' began an article on ''The Role of Spiritualism in Suicide.''

Research *is* limited by the fact that studies are post facto. ''There is very little writing about completed suicides,'' says John Mack, whose *Vivienne* was a recent exception. ''Clinicians are reluctant to write about it because it's so disturbing.'' Suicide is a statistically rare event, and it can take years to accumulate a statistically significant sample. Much research involves as few as six subjects, and the same case studies reappear like prize pupils, often to illustrate different points.

The movement's initial optimism underscored its apparent failure. In Los Angeles, site of the flagship prevention center, the suicide rate jumped from 14.8 in 1961 to 17.6 in 1965; in California, the rate rose from 15.9 to 18.8 during the Sixties. In the nation, the rate rose from 10.6 to 11.6. ''The whole idea was sold as though it was going to change the suicide rate,'' says psychiatrist Herbert Hendin. ''The trouble with the hypersell was that there was bound to be a reaction.'' The government, which in ten years had spent more than $10 million on suicide research, decided it was a bad investment. In 1975 the CSSP was dissolved.

Since then, ''the commitment of the federal government to the cause of suicide prevention has been minimal,'' said psychologist James Selkin, former president of the American Association of Suicidology, in 1982. ''A loss that approaches 30,000 lives per year is practically unrecognized.'' There is no federally funded research on suicide; organized efforts in prevention rest largely with AAS. But as an interdisciplinary organization, AAS reflects continuing conflicts in the field. ''Suicidologists'' are psychologists, social workers, psychiatrists, and others whose perspectives are often at odds. At one point, mental-health professionals and prevention-center volunteers threatened to split the association in two. But the most serious rift may be between members of the AAS and mental-health professionals who remain outside the movement. There is an unspoken feeling that the movement is too lay; it doesn't help that the AAS journal is scorned for its lack of academic rigor. ''I don't want to be described as someone who's *only* interested in suicide,'' says one psychologist. Some express embarrassment over the word *suicidology* and accuse the leadership of self-interest. ''The suicide industry is Shneidman's baby—he popularized it,'' says one well-known psychiatrist. ''Shneidman's empire,'' scoffs another, who has his own theories about Shneidman's devotion to the subject. ''People who think they know

THE MOVEMENT'S initial optimism underscored its apparent failure. The government, which had spent more than $10 million on suicide research, decided it was a bad investment.

me say I'm running from death," says Shneidman. "I don't think so. . . . If you'd let me discover a roomful of personal documents of alcoholics or delinquents, I would have spent my life studying alcoholics or delinquents." He looks tired; the next three days he'll be pushing suicide prevention in three states. "It was a virgin field. My brightness was in recognizing what powerful possibilities lay there."

THE NUMBERS GAME

The main approach to suicide prevention has been demographic: predicting who's going to do it by analyzing who has done it. And who does it? "Any of us, if hit hard enough," says Douglas Jacobs, former director of Emergency Room Services at the Cambridge Hospital. Some are more likely than others. Three times more men than women kill themselves, though women make three times as many attempts. Whites commit suicide more than nonwhites. Suicide rates increase rapidly in the teen years, peak in the twenties, taper off in the thirties and forties, rise rapidly in the fifties and sixties. Divorced men are three times as likely to kill themselves as married men. In general, two out of three suicides are white males. Other factors identified with suicide include alcoholism, homosexuality, and family history. Using these and other factors, a composite emerges, like a police artist's sketch, of an older, white, divorced, unemployed male who lives alone and is in poor health—the "high-risk paradigm" described by clinicians.

Using this information, researchers have devised scales to quantify suicide risk, so that clinicians will have something "objective" to help them in making decisions. Pointing out that test chestnuts like the Rorschach, TAT, and MMPI have been no help in predicting suicide risk, Aaron Beck developed "The Beck Depression Inventory," a "Hopelessness Scale," and a scale to measure the intentions of suicide ideators, with nineteen questions including "Wish to Die," (scored "moderate to strong," "weak," or "none") and "Wish to Live" (which offers the same options). Psychiatrist Robert Litman, reviewing 26,000 cases at LASPC, devised the "Suicide Potentiality Rating Scale," now used in both telephone screening and clinical interviews.

While these scales provide useful checklists for clinicians, playing "the numbers game," says Jacobs, can be dangerous. "You have to be careful. On a percentage basis, young people are less likely to kill themselves, but if one does, for that person, it's a hundred percent." The categories of risk that most scales use have become less reliable. Suicide has increased dramatically among the young; the rate for women is increasing faster than that for men; for young blacks, faster than for whites.

If the scales have not proved their worth, neither has clinical intuition. In a 1974 study, a computer proved more accurate than experienced clinicians in predicting suicide attempters. In addition, more than half the patients preferred computer to doctor as interviewer.

Ed Shneidman looks for other signs; he believes suicidal people communicate their intentions, but most people don't know how to listen. From "psychological autopsies" conducted with family and friends, he claims that four of five suicides give "clues" to their intentions, verbally ("You won't see me around much longer") or by their behavior (giving away prized possessions, putting their affairs in order). A sixteen-year-old Californian asked a teacher, "Do you have to be crazy to kill yourself?", wrote the word "death" on his hand, and one morning told a friend he was "going to heaven very soon." That day in class, he pulled a revolver out of a paper bag and shot himself. He was one of twelve high-school suicides in San Mateo County that year; most had left clues recognizable to "trained suicidologists."

"Education is the single most important item in lowering the suicide rate," says Shneidman. "I don't just mean suicide prevention classes. I mean a general heightening of awareness, so that if I give you my watch, you won't simply take it and thank me. You ought to say 'Ed, sit down, tell me what is happening.' " Shneidman advocates mass-media campaigns like that which helped 30 million Americans give up smoking in the past decade. "It's like V.D. and cancer," he says. "Education is more important than crisis intervention. I don't like putting out fires; I think it's more important to build a hotel where fires won't occur."

Yet even to trained suicidologists, clues are often recognizable only in retrospect—and in hindsight, almost anything can look like a clue. "I'm sure if you or I went out the window right now, somebody might say, 'I knew that was going to happen someday,'" says Harvard psychologist Douglas Powell. Powell did counseling work in 1975 when a Harvard junior ran through a dormitory window to his death during reading period. For weeks, the boy's friends wondered why, agreeing there had been no apparent reason, no clues. Then his roommate recalled one detail: Eric* had always set ashtrays, mugs, and postcards on the window sill. And, for several weeks before his death, each time Eric sat in front of that window, another object had been removed.

Clues are especially well camouflaged in adolescents. In *Vivienne*, the story of a talented, insecure fourteen-year-old who hung herself in her mother's silversmithing studio, psychiatrist John Mack lists warning signs that families and teachers should recognize: suicidal talk, moodiness, loss of self-worth, turning excessively to a diary instead of friends, failure in school, increased drug and alcohol consumption, hypochondria, and philosophical preoccupation with death and dying. But what looks like "presuicidal syndrome" to a psychiatrist may look to parents like "a phase."

People often recognize clues but fail to respond, through fear or ignorance. Nor is it known what proportion of people who leave clues go on to kill themselves.

"All the students come in at some point and talk about suicide," says one high-school social worker. "I can't put them all in the hospital." Her biggest difficulty is persuading parents to let her help: "I get screaming parents who say 'You're nuts—my child's fine, not depressed.'"

"It's often difficult to get parents to acknowledge the problem, because they *are* the problem," says Peter Saltzman, child psychiatrist at McLean Hospital. Even suicide attempts may have no effect. One of Saltzman's patients, following an attempt, was told by his father, "Next time, jump off the Bourne Bridge." Failure to get attention may lead to a "face-saving" success. Leon Eisenberg describes a college student having a turbulent affair with a classmate. "He said, 'If you don't go steady with me, I'll jump off this building.' She said, 'You don't have the guts.'" He did. "He ran right up the steps to the eighth floor, out on the roof, and jumped off," says Eisenberg. "And I might add the young lady showed no remorse at all."

TREATING THE SUICIDAL PATIENT

Extending treatment to all potential suicides is clearly impossible. "We've reached the point of no return in defining vulnerable populations. Finding that vulnerable person amounts to looking for the proverbial needle in a haystack," says Herbert Hendin, who knocks Shneidman's proposed educational campaign. "I don't follow the logic of putting millions into educating the lay public in something that psychiatrists haven't proven *they* can identify. It makes more sense to do something for the people you *do* find. A lot of seriously suicidal people present themselves in ways nobody can miss—they jump from five-story buildings—and nobody does anything for them."

Previous attempt is the highest predictor of suicide risk; between 30 and 40 percent of suicides have tried before, and about 10 percent of those who attempt will succeed within ten years. Yet most attempters, says Hendin, are returned to the community after brief hospitalization, without provision for follow-up treatment. "If you could even identify 20 percent of the seriously suicidal from those who make attempts, and cure 10 percent, you could literally change the suicide rate," he declares.

Can clinicians actually "cure" suicidal people? Patients who have made attempts and entered into treatment have the highest suicide rate of any patient group. Yet many in the prevention movement assume that once we tag people as suicidal, we can help them.

Suicide is a symptom, not a diagnosis. Because a clinician can't treat it, he must treat the "underlying illness"—the patient's closest diagnosable ailment. Some clinicians believe that if they successfully do so, they've treated the suicidal patient, as if suicidality were a nasty side effect of the underlying illness. Yet many suicidal patients have no diagnosable underlying illness, and patients often kill themselves shortly after coming out of depression—or long after depression has lifted. "Suicide proneness is primarily a psychodynamic matter; the formal elements of mental illness only secondarily intensify it, release it, or immobilize it," writes psychi-

atrist John Maltsberger. "The urge to suicide is largely independent of the observable mental state and it can be intense despite the clearing of symptoms of mental illness."

There's little agreement on how to treat any mental illness. A person suffering from some form of depression (an estimated 25 to 30 percent of suicides) may be treated with antidepressant drugs, shock treatment (making a comeback after decades of disrepute), cognitive therapy, Yoga, or any of more than 200 brands of psychotherapy practiced today. Of the 200, only psychoanalysis is agreed to be inappropriate for suicidal patients: "Most are either too anxious, too depressed, or just not well enough put together to stand it," says Hendin. Although suicidal patients come in different diagnoses with different needs, they're likely to get whatever the therapist stocks. "One would hope that clinicians had a number of strings to their therapeutic bow and would change depending on the nature of the problem," says Leon Eisenberg. "Unfortunately, this field is characterized by people who do a type of treatment for every customer

*H*ERBERT HENDIN: "I don't follow the logic of putting millions into educating the lay public in something that psychiatrists haven't proven they can identify."

that comes along." A therapy that works with suicidal patients will be ignored by most clinicians if it's not their *modus operandi*. Group therapy with suicide attempters appears to generate a caring bond which, like Alcoholics Anonymous, can extend beyond formal sessions to times of crisis. Family therapy has been effective, and some therapists won't see a suicidal patient without seeing the family. But many more worry about overstepping patient-therapist confidentiality.

There is no pill for suicide, but there are scores for depression. Most therapists supplement drug therapy with psychotherapy, or vice versa, but drugs are the *only* treatment method for some. "In state hospitals, community mental-health centers, and as practiced by welfare doctors, it is often the sole therapy," writes Jonas Robitscher in *The Powers of Psychiatry*.

Antidepressants have been a red herring in the treatment of suicidal patients. "They do nothing to alter the underlying vulnerability to suicide," says Maltsberger. The same drug that relieves depression may give a patient sufficient energy to act on his impulses. "It's a complicated issue," says Alan Pollack, director of McLean Hospital's outpatient clinic. "You have to judge whether patients can manage medication, how reliable they are in taking it, and how big a supply you can give them without their OD'ing on it." A 1970 study showed that of 200 people who committed suicide with barbiturates prescribed by physicians, over two-thirds had a history of previous attempts. Notes the study, "One might indeed wonder what kind of a nonverbal communication the suicidal person must feel he is receiving when he is handed a prescription for a potentially lethal quantity of drugs." . . .

SOCIAL STUDIES

A suicidologist visiting China asked her guide about suicide. The guide frowned and said, "Oh no, the state would not approve." In some ways, the United States does. Suicidologists point to a number of measures that might lower the suicide rate: gun control (only in the U.S. are guns the primary method of suicide); tighter controls on prescription drugs; availability of abortion and birth control ("An unwanted child is an unhappy child is a depressed child is a suicidal child," reasons Sol Blumenthal of the New York Department of Public Health); safety features—bridges without walkways, nets for observation decks, windows that don't open wide. Opponents of these proposals argue that denying access to one method will merely force the suicidal person to another, but this overlooks the impulsive nature of many suicides who adopt whatever means is at hand. When a plexiglass barrier was erected atop the campanile at the University of California, Berkeley, the campus suicide rate declined. Moreover, would-be suicides are often particular; many women, for instance, choose not to spoil their looks by jumping or shooting. Suicidologists feel that limiting access to lethal methods would lower the rate: from 1971 to 1976 barbiturate prescriptions declined from 40 to 20 million; the number of suicides by barbiturates was halved. Although suicides by other drugs increased during that time, careful prescription practices clearly had a measurable effect. When the suicide rate in England and Wales declined between 1960 and 1975, the drop was attributed to the changeover from coal gas to less toxic natural gas.

San Francisco has one of the highest suicide rates in America, and one of the most picturesque and lethal sites: the Golden Gate Bridge. Since it was built in 1937, 757 have jumped (21 jumped and lived). In 1973, when the 500th suicide was expected, a circus atmosphere of souped-up surveillance and media publicity prevailed. Fourteen attempts were foiled between numbers 499 and 500. Ten years and almost 300 suicides later, the plan for a million-dollar "suicide fence" has been scratched.*

While we cannot make the world "suicide-proof," nor our lives a 24-hour suicide watch, "we should not make it physically easy to commit suicide," argues Sol Blumenthal. Others aren't persuaded. "Ninety-nine percent of us don't need it," says one San Franciscan. "Is it fair to ruin the view for the sake of a few? If they want to die so much, why not let them?"

THE RIGHT TO SUICIDE

In 46 B.C., determined to kill himself before an advancing army could, Cato the Younger threw himself on his sword. He revived to find physicians dressing his wound. Shoving them away, he ripped off the bandages and died.

The right to suicide has been debated ever since; the recent appearance of at least four "how-to" manuals has added urgency to the discussion. "A Guide to Self-Deliverance," published by EXIT, a London-based organization, is a 10,000-word booklet offering four bloodless methods of suicide, and including a table of lethal doses for prescription and nonprescription drugs. Formerly the Voluntary Euthanasia Society, EXIT has expanded its credo to include not only the incurably ill but the right to suicide for everyone, and is now the Society for the Right to Die with Dignity. The book is for members only; membership is a matter of $23. *Suicide, Mode D'Emploi* ("suicide operating instructions") offers recipes for fifty lethal cocktails in its 276 pages. It sold 50,000 copies in the first five months after publication in 1982, made the French best-seller list, and has been linked to ten suicides. Efforts to ban it have failed; American publishers are jockeying for the rights. "I feel no remorse," said the publisher. "This is a book that pleads for life. But it also recognizes that the right to suicide is an inalienable right, like the right to work, the right to like certain things, the right to publish. What use is a right without the means to execute it?"

Courts are struggling with that question. In Ledyard, Connecticut, two young men sawed a shotgun barrel so their best friend, partially paralyzed in an accident, could more easily maneuver the gun to shoot himself. They now face trial. A 22-year-old Oklahoman distributed flyers offering to help people kill themselves if they would split their insurance benefits. Arrested before getting any takers, he faces five years in prison.

With increasing attention to death and dying, and the idea that a "good" death is as important as a "good" life, "suicide will gradually become the culturally sanctioned mode of death," writes Robert Kastenbaum, a former president of AAS. He suggests that suicide "facilitation" be linked to prevention, in one center.

Incidence of suicide in fifteen countries

	(per 100,000)	
	Males	*Females*
Austria	36.7	14.7
Switzerland	34.5	15.4
Denmark	31.8	19.8
West Germany	30.1	15.1
Sweden	28.3	12.9
France	23.3	9.9
Poland	22.8	4.2
Japan	22.6	13.6
Canada	21.2	7.3
UNITED STATES	19.0	6.3
Norway	17.2	7.1
Australia	16.6	6.7
U.K. (England and Wales only)	10.7	6.5
Israel	6.6	4.7
Spain	6.1	2.1

Source: World Health Organization, *World Health Statistics.*

* A study showed that of 1,440 persons prevented from jumping, only 4 percent went on to kill themselves by other means.

GUIDELINES

Like many psychiatrists, John (Terry) Maltsberger took on suicide in self-defense. When he was a resident at Massachusetts Mental Health Center, a woman patient chloroformed herself. Several of Maltsberger's patients were also threatening suicide. He began meeting weekly with Dan Buie, a young colleague who had lost a patient to suicide and felt equally concerned. They worked their way onto a committee whose charges included the "suicide review"; gradually, their combined perspective emerged in professional journals.

Ask about suicide care in Boston today, and Maltsberger's name is almost sure to come up. The man himself is Napoleon-sized, dapper, with a gracious accent an American might import after a week in London. Bright blue eyes peer from a full-moon face under silver hair. The consulting room on the first floor of his Beacon Hill town house is dimly lit but warm, with a working fireplace at one end, and at the other a long drawerless desk on which fifty books—including *The Savage God,* a staple of every therapist's library—are lined. French doors open onto a jungly garden that seems about to entrap Maltsberger in a three-dimensional Rousseau painting. He is explaining the "psychodynamic formulation" of suicide, which he and Buie evolved.

"It boils down to finding what a person has to live for," says Maltsberger. "Most people live for all sorts of things—friends, a special person, work—and if they lose something on one front, they pick it up on another. But suicidal people are quite deficient in any capacity to keep themselves afloat, on the basis of inner resources. Once somebody threatens suicide, you start looking at what resources the person has."

Maltsberger and Buie specify three areas people may live for: other people, work, their own bodies. "Obviously, when someone who is dependent and depressive loses a girlfriend or a husband, it can precipitate a suicidal crisis. There are people who never have relationships, who lock themselves in the library and devote themselves to scholarship. But when they retire, or can't work anymore, they may kill themselves. A surgeon may live only to operate; if he loses the use of his hands, he may do away with himself. And there are people who may be very dependent and depressive, but as long as they can jog and look in the mirror and say, 'Gee, I'm in great shape,' they can go on.

"So if someone has relied all his life on some capacity to work at Sanskrit, and he goes blind and can't do it anymore, the task becomes to find what this person can substitute as a life-saving activity. It isn't always possible. Many people are quite indifferent to the love of others, for instance. Others may be indifferent to success at work. Suicidal people are very specialized in what they will accept as a reason for living." At first, says

Maltsberger, the therapist may have to constitute that reason, "until the patient can regain his balance and stand up again." Maltsberger smiles. "It sounds simpleminded, but it really is like that."

The psychodynamic formulation offers therapists a practical way to decide *when* someone is suicidal, what to do for treatment, and whether hospital admittance or discharge is indicated. In their suicide reviews, Buie and Maltsberger found that such decisions were often based entirely on mental-state examinations, assessments of appearance, speech, and behavior—i.e., experienced gut reactions. "When this is the practice," they write, "it is inevitable that a certain number of otherwise preventable suicides will take place." Like one fifty-year-old widow who was released from a hospital after a course of treatment:

The psychiatrist and staff responsible for her care did not appreciate that she had moved into an attitude of quiet resignation and despair. . . . At no time in the course of her treatment had the patient seemed disturbingly depressed to anyone and there had been no suicidal threatening or preoccupation that the patient had made known. She destroyed herself after discharge when her son announced his plans for marriage. Only in retrospect was it noted that this dependent woman was without support in her social and family context.

The approach developed by Maltsberger and Buie requires a therapist to know patient histories well, and to spot events in a patient's daily calendar that might heighten suicide risk. "Treating suicidal people means being available—intensively—from time to time while they're between supporting figures or research projects," says Maltsberger. "I might call them on the telephone every day, perhaps go to their house . . . you have to be there waiting like a net, hoping that as time goes on the person can widen his repertoire and make room for other sustaining influences.

"People who grow up suicide-vulnerable have failed to get the love they ought to have had from their mothers," says Maltsberger. "My approach looks at suicide in terms of developmental failures that make it impossible to maintain a sense of self-worth." In normal development, he explains, capacity for autonomy increases with age, enabling one to endure degrees of loneliness, depression, and anxiety. Those who don't receive enough mother-love fail to develop sustaining inner resources; they must depend on external supports. When external supports fail, suicide is a danger.

Even the psychodynamic formulation offers only temporary relief. Can vulnerability to suicide be altered? Maltsberger sighs, like the Wizard of Oz after giving out heart, brains, and courage, only to find that Dorothy's still in search of a way to get back to Kansas. "That's most ambitious," he says slowly. "That means helping the patient restructure his mind. There are very few patients where that's possible." He pauses. "Some of us believe you can. Often, psychiatrists don't want to try."

"Suicidology would have two distinct aspects—a continuation of the present effort to understand, predict, and prevent self-destructive behavior, and a sensitive new approach designed to help people in certain circumstances attain the particular death recommended to them by cultural idealizations and their own promptings."

"Certain circumstances," for most advocates, means life-support systems and terminal cancer. (In fact, the suicide rate among terminally ill cancer patients is low; they "tend to cling to what life they have left," says Calvin Frederick.) Model cultures include those of the Greeks, who honored suicide, and the Romans, who promoted it to heroic proportions.

The elderly often figure in discussions about the right to suicide. Doris Portwood, in *Common-Sense Suicide*, writes, "When an older woman leaves a social gathering—perhaps an hour after dinner when younger guests are settling down to a game or a fresh drink—no one urges her to call a cab or offers her a lift. She will receive thoughtful words during the process of departure, but no insistence on her staying. There is the assumption that she has, in fact, some good reason for going." Portwood suggests that to decide when to leave, a person should draw up a list of pros and cons. Though "balance-sheet suicide" is an old concept and Portwood insists she's pushing "only for a right of choice—not trying to eliminate the old," some of her pros are cons. She writes of the high cost of subsidizing the elderly, and points out that by the year 2000 there will be more than 30 million American senior citizens. "Those who have the will to opt out may not (yet) get a public vote of thanks. But who can dare say that they will be missed?" she writes.

Concern with adolescent suicide obscures the fact that older people still make up the majority of suicides; white men over fifty, 10 percent of the population, account for 30 percent of all suicides. "What we tend to forget when we talk about rates is that some of the disparities reflect populations we do not like or do not respond to," says psychiatrist Seymour Perlin, citing skid-row alcoholics and the elderly, who are considered undesirable patients. "Too often the old are written off as treatable with pharmacology while younger patients get psychotherapy," said Herbert Hendin, addressing a sparse crowd on "Suicide over Sixty" at an AAS meeting (across the hall, "Adolescent Suicide" was SRO). "The heart of suicide prevention has always been, and remains, suicide among older people."

Underlying is the assumption that older people have less to live for. But the right to live with dignity may be as neglected as the right to die with dignity; if the first were looked into, the second might not seem so pressing. And what is overlooked in Portwood's own fantasy suicide of a peaceful plunge into a lake with a congenial group of friends (besides the consensus that drowning is one of the most difficult and painful deaths) is that many elderly suicides may not be so "rational." Suicidologists worry that easy access to methods and public approval of "rational suicide" may encourage self-destruction by some for whom it may not be a rational choice. Where do you draw the line? Who draws it?

"Most of us don't think about it very much," says psychiatrist Leon Eisenberg. "If a patient says, 'I'm going to kill myself,' we try to stop him." Yet there is the feeling that if someone *really* wants to die, he will.

"Killing yourself is not a great problem," allows Eisenberg. "There are buildings you can jump off, you can throw yourself on subway tracks . . . there's just no way to stop someone unless you lock them off in a corner, and even then people find a way.

"The reason most clinicians don't take the right to kill yourself seriously," continues Eisenberg, "is that all of us have seen people who have failed, been treated for depression, and are now grateful to be alive." (Research shows that 10 to 12 percent of attempters try again within a year; 1 percent succeed; and 10 percent will kill themselves within ten years.) Over the past thirty years, Hendin has interviewed four people who survived six-story jumps. Two changed their minds in midair, two did not, and only one tried again. "It's clear that suicide must not be a mature, thoughtful, highly motivated decision," says Eisenberg, "because once a person made that decision, there's not a damn thing we could do to stop it, because they could keep going after it, after it, after it. And they don't."

To clinicians, there may be no such thing as a "rational suicide," which to them precludes any "right." Countless studies purport to show what percentage of suicides are "mentally ill": Eli Robins in *The Final Months* goes as high as 94 percent. "The argument connecting suicide and mental illness is tautologically based upon our cultural bias against suicide," Zigrids Stelmachers, director of a Minneapolis prevention center, has said. "We say, in essence, 'all who attempt suicide are mentally ill.' If someone asks, how do you know?, the implied answer is, because only mentally ill persons would try to commit suicide." A study at Harvard found that the highest estimate of mental illness when a sample had been diagnosed *before* suicide was 22 percent. Afterward, the highest estimate was 90 percent.

Finding suicidal people mentally ill has practical implications. In recent years, efforts by civil libertarians to abolish involuntary commitment has led to a maze of different standards in different states. Most statutes specify that the individual must be considered dangerous to himself and also mentally ill—criteria determined by the admitting psychiatrist. This gives psychiatrists another reason to staple suicide to a traditional diagnosis. But there is a less tangible, perhaps more dangerous consequence. Writes psychiatrist Thomas Szasz:

In regarding the desire to live as a legitimate aspiration but not the desire to die, the suicidologist stands Patrick Henry's famous exclamation "Give me liberty or give me death!" on its head. In effect, he says, "*Give him* commitment, *give him* electroshock, *give him* lobotomy, *give him* lifelong slavery, *but do not let him choose death!*" By so radically invalidating another person's (not his own!) wish to die, the suicide-preventer redefines the aspiration of the Other as not an aspiration at all. The wish to die thus becomes something an irrational, mentally diseased being displays, or something that happens to a lower form of life. The result is a far-reaching infantilization and dehumanization of the suicidal person. . . .In short, the suicidologist's job is to try to convince people that wanting to die is a disease.

"One of the underlying assumptions in a lot of this work is that there are 'sickies' and 'wellies'—that I must be a 'wellie' but you look sort of like a 'sickie,'" says

Merton Kahne, psychiatrist in chief at M.I.T. "A lot of people in my trade spend a lot of time trying to demonstrate how sick you are, as if you must be sick or you would never kill yourself . . . which I think is nonsense." He adds, "I don't know what a suicidal person is—I thought virtually everyone who's gotten beyond the seventh grade had thought about killing themselves." Calling suicidal behavior "sick" may help distance us from an act that strikes a disturbing chord.

"I remember dealing with my first suicidal patient. I found it very difficult to understand that a person could really choose this," says Nancy Kehoe. "I had to go out for a long walk and try to take in how much pain that person must feel to want to take his own life." Now, dealing with suicidal patients, "I let myself get in touch with the times I've felt pretty desperate, the fleeting moments of driving down the Mass Pike and wishing a truck would hit you. We've all had those moments where we say 'Enough—I can't take it anymore.'"

Perhaps in an ideal world, people would not want to die, but as Stelmachers says, "Some of the things that happen to these people give them pretty rational reasons for ending their lives." If the cry for help can be translated "help me live," it can also be translated "help me die." "A totally open existential therapeutic relationship must make room for everything, including suicide," writes philosopher Peter Koestenbaum. "Only in such a way can the freedom of the patient be recognized and nurtured." Writes Robert Neale, "Suicide can be prevented only by permitting it." Making room for suicide does not mean a clinician must set up facilitation services in a prevention center, or refuse treatment to a ten-year-old who's tried to hang himself, but that he must respect the possibility of suicide at least as much as he fears it. "If the person says, 'I'm going to kill myself,'" says Stelmachers, "one way to respond is to say, 'Well, maybe suicide is the best way out for you, but let's talk about it first.' This says many things—first, 'I am really interested in you and your problems.' It also negates a sneaking suspicion he might have had about himself, that he must be crazy to even consider such an act."

At a recent Harvard conference on suicide, the ethics of prevention came up: "Do we have the right to say no?" wondered an audience member. There were appreciative chuckles; it's the oldest and, by clinicians, least seriously discussed topic in suicide. "Tough question," said John Mack. "Shall we refer that one to God?" More chuckles. "We have a right to take a different position," continued Mack. "Our responsibility as clinicians is to choose life." Someone else spoke up: "I think the philosophical answer is different from the clinical one." Until they are part of the same answer, the study of suicide and its prevention may never be complete.

THE UNANSWERED QUESTION

It was the last day of the fifteenth annual meeting of the American Association of Suicidology. More than 500 suicidologists from dozens of states and countries had gathered at the Vista International Hotel in New York City for a four-day smorgasbord of workshops on "Suicide and the Big City," "Women and Suicide,"

"The Question of the Right to Die When Pain is Very Intense," and 55 other topics.

A who's who of suicide had assembled. Norman Farberow and Ron Maris were there. So were Herbert Hendin, whose *Suicide in America* was about to appear in bookstores, Ari Kiev, and Nancy Allen, the public-health worker who was instrumental in organizing the first "National Suicide Prevention Week" in 1974. And everywhere you looked was Ed Shneidman, peripatetic ringmaster of the suicide prevention movement in America. Heady company; at one point, twelve past presidents of the AAS sat at the dais. Their combined efforts represented over a hundred books, a thousand articles, and more than 200 years of experience in the study and prevention of suicide.

Now, while volunteers took down posters in the lobby (a blank brick wall—"suicide is a dead end") and the silver-haired proprietor of the Thanatology Book Club ("Save time, save money, receive a free book just for joining") closed up shop, the day's first meeting was getting underway downstairs in the Nieuw Amsterdam Ballroom. It was nine o'clock. Less than a third of the registrants were in attendance. Some were recovering from a "Backstage on Broadway" tour arranged by the entertainment committee, while others opted for last-minute sightseeing or confirming flights home, rather than this session on "Borderline Personality Disorders and Suicidal Behavior."

Grisly fare for a Sunday morning. Several people in back slept through presentations by two mildly eminent psychiatrists. (My notes are hieroglyphs: "central organizing fantasy of narcissistic union" and "objective scrutiny of object relations.") As Otto Kernberg, who pioneered the study of the borderline patient, read a dense, theoretical paper, a group of psychiatrists in front gazed up with adoration and a prevention-center volunteer joked about marketing the speech as a sedative.

When Kernberg finished, the moderator, a young psychiatrist who'd been alternating pensive nods with glances at his watch—it was his job to herd everyone upstairs in time for "Is There Room for Self-Help in Suicide Prevention?"—invited questions. Hands shot up in front, and their owners raised progressively complex issues. But a hand in back, belonging to a shabby fellow with a ponytail, persisted. And the moderator finally gave in.

The man stood. His jeans and flannel shirt were worn but not dirty. His ruddy face couldn't decide on a beard or a shave, and his eyes were as cloudy as his question, a stammering ramble proposing meditation as a panacea for suicide. Eyes started to roll in the audience, and there were tolerant chuckles. The moderator flashed the panel an embarrassed collegial smile. When the ponytailed man slowed for a moment, the moderator broke in: "That's an interesting question, but let's move on. We've got time for one more." He looked for another hand; the man remained standing. The moderator began his thank-you-very-much-I'm-sure-we-all-learned-a-lot speech and the man was beginning to sit, bewildered, when Kernberg reached for the microphone, said "I'd like to answer that question," and in his textbook Viennese accent, began responding with care and respect.

Psychic Surrender
America's Creeping Paralysis

Michael Scott Cain

She was twenty-nine, devoted to her husband and their year-old baby and enthusiastic about her career teaching math at a community college. But one night, she drove to Annapolis's Bay Bridge and jumped off. Though I had known her only casually, the news of her death was still shocking enough to make teaching research papers seem unimportant, futile. I told my class what had happened and sent them away.

Within moments, three of my students arrived in my office. The leader, a dark-haired girl with thick glasses, said that they were worried about me. They wanted to make sure that I was all right.

"It seems bad, I know," she continued, "but things work out for the best."

"That's right," her chubby, blonde friend said. "Tell him what you were telling me, Barbara."

Barbara slid her glasses up her nose. "I used to be suicidal. A year ago, I was into drugs, drinking really heavy. Just throwing my life away. I didn't have a thing to live for and I was always thinking of killing myself and just getting it over with. Now I'm a different person. Since I gave myself

to Christ, everything's different for me. I'm not alone anymore. Christ is right beside me all the time. I don't do drugs or anything anymore."

I looked at her, wondering what kind of person would use another's death as an occasion for a testimonial. The happiness she spoke of wasn't reflected in her face or her slumped posture. In fact, she looked defeated.

I was in a room with three cases of psychic surrender, a kind of creeping zombiism peculiar to our times that needs to be more carefully examined.

Psychic surrender surfaced again at a poetry workshop I teach in a state prison. A balding man in his late thirties, with hawk's eyes and an elaborate daisy chain of tattoos on his arms and neck, shuffled up to put a poem in my hand. Usually he sat with his hands folded on his notebook, staring at the rest of us as though he was not quite sure the language was English. His poem, six lines of religious doggerel, had as its final couplet, "and while he hates what we may do/God's love for us is always true."

"I got my head turned around a while back," he

Chances are you know someone who has totally abdicated
control of his or her life to someone or something.
The extent of such surrender is becoming alarming.

 This article first appeared in the HUMANIST issue of September/October 1983, pp. 5-11, 32, and is reprinted by permission.

said. "I saw how big God is, how powerful he is, and how little and unimportant I am. If I'd given myself to God a lot earlier, things wouldn't be like this now." He indicated the stone room with barred windows. "I can't handle all this, you know? It's too much for me. But God can handle it."

Psychic surrender again.

Psychic surrender encompasses born-again Christianity but goes well beyond it. Aggressive fundamentalism has been around as long as this country has, and the current revival it is undergoing simply repeats an established pattern: fundamentalism gains strength in troubled, insecure times. But this time around, there's a difference. Psychic surrender, a malady that's becoming progressively stronger in many aspects of our culture, can be defined as the act of turning complete control of your life, all responsibility for your total existence, over to someone you perceive as stronger and more capable than you. Both Barbara and the prison poet described their conversions as a surrender, a rejection of their own lives. And although the charismatic fundamentalist movement is the most visible example of psychic surrender, it is by no means the only one. In addition to radical perceptions of God, our people surrender to trendy therapies, gurus, and top gun psychics of the "New Age."

AN ABDICATION OF LIFE

Although psychic surrender looks superficially like just another aspect of the search for a better way to live—a search that has occupied us for as long as we've been on the planet—psychic surrender is a recent twist. Only for a little more than a decade has a sizable portion of our population decided that the search isn't worth it and turned their lives over to others.

It's important to realize before we go further that all excursions into religion or New Age wisdom do not represent psychic surrender. A decision made from a position of strength to explore these areas is a positive step. But to turn to them out of weakness, from a stance of insecurity and fear, which leads to an abdication of life, can in no way be seen as positive. It places responsiblity for a life into the hands of someone else; the goal is not to seek answers but to put an end to the questions.

How does psychic surrender work? On "The PTL Club," a born-again "Tonight Show" complete with host and fawning sidekick, the wife of a former cowboy star,

herself once a queen of the cowgirls, made a statement that contains the key. Her life had been a shambles, sheer turmoil; she had been unable to find the handle. Finally she lost control completely. "I can't take it, Lord. I can't do it," she said, "*You* take it." From that moment on, from the very second that she decided to "give my life to God," she experienced nothing but bliss, she says.

The key is in the language. One does not accept God, acquire religion, or gain a set of beliefs. One quits. As Watergater Charles Colson, faced with disbarment and prison, put it:

"God, I don't know how to find you, but I'm going to try. I'm not much the way I am now, but somehow I want to give myself over to you." I didn't know how to say more, so I repeated over and over the words *"Take me, take me."*

The language of New Age Eastern religion converts is similar. Rennie Davis, former anti-war activist and social reformer who became a follower of the then teen-age wonder guru, Maharaj-ji, described his relationship with his leader this way: "All I want to do is put my forehead right on his boot for as long as he'll let me. Sometimes he lets me."

Davis's submission might be melodramatic, but among the faithful it is not extreme. A friend of mine attended a weekend seminar with another visiting guru. When he was ushered into the guru's presence, he looked into his eyes and "fell in love."

The guru said, "Why is your hair long?"

Gurus often ask nonsensical questions, designed to get their students to look at themselves from fresh perspectives. The obvious answer to this one would be "I prefer it that way." The zen answer to this might be "why is yours short?" But my friend, starry-eyed in surrender, was incapable of introspection; he could only react literally. He cut his hair.

When they next met, the guru asked, "Why did you cut your hair off?"

This case shows that, once surrender occurs, critical thinking is thrown off as though it were spoiled meat. Therefore, the charismatic personality's hold over his zombies can strike rational people as evidence of madness. Once-normal people delight in begging in airports for sixteen hours a day while others work for slave wages—or for no wages at all—in their master's restaurants. Converts grow suspicious of the outside world, viewing it as a foreign nation, a threatening place good only for quick infusions of cash and a storage shed for souls that need saving. The converts are as unconcerned about the wealth their masters accumulate as they are about the rest of reality; they simply do not perceive the Rolls Royces, Lear jets, and lavish estates. If you call it to their attention, they express naïve inability to understand.

An example of such naïveté on a massive scale occurred recently when hundreds of thousands of followers of the Bhagwan Shree Rajneesh descended on what used to be the small town of Antelope, Oregon (only a handful of the original residents are left after the Bhagwan's followers took over the town council and, among other actions, raised taxes), over the Fourth of July weekend just to see their leader ride by in one of his many Rolls Royces, the purchase of which was made possible by the material renunciation of his followers. When asked by "Sixty Minutes" reporter Ed Bradley why the Bhagwan needs or desires these kinds of material embellishments himself, the Bhagwan's secretary and business advisor answered, "Why not? Such things are not important." The acceptance of such answers by this guru's mostly educated flock illustrates plainly the extent of their surrender.

This zombiism leads converts to some irrational positions. In 1975, Chogyam Trungpa, a Tibetan Buddhist who, when the Chinese came in, fled his country to teach his brand of "crazy wisdom" in Boulder, Colorado, got a little carried away. After biting a hole in a woman's cheek as a warm-up, he led his core of elites in the stripping and beating of poet W. S. Merwin and his companion, Dana Naone. The poets Allen Ginsberg and Anne Waldman, followers of Trungpa, led a cover-up of the incident. They seemed to find nothing wrong with Trungpa's behavior and implied that Merwin was at fault because he and Naone were "sexually exclusive."

As it is with his other disciples, Trungpa's hold on these two poets is total. Ginsberg showed how incapable he is of critical thought when he defended Trungpa's actions in an interview with poet Tom Clark by saying that the master was challenging the foundations of American democracy by setting up "an experiment in monarchy." Ginsberg, once a respected fighter for personal freedoms, is reduced by psychic surrender to an apologist for a psychic dictatorship.

SECULAR SIDE NO DIFFERENT

The secular side of the New Age is no different. People rush from one sixties-derived therapy to another. Transcendental Meditation gives way to Lifeline, Scientology to *est*, rebirthing to primal therapy—each promising to effortlessly transform us, to open us to peak experiences, bliss, and joy, instantly, without work. We can get it together, find our space, be whole, get it on with ourselves.

But, say the therapists, our thing is different, not at all the same as giving yourself over to a guru or an electro-evangelist. In some cases, they are right. But in many more cases, we surrender not to the movement but the charismatic figure who leads it. *Est* has Werner Erhard, former door-to-door encyclopaedia hustler and used-car salesman. Around one hundred thousand people have learned to experience their lives and get "it" from Erhard, whose personality is to *est* what cheese is to pizza. Scientology offers L. Ron Hubbard, whose public-relations bio changes from week to week while he sails the seas in search of a country that will admit him. Co-counseling is dominated by its founder, Harry Jackins, a labor organizer from Seattle. And can we imagine primal therapy without Arthur Janov?

The faithful's dependence on the master is so total that disciples can no longer make their own decisions. R. D. Rosen, for example, reports an incident in which a man stood up at one of Werner's sessions and said, "I'm a fifty-three-year-old psychiatrist, and, listening to you, I'm beginning to think I should give up my practice. What do you think?"

The search for a person who can live your life for you also encompasses another aspect of the New Age: the psychic. As Americans rush from guru to guru, looking for someone to whom they can surrender, they invariably touch base with the psy-

Sun Myung Moon marries 2,200 couples: an end to personal choice.

chic circuit. Today, just as these religions and therapies do, the psychic field offers organized churches and organizations, with memberships as committed as Moonies, as off-the-wall as Scientologists. These groups offer half-baked metaphysics, other-worldly helpers, and counseling programs. They play to the same insecurities that lead to psychic surrender. One California-based church states in an ad:

Your days of spiritual loneliness can be ended and your "Quest Eternal" answered. So many questions. Who am I? Have I really lived before? How to discover my real mission? . . . How can I end my karma and gain liberation? How can I find my way out of the labyrinth of my present life?

Another promises nothing less than to "uplift and improve *all* areas of your life."

Can't afford to migrate to a center? Most of them, such as Paradise, California's Magi Center, Inc., offer mail-order tapes, lessons, and memberships "that effectively change your life to whatever you desire."

All are led by the same type of charismatic individuals who dominate the religious and therapeutic groups. The Reverend Brian Seabrook "can reveal the karmic laws of right love, new prosperity, and natural health on all levels of self-expression." Clark and Die Wilkerson of Cosmic Wisdom "guarantee satisfaction" with their lessons on how to make more money, find more love, and expand your soul.

When frightened clients find a psychic who fits their views of the future, they form a bond as tight as Ginsberg's with Trungpa, a dependency as strong as a junkie's on smack. "You have to very gently break their former ties," one psychic told me. "I tell my clients they have to live their own lives."

But most clients, she says, don't want to live their lives. They come to her to give up those lives, to lay them in her more capable hands. The heavily advertised organizations and churches gladly give them every opportunity. They promise miracles, the end to problems, the easy answer.

Certainly the easy answer pervades the psychic books. They emphasize instant relief, new power, and autonomy, without effort. You can overcome your problems, they say; just follow my easy steps.

Suffering from a lack of sex appeal? No problem. Follow *The Magic of Ishtar Power*:

We must recognize that sexual attractiveness may seem to start with the physical, but a good 80 percent of it comes into play with the subtle, often subliminal, energy field that is generally called your aura. When your aura is giving off the special Ishtar Charismatic Energy, literally everyone who comes within ten feet of you will feel titillated, excited, and indeed "turned on" about you.

Need money? Frustrated in your career? *Telecult Power* has the answer:

Many a low-paid worker has risen to great wealth simply by "tuning in" on the mind of his bosses and superiors with Telecult Power. Taking an example from my experience, I was able to double my salary and get rapid promotions with practically no effort at all, simply by tuning in on my bosses' minds. . . .

Why are these books popular? Simply because they offer a promise to desperate people, offer control to those who feel their lives careening away from them, those whose lives are increasingly in despair.

And this desperation is one key to psychic surrender. In their book, *Snapping*, Flo Conway and Jim Siegleman state that under stress people often undergo rapid and unpredictable personality changes. Eldridge Cleaver finds God, Patty Hearst becomes Tania, a group of vacant high school dropouts become instruments of ex-

Hare Krishna school: repeat after me.

ecution for Charles Manson. More subtly, a woman gives up a promising career, consigns away her property, and becomes a member of some weird cult. Conway and Siegleman point to the problem, but psychic surrender goes beyond "snapping." Psychic surrender is a form of symbolic suicide: you lose your life by giving it away. And more and more Americans are "hanging it up" in this manner. Why do they suddenly feel they are incapable of living their lives?

SYMBOLIC SUICIDE

A main reason is that they have discovered their own sense of nothingness and cannot handle it. Michael Novak says that, because of the experience of nothingness, the individual no longer has any self or identity: "In a society like ours, he must constantly be inventing selves." Nothingness, according to Novak, dissolves the pragmatic solidity of the American way of life.

Many of us have felt nothingness. At one time or another we have jerked awake at four o'clock in the morning with the terrifying certainty that we are empty, inauthentic, purposeless creatures simply waiting for final inevitable death. While many of us face the problem and get on with our lives, others are unable to. They see the nothingness at the core of their lives and surrender.

Why are some prone to chuck it while others persevere? Hans Selye's General Adaptation Syndrome offers a clue. When we are hit by stress, our bodies prepare physiologically and psychologically to either fight or flee. But for most of the stress we are likely to encounter in the modern world, the fight-or-flight response

is inappropriate; we can neither fight nor run from a deadline for completing a piece of work, for example, nor can we use either response for the monthly bills. This being the case, we try to adapt, knowing that we cannot be kept continuously in a state of alarm. If the condition cannot be adapted to, exhaustion follows; if we are constantly exposed to the wear and tear of stress without being able to relieve it, we will die. The sense of nothingness removes purpose, meaning, and value from our lives. We can't live so empty a life, and stress results. Psychic surrender is a last desperate attempt to remove the stress. In psychic surrender, we choose a symbolic rather than a literal death. The ultimate effect is the release of stress.

Religion, be it traditional or radical, can provide meaning, purpose, and value. So can the new therapies and the psychic arena. It is a mistake to see every embrace of these areas as a case of surrender. To find meaning in life, integrate it, and use it to develop or refine a value system can be good for us. But that goal simply cannot be accomplished by giving control over your life to a charismatic character. To go into these areas positively, strongly, while retaining and enhancing your identity, is to grow. To retreat into them is to surrender, and surrender can no more be growth than illness can be health.

The question, then, becomes why is the stress of nothingness so prevalent in our culture? Why do we feel so alone and lost? Traditional religion has been in decline since World War II. With its withdrawal as a major force in our lives, we are left with two main socializing agents: the schools and the job scene. Regardless of their intentions, the major effect of each of these institutions is to create nothingness. To understand how this happens, let's take the five factors Michael Novak has identified in the experience of nothingness and relate them to schools and jobs.

Boredom. Schools seem to specialize in producing boredom. Anyone who has seen a group of fifth graders struggling to stay quiet in their seats while a well-meaning teacher drones from the front of the room knows how much of what goes on in our schools is soul-destroying boredom. Children are told where to sit, when to work, and what tasks to perform. Because so much time and energy is devoted to keeping them bored, two important lessons are implanted: the students are not important and the work they do bears about as much resemblance to reality as Wonder Bread does to food.

In psychic surrender, we choose a symbolic rather than a literal death. The ultimate effect is the release of stress.

On the job, the situation is much the same. Rollo May tells the story of a bus driver in New York City who got so sick of driving the same streets that he ordered all his passengers out of the bus and drove it to Miami Beach. When he returned, the bus company tried to fire him, but the outcry—by those who also knew something about boredom—was so strong that they relented.

The collapse of a strongly inculcated set of values. To fully explore the opportunities of our own lives, we must have values, yet our schools are deliberately value-free. The emphasis is on facts, with no attention given to the value perspectives from which those facts originated. Students learn to memorize but not to test an idea against their own important beliefs so that they can intelligently accept or reject it. Indeed, many right-wing groups are attacking the entire idea of teaching either values or concepts in the schools.

The workplace does have a strong value, of course: profits. Workers, doing small and routine tasks for which they cannot even visualize the final goal or product, quickly realize that they are valued only to the extent they contribute to the black ink on the ledger book.

Helplessness. Students have no control over their own education. Thrown by law into huge institutions that are run for the convenience of the administrators, they learn very quickly how little voice they have over the proceedings. When a child has to have permission and a pass to go to the bathroom, he or she knows what helplessness is.

Employees feel their lack of power also. A coal miner or steel worker who knows that he or she may die in an accident is helpless. Also powerless are those who work with chemicals which they know cause cancer. Steel workers and auto workers see their job fields dying and know that they can do nothing about it. Not even white collar workers are safe. They see huge monolithic corporations hurtling

235

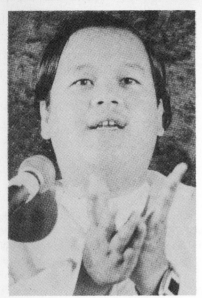

Teen-age wonder guru Maharaj ji: evidence of madness?

forward of their own bulk without control of direction or speed.

To be helpless is frightening, but even worse is to look at those who wield power and to see that even they are powerless to make any long-lasting legitimate changes. The result is to be on a treadmill to nothingness.

The betrayal by permissiveness, pragmatism, and value-free discourse. Young people realize that they truly need to know how to tell right from wrong, true from false, excellent from shoddy. But the critical thinking and valuing processes needed to make these distinctions are not taught in schools. To do so would challenge students, and that is as unacceptable as nudity. Because our educators have a pragmatic, can-it-be-measured approach, they are skeptical of concept learning. They also undervalue the skills and interests of students.

A teacher who offers educational opportunities to his students does so at his own peril. In Montgomery County, Maryland, a teacher was fired for asking his high school English class to read Aristotle to compliment a lesson on tragedy. He was demanding too much of them, the school board insisted when they dismissed him—even though his students testified that he wasn't.

Critical thinking and valuing are not welcome on the job either. That which produces profits is good; falsehoods are used to sell goods and the products are often deliberately shoddy. Manufacturers call it planned obsolescence. Consumers call it a rip-off. The person who makes his or her living producing such goods isn't permitted a voice.

The inevitable result, both from school and work, is cynicism, a deep feeling that nothing is real or worthwhile.

The drug experience. Many drugs are used as casually as coffee in our culture, mostly as a result of the four items listed above. But drugs have a strong, unanticipated effect: for some, they open the mind to the *possibilities* of wholeness, a unity of self with nature and environment, and the certain knowledge that life offers deep love and deeper meaning. They offer a glimpse. But this alone is an empty glimpse, and many of our citizens cannot cope with that glimpse in a drug-free life. They know from the glimpse that the heights are there and are possible to achieve. And having once tasted the joy, they are unable to live without it; joy is as essential as water. Being unable to achieve it results in the experience of nothingness.

We've seen, then, that the two major factors upon which we rely to create fully functioning members of our society create instead the nothingness that leads to psychic surrender. Schools and jobs often destroy both individuality and the social bond, causing our lives to deteriorate in ways we could not have previously imagined. Suddenly we live a lie. The syllogism goes this way: if my life is a lie and I am part of society, then society is a lie. And if I am part of society and I am crumbling, then society is also crumbling.

A SEARCH FOR LESS

And so our people surrender. Psychic surrender is a last-ditch, desperate act. But before we condemn the people who choose it, we might ask what alternatives its victims have. Having dangled from the end of a rope for a lifetime, they made the only decision they could conceive. A person on the edge doesn't ponder how he got there; he simply seeks a way to safety. And certainly psychic surrender is safe. After giving up, you needn't think, question, wonder, or feel. You need only accept the master and his words and reject the rest of the world.

But safety isn't a major goal of life. It's one of the basest of human strivings and, as we shall see, doesn't even satisfy surrenderers for long. We crave more because we sense there is more.

By the time surrender occurs, however, the search is no longer for more but for less. Nothingness has made living a

> *We need to see that the very complexity of life is a source of its beauty, rather than a source of anxiety. We must discover the ingredient that surrenderers lack: self-responsibility.*

forlorn exercise in despair, which the surrenderer desires simply to escape. By the time surrender occurs, therefore, it just might be too late for alternatives. The time to search for alternatives, then, is before surrender occurs.

To escape the possibility of psychic surrender, we need simply to change. If what we are doing leads to nothingness, we should stop what we are doing. If we have no purpose, realness, or value, we must acquire them. The question, of course, is how. Sociologists and philosophers have long said that ours is an other-directed society. Both by our job training and our education, we are taught that what we need to thrive comes from outside us. A refrigerator is acquired by going to a discount house, religious values are picked up by going to a church, purpose comes from a job, and problems are solved by consulting an expert. None of this is true. Both strength and purpose come from within and cannot be injected like a flu shot by an outside agent.

If we are to escape surrender, then, we must learn—and we must teach our children—to discover the inner strength that gives us the will and self-responsibility to go on with a life that very often is going to be anxious and seemingly empty. We need to see that the very complexity of life is a source of its beauty, rather than a source of anxiety. We need to become inner-directed, authentic. We must discover the ingredient that surrenderers lack: self-responsibility.

Authenticity and purpose come from self-responsibility. Responsible people, who feel authentic in good times and bad, do not need to give up their lives. But responsibility requires effort, work, and will—qualities which our society doesn't value highly and which the people to whom we surrender insist are not necessary. This quality also grows out of a sense of self-identity, which is profoundly lacking in psychic surrenderers. Most are as other-directed and lacking in identity as Adelaide Bry, a psychologist who "peaked" in an *est* session: "The high point of the weekend was when a man in charge of logistics said to me, after I'd mapped the shortest and most efficient way to the restrooms, 'Thank you Adelaide, you've done an excellent job with these instructions.' Wow! I was high for hours."

People as hungry for self-validation as Adelaide Bry are not unique in America. The nation turns them out as easily and as quickly as it does TV dinners. As long as this is so, psychic surrender will result. If we are to avoid psychic surrender, we will have to alter both ourselves and our institutions. We must emphasize purpose, self-identity, and responsibility.

Since self-responsibility allows us to recognize the experience of nothingness, it can help us fight it. When he was in his early thirties, psychologist Rollo May caught tuberculosis. At that time, there was no cure; for a year and a half he did not know whether he would live or die. He followed his doctor's advice, which was to rest and "give my healing over to others."

The results were dismal. "I found to my moral and intellectual dismay that the bacilli were taking advantage of my very innocence. . . . I saw that the reason I had contracted the disease in the first place was my sense of hopelessness and defeatism." Only after he renewed his purpose and sought authenticity did May turn the course of the disease around: "Not until I developed some 'fight,' some sense that it was *I* who had tuberculosis, did I make lasting progress."

But many Americans, lacking authenticity and feeling that the self is one more item to be bought with a VISA card, are unable to develop fight. They can only retreat to the safety of surrender. And, as I stated earlier, safety proves not to be enough. The self-society destruction syllogism won't allow a safe rest. It insists that the longed-for death come. Perhaps this explains the death wish so prominent in many cults and born-again sects. Hal Lindsay's widely quoted *The Late Great Planet Earth* and its sequels are devoted to the coming end of the world. Lonnie Frisbee, a San Francisco born-again leader, says the Six Day War, in which Israel regained lost territories, signaled the beginning of the end. The Book of Joel, says Frisbee, prophesies that the earthly return of Christ will be preceded by a great outpouring of the Holy Spirit. The war and the subsequent rise of the born-again movement among West Coast kids were, to him, the great outpouring Joel predicted. Christ's return, incidentally, will be accompanied by apocalypse; he comes not to save us but to destroy us.

New Age and occult wisdom see the same apocalypse coming but from an ecological base. To them the earth is a sick creature which will erupt in an attempt to heal itself by shaking off its parasites—us.

And, of course, the many predictions by psychics of earthquakes, lost continents rising from the ocean and causing vast tidal waves, are simply other forms of the same prophecies of doom, the same death wish.

All of these prophecies present pre-visions of the coming judgment of a larger deity on a wicked world. Unable to live, psychic surrenderers can only project their own wish for death onto the world at large; they make their judgment of themselves a judgment on the entire earth in a last desperate attempt to gain authenticity. They cannot triumph but they can say, "I was right." They are reduced to trying to justify their quitting.

The lack of authenticity they feel becomes clear when you realize that all of the cataclysmic prophecies have one other element in common: the idea that the world at large must and will remain as it is, until it is finally destroyed. They see society as no more capable of change than they are. Society is "them" and psychic surrenderers are "us," unable to make contact with "them." Since psychic surrenderers cannot overcome or unite with their foes, they can only find a charismatic version of themselves, surrender to him, and wait for vengeance on the rest of society. They can only anticipate the apocalypse. What will become of them if the end doesn't come? If neither the earth nor the deities decide to take vengeance?

To whom will they surrender then?

Michael Scott Cain teaches English at Catonsville Community College in Catonsville, Maryland. He has published several books, most recently Book Marketing *from Dustbooks. His essays and articles have appeared in such magazines as* Midwest Review, Writer's Digest, Small Press Review, *and* Publisher's Weekly.

Development During Adulthood and Aging

Developmentalists hold two extreme points of view about the latter part of the life span. One point of view, "disengagement," argues that the physical and intellectual deficits associated with aging are inevitable and should be accepted at face value by the aged. The other point of view, "activity," acknowledges the decline in abilities associated with aging, but also notes that the aged can maintain satisfying and productive lives.

Extreme views in any guise suffer from the problem of homogeneity, which involves stereotyping all individuals within a category or class as having the same needs and capabilities. Whether one's reference group is racial, ethnic, cultural, or age-related, stereotyping usually leads to counter-productive, discriminatory social policy which alienates the reference group from mainstream society.

Evidence obtained during the past decade clearly illustrates the fallacy of extremist views of adulthood and aging. Development during adulthood and aging is not a unitary phenomenon. Although there are common physical changes associated with aging, there also are wide individual differences in the rates of change as well as the degree to which changes are expressed. It is common to think of the changes associated with aging solely as physical. Often the subject of aging is cast in generally negative terms. However, there are psychological changes as well, and, as is the case at all age levels, some individuals cope well with such changes and others do not. New research on the aging process suggests that physical health and mental health changes do not correlate well. Although a variety of abuses can hasten physical and mental deterioration, proper diet and modest exercise can also slow the aging process. One cannot understate the importance of love, social interaction, and a sense of self-worth for combating the loneliness, despair, and futility often associated with aging.

The first article in this section presents information on love, a topic that has received surprisingly little attention from developmentalists although it is a favorite of poets, novelists, and song writers. Although the nature of love differs between siblings, parents and children, and husbands and wives, the measurement of love suggests a common core. Moreover, it is now clear that love is not just for the young. "Never Too Late" contrasts the passionate love of youth with campanionate love of the elderly. It is interesting that despite such differences, the emotional effect of love seems strinkingly similar across the span of adulthood.

Erik Erikson was among the first to draw serious attention to the conflicts associated with each of the age periods in the life cycle. Other investigators have drawn attention to the major problems of middle age—that period of development which marks the transition of maturity, and subsequently, to old age. Popular accounts of the pervasiveness of the "mid-life crisis" have been tempered by more empirical studies which suggest that the mid-life crisis may be a real phenomenon indeed, but real only for a minority of individuals. While the mid-life crisis may not be as widespread as popular reports suggest, this is not the case for loneliness.

Richard Booth focuses on the emotional state of loneliness, a state that nearly every individual experiences one time or another throughout the life-span. However, just as developmentalists have paid little attention to the powerful emotion of love, they have also generally ignored loneliness.

Most adults cope successfully with everyday occurrences of loneliness. However, in its most chronic form, loneliness can lead to a sense of profound isolation, despair, and depression. Depression is a major health problem of adulthood and aging, especially among women. However, it is a problem for which there is cause for optimism. "The Good News About Depression" reveals that a number of therapeutic interventions have proven to be helpful in the treatment of chronic depression.

Behavioral gerontology remains a specialization within human development that is absent from most graduate programs in human development. Perhaps this is due in part to the natural tendency to avoid the negative aspects of aging, such as loneliness and despair over the loss of one's spouse or over one's own impending death. Regardless of the reason, contemporary studies provide fascinating information to challenge many traditional views about brain growth and development, interpersonal relationships, and memory processes among the aging.

Because the proportion of the population represented by the aged is increasing rapidly, it is imperative that significant advances be made in knowledge of the later years of development. Prolonging life and controlling mental and physical illness are only small aspects of all that is involved in the promotion and improvement of the quality of life for the elderly.

Looking Ahead: Challenge Questions

Why has it taken so long for psychologists to attempt a scientific study of love? Do you think they will have as much success as poets and songwriters in detailing the components and dynamics of this pervasive and powerful human emotion?

Many age-related crises seem to be associated with a variety of factors that produce stress. Some stress is situational while other stress is chronic. Loneliness is a stress that can occur at any age and can be either situational or chronic. When chronic it can easily lead to depression and perhaps to suicide. How would you plan programs to deal with the loneliness of old age? Would Erikson's theory provide any guidance?

The pressure and competitiveness of the world of work combined with the marked increase in single-parent and dual-career families almost makes good organizational skills a key to family survival. At this point in your life, which life-space structure holds the most appeal to you? Which do you actually think will be characteristic of your life ten years from now?

Gerontologists suggest that although hormonal influences are an important aspect of aging, significantly greater life expectancies are possible without resorting to medical breakthroughs. Control of childhood disease, better education, better physical fitness, and proper diet are factors that add to increased life span. How would your life differ now if your expected life span was 150 years?

THE MEASURE OF LOVE

LOVE MAY STILL BE A MYSTERY TO POETS, BUT THE SECRETS OF BOTH ITS STRUCTURE AND ITS WORKINGS ARE NOW YIELDING TO A YALE PSYCHOLOGIST.

ROBERT J. STERNBERG

Yale psychologist **Robert Sternberg** created his first intelligence test for a science project in the seventh grade; it ended badly, when he was disciplined for using it to test his fellow students' IQs. A later project, in tenth grade, showed him that music—specifically, the Beatles' rendition of "She's Got the Devil in Her Heart"—increased test scores.

Since then, Sternberg has continued to hunt for the bases of intelligence and the mental processes that underlie thinking. "Many of them are what are sometimes called executive processes," he says, "such as recognizing problems in life, figuring out what steps to take and monitoring how well the solution is going." He is a recipient of many awards for his work and the author of dozens of articles on professional journals.

Love is one of the most important things in life. People have been known to lie, cheat, steal and kill for it. Even in the most materialistic of societies, it remains one of the few things that cannot be bought. And it has puzzled poets, philosophers, writers, psychologists and practically everyone else who has tried to understand it. Love has been called a disease, a neurosis, a projection of competitiveness with a parent and the enshrinement of suffering and death. For Freud, it arose from sublimated sexuality; for Harlow, it had its roots in the need for attachment; for Fromm, it was the expression of care, responsibility and respect for another. But despite its elusiveness, love can be measured!

My colleagues and I were interested both in the structure of love and in discovering what leads to success or failure in romantic relationships. We found that love has a basic, stable core; despite the fact that people experience differences in their feelings for the various people they love, from parents to lovers to friends, in each case their love actually has the same components. And in terms of what makes love work, we found that how a man thinks his lover feels about him is much more important than how she actually feels. The same applies to women.

When we investigated the structure of love, the first question Susan Grajek, a former Yale graduate student, and I looked at was the most basic one: What is love, and is it the same thing from one kind of relationship to another? We used two scales: One, called a love scale, was constructed by Zick Rubin, a Brandeis University psychologist; the other was devised by George Levinger and his colleagues at the University of Massachusetts. (We used two scales to make sure that our results would not be peculiar to a single scale; the Rubin and Levinger scales turned out to be highly correlated.) Levinger's measures the extent to which particular feelings and actions characterize a relationship (see box on page   ). Rubin designed his 13-item scale to measure what he believes to be three critical aspects of love: affiliative and dependent need for another, predisposition to help another, and exclusiveness and absorption in the relationship with another.

Consider three examples of statements Rubin used, substituting for the blanks the name of a person you presently love or have loved in the past. For each statement, rate on a one (low) to nine (high) scale the extent to which the statement characterizes your feelings for your present or previous love.

"If I could never be with ____, I would feel miserable." "If ____ were feeling badly, my first duty would be to cheer him (her) up." "I feel very possessive toward ____."

The first statement measures affiliative and dependent need, the second, predisposition to help, and the third, exclusiveness and absorption.

Validating the Score

Although there is no guarantee that the scale truly does measure love, it seems, intuitively, to be on the right track. What's more important, scores on the Rubin love scale are predictive of the amount of mutual eye gazing in which a couple engages, of the couple's ratings of the probability that they will eventually get married and of the chances that a couple in a close relationship will stay in that relationship. There is thus a scientific basis as well as intuitive support for the scale's validity.

We asked participants to fill out the Rubin and Levinger scales as they applied to their mother, father, sibling closest in age, best friend of the same sex and lover. Thirty-five men and 50 women from the greater New Haven area took part. They ranged in age from 18 to 70 years, with an average

age of 32. Although most were Caucasian, they were of a variety of religions, had diverse family incomes and were variously single, married, separated and divorced.

To discover what love is, we applied advanced statistical techniques to our data and used the results to compare two kinds of conceptions, based on past research on human intelligence. Back in 1927, the British psychologist Charles Spearman suggested that underlying all of the intelligent things we do in our everyday lives is a single mental factor, which Spearman called *G*, or general ability. Spearman was never certain just what this general ability was, but he suggested it might be what he referred to as "mental energy." Opposing Spearman was another British psychologist, Godfrey Thomson, who argued that intelligence is not any one thing, such as mental energy, but, rather, many things, including habits, knowledge, processes and the like. Our current knowledge about intelligence suggests that Thomson, and not Spearman, was on the right track.

We thought these two basic kinds of models might apply to love as well as to intelligence. According to the first, Spearmanian kind of conception, love is a single, undifferentiated and indivisible entity. One cannot decompose love into its aspects, because it has none. Rather, it is a global emotion, or emotional energy, that resists analysis. According to the second, Thomsonian kind of conception, love may feel like a single, undifferentiated emotion, but it is in fact one best understood in terms of a set of separate aspects.

Our data left us with no doubt about which conception was correct: Love may feel, subjectively, like a single emotion, but it is in fact composed of a number of different components. The Thomsonian model is thus the better one for understanding love as well as intelligence.

Although no one questionnaire or even combination of questionnaires is likely to reveal all the components of love, we got a good sense of what some of them are: (1) Promoting the welfare of the loved one. (2) Experiencing happiness with the loved one. (3) High regard for the loved one. (4) Being able to count on the loved one in times of need. (5) Mutual understanding of the loved one. (6) Sharing oneself and one's things with the loved one. (7) Receiving emotional support from the loved one. (8) Giving emotional support to the loved one. (9) Intimate communication with the loved one. (10) Valuing the loved one in one's own life.

These items are not necessarily mutually exclusive, but they do show the variety and depth of the various components of love. Based on this list, we may characterize love as a set of feelings, cognitions and motivations that contribute to communication, sharing and support.

To our surprise, the nature of love proved to be pretty much the same from one close relationship to another. Many things that matter in people's relationship with their father, for example, also matter in their relationship with a lover. Thus, it is not quite correct to say, as people often do, that our love for our parents is completely different from our love for our lover. There is a basic core of love that is constant over different close relationships.

But there are three important qualifications: First, when we asked whom people love and how much, we found that the amounts of love people feel in different close relationships may vary widely. Furthermore, our results differed slightly for men and women. Men loved their lover the most and their sibling closest in age the least. Their best friend of the same sex followed the lover, and their mother and

---

Love for our parents is not so different from love for a lover. A basic core of love is constant.

---

father were in the middle. Women loved their lover and best friend of the same sex about equally. They, too, loved their sibling closest in age the least, with their mother and father in the middle. But whereas men did not show a clear tendency to prefer either their mother or their father, women showed more of a tendency to prefer their mother. These results are good news for lovers and same-sex best friends but bad news for siblings close in age. (Remember, however, that all these results are averages. They do not necessarily apply in any individual case.)

Second, the weights or importances of the various aspects of love may differ from one relationship to another. Receiving emotional support or intimate communication may play more of a role in love for a lover than in love for a sibling.

And third, the concomitants of love—what goes along with it—may differ from one relationship to another. Thus, the sexual attraction that accompanies love for a lover is not likely to accompany love for a sibling. (Although sexual attraction feels like a central component of love, most of us learn, often the hard way, that it is possible to have sexual attraction in the absence of love, and vice versa. As researchers, we decided to keep sexual attraction distinct from our list of central compo-

nents because it enters into some love relationships but not others.)

We did not obtain clear evidence for sex differences in the structure of love for men versus women. However, other evidence suggests that there are at least some. George Levinger and his colleagues, for example, investigated what men and women found to be most rewarding in romantic relationships. They discovered that women found disclosure, nurturance, togetherness and commitment, and self-compromise to be more rewarding than men did. Men, in contrast, found personal separateness and autonomy to be more rewarding than women did. There is also evidence from other investigators to suggest that women find love to be a more integral, less separable part of sexual intercourse than men do.

Some people seem to be very loving and caring people, and others don't. This observation led us to question whether some people are just "all-around" lovers. The results were clear: There is a significant "love cluster" within the nuclear family in which one grows up. Loving one member of this family a lot is associated with a tendency to love other members of the family a lot, too. Not loving a member of this family much is associated with a tendency not to love others in the family much, either. These are only tendencies, of course, and there are wide individual differences. But the general finding is that love seems to run in nuclear families.

Romantic Prediction

These results do not generalize at all outside the nuclear family. How much one loves one's mother predicts how much one loves one's father, but not how much one loves one's lover. So people who haven't come from a loving family may still form loving relationships outside the family—though coming from a loving family doesn't guarantee that you will be successful in love.

Having learned something about the nature of love, we were interested in determining whether we could use love-scale scores to predict and even understand what leads to success or failure in romantic relationships. Because our first study was not directly addressed to this question, Michael Barnes, a Yale graduate student, and I conducted a second study that specifically addressed the role of love in the success of romantic relationships.

In our study, each of the members of 24 couples involved in romantic relationships filled out the love scales of Rubin and Levinger. But they filled them out in four different ways, expressing: (1) Their feelings toward the other member of the couple. (2) Their perceptions of the feelings of the other member of the couple toward them. (3) Their feelings toward an ideal other member of such a couple. (4)

A COST-BENEFIT LOVE SCALE

All relationships are both rewarding and costly; caring for someone, while emotionally satisfying, takes time and can be very taxing. The items that follow are excerpted from a scale George Levinger, of the University of Massachusetts, Amherst, devised to rate negative and positive aspects in any given relationship (with lover, spouse, parent, friend, etc.). In his research, author Robert J. Sternberg adapted the scale to a different purpose: He asked people to rate the importance of each statement in terms of both an actual and an ideal relationship. His findings are discussed in the accompanying story.

____ 1. Doing things together

____ 2. Having no secrets from each other

____ 3. Needing the other person

____ 4. Feeling needed by the other person

____ 5. Accepting the other's limitations

____ 6. Growing personally through the relationship

____ 7. Helping the other to grow

____ 8. Having career goals that do not conflict

____ 9. Understanding the other well

____ 10. Giving up some of one's own freedom

____ 11. Feeling possessive of the other's time

____ 12. Taking vacations together

____ 13. Offering emotional support to the other

____ 14. Receiving affection from the other

____ 15. Giving affection to the other

____ 16. Having interests the other shares

EXCERPTED AND ADAPTED FROM THE INTERPERSONAL INVOLVEMENT SCALE, BY GEORGE LEVINGER, UNIVERSITY OF MASSACHUSETTS, AMHERST

of how love might affect satisfaction in a romantic relationship. According to the first conception, level of satisfaction is directly related to the amount of love the couple feel: The more they love each other, the more satisfied they will be with their relationship. According to the second and more complex conception, the relation between love and satisfaction is mediated by one's ideal other. In particular, it is the congruence between the real and the ideal other that leads to satisfaction. As the discrepancy between the real and ideal other increases, so does one's dissatisfaction with the relationship.

Consider two couples: Bob and Carol and Ted and Alice. Suppose that Bob loves Carol just as much as Ted loves Alice (at least, according to their scores on the love scales). According to the first conception, this evidence would contribute toward the prediction that, other things being equal, Bob and Ted are equally satisfied with their relationships. But now suppose that Bob, unlike Ted, has an extremely high ideal: He expects much more from Carol than Ted expects from Alice. According to the second conception, Ted will be happier than Bob, because Bob will feel less satisfied.

These two conceptions of what counts in a relationship are not mutually exclusive, and our data show that both matter about equally for the success of the relationship. Thus, it is important to remember that although love contributes to a successful relationship, any relationship can be damaged by unrealistically high ideals. At the other extreme, a relationship that perhaps does not deserve to last may go on indefinitely because of low ideals.

In addition to love and the ideal, we found that both how a person feels about his lover and how he thinks she feels about him matter roughly equally to satisfaction in a relationship. But there are three important qualifications.

First, the correlation between the love-scale scores of the two members of a couple is not, on the average, particularly high. In many relationships, the two members do not love each other equally.

Second, it is a person's perception of the way his lover feels about him, rather than the way the lover actually feels, that matters most for one's happiness in a relationship. In other words, relationships may succeed better than one might expect, given their asymmetry, because people sometimes systematically delude themselves about the way their partners feel. And it is this perception of the other's feelings rather than the other's actual feelings that keeps the relationship going.

Third, probably the single most important variable in the success of the relationships we studied was the difference between the way a person would ideally *like*

his partner to feel about him and the way he thinks she really does feel about him. We found that this difference is actually more important to the success of a relationship than a person's own feelings: No variable we studied was more damaging to the success of a romantic relationship than perceived under- or overinvolvement by the partner.

Why might this be so? We believe it is because of the ultimate fate of relationships in which one partner is unhappy with the other's level of involvement. If a person perceives his partner to be underinvolved, he may try to bring her closer. If she does not want to come closer, she may react by pulling away. This leads to redoubled efforts to bring her closer, which in turn lead her to move away. Eventually, the relationship dies.

On the other hand, if a person perceives her partner to be overinvolved, she may react by pulling away—to cool things down. This results in exactly the opposite of what is intended: Her partner tries to push even closer. The relationship becomes too asymmetrical to survive.

All our results suggest that the Rubin and Levinger scales could be useful tools to diagnose whether relationships are succeeding or not. There is one important and sobering fact to keep in mind, though: Scores from a liking scale devised by Zick Rubin were even better predictors of satisfaction in romantic relationships than were scores from the love scales, especially for women. Thus, no matter how much a person loves his partner, the relationship is not likely to work out unless he likes her as well.

An End to a Mystery?

Despite its complexity, love can be measured and studied by scientific means. Those who believe that love is, and should remain, one of life's great mysteries will view this fact as a threat; I view it in exactly the opposite way. With a national divorce rate approaching 50 percent and actually exceeding this figure in some locales, it is more important than ever that we understand what love is, what leads to its maintenance and what leads to its demise. Scientists studying love have the opportunity not only to make a contribution to pure science but to make a contribution to our society. At the very least, the study of love can suggest the cause, if not the cure, for certain kinds of failed relationships. I believe that our ability to measure love is contributing to progress both in understanding the nature of love and in suggesting some of the causes of success and failure in close relationships.

Their perceptions of the feelings of an ideal other member of such a couple toward them. These questions dealt with two basic distinctions: the self versus the actual other and the actual other versus an ideal.

The participants, all of whom were students in college or graduate school and none of whom were married, were asked to think in terms of a realistic ideal that would be possible in their lives, rather than in terms of some fantasy or Hollywood ideal that could exist only in movies or other forms of fiction. In addition to filling out the love-scale questionnaires, participants also answered questions regarding their satisfaction and happiness with their present romantic relationship.

Between Real and Ideal

We compared two different conceptions

The Great Balancing Act

UNSEEN WAYS OF STRUCTURING OUR LIVES CAN TIP THE SCALES BETWEEN FRUSTRATION AND FULFILLMENT.

MARY DEAN LEE

Mary Dean Lee is an assistant professor in the Faculty of Management, McGill University, Montreal, Quebec.

For most of us, the dawn of a new day is the continuation of familiar, well-worn patterns of living. Each of us organizes the key resources in our life—time, space, people and activity—in characteristic ways. But I have found, after studying the individual patterns of about 120 working adults, four main patterns, which I call "life-space structures." People who have "home-based nuclear" or "work-based nuclear" structures organize their lives around that single focal point. Those who have "conjoint" structures focus equally on home and work, while those with "diffuse" structures spend much of their time and energy in organizations and social activities outside both home and work (see "Key Life-Space Structures" box).

Using interviews, questionnaires and people's self-recorded behavior, I have looked at what makes these structures work well or poorly for people.

Life-space structures are seldom consciously or deliberately constructed;

they simply emerge as individuals lead their daily lives, influenced by conflicting demands on their time, energy and attention and by their personalities, needs, desires and values. Although these patterns are usually stable, they can shift greatly through conscious choice or when circumstances alter work or living situations.

We can judge the effectiveness of people's life-space structures by how well they help meet three goals: Do they allow people to accomplish critical tasks at home and work? Do they provide an adequate amount of pleasure or individual fulfillment? Do they protect people against undue stress? Often it is not possible to meet all these goals. But people can try to create the best structure that their circumstances permit.

To meet the first goal—getting critical tasks done at home and work—many people focus exclusively on time, a limited resource, without also considering space, people and activity, all of which can be expanded or contracted largely without regard to time. For example, people concerned that their children are not getting enough attention may have trouble finding addition-

al time for them. But they might think about taking work home, bringing the kids to work or hiring someone to give them extra attention.

Meeting the second goal—creating opportunities for pleasure or fulfillment—is essential for mental and physical health. Many people blame their lack of pleasure on lack of time. But the real problem may be failure to recognize and build into their daily routine the activities, people or places they find fulfilling.

Meeting the third goal—avoiding strain (which I view as undue or negative stress)—is a complex issue, and even experts disagree about what stress is and what causes it. I see strain as stemming chiefly from doing things at odds with your feelings or sensing that you don't have enough time to do what you want to do, rather than what you have to do. Here, too, if people look at their full range of options and resources, there's more flexibility than many imagine.

Some structures seem to work better than others for people in particular circumstances, such as dual-career parents or working people without

From *Psychology Today*, March 1986, pp. 48-56. Reprinted by permission.

Key Life-Space Structures

To identify the life-space structures discussed in this article, I compared many aspects of how people organize their lives, such as where they spend most of their time, where they put out their best efforts, where they enjoy themselves and where they feel they have

Home-Based Nuclear Structures

For people with these structures, there's no place like home. It's the center of their lives, the place where they spend most of their time and energy and experience high levels of activities that they enjoy and/or choose to do. Such people do not necessarily do less work than those with other kinds of structures.

Work-Based Nuclear Structures

For people with these structures, work is clearly the center of their universe. Their time at work is high in quality and quantity, and they find their work highly enjoyable.

Homesteaders

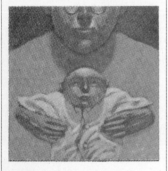

These people, usually married with children, spend high amounts of time at home and have high levels of enjoyable activities there, even though they don't have much choice about what they do there. Their time at work is about average, but it's not high in quality. Their lives are rather slow-paced; they go few places and have little interaction with nonfamily people. They are usually happier and more satisfied with their lives than are other groups.

Architects

These people spend very high amounts of time at home, where they have high amounts of pleasurable and high-choice activity. They tend to be self-employed or in occupations that allow them to do much of their work at home. The amount of time spent at work is low. They move around very little and interact with relatively few people. Although they usually earn less money than most groups, they are likely to be happier and more satisfied with their lives than most groups, including Homesteaders.

Workhorses

Often single and male, they spend extremely large amounts of time at work. Although they don't have much choice in what they do there, they enjoy their work and socialize mostly with people from work. They are less happy with their lives than are most of the other groups.

Work Revelers

These people, often married men with nonworking wives, spend high amounts of time at work, which offers them a great deal of choice and a high percent of especially enjoyable activities. Their lives are fast-paced, with a lot of movement from place to place and interaction with people, usually family members or friends not associated with work. They make more money than most other groups and tend to have high levels of life satisfaction and happiness.

the most choice over what it is that they do.

After looking at all the data, I focused on clusters of responses that were higher or lower than the average for the group as a whole. For example, I found that some people who spent above-average time at home had other life-space characteristics in common. These clusterings provided the basis for the typology below. (For the sake of simplicity, when I refer to high or low levels, I mean higher or lower than the average for the total group of respondents.)

Conjoint Structures

The lives of people with these structures focus equally on home and work, and they spend little time anywhere else. Their time both at home and at work is high in quantity and quality, and they have high levels of choice or pleasurable activity in both places.

Diffuse Structures

The people whose lives are structured these ways are on the move; no place or places form the center of their world. They do not spend both high-quantity and high-quality time at both home and work. Rather, their time and activity are focused on other locales, which are numerous and varied.

Straddlers

These people, mostly married men and women with children, tend to spend high amounts of time at home and work and have high levels of pleasure in both places, although in both they have lower levels of choice than do Jugglers. Straddlers do not show a consistent pattern of life satisfaction.

Jugglers

Found among many occupations and marital situations, they spend high amounts of time at work and at home, having considerable freedom and pleasure in both locales. Jugglers usually socialize with work-related friends and family members. They are generally happier than most other groups but not necessarily more satisfied with their lives.

Gypsies

Usually single or married without children, they spend low-quality or low-quantity time at work and at home and have one or more other types of places that are important to them. Gypsies lead a fast-paced life with extensive social interaction and movement from place to place. They tend to be less satisfied and less happy with their lives than are most other groups.

Satellites

These people, usually married with children, are away from home and work a lot yet have strong home connections. Like Gypsies, Satellites invest their time and energy in places outside work or home. Yet their close interactions occur with family members, even though their focus is not on home life. Compared with most other groups, they are generally not very happy with their lives.

partners. In explaining why, I've found it useful to distinguish between "high-demand" and "low-demand" lives.

Many of the men and women I studied were operating at high gear at home and work and feeling stretched by their responsibilities in both arenas. These high-demand conditions arose most often among men and women in dual-career or dual-earner households who had young children at home. Many were in professions in which very high commitment was necessary for success. Some seemed to meet high expectations at home and at work, gain some pleasure and not feel too strained. Others fulfilled their dual commitments well but clearly with some psychological costs.

Home-based and conjoint structures seem to have a special appeal for people living under high-demand conditions and often work well for them. Eleven of the 14 high-demand people I interviewed in one study had such structures, and seven of the eight in the high-demand group who described themselves as being particularly happy had such structures. But neither structure guarantees happiness for everyone, as the following examples show.

• Fran, married and in her early 30s, has a 1-year-old child and works as a book editor at a university press. Her husband, Chris, a top executive for a company in a city 500 miles away, is out of town each Monday through Thursday, but he and Fran talk on the phone every night. A live-in helper cares for the baby, but Fran also does much of her work at home so she can be around her child. Fran and Chris have many friends and enjoy going to cultural and social events on the weekends. Fran has created a home-based nuclear structure, which so far seems to be working well on all fronts.

• Fred is an accountant for a progressive, high-growth company in a New York suburb. He has two children, 10 and 12, and his wife has just started a career after completing an engineering degree. During the past few years Fred has been taking over a lot of responsibility for the children; both he and his wife spend most of their nonwork time with them. Fred feels that he's handling all his responsibilities adequately. But he has developed stomach trouble and high blood

ALWAYS HAVING TO DECIDE WHAT TO DO, WHERE TO GO AND WITH WHOM MAKES SOME SINGLES ANXIOUS.

pressure since his wife started her new job full-time.

Why did the home-based structure work well for Fran but not Fred? First, she spends more time with friends and goes out more for fun. Second, Fred is quite concerned about his children's being home alone after school and believes someone should be there with them. This creates strain because his feelings and behavior are inconsistent. Perhaps if he found someone to look after them, or spent more time at home when he felt the kids needed it and then went out or socialized more, he would have less strain and more pleasure.

• Anne has a booming career as a manager for a large international consulting firm. She has had five promotions in seven years and now runs a department that has extensive daily pressures. In the past six years, Anne has also had three babies and relocated twice for promotions. Her oldest child is in nursery school, and a neighbor takes care of the other two children four days a week. Her husband, a systems analyst, cares for the children on the fifth day and also puts a great deal of energy into the small farm where they live. Anne sees almost no friends other than the babysitting neighbor. She has created a conjoint structure in which her activity is almost totally devoted to home and work, and she enjoys both immensely.

• Jessie, married to a doctor, went to law school after her children, now 10 and 12, reached school age. She is now a junior attorney with a prestigious law firm in which young lawyers typically work 12-hour days or longer. She manages to get everything done in less time by "taking shortcuts" and

brown-bagging at her desk instead of joining her colleagues for long lunches out of the office. While she likes her work and wants to stay with the firm, she feels her personal life is suffering. She'd like to cut back to part-time work yet knows she must "prove herself" first. She schedules her time efficiently so she can do her work and still have ample time to be with her children. But she rarely enjoys being with them, because she is exhausted. Jessie gets her critical tasks done, although she is clearly under a lot of strain and has found little room for pleasure.

Anne's and Jessie's conjoint structures seem very similar, but subtle differences may account for their contrasting satisfaction. For example, Anne's husband, less career-oriented than Jessie's, has more time to devote to the family. Also, Anne has a successful child-care arrangement, while Jessie does not. Jessie may also be struggling more than Anne with part-time parenthood, still expecting to relate to her children as she did as a full-time mother. Anne, who was home for only a month after each of her children was born, expects to work full-time the rest of her life.

People living under high-demand conditions need to make the best use of their resources. For that reason, either home- or work-based structures in which many activities can overlap in a single location are particularly efficient.

Since conjoint structures stretch resources between work and home, they leave few for other areas and create greater potential for strain. But people with such structures obviously enjoy spending more-than-average amounts of time at home and work. Conjoint structures seem to work best for high-demand people who are very energetic and like a fast-paced life.

Fran, Fred, Anne and Jessie were all trying to meet their own exacting standards for performance at work and at home. Only Jessie ever talked of compromising or altering either expectations or behavior. Some other high-demand people I interviewed clearly had decided that they needed to make some trade-offs. They deliberately gave up some things at work or at home to meet demands in both areas and maintain their health and well-being.

• Mark, a social worker married to

a high school teacher, has two pre-school children. He has decided that family is "number one" and has built a home-based structure that supports and rewards that choice. He has passed up promotion opportunities that involved relocation and has limited his circle of friends to people who tie into his family and community activities. But friends often drop by his house, and he really enjoys his family.

• Laura, a well-known scholar and university professor of English literature, has three children. She and her husband, a research scientist, are seeing a marriage counselor. Although she seems to handle her domestic role competently, Laura finds her work more exciting and satisfying than family and home and has chosen to put relatively more energy into work. An important source of pleasure for Laura is regular contact with her small circle of friends, many of them colleagues.

The diffuse type of structure does not seem to work well under high-demand conditions. It diverts needed energy from home and work, and most high-demand people don't have the time to spend in other locales. In general, people under high-demand conditions are most likely to be happy with home- or work-based structures and least likely to be happy with diffuse structures.

Working people living alone—let's call them "single people" for the sake of simplicity—have a very different set of challenges from those in dual-career families. Their "low-demand" lives lack the home responsibilities that partially dictate the daily patterns of others. Some of the single people I interviewed saw their work as so very demanding that they felt they were functioning under high-demand conditions, while others viewed both work and home as asking little of them.

Single people most often have a work-based structure. In two studies combined, for example, I found that of 26 singles, 10 had this structure. However, all but 1 of the 10 reported happiness levels that were less than average for the group of 80 in these two studies, perhaps because they did not truly choose to be work-based. Eight of the 10 attributed their pattern to the demands of the job rather than to their preference.

Work is an easy organizing focus for single people, especially if they

I FOUND MANY PEOPLE WHO WERE INCREDIBLY EFFICIENT AT DOING EVERYTHING BUT ENJOYING THEMSELVES.

want to contribute to society or be involved in a meaningful career. Furthermore, many young, career-oriented single people believe that sacrificing some fun now in favor of work will pay off for them.

The diffuse structure was the second-most common, found among 9 of 26 single people in the two studies. Spending extensive and enjoyable time in restaurants, bars, friends' homes and other places is a natural choice for many single people, especially those with low demands at work and at home. But only three of the nine with diffuse structures had happiness levels above average, while six were below average. Seven of the 26 single people had home-based or conjoint structures. All the home-based singles had occupations that allowed them to do most of their work there. Those who developed conjoint structures enjoyed both the home and work arenas and preferred to center their lives around both, despite low demands at home. Of the nine single people whose happiness levels were above the average for the total group of 80, five were in home-based or conjoint structures.

Single people with low demands at home sometimes make work their focus by default, as a way of creating a sense of meaning, certainty and stability in their lives. Or they establish a diffuse structure to meet potential mates or make up for what's missing at home or work rather than as an enjoyable way of life in and of itself.

Single people face the additional problem of continually deciding where to go, whom to be with and what to do. For some, this presents too many choices and creates anxiety. Married people or those with major commit-

ments to others have fewer choices but can often gain a valuable sense of efficacy and self-determination choosing among the options that remain.

Some single people I interviewed deliberately created demands at home by having one or more roommates and sharing meals and chores. While economic factors may have influenced this decision, shared living arrangements also provided a way to make home more desirable and compelling. One person forged close links with fellow-renters in an old house with several apartments, turning it into an informal cooperative in which all the renters ate together once a week, worked together in a backyard garden and shared or swapped cars and tools.

Regardless of their life-space structure, single people generally seem less happy than the other people I studied. They often compound their problems by choosing work-based or diffuse structures, which usually yield less happiness than do home-based or conjoint structures.

Some people, both single and married, create structures that allow no opportunity for retreat or repose. I found many who were incredibly efficient at doing everything but enjoying themselves. Yet others leading similarly pressured lives somehow arranged to build in effective ways to relax, often establishing regular patterns or rituals for doing just that. For some, these were associated with a particular location. For example, one very busy graphics designer in a difficult marriage who had three teenage children described her health club as her refuge, the only place where she could relax and feel she had a great deal of choice about what she did.

A professor teaching evening composition classes to a melting pot of striving students in an inner-city university used a bar in an elegant old downtown hotel as his "other place." A couple of times a month he'd treat himself to a leisurely afternoon of good drinks, hors d'oeuvres and stimulating conversation with a friend or colleague in an atmosphere totally unlike his work environment.

Times of release may also be associated with particular people or activities. One young, ambitious executive had a few buddies he could call anytime, just to let off steam or arrange to meet for a drink after work or an afternoon of golf. Another man, a teacher at a troubled urban high school, took hour-long hot baths after work to help him unwind.

No particular kind of life-space structure ensures that people will find adequate release and relaxation. But by paying attention to where, how and with whom they enjoy themselves, most people, however busy they may be, can make room for the necessary ingredients.

At times people may have to live with unsatisfying situations because they are unwilling or unable to make major changes in the status quo at home, at work or in both places. In these cases, they can often adjust their life-space patterns to make the best of the situation, although some of these adjustments have higher costs than others.

A person unhappy at home might try to improve the quality of time there by increasing pleasurable, high-choice activities, such as entertaining or spending focused time one-on-one with the kids, or might shift to structures that focus energy outside the home. One of the people I interviewed who used this approach was Shelley, a 35-year-old TV producer who is married to a lawyer and has a 2-year-old child. She loves her work and her child but says her marriage is in very bad shape. Shelley sees a breakup as inevitable and is already planning how to get custody if her husband sues for divorce. But she wants to postpone this as long as possible because she thinks that her marriage is the best way to get the most out of work and home. She tries to increase pleasure at home by entertaining friends and relatives. Although family life is fine on the sur-

WHEN BAD BUT TRANSIENT TIMES MUST BE ENDURED, THE BEST STRATEGY MAY BE TO PULL BACK FROM HOME, WORK OR BOTH AND DEVELOP A THIRD ARENA FOR SATISFACTION AND FULFILLMENT.

face, Shelley is worried by some of her child's recent behavioral problems, fearing that they may be caused by the strained marriage situation.

Shelley has created a conjoint structure, spending high-quantity and high-quality time at work and at home despite her unsatisfying relationship with her husband. She has clearly taken the path of least resistance, at what might be a high cost to her child and husband. Yet she believes that any other life-space structure would involve even greater compromises.

Similarly, those unhappy both at home and work can try to increase the quality of time in both places or, after meeting their obligations, can seek out other places and people for stimulation, enjoyment and involvement. Stan, 35 and married, with two children, tried this approach. Although he makes good money welding on a night shift, he hates his job. He and his wife, a secretary, rarely see one another since their work hours conflict, and when they are together they don't get along. Stan has recently taken on a second job managing an apartment complex. He loves this work, which has provided him with new friends and career aspirations.

Stan's unusual two-job work-based structure allows him to deal with unsatisfactory conditions at home and work. He may not be meeting all his obligations at both jobs and at home with his kids, and he has difficulty getting enough sleep. But the general principle Stan is following is sound. When bad but transient times must be endured, sometimes the best strategy is to pull back from home, work or both and develop a third arena for satisfaction and fulfillment, as Stan did in taking a second job.

The study of life-space structures is still new and exploratory. As I learn more about how these patterns work for or against people in particular situations, I hope to learn ways to help people coordinate their work and personal lives more effectively at different points in life.

Even at this early stage, however, understanding that these various structures exist as a matter of choice can help people to think about their lives and articulate their concerns in a new way. People who realize that they can exercise some choices over their life space can consciously design it to yield more day-to-day fulfillment, both at work and in their personal lives.

Toward an Understanding of Loneliness

Richard Booth

Richard Booth, MSW, is Associate Professor, Department of Psychology, Sociology, and Social Work, Quad-Cities Campus, Black Hawk College, Moline, Illinois, and a doctoral candidate at the University of Iowa.

WESTERN LITERATURE is filled with themes of people in the grips of emotional turmoil. Sartre carved the image of the alienated individual who must struggle with the essential meaninglessness of existence—alone and with deliberation.[1] One reads of the "marginal man" displaced, suspended from the fringes of the system, struggling not merely for survival but for fundamental personal and social meaning.[2] Romeo and Juliet eagerly offered their lives rather than be separated from each other, and Oedipus punished himself severely for a mistake he could hardly have avoided making. The emptiness of Willy Loman, the curiosity of Huck Finn, the spontaneity of Marie Antoinette, the vanity of Dorian Gray, the misery of Oliver Twist, the persistence of Dorothy in Oz are but a few of the literary examples of the human element coming to terms with life, whether the characters are fictional or real. No matter how it is described, there is a common theme of people attempting to bring emotion under rational control and to eke from their lives some good feeling about themselves and those with whom they interact.

Loneliness is an emotional condition that affects virtually all people from time to time and which is accompanied by such intensity that people sometimes do regrettable things to offset its influence. Loneliness can be found in every therapist's office, every classroom, every workplace, and every clinic and hospital. Every family system experiences it, somehow deals with it, and proceeds with the rest of its life. Yet, how many people have the skills necessary to confront the condition straightforwardly in their own lives, not to speak of the lives of family members, students, clients, and colleagues? Indeed, when this vague although pervasive phenomenon pushes itself to the fore, who can identify it so readily and correctly that it is not confused with moodiness, fleeting bad temper, depression, and even anger or global negativity? What is loneliness, and is it a condition apart from the other conditions just mentioned?

Although literary and psychoanalytic traditions have contributed substantially to knowledge about loneliness, little experimental work has been done in this area until recently. This article suggests a cognitive-emotional approach to understanding loneliness, explores some empirical work on the subject, and offers workers some possible intervention strategies.

CHARACTER OF LONELINESS

Despite its probable high frequency in the population and its compelling intensity, there has been little research on this affective state. Although not much empirical work has been done, profiles of loneliness and of the lonely person have emerged. The bulk of the work has been largely clinical and has come from the psychoanalytic tradition. Horney, Sullivan, and Fromm all speak to the basic need to seek and engage in meaningful relationships.[3] These theorists perceive people as fundamentally interactional and interpersonal and outline the difficulties that may arise if human relationships become disturbed. Fromm goes so far as to suggest that if an individual's basic needs (two of which are rootedness and relatedness) remain unfulfilled, either the person will die or become insane. Even if the individual does not fall prey to those extremes, he or she will become power-less. The need for relatedness about which these theorists speak has long been a central concern of the social work profession. When a person is out of step with his or her context and begins to march to a different drummer, as Slater suggests, problems inevitably will follow. A possible one is loneliness.[4]

From a sociological perspective, Durkheim studied the concept of anomie—that state of "normlessness" in which people are essentially lost in the system and confused about their "fit."[5] Peplau, Russell, and Heim came closer to a definition of loneliness by outlining the conditions within which loneliness is likely to emerge. They stated: "Loneliness occurs when a person's network of social relationships is small or less satisfying than the person desires."[6] From this perspective, it is conceivable that a person with twice the "normal" size friendship group could, nonetheless, be lonely, while the person with a small but satisfying friendship group may not be lonely at all. Peplau, Russell, and Heim seem to be suggesting that loneliness is, to some extent at least, a matter of cognitive expectations rather than merely an affective experience. Weiss agreed with them, saying, ". . . an individual when lonely maintains an organization of emotions, self-definitions, and definitions of his or her relations to others, which is quite different from the ones he maintains when not lonely."[7] Thus, loneliness appears to be a subjective notion— a cognitive-emotional phenomenon that often is of such intense proportions that it may influence overt behavior. Drawing still closer to a definition but remaining general, Weiss added that loneliness is "a gnawing . . . chronic distress without redeeming features."[8] In essence, the lonely individual's interpersonal reality does not fit interpersonal expectations. Why this occurs is unclear.

The Lonely Character

Viewing the situation from a more clinical perspective, one may ask

From *Social Work*, March/April 1983, pp. 116-119. Reprinted by permission.

whether there are any clues by which to identify the condition—any really reliable ways to find the lonely student in the classroom, the lonely family member at home, or the lonely colleague in the agency. In other words, are there any recognized correlates of the condition that may help social workers to intervene early and perhaps counteract potential disabling chronicity? Jones, Freeman, and Goswick suggested that it is difficult to identify a lonely person and found that naive judges tend to rate lonely and nonlonely persons similarly. However, the literature has provided a broad descriptive analysis of the lonely individual. In certain groups of lonely people, high positive correlations have been found between alcoholism and suicide and loneliness. Moreover, lonely people have been found to have significantly lower self-esteem than nonlonely persons and to be significantly more sensitive to rejection by others.[9]

Cutrona found that lonely persons who have passed through situational loneliness into chronicity tend to view themselves as personally responsible for their plight. They appear to have become involved in a cyclic situation in which they reinforce their loneliness through its chronicity—a situation from which they are unable to extricate themselves. Cutrona also found that measures on the Beck Depression Inventory were significantly and positively correlated with loneliness in this group.[10] It is interesting to note that both situationally and chronically lonely people were found to feel negative about human relationships in general. They also tended to reject others with greater frequency than others rejected them. Russell, Peplau, and Cutrona also found that the lonely subjects they studied showed signs of general restlessness, were more self-enclosed, and felt more "empty" than nonlonely persons. These findings were statistically significant.[11]

Jones, focusing specifically on how lonely people report they feel, identified an array of negative emotions. Generally, his subjects indicated they felt helpless, misunderstood, and unloved.[12] They also tended to feel unacceptable to others, separated from meaningful interaction with others, and generally worthless. They described themselves as being bored with life and as feeling unhappy. Although a number of self-concept factors were involved, the strength of the individual factors and the exact nature of their interaction remain far from clear.

Impact on Others

As was noted, the judges used by Jones, Freeman, and Goswick had some difficulty in consistently differentiating the lonely from the nonlonely in their ratings of subjects.[13] However, the judges used by Solano and those used by Sansone, Jones, and Helm had different types of problems in identifying lonely people.[14] Thus, the literature clearly is inconsistent on this matter. Nevertheless, it seems safe to hypothesize that because they are unable to differentiate, completely and consistently, lonely from nonlonely persons through casual interaction with them, observers would tend to deal similarly with both types of persons.

What effect, if any, would knowing that another person is lonely have on the behavior of observers? Fromm-Reichman argued that nonlonely people tend to construct walls of defense against lonely people in an attempt to ward off their own fears.[15] In other words, lonely people seem to frighten nonlonely people and perhaps even threaten the condition of nonloneliness; thus, nonlonely people feel the need to defend against lonely people for their own well-being. Wimer and Peplau found that whether nonlonely people enjoy dealing with lonely people largely depends on what they perceive is causing and maintaining the loneliness. Specifically, when nonlonely people think that lonely people are trying to help themselves get away from their loneliness, they are disposed to feel much more positively toward them than when they perceive that lonely people are not even attempting to rid themselves of their condition. Further, if nonlonely people think that others are lonely through no fault of their own, they are more likely to have positive feelings than if they think that the lonely people are the source of their own problems.[16] Thus, although people may generally find it difficult to identify a lonely person unless the individual tells them, there is some evidence to suggest that once they are aware of the condition in another person, they begin to look for the cause of the loneliness. It follows that whether they are unaware of the condition or whether they perceive the lonely person as perpetuating his or her own dilemma, they are unlikely to intervene spontaneously in the condition of loneliness.

IMPLICATIONS FOR PRACTICE

In linking the descriptive and correlational findings of the literature to practice, the author will consider five of the most central themes. They are

described in the following material.

Helplessness

The first theme is the fundamental feeling of helplessness that lonely people report about themselves. As was already stated, some lonely persons do not feel in control of their lives. As with the construct of learned helplessness discussed by Seligman, people with an external locus of control tend to blame the environment for their difficulties and failures.[17] In the extreme, they could believe that no matter what they might do to bring about a change, they would have no influence. Peplau, Russell, and Heim found that if external locus of control and loneliness exist in the same person, chances are good that the loneliness is situational rather than chronic. In cases like these, they suggest, clarification of the perceived causes of loneliness is paramount in the helping process. That is, the client and the therapist should identify what most probably is responsible for the client's condition, focusing on the potential alterability of the "causal" situation. It is possible that lonely clients, because of their preoccupation with their own condition and the pain that accompanies it, pay too little attention to those aspects of loneliness over which they could learn to exert control. They may be overstating the actual degree of their helplessness to the detriment of ameliorating their own condition. They can be taught and helped to explore the possibility that part of their loneliness may be due to their own resistance to taking control of it—a control which they can find within themselves with therapeutic assistance.[18]

Unrealistic Expectations

The second theme is that the expectations of some lonely clients are not realistic. Therefore, it may be helpful in certain cases of loneliness for the therapist to help the client scale down his or her expectations. In reassessing their expectations of interpersonal relationships, clients can be helped to avoid frustration and loneliness by coming to understand that all is not lost if there really is no ideal mate, no infallible spouse, no perfect lover, or no all-fulfilling friendship group. Sometimes clients—and therapists as well—may do well to learn that Santa Claus (the personification and deliverer of one's dreams) is but a limited reality.[19] That is, no one can expect to receive all that he or she may wish to receive.

Overcoming the Past

The third theme is that some lonely

people focus their energy and attention on the events that precipitated their loneliness rather than on the "maintenance factors" of the condition.[20] Such clients need help in overcoming a painful event in the past. For example, lonely single parents may have been deeply affected by the divorce from or the death of a spouse. Even after a considerable time, they may be preoccupied with the possibility of a reunion (in the case of divorce) or may continue to deny the death of the beloved. In either event, these persons may have shut themselves off from possible ameliorative efforts such as dating or other types of social interaction. Ignorant of the effects of this counterproductive process, they may blame their children or other external factors for occupying all their time.

These persons may continue to blame their spouses for divorcing them or for dying and leaving them alone. Thus, they maintain their loneliness, desiring meaningful interaction but disallowing it by focusing on "what could have been if" or "what could be if" or "if only" kinds of statements. With help, such clients can be redirected to recognize the difference between the initial conditions out of which their loneliness emerged and the factors that perpetuate the condition. It is those maintenance factors of the condition that are frequently amenable to change.

Loneliness vs. Depression

The fourth theme is the confusion of loneliness with depression. Because of this confusion, more research is needed to identify the differences and similarities between the two states. As was mentioned earlier, those who pass from situational to chronic loneliness tend to blame themselves for their condition and score higher on the Beck Depression Inventory. It may be that the transitory loneliness—the type of loneliness that comes and goes in the lives of all people—is largely situational, while depression is related more to internal-blame factors. In this regard, Peplau, Russell, and Heim found that depression scores were highest for subjects who felt they were to blame for their loneliness and that the loneliness was not likely to go away.[21] There still remains, however, enough confusion between these two concepts to justify further work at clarification.

Avoidance of Clients

The fifth theme is that in their efforts to defend against loneliness, some therapists may shut out significant cues from clients which may indicate that the clients are lonely even though the clients are not telling them so directly. Therefore, it behooves clinicians in all fields of practice to be alert to possible tendencies to avoid clients who present themselves for help. It seems safe to suggest that all clinicians are on the alert for potential psychological as well as physical dangers and hence may avoid certain aspects of a client's problem because they "sense" that he or she is lonely. Clearly, this idea requires further research. Nevertheless, clinicians should bear in mind that clients, students, friends, and family members who manifest general negativity, poor self-esteem, pessimism, social isolation—conditions that make spending time with them less desirable—may be, in fact, lonely. The perceptive clinician will pick up on these correlates and examine the possibilities.

CONCLUSION

The common social and personal problem of loneliness is multifaceted and requires much more empirical investigation. Researchers need to examine more thoroughly the relationship between loneliness and depression in normal persons to separate out the clinical similarities and differences of these two conditions. Further work is necessary to shift from an analysis of correlations to the realm of cause and effect. However, the literature seems clear about one thing: lonely people are negative about social relationships in ways that nonlonely persons do not share. Thus, another significant question for further study is this: Does negativity lead people to be lonely or does loneliness lead to negativity? This question has serious implications for the socialization of children; it may be reasonable to hypothesize that children who are taught to view life negatively may be prime candidates for loneliness in adulthood as well as for the many ripple effects accompanying that condition.

Notes and References

1. Jean-Paul Sartre, *Existentialism and Human Emotions* (New York: Book Sales, 1957).

2. Charles C. McCaghy, James Skipper, Jr., and Mark Lefton, eds., *In Their Own Behalf: Voices from the Margin* (New York: Appleton-Century-Crofts, 1968).

3. See Karen Horney, "Alienation and the Search for Identity," *American Journal of Psychoanalysis*, 21 (1961); Harry Stack Sullivan, *The Interpersonal Theory of Psychiatry* (New York: W. W. Norton & Co., 1953); and Erich Fromm, *Man for Himself* (New York: Fawcett Books, 1947).

4. Philip Slater, *The Pursuit of Loneliness: American Culture at the Breaking Point* (Boston: Beacon Press, 1970).

5. Emile Durkheim, *Suicide: A Study in Sociology*, John Spaulding and George Simpson, trans. and eds. (New York: Free Press, 1951).

6. Letitia Anne Peplau, Daniel Russell, and Margaret Heim, "An Attributional Analysis of Loneliness," in Irene H. Frieze, Daniel Bar-Tal, and John S. Carroll, eds., *Attribution Theory: Application to Social Problems* (New York: Jossey-Bass, 1979), pp. 53–78.

7. Robert Weiss, ed., *Loneliness: The Experience of Emotional and Social Isolation* (Cambridge, Mass.: M.I.T. Press, 1973), p. 11.

8. Ibid., p. 15.

9. Warren H. Jones, J. E. Freemon, and Ruth Ann Goswick, "The Persistence of Loneliness: Self and Other Determinants," *Journal of Personality*, 49 (March 1981), pp. 27–48. See also W. H. Jones, "Loneliness and Social Behavior." Unpublished manuscript, University of Tulsa, 1980.

10. For further insight into the clinical aspects of loneliness and their relationship to depression, see Carolyn E. Cutrona, "Transition to College: Loneliness and the Process of Social Adjustment," in Letitia Anne Peplau and Daniel Perlman, eds., *Loneliness: A Sourcebook of Current Research, Theory, and Therapy* (New York: Wiley Interscience, 1982).

11. David Russell, Letitia Anne Peplau, and Carolyn Cutrona, "The Revised UCLA Loneliness Scale: Concurrent and Discriminant Validity Evidence," *Journal of Personality and Social Psychology*, 39 (September 1980), pp. 472—480.

12. Jones, op. cit.

13. Jones, Freemon, and Goswick, op. cit.

14. See especially Cecilia Solano, "Two Measures of Loneliness: A Comparison," *Psychological Reports*, 46 (February 1980), 23—28; and C. Sansone, Warren Jones, and Bob Heim, "Interpersonal Perceptions of Loneliness," paper presented at the annual meeting of the Southwestern Psychological Association, San Antonio, Tex., April 1979.

15. Frieda Fromm-Reichman, "Loneliness," *Psychiatry*, 22 (February 1959).

16. Scott Wimer and Letitia Anne Peplau, "Determinants of Reactions to Lonely Others." Paper presented at the annual meeting of the Western Psychological Association, San Francisco, April 1978.

17. Martin Seligman, *Helplessness* (San Francisco: W. H. Freeman 1975).

18. Peplau, Russell, and Heim, op. cit.

19. See Eric Berne, *What Do You Say After You Say Hello?* (New York: Bantam Books, 1975).

20. See Cutrona, op. cit., for a discussion of maintenance factors.

21. Peplau, Russell, and Heim, op. cit.

The Good News About Depression

LAURENCE CHERRY

There are so many resources available (including light therapy, new antidepressants, and short-term counseling) that, a noted psychiatrist promises, "80 percent of the depressed can be significantly helped."

THE ELEGANT YOUNG WOMAN, TOO depressed to deal with the demands of her fast-paced job, has not been to work in three days; the deep-voiced man in his fifties, still troubled by a recent divorce, has not had the energy to open his mail in weeks. "I just watch it piling up, bills and all," he says indifferently. "I suppose I'll mind when Con Ed shuts off my electricity, but right now I don't give a damn."

An evening storm splatters the city but draws no comment from the people in the cozy auditorium on East 62nd Street. At this weekly meeting of Depressives Anonymous, one of several groups formed in recent years to help people cope with the blues, conversation centers on more pressing problems: how to make it through the night, how to summon the energy to arrive at work the next morning.

Depression has been with us for centuries, of course. Its celebrated victims have included the Bible's King Saul, England's first Queen Elizabeth, Abraham Lincoln, and, more recently, Ernest Hemingway, the poets Anne Sexton and Robert Lowell, and First Lady Betty Ford. Depression was the "black dog" that shadowed much of Winston Churchill's adult life.

What's more, the number of sufferers is growing. This past year, scientists at the National Institute of Mental Health reported that the incidence of depression of all kinds has "shockingly" increased. "This is particularly true among people born since 1940," says Dr. Elliot S. Gershon of the Clinical Neurogenetics Branch of NIMH. "More people are becoming depressed, and at much younger ages." The ailment is so prevalent, in fact, that some NIMH scientists privately speculate that an unknown "Agent Blue" may be spurring its spread.

Today, some 14 million Americans suffer from prolonged depression; one in four women and one in ten men can expect to suffer a serious bout at some point in their lives. The average age at the onset of the disease has dropped from about 40 a generation ago to the mid-twenties today, and some experts now estimate that depression affects 10 percent of those under 12. Between 1980 and 1984, adolescent admissions to private psychiatric hospitals increased more than 350 percent, with depression often cited as one of the main reasons for the rise.

But there's some cheering news. New antidepressants, as well as older drugs being used in novel ways, are producing dramatic relief. Research into light therapy and sleep cycles is showing great promise. And new psychiatric services at medical centers—as well as local support groups—have proliferated, helping the depressed deal better with their condition. "Well over 80 percent of people can be significantly helped," says Dr. David J. Kupfer, chief of the Department of Psychiatry at the University of Pittsburgh School of Medicine at the Western Psychiatric Institute and Clinic, who was chairman of a special 1984 consensus conference on mood disorders sponsored by the National Institutes of Health. "We've accomplished a lot."

To publicize psychiatry's increasing ability to relieve depression, NIMH is organizing a national Depression Awareness, Recognition and Treatment campaign (D/ART), which it hopes will be under way by the fall. Brochures for the public and special seminars for general practitioners and mental-health professionals—"often abysmally ignorant about effective treatment," says one top NIMH official—will emphasize that today, depression *can* be controlled. Astonishingly, "fewer than 20 percent of seriously depressed people in this country are being properly treated," says Dr. Robert M. A. Hirschfeld, chief of the Affective and Anxiety Disorders Research Branch at NIMH and the D/ART campaign's clinical director. "But there's no reason why. No depressed person nowadays should have to listen to someone—a relative or a professional—say, 'Come on now, just pull yourself together.' We've gone way past that in what we know and what we can offer, and we have to get the word about that out."

TREATING CHRONIC MILD DEPRESSION

DEPRESSION COMES IN MANY GUISES. According to psychiatry's bible, the *Diagnostic and Statistical Manual of Mental Disorders III,* issued in 1980 and currently being revised, there are at least a half-dozen subtypes, from major *unipolar* depression—depression that usually lingers for months before lifting—to the seesaw ups and downs of manic-depressive (*bipolar*) illness. "We've acquired a welter of labels," concedes Dr. Frederic Quitkin, head of the Depression Evaluation Service at the New York State Psychiatric Institute. "Probably the most important thing to remember is that depression occurs on a *spectrum,* from the ordinary down moods that everyone experiences to the major, crippling kinds of depression that may require hospitalization."

Scientific attention once focused almost exclusively on one end of the spectrum—depression that is almost totally incapacitating. But now scientists are finally paying attention to a long-ignored, less flamboyant form of the illness—chronic mild depression. Some 3 to 6 percent of people in the U.S. experience this kind of depression, according to NIMH's current Epidemiology Catchment Area Study, which involved 20,000 people in five cities (Baltimore, Los Angeles, St. Louis, New Haven, and Durham, North Carolina). These are not the people rushed to hospital emergency rooms with stomachs needing to be pumped, or slashed wrists. "Chronic mild depressives go to work and maintain relationships, but they never seem to get out from under the black cloud hanging over them," says Dr. James Kocsis, director of the Clinical Inpatient Research Unit at the Payne Whitney Clinic of New York Hospital—Cornell Medical Center. "Often they tell us they don't answer their phones because they're too down." This depressed mood almost never abates; instead, chronic mild depressives consider their gloominess to be the normal way of experiencing life and can't quite fathom why other people seem so energetic or buoyant.

In a certain subgroup of these mild depressives—"atypical" depressives—the bleak mood can sometimes briefly improve, doctors at the New York State Psychiatric Institute have noted. "A compliment, a success, an unexpected phone call from an acquaintance can cheer them up immensely," says Quitkin. "But their good mood doesn't last long. A few hours later, they're back in the dumps." These atypical victims reverse some of the symptoms of major depression: Instead of suffering from insomnia, for example, they oversleep (they may occasionally sleep as much as twelve to fifteen hours a day); rather than avoiding food, they overeat; and they feel

worst in the evening, rather than in the morning. "There's another marked trait—hypersensitivity to rejection," says Quitkin. "Trivial slights can devastate victims of atypical depression."

Only a few years ago, chronic mild depressives would not have been considered suitable candidates for antidepressants; their persistently low mood and social ineptitude would have been regarded as personality flaws that only long-term psychotherapy could, possibly, cure. But at the New York Psychiatric Institute, a recent twelve-week study of 120 atypical depressives revealed that over 75 percent responded to the antidepressant Nardil (phenelzine); a mere 24 percent improved on a placebo. "And within a few months, if you give these people standard psychological tests, including social-adjustment scales, you see striking changes," says Quitkin. "They're making friends, getting along better at work. That's amazing, since improving patients' social skills is one of the hardest things to achieve. But this happens so quickly—usually within six weeks—that we now believe these people have a flaw in their neurochemistry that kept their mood down and made them so clumsy around others."

Celia Burke (not her real name), 47, a former copy editor at a Manhattan publishing house, has worked on and off (mostly off) as an office temp for the past five years. For most of her life, she's lived with her mother. She'd already been in therapy for years, Celia told researchers at the New York State Psychiatric Institute last fall, but had not been helped. She was enrolled in a six-month study at PI and was informed she would be given Nardil or a placebo. For weeks, Celia felt nothing. "But then in the fifth week, I suddenly felt a *surge* of energy," she recalls. "I bounded up subway steps. And my life's turned around. I'm dating again for the first time in years, I'm going on job interviews, I'm full of goals."

Other kinds of antidepressants seem able to help victims of both chronic mild depression and major episodic depression—if doctors can only be motivated to write the prescription. (At Payne Whitney, researcher Kocsis found that despite the fact that three quarters of patients studied had been in psychotherapy, only 15 percent had ever received a trial course of antidepressants. Well over half responded "beautifully" when the Payne Whitney team prescribed them.) Unfortunately, the old split between the drugs-only or psychotherapy-only camps persists in depression treatment, although the gap is narrowing. "Part of the message we're trying to get out to practitioners around the country is that flexibility's the key," says psychologist Harold Goldstein, coordinator of training of NIMH's D/ART campaign. "If a month or two has gone by and talk therapy isn't helping your patient's depression, it may be time to try drugs, just as psychotherapy can often help people taking antidepressants recover even more quickly."

Indeed, the impressive ability of two new types of brief psychotherapy to help the depressed was underscored only a few weeks ago, when scientists announced the results of a long-awaited major NIMH-supported study. It turns out that both cognitive behavior therapy (which teaches patients to identify and change unrealistic, negative, or pessimistic views of the world and themselves) and IPT, interpersonal psychotherapy (which focuses on helping the depressed improve their dealings with others), ease depression as effectively as the standard antidepressant drug Tofranil. All three approaches dramatically improved the mood of depressed subjects within the startlingly short span of sixteen weeks.

Along with more rapidly effective forms of psychotherapy, psychiatrists can now offer the depressed an array of new drugs (page 42) as well as better strategies for using some older ones. The new drugs have fewer side effects (such as the all-too-common blurry vision, constipation, or dizziness) than the older drugs. And some begin to work more quickly—often within days, rather than weeks—thereby shortening the "suicide watch" period, when doctors must anxiously watch very depressed patients to be sure they don't harm themselves. "Today, up to three quarters of the depressed can be helped with the right medication," says Dr. Leslie L. Powers, former director of Group Psychotherapy at St. Luke's-Roosevelt Hospital Center. "We can tell patients who may have tried and failed before, 'Look, there are new drugs, new combinations, that can help you now. We know a lot more about dosages that work, and which drugs may succeed when others have not. It's a different ball game.'"

'In my fifth week on the antidepressant, I felt a *surge of energy*," one patient recalls. "I bounded up the steps. I'm dating again for the first time in years. My life's turned around."

IMPROVING MOOD WITH LIGHT

'IT'S THIS DREARY WINTER—THAT'S WHY I'm depressed," people have been complaining for years. And for years experts have dismissed the idea as an old wives' tale. But it seems dreary winters *can* cause prolonged depression—at least for victims of a newly discovered kind of depression called Seasonal Affective Disorder (SAD). No one yet knows how many people suffer from the syndrome. "My guess is that there are many thousands in the New York metropolitan area alone," says Dr. Michael Terman, research psychologist at the New York State Psychiatric Institute and head of its Light Therapy Program. These are the people who grow depressed as the long hours of summer light begin to give way to the shortened days of fall and winter. They become lethargic, sleepy, and begin to gorge themselves on carbohydrates. Their ability to concentrate, work, and enjoy sex begins to fade as well.

Dr. Norman Rosenthal, chief of outpatient services at the Clinical Psychobiology Branch at NIMH, was the man chiefly responsible for identifying the disorder in the late seventies. In 1980, basing his work on other research at the institute, he and his team came up with a tentative explanation of how the disorder might be triggered. Although biologists have long known that light affects animal behavior (as in determining reproductive cycles), only recently were they able to prove that the pineal gland, a tiny protuberance at the base of the brain, is not a vestigial organ (like the appendix) but the body's "Dracula" gland. Coming to life each night, it

secretes melatonin, a hormone that seems to play a key role in maintaining the biological clock that keeps our body rhythms running smoothly on their daily cycles. Taken orally, the hormone makes subjects drowsy, lethargic, drained. NIMH researchers proved that light suppresses melatonin production.

Based on hundreds of cases, says Rosenthal, the profile of a typical SAD victim has gradually emerged. She (female SAD victims far outnumber male victims) is a woman in her early thirties who has suffered from the syndrome, often without understanding what was wrong, for years.

Treatment for SAD is remarkably simple: exposure to a two-by-four-foot rectangular fixture studded with Vita-Lites—special fluorescent lights devised in the mid-seventies that include all the colors found in natural daylight, from far red to ultraviolet. Vita-Lites are manufactured by the Duro-Test Corporation of North Bergen, New Jersey; the fixture it designed produces 2,500 lux—slightly less than the light outside just when the sun is over the horizon on a clear day. (To compare: The standard fluorescent lighting in an office is 500 lux; the light outdoors at noon on a sunny June 21 is about 113,000 lux.)

Those who respond to treatment do so remarkably quickly—usually within four days, report Rosenthal and Terman. The turnaround can be dramatic. Last November, RoseAnne Tockstein, 42, of Franklin Lakes, New Jersey, happened to hear a friend mention the research on winter depression being conducted at the New York State Psychiatric Institute. Raised in Wisconsin, where winters are long and cloudy, she remembers her strange seasonal slumps. Her husband's career required the couple to relocate to New York. "It's a little better than Wisconsin, but come October, just about the time we go off daylight saving time, the same old depression begins," she says. "In spring, summer, I'm a go-getter—a hospice volunteer, industrial photographer, floral designer. But in winter, during those inevitable stretches of overcast weather, I'll just get up, stare out the window, and spend the rest of the day in bed. I don't function." Tockstein saw various doctors and described her symptoms; most were skeptical, and none could offer her any help.

After interviews and tests by the light-therapy staff at the New York State Psychiatric Institute, Tockstein began her first light treatment last December 6, sitting in front of a light fixture that resembles a large, glowing mirror. (The lights hurt her eyes, but now she does office work in front of them without any discomfort.)

At first, she was disappointed; she saw no change in her mood. But after three days of treatment, Tockstein's depression began to lift. The institute lent her a Vita-Lite unit, which she installed in her family room. She sat in front of it every day from 6 to 8 A.M. and 6 to 7 P.M.

The log Tockstein has kept shows her steady improvement. Last December 11, she wrote: "Today is gray, foggy, overall dreary, the kind that always put me into a depression. . . . But I don't have to head for bed." On December 29: "I woke up tired, but felt emotionally well. Wonderful day. I'm much better—near my old summer energy levels."

Children, too, seem to suffer from SAD. "This is an area we're just beginning to explore," says Rosenthal. His team has treated four boys and two girls, ages six to fourteen, who began to complain of being tired and unhappy in school in the late fall; they slept later as the weeks passed, had difficulty waking up, cried, and complained of headaches; January and February were their worst months. When Rosenthal put the children on a light-therapy regimen, almost all of them rapidly improved: One, a champion swimmer whose times had always mysteriously dropped in the winter, broke his previous records; another, a thirteen-year-old boy, made the honor roll at school for his first winter ever.

Some depression researchers even suggest now that millions of adults who don't have a full-blown case of SAD may nevertheless be suffering from a wintertime loss of emotional equilibrium. "They still function, but without the ease or efficiency that they have in other seasons," says Michael Terman. (Those wishing to take part in light-therapy studies at the New York State Psychiatric Institute are welcome to call 212-960-5714.) Other investigators theorize that "for one reason or another, a large number of depressed people are light-starved *throughout* the year," says Dr. Jack D. Blaine, a psychiatrist at NIMH.

Not surprisingly, as news about SAD has spread, Duro-Test reports soaring sales of its light units, available for $477. But most experts caution against self-treatment. "People may wrongly diagnose themselves as being SAD sufferers when in fact they're suffering from another kind of depression," says Dr. Boghos Yerevanian, head of the Affective Disorders Program at the University of Rochester Medical Center, where light therapy for depression is also being studied. "No eye damage has yet been reported, for example, but obviously we can't yet predict long-term consequences; this is still too new." For SAD victims with manic-depressive illness as well, says Yerevanian, "the lights may push them into a manic episode; they may become agitated and grandiose. We've heard of cases where secretaries who used the units went out and bought $25,000 cars they couldn't possibly afford. Light therapy appears to be generally safe, but it isn't something you should trifle with."

RESETTING THE BIOLOGICAL CLOCK

LIGHT THERAPY FOR DEPRESSION HAS helped to illustrate the important role of the body's daily rhythms: Light treatments in the morning, for example, are generally more effective than those given at night. Moreover, victims of unipolar depression generally feel worse in the morning; manic-depressives are more gloomy as the day wears on; most atypical depressives are at their lowest in the evening.

Depression plays havoc with sleep. "The sleep cycles of depressed people are disordered in all kinds of ways," says Dr. Neil Kavey, head of the Sleep Disorders Center at Columbia-Presbyteri-

an Medical Center. Most depressed people take a long time to fall asleep and wake hours earlier than others. In the non-depressed, the first REM period of the night—when dreaming usually occurs—generally begins 90 minutes after falling asleep, but in the severely depressed, it may begin a mere 20 minutes after falling asleep. Moreover, the REM activity of the severely depressed is abnormal, with unusually intense bursts of eye movement. The pattern of healthy sleep is altered: The depressed generally have the most REM activity during the first third of the night; the healthy usually experience the most REM periods during the last third of the night.

In many sleep labs, such as those at Columbia-Presbyterian and Montefiore Medical Center, doctors can now routinely use these striking differences to decide whether a person is clinically depressed or suffering from an illness—such as Alzheimer's disease—that can mimic true depression. "Testing to see if REM periods occur abnormally early in the night is usually a neat, easy way of confirming depression," says Kavey.

As useful is a sleep-monitoring technique that can predict whether a particular antidepressant will work. If it is successful, it will usually move REM patterns closer to normal within two nights. "This can save you precious time—weeks, in fact—by preventing you from keeping the patient on a medication that isn't going to help him," says one expert. "You can then turn to another antidepressant that *will* work."

It was not a very long step from using sleep patterns to diagnose depression and predict a drug's potency to actually modifying sleep to improve mood. At the Sleep-Wake Disorders Center at Montefiore, researchers are doing just that with patients whose body clocks have gone awry. The human organism seems naturally set to a 25-hour cycle; most of us use time cues from light and clocks to adapt to a 24-hour day. "But in some patients—mostly adolescents—the ordinary cues just aren't enough," says psychologist Paul Glovinsky, a Montefiore sleep expert. "The result is someone who can't function in sync with a normal day, and as a result may become progressively more depressed."

Christopher (not his real name), a sixteen-year-old student at a New England prep school, came to Montefiore last August, referred by a psychiatrist whom his parents in Westchester had consulted about his worsening depression. Once an honor student and star athlete, he fell asleep in class five times last year, annoying himself and angering his teachers. "He was on academic probation, had lost his enthusiasm for sports, and talked openly about suicide," says Glovinsky. His sleep habits had changed drastically during the past year. Where once he had gone to sleep at 11 P.M., he was now unable to fall asleep until 3 A.M.

Monitoring confirmed his disordered sleep patterns. "We put him on a 27-hour day for a week, in effect resetting his body clock," says Glovinsky. The first night of treatment, Christopher went to bed at 3 A.M. and was awakened at 10:30 in the morning; the following day he went to bed at 6 A.M. and was roused at 12:30 in the afternoon. Within a week, he had been moved

In fall and winter, many people become lethargic and lose their ability to concentrate, work, and enjoy sex. A new therapy—exposure to special lights—may improve their mood within three days.

back to a falling-asleep time of 11 P.M.—and was able to maintain it at that hour. His fatigue and depression soon vanished. Back at school, the eleventh-grader is doing well; his mother recently phoned the sleep center to thank the staff for performing "a miracle." The Montefiore team has treated almost two dozen other depressed young victims of this delayed-sleep-phase syndrome; although the specialists are still unsure exactly why their chronotherapy is so successful, many of their patients report improvement in mood.

The striking connection between sleep and depression has been demonstrated even more graphically by another discovery: Depriving depressed patients of sleep, even for one night, can often immediately buoy their spirits. The key seems to be not lack of sleep but elimination of REM periods. Unfortunately, the effect lasts only a few days—but frequently that's all that's required before an antidepressant can take effect. "We use this to tide some patients over until their drugs begin to act," says Glovinsky. "It usually works very well indeed."

SHOCK THERAPY: ENLIGHTENED USES

A MIDDLE-AGED WOMAN LIES IN AN anesthesia-induced sleep in a narrow, brightly painted room in the New York State Psychiatric Institute. Her scalp is wreathed with tiny electrodes that look like a tangle of hair curlers; the only sound is the low, persistent thump of monitors tracking her vital functions. As a psychiatrist applies electric current to her temples, her legs and arms jerk almost imperceptibly; five minutes later, wheeled into a makeshift recovery cubicle in an adjoining room, she is awake, still a bit groggy, eager to know about her breakfast. The entire procedure has taken no longer, and has been barely more dramatic, than a session in a dentist's chair.

ECT—electroconvulsive (electroshock) therapy, the oldest of the treatments in use for depression—has never quite lived down its spooky reputation as a psychiatric torture callously practiced on the poor and helpless, an indelible image left behind by popular movies like *One Flew Over the Cuckoo's Nest*. Groups of former mental patients in California and Vermont have recently tried to have the treatment banned altogether. "But ECT bears little similarity to the old caricature," says Dr. Arnold J. Friedhoff, professor of psychiatry at NYU School of Medicine and a member of a panel of experts who participated in a special NIMH conference on electroconvulsive therapy held last July. "The procedure's become both acceptable *and* respectable again in depression treatment." Between 10 and 20 percent of depressed patients do not respond to drugs or cannot be given them because of medical conditions (such as certain serious heart problems). Moreover, in cases where patients are

violently suicidal, ECT can often offer prompt relief.

Many experts admit that in the past, ECT was indeed wrongly administered to patients who could not benefit from it, such as those with vague anxiety disorders. Today, only about 60,000 to 100,000 people a year receive ECT treatments, about half the number getting them twenty years ago. Contrary to the stereotype, the typical ECT patient is well-to-do and receiving treatment for major depression in a university teaching hospital after drug therapy has failed. Public hospitals avoid giving ECT; it's too expensive.

Memory loss is one of the main complaints of patients who've undergone ECT, and the recent NIMH conference agreed that the loss is real, generally involving poor recall of events that occurred six months before and two months after treatment. Partly to lessen the impact on memory, low-dose ECT—about one third as strong as the standard dose—is becoming more popular. A five-hospital study in the New York area is evaluating its effectiveness.

But Dr. Harold Sackeim of the New York State Psychiatric Institute admits that even in its improved version, ECT is most often used only with the small percentage of the depressed for whom other treatments haven't worked, or probably won't. As researchers unravel the biochemical and other mysteries of depression, acquiring an ever-sharper skill in dealing with the complexities of this most common of mental ailments, the role of ECT is likely to shrink even more. "Only three decades ago, this was about all we could offer the depressed," Sackeim says. "It's a good measure of the astonishing progress we've made in recent years that it's now just one option out of several. And we're delighted that's so."

DEPRESSION CLINICS

O BVIOUSLY, NO PLACE SEEMS LIKE A very lucky place in which to be depressed. But New Yorkers *are* lucky in that this city offers a diversity of specialized services and support groups—many of them newly formed—to help the troubled. "Outpatient mood-disorder programs" have become so trendy, in fact, that it would make sense to check with your local hospital to see if it has already established one. Here's a sampling of the many programs for the depressed available in the metropolitan area. Unless otherwise noted, all are open Monday through Friday between 9 A.M. and 5 P.M.

The **Depression Evaluation Service, New York State Psychiatric Institute,** 722 West 168th Street (960-5734). Few in the field would deny that PI, with its host of ongoing depression studies, is the leading research and treatment center for depression in the metropolitan area. Its Depression Evaluation Service, an outpatient clinic, offers free diagnosis and treatment to all those accepted into its research projects. "Subjects really get about $3,000 worth of treatment free, and the only risk they face is possibly being exposed to a harmless placebo for six weeks," says director Frederic Quitkin. "If a patient doesn't respond to the drug we're investigating, he or she doesn't continue in the study, but we offer access to all other possible forms of treatment [at no cost]." Quitkin notes that PI is now studying and treating people with alcohol problems and a history of depression, anxiety, or panic. Along with the main facility at 168th Street, the institute maintains a satellite depression clinic at 79th Street and Third Avenue. A few sessions are scheduled until 8 P.M. on Mondays.

Child and Adolescent Depression and Suicidal Disorders Clinic, Columbia-Presbyterian Medical Center, 622 West 168th Street (305-3093). In 1984, Presbyterian Hospital established this adolescent clinic to deal with the growing numbers of depressed youngsters. A staff of child psychiatrists, psychologists, and nurse clinicians diagnoses and treats about 200 children and adolescents a year—some as young as six years old. "We evaluate them and then decide which treatment would be most effective," says the director, Dr. Paul Trautman. "Medication, family therapy, and group therapy are all options." Each session is $50; some patients may pay on a sliding scale.

Depression Clinic and **Post-Schizophrenia Depression Program, Mount Sinai Medical Center**, 1450 Madison Avenue, at 99th Street. These clinics treat about 50 patients a year and evaluate many more, says Dr. Samuel Siris, director of outpatient psychiatry at Mount Sinai. Therapy ranges from drugs to group therapy, individual counseling, or a combination of techniques. Associated with the clinic is a special program—reportedly unique in the country—for the treatment of depression in schizophrenics whose illness has stabilized, usually thanks to medication. "Although these patients aren't hearing voices anymore, they're often very, very down," says Siris. "Sometimes this is a result of the antipsychotic drugs they have to take, but in other cases it seems to be quite a separate problem. But no matter what the cause, we've had good success in treating their depression in many cases." Fees are based on the patient's ability to pay. For more information, call the Depression Clinic (650-7191) or the Post-Schizophrenia Depression Program (650-7192).

The **Adolescent Health Center at Mount Sinai Medical Center**, 19 East 101st Street (650-6016), deals with the gamut of adolescent medical problems. "But frequently you find that teenagers who come with headaches, stomach pain, or chronic fatigue are really suffering from depression," says Dr. Richard Wortman, a psychiatrist and director of the center's Mental Health Unit. After diagnosis, the hundreds of depressed adolescents who come to the center each month

N ew York City offers a diversity of services and support groups to help the troubled.

YOUTH SUICIDE

PSYCHIATRIC DOGMA ONCE HELD THAT ONLY adults, never children or adolescents, could be "truly" depressed. "We've learned that was nonsense," says Charlotte Ross, director of the newly established Youth Suicide National Center in Washington, D.C. Within the past two decades, the suicide rate among children and adolescents has increased threefold; suicide is now the third leading cause of death for those between 15 and 24, with 5,000 such deaths verified in 1984 and an estimated 120 attempts for each death.

Very few studies on youth suicide have been conducted; all have presented only rudimentary findings. One study, begun in 1984 and headed by Dr. David Shaffer, director of child psychiatry at Columbia University, is examining all suicides under nineteen in the New York metropolitan area. Almost two thirds of the teens studied had seen a mental-health professional, but only one fifth had shown clear-cut signs of suffering major depression. Two thirds had a history of antisocial behavior or drug or alcohol abuse; 14 percent were homosexual, and often under severe social stress. "But the most striking characteristic we've seen in young suicides and suicide attempters is that they tend to be both impulsive and poor problem solvers," says Paul Trautman, head of the adolescent clinic at the Columbia-Presbyterian Medical Center. "Suicide isn't something they plan, it's just something they do—usually on the spur of the moment." One sixteen-year-old boy Trautman treated had a fight with a friend over a radio, went home, and swallowed 60 sleeping pills; his parents found him unconscious a few hours later when they returned home, and rushed him to the hospital, where his stomach was pumped. "This wasn't the result of premeditation, or even prolonged depres-

sion," says Trautman. "The boy was angry, happened to see the pills in his parents' bedroom, and thought, 'Why not?'"

Other experts don't feel that the Columbia study reflects the truth about adolescent suicide. "Yes, some attempters are impulsive, antisocial types," says Ross. "But that's only one subgroup. Just as often, these are cream-of-the-crop kids—class presidents, top honor students—who are simply very unhappy behind their façade of unruffled success."

Treatment for suicide attempters usually involves psychotherapy and family therapy. Most medical centers reserve ECT only for those over eighteen because of the memory losses it can cause; until recently, even giving antidepressants to adolescents was unusual. "But that's become much more common within the past five years," says Trautman; some pilot studies have even reported success at treating very depressed or suicidal children under age twelve with drugs.

Prevention obviously is the main goal in coping with teen suicide. But Ross doubts if conventional psychiatric treatment is an effective preventive measure, because "most kids hate therapy—there's too much stigma attached to it." One of the aims of the Youth Suicide National Center is to bring suicide prevention into the classroom, where until recently it was usually a taboo subject. "Suicidal kids don't turn to counselors, or their parents—over 90 percent look to their friends for help," she says. "We're promoting efforts in which people go into schools and teach kids the warning signs of depression and suicide that they can spot in their classmates." This has already worked in California, which has one of the most active state teen-suicide-prevention programs, says Ross. Governor Mario Cuomo's task force on teen suicide is studying a similar approach for dealing with the problem in New York. **—L.C.**

typically undergo a one-month trial of psychotherapy before drug treatment (if any) is attempted. Self-referral by teenagers is welcomed; parents need not be informed, although the staff usually encourages family involvement. A few patients pay nothing, if their financial circumstances warrant that; others pay from $5 to $30 per session. The center is open Monday through Friday from 8:30 A.M. to 5:30 P.M.

Anxiety and Depression Clinic, New York Hospital–Cornell Medical Center, Westchester Division, 21 Bloomingdale Road, White Plains, New York (914-997-5967). Established only this past March, the clinic emphasizes a multidisciplinary approach to anxiety, phobia, and depression treatment, says its chief, Dr. Joseph Deltito. That includes drug therapy, psychotherapy, stress management, hypnosis, and behavior therapy. "We think our strength is that we don't favor one therapeutic approach over another," Deltito says. The clinic draws most of its patients (who must be eighteen or older) from the Bronx, and Westchester, Fairfield, and Rockland counties, but a few come from as far away as Brooklyn and New Jersey. Fees range from $5 to $100 per session, based on the patient's ability to pay.

At the **Mood Disorder Clinic at Hillside Hospital,** a division of Long Island Jewish Medical Center at 76th Avenue and 266th Street, Glen Oaks, Queens (718-470-8151), the depressed are offered both drug treatment and psychological support. Current research projects include the use of the new but not yet marketed antidepressant Wellbutrin; low-dose lithium for manic-depressives; Pramircetam, a new drug that *may* help reduce memory loss after ECT; and other drugs. Fees are based on the patient's ability to pay.

Affective Disorders Clinic, Montefiore Medical Center, 111 East 210th Street, the Bronx, New York (920-4596). Although all kinds of depression are treated at Montefiore, the main activity is investigating the biology of depressive illness and how patients respond to one tricyclic antidepressant, desipramine (Norpramin, Pertofrane). Those enrolled in the research project are treated with the drug or a placebo for six weeks; subjects who benefit from the drug are continued on it for six months, while others are referred for different kinds of treatment. Fees are levied on a sliding scale. "Self-referrals are very welcome," says the director, Dr. Gregory Asnis. The clinic is open Monday through Friday from 9 A.M. to 7 P.M.

The **Affective Disorders Clinic at Beth Israel Medical Center,** First Avenue near 15th Street (420-4135), opened last month. The clinic mostly treats depressed patients ranging from pre-teenagers to people in their late seventies. "We offer the full range of antidepressant drugs but especially emphasize short-term interpersonal psychotherapy, lasting no longer than ten to fifteen weeks," says psychologist Steven Klee, the program director. Each session costs $50, but there's a sliding-scale fee applied in some cases.

Depression Studies Program, New York University Medical Center, 560 First Avenue, at 32nd Street (340-5705). "Our emphasis here is on a sadly neglected problem—depression in the el-

derly," says staff psychiatrist Dr. Robert McCue, clinical instructor of psychiatry at the NYU School of Medicine. "Although geriatric depression is the most common psychiatric illness among the aged, we still know little about it." What *is* known is that older patients often take longer than younger patients to respond to antidepressants—sometimes several weeks longer. "The elderly are often misdiagnosed as hopeless, when in fact they would respond in time if their doctors were more patient," says McCue. "We've seen tragic mistakes. In our experience, most older people *do* eventually respond well to antidepressants." Those who improve are followed for up to three years. "We try not to give up on anyone," says McCue. Diagnosis and treatment are free to those 55 and older.

SUPPORT GROUPS

IN ADDITION TO THESE PROFESSIONAL services, a wide variety of nonprofit support groups for the depressed—and their relatives—have also sprung up in New York. Self-help is the theme, and in many cases these groups can indeed be helpful.

Depressives Anonymous. Founded in 1977 by psychiatrist and author Dr. Helen DeRosis (*The Book of Hope*), Depressives Anonymous meets most Wednesday evenings in the auditorium of the Karen Horney Clinic, 329 East 62nd Street. "This is a self-help group where the focus is strictly on the depression, the stress associated with it, and what concrete steps the person can take to get out of it," says DeRosis, a motherly-looking woman who occasionally serves as moderator. At a recent meeting, for example, a woman depressed about her unfinished graduate thesis agreed she could probably manage to write a page a day; an older woman distraught over the end of a romantic relationship agreed that she could get in touch with several acquaintances she hadn't seen in some time. "These small steps can help you out of your depression," urged DeRosis. For information, send a stamped envelope to Depressives Anonymous, 329 East 62nd Street, New York, New York 10021.

The **Manic and Depressive Support Group** (924-4979) was founded in 1981 by depressives, manic-depressives, and some of their family members. There are chapters in Manhattan, Long Island, and New Jersey; groups are also forming in the Bronx and Westchester. Separate lecture-meetings for depressives and manic-depressives are held once a month at Beth Israel Medical Center; the meetings, involving a talk and question-and-answer session, are open to patients and their families and friends. These are large gatherings; as many as 250 people turn out. At other monthly meetings, members form into small groups of a half-dozen or so and share advice, complaints, and encouragement about the difficulties—often medication-related—of understanding their ailment. Membership is $25 per year (individual) or $40 (family). Non-members pay $3 per session. For information, write the Manic and Depressive Support Group, 15 Charles Street, 11H, New York, New York 10014.

'F requently," says the director of Mount Sinai's Mental Health Unit, "teenagers who come in with headaches, stomach pain, or chronic fatigue are really suffering from depression."

At **Recovery, Inc.**, participants are gently taught not to use the word "depressed"; the less dramatic term "lowered feelings" is preferred. "But obviously, depression is one of the problems we see most often," says spokeswoman Marion Zukoff. Recovery is an international nonprofit support network that subscribes to the tenets of Dr. Abraham A. Low, the late Chicago psychiatrist. In the 1930s and 1940s, Low was a pioneer in advocating and organizing self-help groups for those with nervous problems, including former mental patients. At the amiable Friday-evening Recovery meeting I attended at St. Vincent's Hospital, Dr. Low's writing and common-sense dictums served as the focus for a discussion of problems; afterward, most of the dozen participants adjourned to a coffee shop on Sixth Avenue for more informal group support. (Low believed that after-meeting socializing is often crucial for troubled people, who frequently isolate themselves.) Recovery meetings are held nightly somewhere in the metropolitan area—in churches, synagogues, Ys, mental-health clinics, and other public spaces. Contributions are requested. For more information, call 718-266-4715; in Westchester, 914-968-6021; in Nassau, 516-333-6500; in Suffolk, 516-289-3071.

New Images for Widows (972-2084) was formed in 1983 by author Lynn Caine (*Widow*) and social worker Pat Bertrand. Its aim is to help widows and widowers in the metropolitan area through the trying initial weeks of bereavement and then through the slow process of readjustment to the social world. "The standard rule of thumb among psychiatrists has been that widows experience depression in some form for at least a year following the death of a spouse," says Dr. Ronee I. Herrmann, medical adviser to the group. "In fact, in many cases the depression continues for three to five years." The subgroup Social Networks for the Widowed, which has 85 members ranging from women (and some men) in their twenties to those in their seventies, holds monthly meetings and regular outings. Membership is $15; meetings are $5.

Families of Depressives. "I started the group in 1983 because there was nowhere I could turn," Myrna explained to me over the phone. When her husband was being treated for depression, she found herself in a quandary familiar to relatives of the depressed: how to cope with the new situation and how to deal with a painful lack of understanding from relatives and friends. "I did my crying and said, 'Enough of that,'" she says. "I was sure that there must be others in the same boat." There were. She got in touch with the New York City Self-Help Clearinghouse and soon had formed a group. Now composed of about a dozen members—its numbers constantly fluctuate as relatives improve—it meets every other Wednesday evening at the Brotherhood Synagogue at 28 Gramercy Park South.

The depressed relative is deliberately *not* invited to attend; this is the time when spouses, parents, children, or siblings can air all their accumulated grievances and complaints. "You walk a thin line between love and resentment," one woman admitted in their free-ranging discussion. The resentment is directed at doctors who prescribe medications that may not work as quickly as hoped or cause troublesome side effects; at uncooperative friends and relatives; at the demands of the depressed person himself (most of the Families of Depressives participants are women). But the deep ties of affection were as obvious as the understandable irritation. "Thank God for Wednesday evenings; we laugh and cry and go back to our posts in a decent mood," said a woman whose aged mother has been intermittently depressed for the past two years. "This is what keeps *us* from going down the tubes ourselves." Information about doctors and new treatments is exchanged; in between the weekly meetings, many members participate in a telephone support network. Contributions ($2) for rental of the room are required. For more information, get in touch with the New York City Self-Help Clearinghouse, 186 Joralemon Street, Suite 1100, Brooklyn, New York 11201, or call 718-352-4290.

NEW, FAST-ACTING ANTIDEPRESSANTS

TRADITIONAL TRICYCLIC DRUGS (THE name derives from their three-ring chemical structure), such as Tofranil and Elavil, have been on the market since the early sixties. They continue to be prescribed, along with lithium, for manic-depressive illness. But doctors can now choose among several newer drugs as well as older ones being used in new ways. Here are several, most of which have come on the market in the last six years:

Asendin (amoxapine). Introduced in 1980, Asendin has a mild sedative effect along with its antidepressant properties; its mechanism of action is still not well understood.

Advantages: "Asendin's biggest asset is that it works quickly," says Dr. Arnold Friedhoff, professor of psychiatry at NYU's School of Medicine and director of its Millhauser Laboratories, who participated in researching the drug. "According to our studies, it was effective within four days—or earlier."

Drawbacks: It may be harmful to patients with cardiovascular problems. It causes drowsiness in 14 percent of those taking it, constipation in 12 percent, blurred vision in 7 percent.

Desyrel (trazodone). Introduced in 1982, it is unrelated to tricyclics or other antidepressants.

Advantages: Rarely causes the dry mouth, constipation, urinary retention, or cardiac problems common with older drugs.

Drawbacks: In some male patients, Desyrel has caused priapism—a disorder that involves prolonged erection of the penis; this sometimes requires surgery and has led to permanent impotence. Mead Johnson (the manufacturer of the drug), while stating that such cases are "extremely rare," strongly advises male patients with "prolonged or inappropriate erection" to immediately discontinue use and consult their physician.

Participants adjourn to a nearby coffee shop after the amiable Recovery meetings; socializing is crucial for troubled people, who often isolate themselves.

The depressed relative is *not* invited to attend meetings of the support group Families of Depressives; this is the time when spouses, parents, and children can air their grievances. "You walk a thin line between love and resentment," one woman admits.

Ludiomil (maprotiline). *Advantages:* Fewer cardiovascular side effects are reported with Ludiomil than with older antidepressants.

Drawbacks: The drug causes dry mouth in 22 percent of patients, drowsiness in 16 percent.

Tegretol (carbamazepine). *Advantages:* It has been used as an anti-convulsant drug for epileptics; researchers at NIMH have found that some 60 percent of manic-depressives who do not do well on lithium do, in fact, do well on Tegretol. Patients who have many seesaw mood swings during a year appear to do better on Tegretol.

Drawbacks: The drug may impair bone-marrow function and cause blood abnormalities.

Xanax (alprazolam). *Advantages:* Used primarily for victims of anxiety disorders—especially for those who suffer panic attacks—Xanax is now also being employed as an antidepressant, particularly for those who suffer from depression combined with agitation. It is especially helpful in mild to moderate depressions.

Drawbacks: Chemically related to Valium, Xanax poses similar problems of dependence with prolonged use.

Nardil (phenelzine). "Nardil is two decades young," says Dr. Frederic Quitkin of the Depression Evaluation Service of the New York State Psychiatric Institute. A member of the monoamine oxidase inhibitor (MAOI) family (which also includes Parnate and Marplan), it was largely discarded in the 1970s because of the strict diet that patients taking it must follow. Within the past two years, the drug has been increasingly prescribed.

Advantages: According to studies at PI, Nardil is the most effective drug yet tested for atypical depression.

Drawbacks: When combined with the substance tyramine, found in some food, beverages, and drugs, Nardil can cause blood pressure to soar. Patients must therefore follow a careful regimen. Prohibited items include processed meats, aged cheeses, pickled foods, beer, red wines, sherry, amphetamines, barbiturates, some cough medicines, and nasal decongestants. Research continues into MAOI drugs that will be as effective as Nardil without its potential dangers; one, deprenyl, is currently being tested in New Jersey and at the New York State Psychiatric Institute and, according to experts, may be marketed within the next two to three years.

Aging

What Happens to the Body as We Grow Older?

Given our medical ignorance and the fact that the body does not age all at once—we can have a young kidney and an old heart—the whole concept of aging needs careful re-examination.

Is it possible to describe "typical" aging? Not really. We talk about someone who has begun to be forgetful, whose skin shows signs of losing elasticity, whose lung capacity is diminished because of emphysema, whose cardiac reserve is diminished because of atrophy of the heart muscle, whose organ functions (for instance, kidney or liver function) are at a fraction of what they once were, whose skeletal structure is softened, whose hair is grey, whose eyes are clouded by cataracts and whose hearing is diminished—that is a caricature of a typical old person.

This stereotypical image of aging, however, does not hold true among all individuals. The debate continues over how much of elderly appearance is the result of natural aging and how much is the result of abuse of the body. According to William Kannel, one of the principal investigators of the Framingham Heart Study, "The issue of what constitutes aging and normal aging is an enigma that has never been satisfactorily solved." In epidemiology, there are several approaches that try to answer the question.

One approach is employed by the Veterans Administration Normative Aging Study, which seeks to find people who might eventually develop certain diseases but, at the beginning of the study are free of any ailment. By studying what happens to people over time, project director Pantel Vokonas and co-workers are trying to identify the effects of age. The researchers are looking at the signs of growing older that can be attributed to age rather than illness or disease.

The normative aging approach is, in a sense, a quest for immortality. The assumption is that if we could remove all these diseases, people would live, if not forever, at least for much longer than they now do. Unfortunately, it is virtually impossible to find people totally free of the disabilities sooner or later associated with old age. Even if a person seems to be free of heart or kidney damage, for example, there is no way of being sure that these organs are still in their pristine state.

Another approach to aging is that taken by the Framingham Heart Disease Epidemiology Study, underway since 1949, in which a whole population is followed as they age to see what problems they encounter. "Ours is a more pragmatic approach," said Kannel. "We are interested in seeing what kinds of things cause people who reach advanced age to no longer have much joy in living. We don't care whether it is cardiovascular disease, opaqueness of the lens, poor hearing, soft bones, arthritis, strokes, mental deterioration or normal aging. We're studying the ailments that afflict an aging population and take the joy out of reaching a venerable stage in life."

According to Kannel, "The reward for reaching a venerable stage of life is too often a cardiovascular catastrophe." Cardiac function, muscle, skeleton and so on all decline with time, although recent evidence suggests that cardiac function in a non-diseased heart remains amazingly stable well into old age. Some of this decline must be due to wear and tear, but according to Kannel, "it is just too difficult to dissociate from the long-term effects of noxious influences."

With respect to cardiac function, for example, decline is not necessarily unpreventable. It has been shown that 65-year-olds can be trained to improve their levels of performance. It is easy, for example to train somebody to restore his or her exercise capacity and measure cardiac function and oxygen utilization. We have the technology to measure these things. But how do you train a kidney? Moreover, it is still not clear what noxious influences cause the decline in function in most of the organ systems. If we did know, we would be able to remove the noxious influences and watch the recovery. "For many of the organ declines, we really have poor information," said Kannel. "It just so happens that for cardiovascular disease, we have a good body of data on what risk factors there are. And it turns out that many of these are modifiable."

Given our medical ignorance and the fact that the body does not age all at once –we can have a young kidney and an old heart–the whole concept of aging needs careful re-examination. The assumption that all the organs fail in concert is not borne out by experience. There are many people who are alert and showing few signs of diminished intellectual capacity, but have a failing heart or damaged liver. People have different rates of organ decline.

The Framingham Heart Study, however, is showing that many of the risk factors for the young are still operative in the elderly. Even though it may take decades for the disastrous effects of a habit like smoking to show up, there is still good reason to quit. One might think that once a lifetime of smoking has put someone on the track for cancer, eliminating smoking in advanced years will not remove that risk of cancer. The risk does remain, but there is no good reason to multiply the risk by continuing to smoke.

But beyond that, there are other good reasons for elderly people to give up smoking. Smoking contributes to chronic bronchitis, emphysema and may precipitate coronary attacks, peripheral vascular disease and perhaps even stroke. Quitting smoking will not bring the person back to total normal function; but it helps slow deterioration. With coronary disease, in particular, the advantage seems to occur regardless of how long one has smoked. According to Kannel, the data show an immediate 50 percent reduction in the risk of coronary disease whether

> "The issue of what constitutes aging and normal aging is an enigma that has never been satisfactorily solved."
>
> WILLIAM KANNEL

the person has smoked 10 years, 20 years, 30 or 40 years. In terms of coronary attacks, there is trouble showing the benefit of quitting. But for coronary *deaths* and peripheral vascular disease, there is no difficulty at all showing that quitting helps.

"I think that the elderly are becoming increasingly health conscious," said Kannel. "It's curious: One would think that young people, who have so much life ahead of them, would take things more seriously. But the elderly, they are the ones who are driving more carefully, avoiding doing stupid and reckless things because they more acutely feel the approach of the grim reaper."

Some of the other effects commonly attributed to aging may well be preventable. Data from the Framingham studies show that a great deal of the high-frequency hearing loss in the elderly *male* can be laid at the door of noisy industries. We can predict, from the popularity of loud music among the young today, a generation of deaf elderly in 50 or 60 years. Osteoporosis, too, could be reduced if people were kept more active and ate more calcium-rich foods. In other words, the conclusions drawn from the Framingham Heart Study are that prevention is possible, that it must be started early and that it takes sustained effort. The burden of these common and disabling conditions, whether or not we term them aging phenomena, are the sources of a great deal of discontent in the elderly.

There is also a strong genetic component in the process of aging. "People with superb genes are able to withstand a lifetime of abuse because they may be better able to cope with an overload of fat in the diet, too many calories, too much salt, too much trauma, too little exercise, smoking. If they have been blessed with superb metabolic machinery, they somehow survive. Other people, with inferior metabolic machinery, may avoid all these risks and live longer than the great risk taker. There is a lot to be said for genetics."

"Even with bad genes, one can do something effective to reduce the liability," stated Kannel.

There is also a mistaken notion, he continued, that following a healthy lifestyle entails considerable sacrifice. "Diet, exercise and the like need not be so austere that they are painful. We are only recommending a Mediterranean or Asian diet. If you follow the specifics of those, you get the fat content, lower cholesterol, lower calories that you need. That is hardly a gastronomic nightmare. These are good foods. Ham and eggs need not be the epitome of gastronomic experience. One can eat very well following a prudent diet, as recommended by the American Heart Association."

"Exercise is something we need to build back into daily living," added Kannel. "We have taken it out by all the modern conveniences. A better way of living is to exercise naturally without the contriving. If you can walk to work instead of driving and parking right next to the door, you are better off. Try not to use escalators in two-story buildings. We have engineered exercise out of our lives; the time has come to engineer it back."

The Effects of Aging

I t is difficult to measure the rate of aging. One study (Hodgson and Buskirk) has shown that the maximal oxygen intake declined after age 25 at the rate of 0.40 to 0.45 ml of oxygen taken in per minute per kilogram of body weight each year. Grip strength went down about 0.20 kg per year. The investigators also found, however, that training at age 60 could improve the maximal oxygen intake by about 12 percent.

Average decline in human male, from age 30 to age 75

Factor	% Decline
Brain weight	44
Number of axons in spinal nerve	37
Velocity of nerve impulse	10
Number of taste buds	64
Blood supply to brain	20
Output of heart at rest	30
Number of glomeruli in kidney	44
Vital capacity of lungs	44
Maximum oxygen uptake	60

*There is controversy about how much of the decline reported is a result of aging as opposed to disease.

Of all the biological changes associated with growing old, young people are probably most acutely aware of the cosmetic changes. Hair greys, wrinkles become pronounced and shoulders tend to narrow with advancing age.

In American society, these changes are greeted with less than enthusiasm because, in a society that seems to cherish youth, these changes make one look old.

Cosmetic changes affect the sexes differently. Women may be outraged or humiliated by physical changes that lead to what Susan Sontag has called the process of "sexual disqualification." Women may be forced into roles of helplessness, passivity, compliance and non-competitiveness. Men, on the other hand, may

enjoy the assertiveness, competency, self-control, independence and power—all signs of "masculinity"—that come with age.

What creates the common cosmetic changes in aging? Wrinkles begin below the skin's outer layer (epidermis) when the dermis, a layer of tissue filled with glands, nerve endings and blood vessels, begins to shrink. At the same time, the dermis begins to atrophy, changes in the fat, muscle and bone create the deep wrinkles. Other factors—exposure to the sun, environmental toxins, heredity and disease—also affect the wrinkling of the skin.

Greying of the hair is the result of progressive loss of pigment in the cells that

give hair its color. Age spots are caused by the accumulation of pigment in the skin. But the shortened stature and flabby muscles are the result of lack of exercise and other behavioral factors. These, and many other so-called effects of aging, may be reversed.

After a while, time does take its toll. Between the ages of 35 and 80, the maximum work a person can do goes down by 60 percent. The strength of the grasp by the dominant hand (right in right-handers) goes down 50 percent and the endurance to maintain the strongest grasp goes down 30 percent. For some reason, not understood, the other (subordinate) hand, which was weaker to begin with, does not lose as much strength and

Alzheimer's Disease: The Search for a Cure Continues

It is difficult to paint anything but a bleak picture of Alzheimer's disease.

Named after the German neurologist Alois Alzheimer (1864–1915), it is a relentless and irreversible form of dementia that has been known to strike adults as young as 25, but most often appears in people over 70. For the estimated one to three million afflicted Americans, the early symptoms often involve memory loss, apathy and difficulties with spatial orientation and judgment. As these problems worsen, victims become increasingly depressed, confused, restless and unable to care for themselves.

In the final stages, Alzheimer patients may become so helpless that they are bedridden until they die from secondary problems such as pneumonia caused by accidentally inhaling food. Unfortunately, the victims of Alzheimer's often include the patient's stressed family and caregivers who, despite their devoted efforts, must watch loved ones turn into unmanageable strangers.

At this juncture, the exact cause, diagnosis and treatment of Alzheimer's continues to elude medical researchers. An abundance of new clues and insights into this mysterious disease is being uncovered, however. In addition, growing public awareness has led to the development of support groups and experi-

mental programs to help families and caregivers cope with caring for Alzheimer patients. "I'm amazed that researchers have progressed so far in such a short time," noted Professor Marott Sinex, from the School of Medicine's Department of Biochemistry, who has been researching the disease for the past 12 years.

Sinex pointed to many advances in the past decade, including improvements in drug therapy that ease symptoms. There is an increased understanding of possible genetic causes, of the subtle differences in neurotransmitters in the brains of victims and how they change over time and of anatomical changes in the brain that relate to memory loss and other cognitive problems. In addition, improved medical technology—such as the PET scan, which allows scientists to visualize the living brain—is adding new insights. "We now have a dynamic anatomy of the disease," said Sinex. "We can visualize its progression, which we couldn't do before. In other words we're dealing with a real, three-dimensional problem now, whereas before we only had two-dimensional understanding."

Although the exact cause of Alzheimer's is not known, Sinex noted that researchers now know it is associated with an excess of genetic material on a particular chromosome—an abnor-

mality that, interestingly, has already been linked to Down's syndrome. (Several Boston University researchers are independently investigating biochemical, genetic and physical similarities seen in Down's syndrome—a congenital disease whose victims are born moderately to severely retarded with distinctive physical traits—and Alzheimer's disease.) Scientists also know the disease can be inheritable and that—despite the fact that it can strike relatively young adults—it is "strongly age-dependent," with most cases not appearing until people are in their 70s. In addition to these factors, researchers are looking at other possible causes; for example, Sinex is investigating the possiblity that a virus may be involved.

As for diagnosis, the best physical evidence researchers have is the "plaques and tangles"—filamentous material whose nature and origin scientists are not exactly sure of—found in the brains of autopsied victims. (Although similar plaques and tangles are found in the brains of normally aging people, in Alzheimer victims the structures appear more frequently and in specific areas.) Sinex pointed out, however, that today "a really good clinic" can accurately diagnose the disease about 85 percent of the time by eliminating other problems such as stroke, a tumor,

drug poisoning or unrelated depression. Fortunately, research is advancing in this area as well. For example, Mark Moss, an assistant research professor in the School of Medicine's Department of Anatomy, recently developed a "gamelike" diagnostic test that the National Institute on Aging has recommended for clinical use and that "will help us understand the brain structures responsible for memory impairment."

The best method of treating Alzheimer victims today, explained Sinex, involves prescribing medications that "fall in the general category of anti-depressants." Apart from drug therapy, Sinex pointed out that organizations such as the Eastern Massachusetts Chapter of the Alzheimer's Disease and Related Disorders Association, of which he is president, can be helpful to both victims and their caregivers by providing information and connecting people with support groups.

Given this variety of recent advances, Sinex concluded optimistically, "It's simply a lot less traumatic to have Alzheimer's now than it was 10 or 15 years ago."

Editor's Note: Researchers in Boston University's Medical Center—including the School of Medicine and several affiliated hospitals—comprise one of the largest Alzheimer's disease research efforts in the country. *Bostonia* will take a closer look at these projects in an upcoming Insight issue.
JON QUEIJO

endurance. The speed of nerve conduction is slower. The volume of blood pumped throughout the body goes down 50 percent. The maximum volume of air a person can inhale goes down 50 percent and oxygen diffuses from the lungs to the red cells of the blood 30 percent more slowly. Blood flow to the kidneys at age 80 is considerably less than that of age 20 in many, but not all. And, by age 70, the bones of the coccyx (the "tail bone" at the base of the spine) fuse. In short, our bodies slow down and stiffen as we age. This is natural and occurs even in the absence of disease.

Vision, hearing, taste, smell and touch have all been reported to change with age. New research suggests, however, that the effects of aging per se may not be as major a factor as originally believed on declining senses.

Probably the most familiar changes are those in seeing. Presbyopia (presby = old + opia = vision) is a sign of the gradual inability of the lens of the eye to focus on near objects—hence, the growing need for "reading glasses" or bifocals as people age. There are other relatively harmless changes, as well. Almost from birth, the lens of the eye begins to get more rigid. By around age 45, printed pages must be held at arm's length or farther to get them into focus. But, of course, at that distance, the letters are usually too small to read. As people age, their eyes may become more sensitive to glare and bright lights. They

> Smoking contributes to emphysema and may precipitate coronary attacks. Quitting smoking will not bring the person back to absolutely normal function; but it helps slow the process down.

may also be less able to discriminate between gradations of color.

More serious eye conditions increase with age as well. Approximately seven percent of all people between the ages of 65 and 74 have serious visual deficits. After age 75, the proportion more than doubles to 16 percent. Approximately two-thirds of all severe visual impairments occur in people 65 or older.

Macular degeneration is the most serious cause of low vision in the elderly in the United States. The macula is a spot on the retina of the eye needed for very fine focusing. With age, this region can degenerate and become obstructed with fine blood vessels.

Other visual impairments include: glaucoma, which is a dangerous and

painful increase of pressure within the eye; diabetic retinopathy, which is a destruction of the fine blood vessels in the eye, destroying parts of the retina, associated with uncontrolled diabetes; and cataracts, which are cloudy eye lenses.

Just as presbyopia is a vision deficit associated with advancing age, so presbyacusis (presby = old + acousis = sound) is a progressive hearing loss associated with aging. This is especially true for the higher frequencies. Of the estimated 14.2 million Americans with measurable hearing loss, about 60 percent of those with the most severe hearing problems are older than 65.

Hearing loss associated with aging may be of several causes. Genetic factors, infection and a lifetime of noise certainly contribute. Poor personal hygiene, build-up of ear wax, may also reduce hearing. Certain medications (for example, drugs of the streptomycin group) can injure the hair cells in the ear and interfere with both hearing and the sense of balance.

While these physical and sensory changes occur in all people as they age, keep in mind that the degree to which they affect individuals varies greatly. The key is how we take care of ourselves. The pleas from the medical profession to cut down smoking, drinking and to exercise are grounded in heavy evidence. Keeping yourself active and healthy throughout your life can result in an old age that is productive and rewarding.

Osteoporosis
The Stooping Disease

An estimated five million people in the United States are afflicted with osteoporosis—a disease marked by loss of bone mass that weakens the skeleton and may result in spontaneous fractures. Although everyone loses bone mass as they age, this process is accelerated in women. In the first 10 years after menopause, women lose bone at twice the rate men do.

Healthy bone is constantly changing. In childhood and adolescence, bones form at a faster rate than they are reabsorbed. In healthy adults, up until age 40, peak bone mass is maintained through a balance between the processes of bone formation and loss. Peak bone mass, which usu-

ally occurs around age 30, is influenced by hormones, calcium intake, level of physical activity and the stress of weight bearing. Heredity also plays a role. From about age 40, bone absorption is more rapid than bone formation.

There is no way to diagnose the early stages of osteoporosis. The bone loss that characterizes the disease does not show up on x-rays until a substantial portion of the bone is already lost. At its later stages, however, osteoporosis produces extreme visible changes: loss of height, rounding of the upper back ("dowager's hump"), forward thrust of the head, protruding abdomen and expansion of the chest. These symptoms are all due to

the collapsed vertebrae weakened by osteoporosis.

What causes osteoporosis is not yet understood. Because post-menopausal women are at elevated risk, some investigators have suggested that it involves estrogen deficiency and should be treated with hormones. This cannot be the case, however, because not all post-menopausal women develop osteoporosis.

At highest risk are white women with a family history of osteoporosis, of northern European descent, with small bone frames and of normal or less than average weight. Certain dietary and behavioral factors can increase the risk: drinking more than four to six cups of coffee a day, smoking, heavy use of alcohol and lack of calcium in the diet all increase the risk. On the other hand, physical exercise

decreases the risk.

University Hospital is currently planning to open a clinic at the end of February to treat osteoporosis victims. The clinic will be multidisciplinary involving orthopedic surgery, endocrinology, nutrition and internal medicine. All patients referred to the clinic will be prescreened by a special x-ray test which will determine bone mineral content. A specialized blood work-up will be done to rule out metabolic causes for metabolic bone disease and other appropriate studies as indicated by each case. Also a program of functional bracing will be started. Questions concerning the clinic should be directed to the Department of Orthopedic Surgery, University Hospital, (617) 638-8905.

New Evidence Points to Growth of the Brain Even Late in Life

Daniel Goleman

Evidence is building that development and growth of the brain go on into old age. It was once thought that the brain was fixed by late childhood, according to innate genetic design.

As long ago as 1911, however, Santiago Ramón y Cajal, a pioneering neurobiologist, proposed that "cerebral exercise" could benefit the brain. But a scientific consensus that the brain continues to bloom if properly stimulated by an enriched environment was long in coming.

"Over the last decade, neuroscientists have become impressed by the degree to which the structure and chemistry of the brain is affected by experience," said Floyd Bloom, director of the division of neuroscience and endocrinology at the Scripps Clinic and Research Foundation in La Jolla, Calif. The new research seeks to provide a more detailed understanding of that phenomenon.

Investigations at several different laboratories have shown that environmental influences begin while the brain is forming in the fetus and are particularly strong in infancy and early childhood.

Among the most striking new evidence is a report published in a recent issue of Experimental Neurology showing that even in old age the cells of the cerebral cortex respond to an enriched environment by forging new connections to other cells. Marian Diamond, a professor of physiology and anatomy at the University of California at Berkeley, led the team of researchers who did the study.

In Dr. Diamond's study, rats 766 days old—the equivalent in human terms to roughly 75 years—were placed in an enriched environment and lived there until they reached the age of 904 days. For a rat, an impoverished environment is a bare wire cage a foot square with a solitary occupant; an enriched one is a cage a yard square where 12 rats share a variety of toys, such as mazes, ladders and wheels.

The elderly rats, after living in the stimulating environment, showed in-

Physical changes linked to enriched environment.

creased thickening of the cortex. This thickening, other research has shown, is a sign that the brain cells have increased in dimension and activity, and that the glial cells that support the brain cells have multiplied accordingly.

The brain cells also showed a lengthening of the tips of their dendrites, the branches that receive messages from other cells. This increase in the surface of the dendrites allows for more communication with other cells.

Previous studies have shown that enriched environments changed brain cells in a number of ways, these among them. While the specific effects differ from one region of the brain to another, in general the enriched environment has been generally seen to result in growth in the bodies of nerve cells, an increase in the amount of protein in these cells

and an increase in the number or length of dendrites. In more fully developed dendritic spines, a part of the dendrite that receives chemical messages from other brain cells is induced to further growth.

Moreover, as in the new study, the thickness of the cortex was seen to increase, in part because of an increase in the numbers of glial cells needed to support the enlarged neurons. Dr. Diamond's studies on the older rats show that many, but not all, of these effects continue into old age.

These changes, in Dr. Diamond's view, mean that the cells have become more active, forming new connections to other brain cells. One sign of what the increased brain cell activity signifies for intellectual abilities is that the rats in the enriched environment became better at learning how to make their way through a maze. Indeed, Dr. Diamond and other researchers recently examined speci-

BRAIN GROWTH IN RICH ENVIRONMENT

Cells from cerebral cortex of rats put in an enriched environment show more numerous branching of dendrites that stretch out to other brain cells and more fully developed dendritic spines, which receive chemical messages from the other cells.

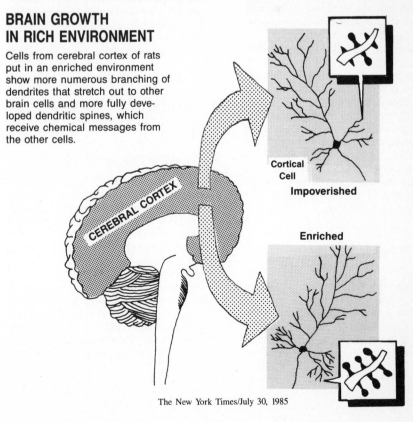

Cortical Cell

Impoverished

Enriched

The New York Times/July 30, 1985

mens from Einstein's brain. The tissue samples, from parts of the cortex presumed critical for mathematical skills, seemed to have unusually large numbers of glial cells.

MORE NEURAL FLEXIBILITY

What does all this mean for the aging brain? "There is much more neural flexibility in old age than we had imagined," said Roger Walsh, a psychiatrist at the University of California medical school at Irvine, who has done research similar to Dr. Diamond's. "The changes in brain cells have been found in every species investigated to date, including primates. They certainly should occur in humans as well."

"In my work," Dr. Walsh added, "I've found that an enriched environment in late life can largely compensate for brain cell deficiencies from earlier deprivations."

"We've been too negative in how we view the human brain," Dr. Diamond said in an interview. "Nerve cells can grow at any age in response to intellectual enrichment of all sorts: travel, crossword puzzles, anything that stimulates the brain with novelty and challenge."

Still, there seem to be limits to the degree to which the brain can respond to experience. Richard Lerner, in "On the Nature of Human Plasticity" (Cambridge University Press), notes, for example, that the impact of environmental enrichment on brain cells seems to diminish with age, although it continues into old age, an effect Dr. Diamond has noted in her research.

The effects of enriched environments on the brain are but part of a larger investigation of the impact of life's experiences on the brain, and the picture is not always positive.

"Brain plasticity can operate for better or for worse," said Jeannine Herron, a neuropsychologist at California Neuropsychology Services in San Rafael. Dr. Herron has organized a conference to be held later this month at which Dr. Diamond and other researchers will describe their findings.

TESTS ON VISION OF KITTENS

Perhaps the most frequently cited example of how experience—or the lack of it—can have a negative effect on the brain is the work of David Hubel. Dr. Hubel, who will also speak at the California conference, won a Nobel Prize for his research on the visual cortex.

As part of his research, Dr. Hubel showed that if the eye of a growing kitten is kept shut so that it is deprived of its normal experience, the cells that would ordinarily register what that eye sees will develop abnormally.

"There is much more neural flexibility in old age than we had imagined."

The notion that certain experiences go hand in hand with the growth and development of the brain has been demonstrated in other research, as well. For example, Arnold Scheibel, a professor of anatomy and psychiatry at the University of California at Los Angeles, has found that the cells in the speech centers of infants undergo a growth burst, in which they form many new connections to other cells, just at the time the infant is beginning to respond to voices, between 6 and 12 months. Between 12 and 18 months, as the infant begins to grasp that words have meanings, this growth accelerates.

Part of this explosion of growth, Dr. Scheibel proposes, may be primed by the infant's interactions with adults, who stimulate the centers for speech by talking to the infant.

The main changes that occur during this growth in the cells of the speech centers are in the dendrites ensemble, the projecting branches of the cell that spread to send and receive messages from other cells. "The dendritic projections are like muscle tissue," Dr. Scheibel said. "They grow more the more they're used."

"Even in adulthood," he added, "if you learn a new language, it's dendritic fireworks."

RESPONSES TO INJURY

The brain's ability to adapt to circumstances can also be seen in its response to injury. Patricia Goldman-Rakic, a neuroanatomist at Yale University medical school, is one of many researchers who have shown that brain cells, within limits, can rearrange themselves to compensate for a brain injury.

"The new connections that occur after an injury to the brain show that the brain's anatomy is not rigidly fixed," Dr. Goldman-Rakic said in an interview. "The uninjured cells reroute how they grow and interconnect. This ability is

most prominent during infancy, when neurons are still growing. It doesn't go on forever, but we don't yet know precisely at what point in later life the brain no longer can compensate in this way. We need to more fully understand normal brain maturation first."

Norman Geschwind, a noted neuroanatomist at Harvard medical school who died [in 1985], had been pursuing evidence suggesting that the experiences of a mother can have a lasting effect on the structure of the developing fetus's brain.

In a series of articles published posthumously in the most recent issues of Archives of Neurology, Dr. Geschwind, with Albert Galaburda, a colleague at Harvard Medical School, proposes that the infant brain is shaped in crucial ways by the level of testosterone, a male sex hormone, present in the intrauterine environment at different stages of fetal development.

At crucial points in the growth of the fetus, brain cells are formed and then migrate to the part of the brain ordained by a genetic plan. In certain parts of the brain these patterns of migration can be affected by the presence of sex hormones, particularly testosterone.

Testosterone levels in the fetus can vary with such factors as the amount of psychological stress the mother feels, maternal diet and possibly even the season of the year.

The main effects of testosterone, according to Dr. Geschwind and Dr. Galaburda, are in the areas of the brain that control such skills as speech, spatial abilities and handedness. One of the key effects of testosterone is in determining the side of the brain on which the centers that control such skills will be located.

When the process goes awry, according to the theory, the result can be problems such as dyslexia, on the one hand, or unusual talents, such as mathematical giftedness, on the other.

These effects are more marked among males, in part because the brains of males develop more slowly than those of females, and in part because testosterone plays a direct role in the growth of certain areas of the male brain. The unusual patterns of brain formation are most common, the theory holds, among left-handed males.

Before his death, Dr. Geschwind found from autopsies of people who had severe dyslexia in childhood that the parts of the cortex that control speech had abnormal cell development along the lines predicted by his theory.

Never Too Late

SINGLE PEOPLE OVER 65 WHO ARE DATING
AND SEXUALLY ACTIVE BELIE THE NOTION THAT
PASSION AND ROMANCE ARE ONLY FOR THE YOUNG.

KRIS BULCROFT AND
MARGARET O'CONNER-RODEN

Kris Bulcroft is a sociologist at St. Olaf College in Northfield, Minnesota. Margaret O'Conner-Roden is a sociology doctoral candidate at the University of Minnesota in Minneapolis.

What is the age of love? The star-crossed lovers Romeo and Juliet were teenagers; Anthony and Cleopatra's torrid affair occurred at the prime of their health and beauty; Lady Diana Spencer was barely 20 when she married her Prince Charming. How old is too old for the sparkle in the eye and the blush in the cheek?

The message our culture often gives us is that love is only for the young and the beautiful—people over 65 are no longer interested in or suited for things such as romance and passion. Few of us imagine older couples taking an interest in the opposite sex other than for companionship—maybe a game of bridge or conversation out on the porch. But, in fact, there are quite a few older single people who not only date but are involved sexually with someone.

Statistically there are good reasons for older people to be dating. At the turn of the century only about 4 percent of the total American population was 65 years of age or older. Today that number has soared to approximately 11 percent, with the total expected to increase to about 20 percent by the year 2050. In addition, older people are living longer and staying healthier, and they are less likely than before to have children living at home. And an increasing number of divorces among the elderly is casting many of these older people back into the singles' pool. All of these factors create an expanded life stage, made up of healthy and active people looking for meaningful ways to spend their leisure.

The question of whether older people date, fall in love and behave romantically, just as the young do, occurred to us while we were observing singles' dances for older people at a senior center. We noticed a sense of anticipation, festive dress and flirtatious behavior that were strikingly familiar to us as women recently involved in the dating scene. Although our observations indicated that older people dated, when we looked for empirical research on the topic we found there was none. We concluded this was due partly to the difficulty in finding representative samples of older

daters and partly to the underlying stereotype of asexual elders. So we decided to go out and talk to older daters ourselves. Once we began looking, we were surprised at the numbers of dating elders who came forward to talk to us. We compared their responses to those from earlier studies on romance and dating, in which the people were much younger.

Dating, as defined by our sample of older people, meant a committed, long-term, monogamous relationship, similar to going steady at younger ages. The vast majority of elderly daters did not approach dating with the more casual attitude of many younger single people who are "playing the field." All respondents clearly saw dating as quite distinct from friendship, although companionship was an important characteristic of over-60 dating.

One of our major findings was the similarity between how older and younger daters feel when they fall in love—what we've come to call the "sweaty palm syndrome." This includes all the physiological and psychological somersaults, such as a heightened sense of reality, perspiring hands, a feeling of awkwardness, inability to concentrate, anxiety when away from the loved one and heart palpitations. A 65-year-old man told

When they fall in love, older daters experience the same emotional somersaults, sweaty palms and beating hearts as do younger couples.

us, "Love is when you look across the room at someone and your heart goes pitty-pat." A widow, aged 72, said, "You know you're in love when the one you love is away and you feel empty." Or as a 68-year-old divorcée said, "When you fall in love at my age there's initially a kind of 'oh, gee!' feeling . . . and it's just a little scary."

We also found a similarity in how both older and younger daters defined romance. Older people were just as likely to want to participate in romantic displays such as candlelight dinners, long walks in the park and giving flowers and candy. Older men, just like younger ones, tended to equate romance with sexuality. As a 71-year-old widower told us, "You can talk about candlelight dinners and sitting in front of the fireplace, but I still think the most romantic thing I've ever done is to go to bed with her."

A major question for us was "What do older people do on dates?" The popular image may suggest a prim, card-playing couple perhaps holding hands at some senior center. We found that not only do older couples' dates include the same activities as those of younger people, but they are often far more varied and creative. In addition to traditional dates such as going to

the movies, out for pizza and to dances, older couples said they went camping, enjoyed the opera and flew to Hawaii for the weekend.

Not only was the dating behavior more varied, but the pace of the relationship was greatly accelerated in later life. People told us that there simply was "not much time for playing the field." They favored the direct, no-game-playing approach in building a relationship with a member of the opposite sex. As one elderly dater commented, "Touching people is important, and I know from watching my father and mother that you might just as well say when lunch is ready . . . and I don't mean that literally."

Sexuality was an important part of the dating relationship for most of those we spoke to, and sexual involvement tended to develop rapidly. While sexuality for these couples included intercourse, the stronger emphasis was on the nuances of sexual behavior such as hugging, kissing and touching. This physical closeness helped fulfill the intimacy needs of older people, needs that were especially important to those living alone whose sole source of human touch was often the dating partner. The intimacy provided through sex also contributed to self-

esteem by making people feel desired and needed. As one 77-year-old woman said, "Sex isn't as important when you're older, but in a way you need it more."

A major distinction we found between older and younger daters was in their attitudes toward passionate love, or what the Greeks called "the madness from the gods." Psychologists Elaine Hatfield, of the University of Hawaii in Manoa, and G. William Walster, of Los Gatos, California, have similarly defined passionate love as explosive, filled with fervor and short-lived. According to their theory of love, young people tend to equate passionate love with being in love. Once the first, intense love experience has faded, young lovers often seek a new partner.

For older daters, it is different. They have learned from experience that passionate love cannot be sustained with the same early level of intensity. But since most of them have been in marriages that lasted for decades, they also know the value of companionate love, that "steady burning fire" that not only endures but tends to grow deeper over time. As one older man put it, "Yeah, passion is nice . . . it's the frosting on the cake.

But it's her personality that's really important. The first time I was in love it was only the excitement that mattered, but now it's the friendship ... the ways we spend our time together that count."

Nonetheless, the pursuit of intimacy caused special problems for older people. Unlike younger daters, older people are faced with a lack of social cues indicating whether sexual behavior is appropriate in the dating relationship. Choosing to have a sexual relationship outside of marriage often goes against the system of values that they have

WHEN MY GIRLFRIEND SPENDS THE NIGHT SHE BRINGS HER CORDLESS PHONE, JUST IN CASE HER DAUGHTER CALLS.

followed during their entire lives.

Older couples also felt the need to hide the intimate aspects of their dating relationship because of a fear of social disapproval, creating a variety of covert behaviors. As one 63-year-old retiree said, "Yeah, my girlfriend (age 64) lives just down the hall from me ... when she spends the night she usually brings her cordless phone ... just in case her daughter calls." One 61-year-old woman told us that even though her 68-year-old boyfriend has been spending three or four nights a week at her house for the past year, she has not been able to tell her family. "I have a tendency to hide his shoes when my grandchildren are coming over."

Despite the fact that marriage

WHO'S WHO IN THE SAMPLE

For our study we interviewed 45 older people in a Midwestern metropolitan area who were widowed or divorced and had been actively dating during the past year. Fifty-four percent were men and 46 percent were women; all were white. The age of the subjects ranged from 60 to 92; the average age was 68. Although most of the group was middle-class, some were affluent and others lived solely on Social Security. Names were obtained through a variety of methods, including a membership list of a singles' club for older persons, senior citizens' centers, newspaper ads and word of mouth. The face-to-face interviews were, for the most part, conducted in the home of the older person. We asked people questions about how they met, what they did on a date, how important sexuality was in their relationship and what family and friends' reactions were to their dating.

would solve the problem of how to deal with the sexual aspects of the relationship, very few of these couples were interested in marriage. Some had assumed when they began dating that they would eventually marry but discovered as time went on that they weren't willing to give up their independence. For women especially, their divorce or widowhood marked the first time in their lives that they had been on their own. Although it was often difficult in the beginning, many discovered that they enjoyed their independence. Older people also said they didn't have the same reasons for marriage that younger people do: beginning a life together and starting a family. Another reason some elders were reluctant to marry was the possibility of deteriorating health. Many said they would not want to become a caretaker for an ill spouse.

Contrary to the popular belief that family would be protective and jealous of the dating relative, family members tended to be supportive of older couples' dating and often included the dating partner in family gatherings. The attitude that individuals have the right to personal happiness may be partially

responsible for families' positive attitudes. But more importantly, many families realize that a significant other for an older person places fewer social demands on family members.

Peers also tended to be supportive, although many women reported sensing jealousy among their female friends, who were possibly unhappy because of their inability to find dating partners themselves and hurt because the dating woman didn't have as much time to spend with them.

Our interviews with older daters revealed that the dating relationship is a critical, central part of elders' lives that provides something that cannot be supplied by family or friends. As one 65-year-old man told us, "I'm very happy with life right now. I'd be lost without my dating partner. I really would."

Our initial question, "What is the age of love?" is best answered in the words of one 64-year-old woman: "I suppose that hope does spring eternal in the human breast as far as love is concerned. People are always looking for the ultimate, perfect relationship. No matter how old they are, they are looking for this thing called love."

The Reason of Age

WE LOSE SOME MENTAL SPEED
WITH THE YEARS, BUT WE CAN OFTEN
SUBSTITUTE EXPERIENCE FOR QUICKNESS.

JEFF MEER

Jeff Meer, 26, is an assistant editor at Psychology Today *whose grandparents are the wisest people he knows.*

The golden years are making a comeback. As researchers spend less time looking at what we lose as we get older and more at what we keep or gain, aging is looking better.

Consider Andrés Segovia, still giving acclaimed concerts on the classical guitar at age 92 ... Claude Pepper, who came in with the 20th century and has served in Congress for most of the past 50 years ... Bob Hope, entertaining and golfing his way around the world 82 years after his birth in Eltham, England.

But aren't these people exceptions? Of course they are. Men and women with unusual abilities are always exceptions, whatever their age. Ability and activity vary among people in their 70s, 80s and 90s just as they do earlier in life.

Evidence is piling up that most of our mental skills remain intact as long as our health does, if we keep mentally and physically active. Much of our fate is in our own hands, with "use it or lose it" as the guiding principle. We are likely to slow down in some ways, but there is evidence that healthy older people do a number of things better than young people.

Psychologist James Birren, dean of the Andrus Gerontological Center at the University of Southern California, is one of many researchers to show that older people perform tasks more slowly, from cutting with a knife and dialing a telephone to remembering lists. There are numerous theories about what body changes are responsible but no conclusive answers.

More important, slowing down doesn't make much difference in most of what we do. Slower reflexes are certainly a disadvantage in driving an automobile, but for many activities speed is not important. And when it is, there are often ways to compensate that maintain performance at essentially the same level. "An awful lot of what we can measure slows down," says psychologist Timothy Salthouse of the University of Missouri at Columbia, "but it isn't clear that this actually affects the lives of the people we study in any significant way."

As an example, Salthouse cites an experiment in which he tested the reaction time and typing skills of typists of all ages. He found that while the reactions of the older typists were generally slower than those of younger ones, they typed just as fast. It could be that the older typists were even faster at one time and had slowed down. But the results of a second test lead Salthouse to believe that another factor was at work.

When he limited the number of characters that the typists could look ahead, the older typists slowed greatly, while the younger ones were affected much less. "There may be limits, but I'm convinced that the older typists have learned to look farther ahead in order to type as quickly as the younger typists," Salthouse says.

A similar substitution of experience for speed may explain how older people maintain their skills in many types of problem-solving and other mental activity. Because of this, many researchers have come to realize that measuring one area of performance in the laboratory can give only a rough idea of a person's ability in the real world.

As an example, psychologist Neil Charness of the University of Waterloo in Ontario gave bridge and chess problems to players of all ages and ability levels. When he asked the bridge players to bid and the chess players to choose a move or remember board positions, the older players took longer and could remember fewer of the chess positions. But the bids and the moves they chose were every bit as good as those of younger players. "I'm not sure exactly what the compensatory mechanisms are," Charness says, "but at least until the age of 60, the special processes that the older players use enable them to make up for what they have lost in terms of speed and memory ability."

Many researchers now believe that one reason we associate decline with age is that we have asked the wrong questions. "I suspect that the lower performance of older people on many of the tasks we have been testing stems from the fact that they have found that these things are unimportant, whereas young people might enjoy this kind of test because it is novel," says psychologist Warner K.

Schaie of Pennsylvania State University. Relying on their experience and perspective, he says, older people "can selectively ignore a good many things."

Memory is probably the most thoroughly studied area in the relationship between age and mental abilities. Elderly men and women do complain more that they can't remember their friends' names, and they seem to lose things more readily than young people. In his book *Enjoy Old Age,* B.F. Skinner mentions trying to do some-

AN AWFUL LOT OF WHAT WE MEASURE SLOWS DOWN, BUT IT ISN'T CLEAR THAT THIS AFFECTS PEOPLE'S LIVES IN ANY SIGNIFICANT WAY.

thing that one learned to do as a child—folding a piece of paper to make a hat, for example—and not remembering how. Such a failure can be especially poignant for an older person.

But the fact is that much of memory ability doesn't decline at all. "As we get older, old age gets blamed for problems that may have existed all along," says psychologist Ilene Siegler of the Duke University Medical Center. "A 35-year-old who forgets his hat is forgetful," she says, "but if the same thing happens to grandpa we start wondering if his mind is going." If an older person starts forgetting things, it's not a sure sign of senility or of Alzheimer's disease. The cause might be incorrect medication, simple depression or other physical or mental problems that can be helped with proper therapy.

Psychologists divide memory into three areas, primary, secondary and tertiary. Primary or immediate memory is the kind we use to remember a telephone number between the time we look it up and when we dial it. "There is really little or no noticeable decline in immediate memory," according to David Arenberg, chief of the cognition section at the Gerontology Research Center at the National Insti-

tute on Aging. Older people may remember this type of material more slowly, but they remember it as completely as do younger people.

Secondary memory, which, for example, is involved in learning and remembering lists, is usually less reliable as we get older, especially if there is a delay between the learning and the recall. In experiments Arenberg has done, for example, older people have a difficult time remembering a list of items if they are given another task to do in between.

Even with secondary memory, however, where decline with age is common, the precise results depend on exactly how memory is tested. Psychologist Gisela Labouvie-Vief of Wayne State University in Detroit has found that older people excel at recalling the metaphoric meaning of a passage. She asked people in their early 20s and those in their 70s to remember phrases such as "the seasons are the costumes of nature." College students try to remember the text as precisely as they can. Older people seem to remember the meaning through metaphor. As a result, she says, "they are more likely to preserve the actual meaning, even if their reproduced sentence doesn't exactly match the original." In most situations, understanding the real meaning of what you hear or read is more important than remembering the exact words.

Part of the problem with tests of memory is that most match older people against students. "As long as we accept students as the ideal, older people will look bad," Labouvie-Vief says. Students need to memorize every day, whereas most older people haven't had to cram for an exam in years. As an example of this, psychologist Patricia Siple and colleagues at Wayne State

University found that older people don't memorize as well as young students do. But when they are matched against young people who are not students, they memorize nearly as well.

The third kind of memory, long-term remembrance of familiar things, normally decreases little or not at all with age. Older people do particularly well if quickness isn't a criterion. Given time and the right circumstances, they may do even better than younger men and women. When psychologist Roy Lachman of the University of Houston and attorney Janet Lachman tested the ability to remember movies, sports information and current events, older people did much better, probably because of their greater store of information. Since they have more tertiary memory to scan, the Lachmans conclude, older people scan that kind of memory more efficiently.

Psychologist John Horn of the University of Denver and other researchers believe that crystallized knowledge such as vocabulary increases throughout life. Horn, who has studied the mental abilities of hundreds of people for more than 20 years, says, "If I were to put together a research team, I'd certainly want some young people who might recall material more quickly, but I'd also want some older crystallized thinkers for balance."

Researchers often echo the "use it or lose it" idea. When psychologist Nancy Denney of the University of Wisconsin-Madison uses the game "20 questions" in experiments, she finds that the needed skills are not lost. "The older people start off by asking inefficient questions," she says, "but we know that the abilities are still there because once they see the efficient strategy being used by others, they learn it very quickly."

Psychologist Liz Zelinski of the University of Southern California makes a similar point when she tests the ability to read and understand brief passages. People in their 70s and 80s show no significant decline in comprehension. "Our tests don't involve the kind of questions that require older people to store information temporarily in memory," she cautions. "Tests like that might show declines." Zelinski has also found that older men and women read her tests just as fast as younger people do. "It is a good guess that they maintain the ability to read

Most of our mental skills remain intact as long as we remain healthy. In order to stay healthy, we must be mentally and physically active. The above couple in Beverly, Massachusetts, use gardening to continue an energetic life style—she is 95 and he is 100.

quickly because they do it all the time," she says.

Even when skills atrophy through disuse, many people can be trained to regain them. Schaie and psychologist Sherry Willis of Pennsylvania State University recently reported on a long-term study with 4,000 people, most of whom were older. Using individualized training, the researchers improved spatial orientation and deductive reasoning for two-thirds of those they studied. Nearly 40 percent of those whose abilities had declined returned to a level they had attained 14 years earlier.

Mnemonics is another strategy that can help people memorize something as simple as a shopping list. Arenberg has found that older people are much better at remembering a 16-item list if they first think of 16 locations in their home or apartment and then link each item with a location. With practice, they master this technique very easily, Arenberg says, "and become very effective memorizers."

When it comes to aging's effect on general intelligence, as measured by standard IQ tests, the same questions of appropriateness, accuracy and motivation complicate the findings. Psychologist Paul Costa, chief of the laboratory of personality and cognition of the Gerontology Research Center at the National Institute on Aging in Baltimore, points out that many early studies on aging tested older people and younger people at the same time, instead of testing the same people over a period of years. These studies were, in effect, measuring the abilities of older people, largely lower-income immigrants, against the generations of their children and grandchildren. "The younger people enjoyed a more comfortable life-style, were better educated and didn't face the same kind of life stresses," he says, "so comparisons were mostly inappropriate."

Most researchers today are uncomfortable with the idea of using standard intelligence tests for older people. "How appropriate is it to measure the 'scholastic aptitude' of a 70-year-old?" asks University of Michigan psychologist Marion Perlmutter.

She and others, including Robert Sternberg at Yale, believe that aspects of adult functioning, such as social or professional competence and the ability to deal with one's environment, ought to be measured along with traditional measures of intelligence. "We are really in the beginning stages of developing adequate measures of adult intelligence and in revising what we think of as adult intelligence," Perlmutter says. "If we had more comprehensive tests including these and other factors, I suspect that older people would score at least as well and probably better than younger people."

Erroneous ideas about automatic mental deterioration with age hit particularly hard in the workplace. Although most jobs require skills unaffected by age, many employers simply assume that older workers should be phased out. Psychologists David Waldman and Bruce Alvolio of the State University of New York at Binghamton recently reviewed 13 studies of job performance and found little support for deterioration of job performance with increasing age. Job performance, measured objectively, increased as employees, especially professionals, grew older. The researchers also discovered, however, that if supervisors' ratings were used as the standard, performance seemed to decline slightly with age. Expectation became reality.

Despite all the experiments and all the talk about gains and losses with age, we should remember that many older people don't want to be compared, analyzed or retrained, and they don't care about being as fast or as nimble at problem solving as they once were. "Perhaps we need to redefine our understanding of what older people can and cannot do," Perlmutter says. Just as children need to lose some of their spontaneity to become more mature, perhaps "some of what we see as decline in older people may be necessary for their growth." While this does not mean that all age-related declines lead to growth or can be ignored, it does highlight a bias in our youth-oriented culture. Why do we so often think of speed as an asset and completely ignore the importance of patient consideration?

SPECIAL CASES: THE OLDEST

One group of particular interest to researchers on aging is those more than 85 years old. There are at least two million Americans in this category, more than 1 percent of the population, and growing faster than any other segment.

Depending upon whom you talk to, this group is called the old old, the oldest old or the extreme aged. More than half live independently, by themselves or with a spouse. And many do more than just live. History and current headlines tell us of extraordinary individuals who have done important work in their ninth decade (see "The Art of Aging," *Psychology Today,* January 1984). Mystery writer Agatha Christie, statesman Konrad Adenauer and cellist Pablo Casals are only three well-known examples.

Other less famous but equally industrious men and women of similar advanced age contribute in their own fields. During the 1950s, gerontologist S.L. Pressey studied the lives of 313 people more than 80 years old whose names he found in newspaper clippings, nursing home records and other random sources. He learned that most were working at least part-time. Two men past the age of 90 were presidents of small-town banks and one nonagenarian woman ran an insurance business. If people are given the opportunity to continue making contributions, Pressey concluded—especially if they work in professional fields or are self-employed—they are likely to do so.

Psychologist Marion Perlmutter has begun to study a group of 80-year-olds to see what keeps them going and what we can learn from them to help others. The first interviews suggest that the abilities to be open to new situations and cope with challenges distinguish people who grow during adulthood from people who stabilize or decline. "One reason to study these people is because they're successes. If you make it to 80, you must be doing something right."

Index

hearing: effect of aging on, 264; fetal, 25
helplessness: learned, 72-73; and loneliness, 250; and psychic surrender, 235
Heredity Choice, 19, 22, 23
heredity vs. environment, *see* nature/nurture debate
Holocaust twins, 204-210
home-based nuclear structures, 243, 244, 247
homosexuality, role of sex hormones in, 9
hormones: and behavior, 9-11; as hereditary, 6-12; and sex differences, 212, 213
hospitalization, and children's stress, 74
household density, and stress in children, 73

infant education, 44
infant(s): brain development of, 36-39; cognitive development in, 44-45; function of crying in, 46-50; imitation of facial expressions by, 41-42, 101-103; interaction with fathers, 147; interaction with mothers, 44, 67; use of memory by, 111; perceptual development of, 41-42; research, 40-41
infantile amnesia, 111, 112
inferiority, intellectual, and blacks, 176, 177
infertility: causes of, 13, 14, 16; incidence of, in America, 13, 16; new solutions to, 13-18; *see also,* sterility
intellectual development: and black Americans, 174, 175, 176; *see also,* intelligence
intelligence: effect of age on, 39, 273; effect of fathering on children's, 147, 148; and "g," 123; and genetics, 19-24; in infants, 44-45; and debate over IQ tests, 119, 120; nature/nurture debate over, 123, 124; and private speech of children, 135; and self-esteem in children, 53-56
Intelligence Quotient, *see* IQ
intensive care, for premature babies, 27, 28, 29
intermittent positive pressure ventilation, 32
intimacy, need for, by elderly, 268
in vitro fertilization, 13-16
IQ: 119-122; sperm from donors with high, 19-24
IVF, *see* in vitro fertilization

Japan, and teaching discipline to children, 144, 145; and youthful alienation, 181

"Kilogram Kids": 28, 29; *see also,* premature babies
Kohlberg, Lawrence: and moral reasoning in children, 58-59, 169, 170; and private speech, 133, 134

language, and private speech behavior, 132-137
lateralization, and male vs. female brain, 216, 217
law, lack of understanding of, by adolescents, 187, 188, 189, 190
"learned helplessness," and coping with stress in children, 72-73
learning, critical periods of, 37-39
left-handed people, and sex differences in brain organization, 221

left hemisphere, of brain, and male vs. female organization of, 216-221
life-space structures, 243-248
life styles: alternate, and child rearing, 149, 157-161; and life-space structures, 243-248
light therapy, for depression, 254, 255
loneliness, 249-251
love: measuring of, 240-242; and older people, 267, 268

male(s): effect of hormones on behavior in, 9-10; infertility, 16-17; physiological requirements for development of, 7-9; *see also,* men
manic-depressive behavior, 253, 255
memory: of childhood, 110-113; and elderly, 271; and infants, 111, 112; semantic, 116
men: and sex differences, 212-221; *see also,* males
mental illness, and suicide, 230, 231
mental skills, and elderly, 270-273
monologue, and private speech of children, 132, 133
Montagner, Hubert, on nonverbal communication in children, 64-68
moral development, 57-61, 146, 169-173
moral principles, and adolescents' lack of understanding, 188, 189
mother-infant relationship, and infant competence, 44
mothering: vs. fathering, and interaction with infants, 147; quality of, and working mothers vs. nonworking mothers, 152-156
Mullerian regression factor (MRF), 8, 9

nature/nurture debate: and aggression, 85-88; and depression, 89-92; and gestural language, 66; and infant competence, 44-45; and intelligence 19-24; and moral development, 57-61; and sexual identity, 11-12
Neonatal Behavioral Assessment Scale, 43
neonatal intensive care, for premature babies, 27-29
nervous system: crying behavior as signal of developing, 46-48; development of, 38
newborns: and facial expression, 101; preference of, for mothers' voices, 26; *see also,* babies, infants
Nobel sperm bank, *see* Repository for Germinal Choice
nonverbal communication, in children, 64-68
nuclear war, threat of, and stress in children, 75

osteoporosis, 262, 264

pacifying-or-attaching gestures, in children, 65-66
parent-child interaction: 66-67, 146-151; *see also,* fathering; mother-infant interaction; mothering; parent-infant interaction; parents
parent-infant interaction: and infant crying behavior, 46-50; with premature infants, 28-33; and self-esteem development, 54-56; *see also,* fathering; mothering; mother-infant interaction; parents

parents: in alternative life styles, 157-161; and crying behavior of infants, 48-50; death of, and stress in children, 76-77; emotions of infants and behavior of, 51-52; ;and importance of fathering to children, 146-151; and moral development in children, 171, 172; of premature babies, 28-33; and punishment vs. discipline of children, 140-145; and self-esteem in children, 54-56; and helping children cope with stress, 69-81; and working mothers, 152-156; and youthful alienation, 179-183
patent ductus arteriosis, 33
peer interaction, 68, 181, 182
pelvic inflammatory disease, and infertility epidemic, 16
personality, and use of private speech by children, 135
Piaget, Jean, 41, 58, 65, 110, 126, 132-134, 169
play, 135, 136, 147
poverty: and blacks, 174; and stress in children, 73-74; and youthful alienation, 181, 182
prejudice: as learned, 82-84; *see also,* discrimination
premature babies, 28-33, 72
prenatal development, 9, 25-26
presbyacusis, and effect of aging on hearing, 264
presbyopia, and effect of aging on vision, 264
preschool children: and appearance-reality competence, 108, 109; private speech of, 132-137
preterms, *see* premature babies
primates, vocal apparatus of, 128, 129
private speech, of children, 132-137
problem solving: by children and private speech, 136; and elderly, 270; and male vs. female brain, 216, 219
programmed instruction, need for, in American education, 162-168
pronatural family, 158-161
prosopagnosia, 106
psychic surrender, 232-237
psychodynamic formulation, and suicide, 229
psychotherapy: and depression, 253, 254; and siblings, 202
puberty, and sex differences, 213
punishment: vs. discipline of children, 140-145; and moral development, 169

quality time, with children, and working mothers vs. at-home mothers, 153

race, and intelligence, 120, 123, 124
racism: 175; as learned, 82-84
religion, and psychic surrender, 232-235
Repository for Germinal Choice, 19, 21
repression, and self-deception, 115, 117
reproduction: and use of sperm from high IQ donors, 19-24; new techniques for, 13-18
"required helpfulness," 94, 97
resilient children, 93-98
responsivity, importance of, to child's competence, 55-56
right to suicide, 228, 230, 231
romantic relationships, role of love in, 241, 242

Credits/ Acknowledgments

Cover design by Charles Vitelli

1. Development During the Prenatal Period
Facing overview—WHO photo.
2. Development During Infancy
Facing overview—T. Polumbaum. 51—Beth Baptisle. 59—Susie Fitzburg.
3. Development During Childhood
Facing overview—K. Amye Steiner. 82—Ken Spencer. 96—UN photo. 100-101—Nancy Burson with Richard Carling and David Kramlich. 102—Paul Ekman. 103—A.N. Meltzoff and M.E. Moore "Imitation of facial and manual gestures by human neonates," *Science,* 1977, 198, 75-78. 104—Nancy Burson with Richard Carling and David Kramlich. 105—Mark Kauffman.

4. Family, School, and Cultural Influences on Development
Facing overview—United Nations/W.A. Graham.
5. Development During Adolescence and Early Adulthood
Facing overview—WHO photo/J. Mohr. 193—Dover *Pictorial Archives* Series. 201—UN photo/Derek Lovejoy. 205-207, 209—Courtesy Nancy L. Segal.
6. Development During Adulthood and Aging
Facing overview—UN photo by Jeffrey J. Foxx. 244-245—Bart Goldman. 268—Susan Gilmore. 272—UN photo by Shelly Rotner.

ANNUAL EDITIONS:
HUMAN DEVELOPMENT 87/88

Article Rating Form

Here is an opportunity for you to have direct input into the next revision of this volume. We would like you to rate each of the 47 articles listed below, using the following scale:

1. **Excellent: should definitely be retained**
2. **Above average: should probably be retained**
3. **Below average: should probably be deleted**
4. **Poor: should definitely be deleted**

Your ratings will play a vital part in the next revision. So please mail this prepaid form to us just as soon as you complete it.
Thanks for your help!

Rating	Article	Rating	Article
	1. Genes: Our Individual Programing System		26. The Importance of Fathering
	2. The New Origins of Life		27. Women at Odds
	3. Are the Progeny Prodigies?		28. The Children of the '60s as Parents
	4. Life Before Birth		29. The Shame of American Education
	5. Before Their Time		30. Moral Education for Young Children
	6. Making of a Mind		31. Rumors of Inferiority
	7. What Do Babies Know?		32. Alienation and the Four Worlds of Childhood
	8. There's More to Crying Than Meets the Ear		33. Rites of Passage
	9. How to Understand Your Baby Better		34. The Sibling Bond: A Lifelong Love/Hate Dialectic
	10. Your Child's Self-Esteem		35. Holocaust Twins: Their Special Bond
	11. The Roots of Morality		36. Men and Women: How Different Are They?
	12. Children's Winning Ways		37. Male Brain, Female Brain: The Hidden Difference
	13. Stress and Coping in Children		38. Suicide
	14. Racists Are Made, Not Born		39. Psychic Surrender: America's Creeping Paralysis
	15. Aggression: The Violence Within		40. The Measure of Love
	16. Depression at an Early Age		41. The Great Balancing Act
	17. Resilient Children		42. Toward an Understanding of Loneliness
	18. Face to Face, It's the Expression That Bears the Message		43. The Good News About Depression
	19. Really and Truly		44. Aging: What Happens to the Body as We Grow Older?
	20. Voices, Glances, Flashbacks: Our First Memories		45. New Evidence Points to Growth of the Brain Even Late in Life
	21. Insights into Self-Deception		46. Never Too Late
	22. Intelligence: New Ways to Measure the Wisdom of Man		47. The Reason of Age
	23. Origins of Speech		
	24. Why Children Talk to Themselves		
	25. Punishment Versus Discipline		

(continued on back)

ABOUT YOU

Name _____ Date _____

Are you a teacher? ☐   Or student? ☐

Your School Name _____

Department _____

Address _____

City _____ State _____ Zip _____

School Telephone # _____

YOUR COMMENTS ARE IMPORTANT TO US!

Please fill in the following information:

For which course did you use this book? _____

Did you use a text with this Annual Edition?   ☐ yes ☐ no

The title of the text: _____

What are your general reactions to the Annual Editions concept?

Have you read any particular articles recently that you think should be included in the next edition?

Are there any articles you feel should be replaced in the next edition? Why?

Are there other areas that you feel would utilize an Annual Edition?

May we contact you for editorial input?

May we quote you from above?

HUMAN DEVELOPMENT 87/88

BUSINESS REPLY MAIL		
First Class	Permit No. 84	Guilford, CT

Postage will be paid by addressee

The Dushkin Publishing Group, Inc.
Sluice Dock
Guilford, Connecticut 06437

No Postage
Necessary
if Mailed
in the
United States

||||...||..|.||..|.|.||.|.||.|.|.|.|.||.|.|.|.|..|.|.|